Cathodic Corrosion Protection Systems

Cathodic Corrosion Protection Systems

A Guide for Oil and Gas Industries

Alireza Bahadori, Ph.D.

School of Environment, Science & Engineering,
Southern Cross University, Lismore, NSW, Australia

ELSEVIER

AMSTERDAM • BOSTON • HEIDELBERG • LONDON
NEW YORK • OXFORD • PARIS • SAN DIEGO
SAN FRANCISCO • SINGAPORE • SYDNEY • TOKYO
Gulf Professional Publishing is an imprint of Elsevier

Gulf Professional Publishing is an imprint of Elsevier
225 Wyman Street, Waltham, MA 02451, USA
The Boulevard, Langford Lane, Kidlington, Oxford, OX5 1GB, UK

Library of Congress Cataloging-in-Publication Data
Bahadori, Alireza.
Cathodic corrosion protection systems : a guide for oil and gas industries / Alireza Bahadori.
pages cm
Includes bibliographical references and index.
ISBN 978-0-12-800274-2
1. Cathodic protection. 2. Pipelines–Cathodic protection. I. Title.
TA462.B27 2014
660'.28304—dc23
2014021031

British Library Cataloguing in Publication Data
A catalogue record for this book is available from the British Library

ISBN: 978-0-12-800274-2

For information on all Gulf Professional Publishing publications visit our web site at store.elsevier.com

This book has been manufactured using Print on Demand technology. Each copy is produced to order and is limited to black ink. The online version of this book will show color figures where appropriate.

Dedication

Dedicated to the loving memory of my parents,
grandparents, and to all who contributed so much
to my work over the years.

Contents

Biography **xxiii**
Preface **xxv**
Acknowledgments **xxvii**

1 Principle of Electrochemical Corrosion and Cathodic Protection **1**
1.1 Behavior of Buried or Immersed Metals in the Absence of CP **5**
1.1.1 The Nature of Metallic Corrosion **5**
1.1.2 Polarization **6**
1.1.3 Formation of Cells **7**
1.1.4 Passivity **9**
1.1.5 Reactions at Cathodic Areas **9**
1.2 Cathodic Protection **9**
1.2.1 Basis of CP **9**
1.3 Considerations Applicable to Most Types of Structures **10**
1.3.1 Range of Application **10**
1.3.2 Basis of Design **11**
1.3.3 Design or Modification of Structures to be Protected **13**
1.3.4 Comparison of the Various Systems **15**
1.3.5 Special Considerations **16**
1.3.6 Measures to Safeguard Neighboring Structures **21**
1.3.7 Design of CP Installations to Minimize Corrosion Interaction **24**
1.3.8 Measures to Reduce Corrosion Interaction **27**
1.3.9 CP Systems Installed Adjacent to Telecommunication Services **28**
1.3.10 CP Systems Adjacent to Railway Signals and Protection Circuits **29**
1.3.11 Interaction at Discontinuities in Cathodically Protected Structures **29**
1.3.12 Jetties and Ships: Corrosion Interaction at Sea and River Terminals **29**
1.4 Safety Aspects **30**
1.4.1 Danger of Electric Shock **30**

1.4.2 Fault Conditions in Electricity Power Systems in Relation to Remedial and/or Unintentional Bonds **31**
1.4.3 Hydrogen Evolution **32**
1.4.4 Installation in Hazardous Atmospheres **32**

2 Application of Cathodic Protection **35**
2.1 Criteria for Cathodic Protection **36**
2.2 Buried Pipes **37**
2.3 Tanks Exteriors **40**
2.4 Submerged Pipelines **40**
2.5 Offshore Structures and Ship Hulls **40**
2.6 Tank, Pipe, and Water Box Interiors **40**
2.7 Well Casings **40**
2.8 Types of Cathodic Protection Systems **41**
2.8.1 Impressed Current Systems **41**
2.8.2 Galvanic Anode Systems **42**
2.8.3 Galvanic versus Impressed Current System **42**
2.9 Equipment and Facilities for Impressed Current Systems **42**
2.9.1 Cathodic Protection Transformer Rectifiers **42**
2.9.2 Alternative Cathodic Protection Power Sources **43**
2.9.3 Impressed Current Anodes **43**
2.9.4 Cables **45**
2.9.5 Insulating Devices **45**
2.9.6 Anode Beds **45**
2.9.7 Electrical Bonding Station **47**
2.9.8 Test Stations **47**
2.9.9 Connections **47**
2.9.10 Electrical Continuity **47**
2.9.11 Protective Coatings **47**
2.9.12 Insulation **47**
2.9.13 Safety **48**
2.9.14 Site Survey **48**
2.9.15 Provision for Testing **48**
2.10 Cathodic Protection of Buried Steel Pipes **48**
2.10.1 Application **48**
2.10.2 Types of Cathodic Protection Systems **50**
2.10.3 Galvanic Anode Systems **51**
2.10.4 Isolation of Buried Pipes **52**
2.10.5 Test and Bonding Stations **54**
2.10.6 Cased Crossing Test Station **55**
2.10.7 Line Crossing Test Station **56**
2.10.8 Insulated Fitting Test Station **56**
2.10.9 System Design **56**
2.10.10 Particular Considerations **56**
2.11 Cathodic Protection In-Plant Facilities **58**

2.11.1 Application 58
2.11.2 Types of Cathodic Protection Systems 59
2.11.3 Impressed Current System Details 60
2.11.4 Galvanic Anode System Details 61
2.11.5 Test and Bonding Stations 61
2.11.6 Hazardous Locations 61
2.11.7 Oil Storage Tank Bottoms 62
2.12 Cathodic Protection of Vessel and Tank Internals 63
2.12.1 Types of Cathodic Protection Systems 64
2.12.2 Current Density 65
2.12.3 Anode Distribution 65
2.12.4 Reference Electrode Position 66
2.12.5 Protection of Specific Installations 66
2.13 Cathodic Protection of Marine Structures 70
2.13.1 Current Density 71
2.13.2 Marine Structural Zones 72
2.13.3 Zone Protection 72
2.13.4 Protection of Specific Installations Including Vessels 75
2.13.5 Anodes for Impressed Current Systems 77
2.13.6 Anodes for Galvanic Anode Systems 80
2.13.7 Boosted Polarization 81
2.13.8 Cathodic Protection Design for Marine Structures 81
2.13.9 Fixed Potential Monitoring Systems 82
2.13.10 Potential Survey 82
2.13.11 Retrofits 82
2.14 Anodic Protection 83
2.14.1 Principles of Anodic Protection 83
2.14.2 Contrast with Cathodic Protection 85
2.14.3 Effects of Variable Factors on Anodic Protection 86
2.14.4 Applications to Process Plants 88

3 Design Considerations on Cathodic Protection for Buried Pipelines and Marine Structures 91
3.1 Principles of Cathodic Protection 91
3.2 Methods of Applying Cathodic Protection 94
3.2.1 Impressed Current 94
3.2.2 Sacrificial Anodes 96
3.3 Protection Potentials 98
3.4 Current Density 98
3.5 Coatings 99
3.6 Calcareous Scales 99
3.7 Choice of a Cathodic-Protection System 99
3.8 Anode Resistance 100
3.9 Impressed-Current System 102
3.9.1 Current–Voltage Relationships 102

3.9.2 Single Drain Point **105**
3.9.3 Multiple Drain Points **106**
3.9.4 Effect of Coating **107**
3.9.5 Point of Minimum Protection **107**
3.9.6 Bare and Poorly Coated Lines **108**
3.9.7 Voltage Limitations **108**
3.9.8 Current Requirements **109**
3.9.9 Measurements **110**
3.9.10 Sizing of Rectifiers **110**
3.9.11 Calculations **110**
3.10 Galvanic Anode System **116**
3.10.1 Symbols **116**
3.10.2 Current Output **117**
3.11 Design Principle of Cathodic Protection for Marine Structures **118**
3.11.1 Design Calculations **118**
3.11.2 Single Cylindrical Vertical Anode **119**
3.11.3 Group of Vertical Anodes Equally Spaced in a Straight Line **120**
3.11.4 Galvanic Anode Systems **120**
3.11.5 Anode Output Formula **121**
3.11.6 Pipelines and Attenuation of Potential **124**
3.11.7 Sacrificial Anode Materials **124**
3.11.8 The Use of Sacrificial Anodes for Internal Cathodic Protection **124**
3.11.9 The Type and Number of Anodes Required **126**
3.11.10 Anode Location **127**
3.11.11 The Use of Sacrificial Anodes for External Cathodic Protection **127**
3.11.12 The Type and Number of Anodes Required (Full Hull Protection) **127**
3.12 Anode Location **128**
3.12.1 Fitting Out Protection **128**
3.12.2 Tank Descaling **128**
3.12.3 Individual Anode Output Determination and Calculation (Sacrificial Anodes) **129**
3.12.4 Impressed Current **130**
3.13 Anodic Protection **130**

4 Materials and Cathodic Protection Systems **133**
4.1 Impressed Current Anodes **133**
4.2 Manufacturing of Materials **134**
4.3 Anode Casting **135**
4.3.1 Casting of High-Silicon–Chromium–Iron Anode **135**
4.3.2 Casting of Graphite Anode **135**
4.3.3 Casting of Magnetite Anode **136**

4.4	Lead Wire-to-Anode Connection	**136**
4.5	Anode Lead Wire	**137**
	4.5.1 Conductor	**137**
	4.5.2 Lead Wire Insulation	**137**
4.6	Properties	**138**
	4.6.1 Mechanical Resistance Test	**138**
	4.6.2 Electrical Resistance Test	**138**
	4.6.3 Temperature Dependence	**138**
	4.6.4 Fluoropolymer-Insulated Anode Lead Wire Test	**138**
	4.6.5 Radiographic Test	**139**
4.7	Dimensions and Weights	**139**
4.8	Carbonaceous Backfill for Impressed Current Anodes	**140**
	4.8.1 Requirements	**141**
	4.8.2 Bulk Density	**142**
	4.8.3 Resistivity	**142**
	4.8.4 Particle Size	**142**
4.9	Galvanic Anodes for Underground Applications (Magnesium and Zinc)	**142**
	4.9.1 Classification	**143**
	4.9.2 Materials and Manufacture	**143**
	4.9.3 Anode Composition	**144**
	4.9.4 Anode Core	**145**
	4.9.5 Production Testing	**145**
	4.9.6 Requirements	**146**
	4.9.7 Workmanship, Finish, and Appearance	**146**
	4.9.8 Mechanical Resistance	**146**
	4.9.9 Electrical Resistance	**146**
	4.9.10 Anode Dimensions and Weights	**147**
	4.9.11 Zinc Anodes	**147**
	4.9.12 Methods of Attachment of Cable for Cast Anodes	**148**
	4.9.13 Identification of Anodes	**150**
4.10	Chemical Backfill for Galvanic Anodes	**150**
	4.10.1 Backfill Composition	**150**
	4.10.2 Particle Size	**150**
	4.10.3 Backfill Analysis	**150**
	4.10.4 Anode Packaging	**150**
4.11	Method for the Determination of the Anode-to-Core Resistance of Galvanic Anodes	**151**
	4.11.1 Circuit	**152**
	4.11.2 Procedure	**153**
4.12	Galvanic Anodes for Submerged Applications (Magnesium and Zinc)	**153**
	4.12.1 Materials and Manufacture	**153**
	4.12.2 Magnesium Anodes	**154**

4.12.3 Zinc Anodes 154
4.12.4 Production Testing 154
4.12.5 Requirements 154
4.12.6 Quality of Steel Inserts 155
4.12.7 Electrical Resistance 155
4.12.8 Identification of Anodes 155
4.13 Method for the Determination of the Anode-to-Core Resistance of Galvanic Anodes 155
4.13.1 Principle 155
4.13.2 Apparatus 155
4.13.3 Circuit 156
4.13.4 Procedure 156
4.14 Bracelet-Type Galvanic Anodes for Submarine Pipelines (Aluminum and Zinc) 157
4.14.1 Materials and Manufacture 157
4.14.2 Alloy Composition 157
4.14.3 Production Testing 159
4.15 Anode Core Steel Works 160
4.15.1 Insert Material 160
4.15.2 Insert Surface Preparation 161
4.15.3 Welding 161
4.15.4 Requirements 161
4.15.5 Mechanical Resistance 162
4.15.6 Electrical Resistance 162
4.15.7 Consumption Rate of Aluminum Anodes in Seawater 162
4.15.8 Closed Circuit Potential of Aluminum Anodes in Seawater 162
4.15.9 Surface Irregularities in the Anode Casting 163
4.15.10 Cracks in Cast Anodes Material 163
4.15.11 Anode Connections 163
4.16 Method for the Determination of the Anode-to-Core Resistance of Galvanic Anodes 164
4.16.1 Apparatus 164
4.16.2 Circuit 164
4.16.3 Procedure 165
4.17 Method for the Determination of the Consumption Rate of Aluminum Anode Alloys Immersed in Seawater 166
4.17.1 Apparatus 166
4.17.2 Circuit 166
4.17.3 Preparation of Test Anodes 166
4.17.4 Procedure 167
4.18 Method for the Determination of the Closed-Circuit Potential of Aluminum Anodes Immersed in Seawater 168
4.18.1 Apparatus 168

4.18.2 Circuit 168
4.18.3 Preparation of Test Anode 168
4.18.4 Procedure 168
4.19 Cast Galvanic Anodes for Fixed Offshore Installations (Aluminum) 169
4.19.1 Materials and Manufacture 169
4.19.2 Type of Anodes 170
4.19.3 Chemical Composition 170
4.19.4 Production Testing 171
4.19.5 Anode Insert 172
4.20 Requirements 173
4.20.1 Casting Quality 173
4.20.2 Weight and Dimensional Tolerances 174
4.20.3 Performance Requirements 174
4.20.4 Identification of Anodes 175
4.20.5 Anode Protection 175
4.21 CP Cables 175
4.21.1 Conductor 175
4.21.2 Insulation 175
4.21.3 Insulating Materials 176
4.22 Specific Cable Requirements 177
4.22.1 Positive Conductor Cable 177
4.22.2 Negative, Bond, or Test Conductor Cable 177
4.23 Quality Assurance Provisions 178
4.24 Tests 178
4.25 Fabrication and Inspection of Monolithic Insulating Joints 178
4.25.1 Design 179
4.25.2 Materials 180
4.25.3 Manufacture 182
4.26 Approved Welding Processes 183
4.27 Coating 183
4.28 Quality Assurance Provisions 183
4.29 Tests 184
4.29.1 Chemical Analysis 184
4.29.2 Mechanical Test 184
4.29.3 Dielectric Test and Megger Test 185
4.29.4 Hydrostatic Test 185
4.29.5 Low Pressure Leakage Test (Air Test) 185
4.29.6 Prototype Tests 185
4.29.7 Radiographic Test 185
4.29.8 Magnetic Particle and Ultrasonic Test 185
4.29.9 Dye Penetrant Test 186
4.29.10 Visual Inspection and Dimensional Check 186
4.30 Flange Insulation Kit 186

4.30.1 Materials **187**
4.30.2 Requirements **187**
4.30.3 Detail Requirements for Insulating Gaskets and Sleeves **188**
4.30.4 Packaging **190**
4.30.5 Quality Assurance Provisions **190**
4.30.6 Sampling **190**
4.31 Polypropylene Rope **190**
4.31.1 Requirements **191**
4.31.2 Other Characteristics **193**
4.31.3 Quality Assurance Provisions **193**
4.31.4 Splicing Kit **193**
4.31.5 Description **193**
4.31.6 Materials and Manufacture **194**
4.31.7 Properties of Catalyzed Resins **194**
4.31.8 Packaging **195**
4.31.9 Sampling **196**
4.32 Split Bolt Connector (Line Tap) **196**
4.32.1 Materials and Manufacture **197**
4.32.2 Dimensions **197**
4.32.3 Workmanship, Finish, and Appearance **197**
4.32.4 Quality Assurance Provisions **197**
4.32.5 Splice Coating Materials **198**
4.32.6 Electrical Insulating Plastic Tape **198**
4.33 Rubber Splicing Tape **200**
4.33.1 Materials and Manufacture **200**
4.33.2 Requirements **200**
4.33.3 Storage Test **200**
4.33.4 Roll Size **201**
4.33.5 Color **201**
4.33.6 Packaging, and Sampling **201**
4.34 Thermit Weld (Cad Weld) Powder **201**
4.34.1 Sampling **202**

5 Monitoring Cathodic Protection Systems **203**
5.1 Criteria for Cathodic Protection **205**
5.1.1 Buried Pipes **206**
5.1.2 Tanks Exteriors **206**
5.1.3 Submerged Pipelines **206**
5.1.4 Offshore Structures and Ship Hulls **207**
5.1.5 Tank, Pipe, and Water Box Interiors **207**
5.1.6 Well Casings **207**
5.1.7 Potential Limits **207**
5.1.8 Aluminum **207**
5.1.9 Lead **207**

	5.1.10 Stainless Steels	**210**
	5.1.11 Steel in Concrete	**210**
5.2	Periodic Inspection	**210**
	5.2.1 Potential Survey	**211**
	5.2.2 Test Equipment for Potential Survey	**211**
	5.2.3 Procedure to Be Observed	**213**
5.3	Potential Survey of Buried Steel Pipes	**213**
	5.3.1 Instruments	**213**
	5.3.2 Potential Measurements	**215**
	5.3.3 Compensation for the IR Drop Component in Cathodically Protected Pipelines	**216**
	5.3.4 Overprotection	**217**
	5.3.5 Potential Survey at Cased Crossings	**217**
	5.3.6 Potential Tests at Insulating Devices	**218**
5.4	Potential Survey of Cathodically Protected Reinforced Concrete	**219**
	5.4.1 Potential Measurement	**219**
	5.4.2 Electrodes and Probes for Some Concretes	**220**
5.5	Potential Survey of Offshore Structures	**221**
	5.5.1 Reference Electrodes	**221**
	5.5.2 Methods of Measurement	**221**
	5.5.3 Potential Measurements with a Diver-Operated Unit	**222**
	5.5.4 Potential Measurements with Surface Voltmeter, Cable, and Measuring Electrode	**225**
	5.5.5 Subsea Pipeline Potential Survey	**229**
	5.5.6 Time Intervals	**229**
	5.5.7 The Extent of the Potential Survey	**229**
5.6	Inspection of Rectifiers	**230**
	5.6.1 Records	**230**
	5.6.2 Major Points for Routing Inspection	**230**
	5.6.3 Annual Inspection	**230**
5.7	Inspection of Ground Bed	**231**
	5.7.1 On-land Ground Bed	**231**
	5.7.2 Submerged Ground Beds	**232**
5.8	Inspection of Offshore Sacrificial Anodes	**232**
5.9	Current Survey	**233**
	5.9.1 Pipelines	**233**
	5.9.2 Offshore Structures	**233**
5.10	Inspection Following Failure Report	**234**
	5.10.1 Increase in Circuit Resistance	**234**
	5.10.2 Stray Electric Currents	**234**
	5.10.3 Tests for Electrical Continuity	**234**
	5.10.4 Problems Associated with Galvanic Anode Installation	**235**
5.11	Inspection and Survey for Efficiencies of Coatings	**235**
	5.11.1 Coating Resistance Measurement	**235**

5.11.2 Attenuation Test Method **236**
5.11.3 Pearson Method **236**
5.11.4 Coating Inspection by C-Scan **236**
5.11.5 Visual Inspection **236**
5.11.6 Over-the-Line Potential Survey **238**
5.12 Data Recording and Analysis **238**
5.12.1 Data Recording **238**
5.12.2 Analysis of Data **239**
5.13 Interferences **241**
5.13.1 Cathode Field Interferences **241**
5.13.2 Measurements of Cathode Field Interferences **241**
5.13.3 Anode Field Interferences **242**
5.13.4 Measurements of Anode Field Interferences **242**
5.13.5 Corrective Measurers **242**
5.14 Meters and Equipment **242**
5.14.1 Reference Electrode **242**
5.14.2 Copper/Copper Sulfate Electrode **243**
5.14.3 Silver/Silver Chloride Electrode **244**
5.14.4 Pure Zinc Electrode **245**
5.14.5 Calomel Electrode **246**
5.15 Potentiometer and Voltmeter **246**
5.15.1 Conventional Voltmeter (High Resistance) **246**
5.15.2 Potentiometer–Voltmeter **248**
5.15.3 Potentiometer **249**

6 Test Methods **253**
6.1 Test Methods for the Short Circuit Point **253**
6.1.1 Locating Casing Short Circuit **253**
6.1.2 Insulation Tests **255**
6.2 American Society for Testing and Material Standard Test Method for Half-Cell Potentials of Uncoated Reinforcing Steel Bar in Concrete **257**
6.2.1 Scope **257**
6.2.2 Significance and Use **258**
6.2.3 Apparatus **259**
6.2.4 Calibration Standardization **260**
6.2.5 Procedure **260**
6.2.6 Recording Half-Cell Potential Values **262**
6.2.7 Data Presentation **263**
6.2.8 Interpretation of Results **263**
6.2.9 Report **264**
6.2.10 Precision and Bias **264**
6.3 Subsea Pipeline CP Survey Method **264**
6.3.1 Introduction **264**
6.3.2 System Components **264**

6.3.3 Survey Operation **267**
6.3.4 Survey Results **267**
6.4 Line Current Survey Test Method **268**
6.4.1 Test Procedure with the Test Point Consisting of Four Wires **270**
6.4.2 Test Procedure Using the Null Amp Test Circuit for Line Current Measurement **271**
6.5 Computer Modeling of Offshore CP Systems Utilized in CP Monitoring **272**
6.5.1 Computerized Modeling Techniques **273**
6.5.2 Short Theoretical Background **273**
6.5.3 Numerical Solutions **275**
6.5.4 Analysis of Existing CP Systems **275**
6.5.5 Performance of Sacrificial Anodes **276**
6.5.6 Analysis of Attenuation Curves at Sacrificial Anodes **277**
6.5.7 Potentials in Nodal Areas—Improved Efficiency in Potential Surveys of Nodes **277**
6.6 Coating Resistance Measurement Method **278**
6.6.1 Current–Voltage Change Method **278**
6.6.2 Pipeline Current with Interrupter On and Off **279**
6.7 Test Method and Calculation for "Attenuation Constant" **281**
6.7.1 Significance of Attenuation Constant **283**
6.7.2 Example Use of Graph Relating to Attenuation Constant **285**
6.8 Coating Inspection by the Pearson Method **285**
6.8.1 Equipment **285**
6.8.2 Procedure **286**
6.8.3 Data Obtained **287**
6.8.4 Presentation of Data **288**
6.9 Coating Inspection by the C-Scan System **288**
6.9.1 C-Scan System Features **289**
6.9.2 Performing Survey by C-Scan **289**
6.9.3 Advantages of the System **290**
6.9.4 Theoretical Background **290**
6.9.5 Operation Principle **291**
6.10 Coating Evaluation by Electromagnetic Current Attenuation Survey **293**
6.10.1 Equipment **294**
6.10.2 Procedure **295**
6.10.3 Data Obtained **296**
6.10.4 Presentation of Data **296**
6.10.5 Criteria and Interpretation **297**
6.11 Close Interval Pipe-to-Soil Potential Survey **298**
6.11.1 Measurement Intervals **300**
6.11.2 Switching Frequency **300**

6.11.3 Distance Measurement 301
6.11.4 Stationary Measurements 301
6.11.5 Equipment 301
6.11.6 Procedure 302
6.11.7 Data Obtained 302
6.11.8 Presentation of Data 303
6.11.9 Criteria and Interpretation 303
6.12 An Example for Cathodic Field Interference Test Method 303
6.13 Tests for Electrical Continuity 306

7 Installation of Cathodic Protection Systems 309
7.1 Galvanic Anodes 309
7.2 Impressed Current Anodes 310
7.3 Transformer/Rectifier Equipment 310
7.4 Excavation and Backfilling 311
7.5 Installation of CP Systems for Buried Pipelines 312
7.5.1 Installation of Impressed Current Systems 312
7.5.2 Installation of Transformer/Rectifier Equipment 316
7.5.3 Cabling 317
7.5.4 Electrical Connections 320
7.5.5 Installation of Test Stations (Test Points) 323
7.5.6 Installation of Test Box(es) 326
7.5.7 Earthing of CP Equipment 326
7.5.8 Fencing 327
7.5.9 Parallel Power Lines 327
7.5.10 Lightning Protection 327
7.5.11 Surge Arrestors 327
7.6 Installation of Galvanic Anode Systems 328
7.6.1 Single Packaged Anode 328
7.6.2 Multiple Galvanic Anodes 328
7.6.3 Extruded Ribbon Anodes 329
7.6.4 Connection of Galvanic Anodes to the Pipeline 329
7.7 Installation of CP Systems for Compact Buried Structures 330
7.7.1 Structure Preparation (to be Considered by the Structural Constructor) 331
7.7.2 Installation of Permanent Reference Electrodes 332
7.7.3 Installation of Insulating Flanges, Joints, and Couplings 332
7.8 Installation of CP Systems for Internal Surfaces 332
7.8.1 Materials and Equipment Acceptance (or Compliance) 333
7.8.2 Installation of Impressed Current Systems 333
7.8.3 Safety Precautions 335
7.8.4 Installation of Galvanic Anode Systems 335
7.8.5 Permanently Installed Reference Electrodes 336
7.9 Installation of CP Systems for Marine Structures 337
7.9.1 Immersed Structures 337

7.9.2 Installation of Galvanic Anode Systems **339**
7.9.3 Electrical Connections **340**
7.9.4 Corrosion Control Test Stations, Connection, and Bonds **341**
7.9.5 Installation of Insulating Joints/Flanges and Devices **341**
7.10 Submarine Pipelines **342**
7.10.1 Installation of Impressed Current Systems **342**
7.10.2 Installation of Galvanic Anode Systems **343**
7.10.3 Corrosion Control Test Stations, Connections, and Bonds **344**
7.10.4 Reinforcement **344**
7.10.5 Pipeline Crossings **345**
7.11 Electrical Measurements and Tests **345**
7.11.1 Potential Measurements **345**
7.11.2 Potential Survey of Internal Protection of the Plant **346**
7.11.3 Determination of Bond Resistance **347**
7.12 Tests Prior to Installation of CP on Buried or Immersed Structures **347**
7.12.1 Soil/Water Evaluation **347**
7.12.2 Structure/Electrolyte "Natural" Potential Survey **349**
7.12.3 Stray Electric Currents **349**
7.12.4 Tests for Electrical Continuity **349**
7.13 Tests during the Commissioning Period **350**
7.13.1 Buried Structures **350**
7.13.2 Fixed Immersed Structures **351**
7.13.3 Internal Protection of Plant **351**
7.13.4 Internal Surfaces **352**
7.14 Specialized Surveys **354**

8 Commissioning of Cathodic Protection Systems **355**
8.1 Precommissioning Inspection and Check **355**
8.2 Hookup and Commissioning **357**
8.2.1 Impressed Current Systems **358**
8.2.2 Sacrificial Anodes **359**
8.2.3 Interference **359**
8.3 Commissioning Survey **360**
8.4 Commissioning Report **361**
8.5 Installation of Electrical Isolation Equipment **361**
8.5.1 Installation **363**
8.5.2 Insulating Joints **363**
8.5.3 Isolated Flange Joints **364**
8.5.4 Protection Against External Moisture Ingress **365**
8.5.5 Pipeline Casing Insulators **365**
8.5.6 High Voltage Protection **366**
8.6 Thermit Welding of CP Leads **366**
8.7 Pipe Preparation **367**
8.8 Thermit Weld Preparation and Procedure **367**

8.9 Control of Interference Currents on Foreign Structures **369**
8.10 Notifying Owners of Other Structures for Interference Testing **370**
8.11 Interference Testing **371**
8.11.1 Stage at Which Interference Tests Shall Be Made **371**
8.11.2 Tests to Assess Interference **371**
8.11.3 Tests After Remedial Measures Have Been Applied **373**
8.12 Criteria for Limiting Corrosion Interaction **374**
8.12.1 Limit of Positive Structure/Electrolyte Potential Changes for All Structures **374**
8.12.2 Negative Changes of Structure/Electrolyte Potential **374**
8.12.3 Control of Interference **374**
8.12.4 Control by the Use of Galvanic Anodes **375**
8.12.5 Control by the Use of Impressed Current CP **375**
8.12.6 Control by Bonding **376**
8.12.7 Fault Conditions in Electricity Power Systems in Relation to Remedial and/or Unintentional Bonds **377**
8.13 Telluric Current **377**
8.14 AC Effects **378**
8.15 Measurement of Soil Resistivity **382**
8.16 Field Procedures **384**
8.17 Frequency of Measurement **384**
8.17.1 Presentation of Results **385**
8.17.2 Criteria and Interpretation **385**
8.18 Measurement of Electrode Resistance **385**
8.18.1 Measurement of Earth Electrode Resistance **386**
8.18.2 Measurement of Resistance of Earthing Conductor **388**
8.19 Current Drainage Survey **389**
8.19.1 Method **389**
8.20 Determination In situ of the Redox Potential of Soil **391**
8.20.1 Apparatus **391**
8.20.2 Materials **392**
8.20.3 Procedure **392**
8.20.4 Calculations and Expression of the Results **393**
8.20.5 Test Report **394**
8.21 Inspection of CP Installations **394**
8.21.1 CP Installations **394**
8.21.2 Test Points, Cased Crossings and Insulating Joints **395**
8.21.3 Coating Inspection **396**
8.22 Installation in Hazardous Atmospheres **397**
8.22.1 Bonds **397**
8.22.2 Isolating Joints **397**
8.22.3 Short Circuits between Points of Different Potentials **397**
8.22.4 Disconnection, Separation, or Breaking of Protected Pipework **397**

8.22.5 Electrical Equipment **398**
8.22.6 Test Instruments **398**
8.22.7 Internal Anodes **398**
8.22.8 Sacrificial Anodes **398**
8.22.9 Instruction of Personnel **399**

Additional List of Reading on Retrofitting **401**
Glossary of Terms **413**
Index **455**

Biography

Alireza Bahadori, Ph.D., is a research staff member in the School of Environment, Science and Engineering at Southern Cross University, Lismore, NSW, Australia. He received his Ph.D. from Curtin University, Perth, Western Australia.

During the past twenty years, Dr. Bahadori has held various process and petroleum engineering positions and was involved in many large-scale projects at National Iranian Oil Co. (NIOC), Petroleum Development Oman (PDO), and Clough AMEC PTY LTD.

He is the author of 250 articles and 12 books. His books have been published by multiple major publishers, including Elsevier.

Dr. Bahadori is the recipient of the highly competitive and prestigious Australian Government's Endeavor International Postgraduate Research Award as part of his research in oil and gas area. He also received a Top-Up Award from the State Government of Western Australia through Western Australia Energy Research Alliance (WA:ERA) in 2009. Dr. Bahadori serves as a member of the editorial board and reviewer for a large number of journals. He was honored by Elsevier to be an outstanding author of the *Journal of Natural Gas Science and Engineering* in 2009.

Preface

The oil and gas industry relies on the strength of steel and other metals to build pipelines, storage tanks, and other infrastructure that stand up to the rigors of industry activity. However, metal has one major weakness: when it comes into contact with water or soil, it can corrode. It is obvious, corrosion in a pipeline or storage tank is not good.

Cathodic protection (CP) is an electrical method of preventing corrosion on metallic structures that are in electrolytes such as soil or water. It has had widespread application on underground pipelines, and ever increasing use as the most effective corrosion control method for numerous other underground and underwater structures in oil and gas industries. It is a scientific method that combats corrosion by use of the same laws that cause the corrosion process.

To protect pipelines and other metal structures from corrosion, the oil and gas industry uses CP. The science of CP is based on electrochemistry. It is complex but in short, CP suppresses unwanted corrosion reactions by applying a protective electrical current.

CP provides an effective method of mitigating the corrosion damage to metal surfaces exposed to a conducting (corrosive) electrolyte. This engineering book provides the design requirements for electrochemical protection (CP) of metals against corrosion. The electrochemical methods of preventing corrosion consist of cathodic and anodic protection.

Anodic protection at this stage of development is applicable to limited combinations of metal and corrosive environment so there has been little applications for it in industries so far. Economics and difficulty in application has also limited its application to metal structures. For this reason, the book has emphasized on CP, which had been used widely and effectively in different industries as well as in oil, gas, and petrochemical industries.

Design requirements for CP systems (impressed and galvanic) for buried and immersed metal structures, such as buried pipelines, distribution pipelines, in-plant facilities, vessels and tanks, and marine structures, are described in this engineering book. The book also provides general guidelines for applying cathodic and anodic protection to metal structures.

Also, this engineering book covers the minimum requirements for anodes (high-silicon iron, graphite, magnetite) for use in impressed current CP systems. It specifies the composition, materials, manufacture, properties, inspection, and testing for high-silicon iron, graphite, and magnetite anodes.

Moreover, this book revised survey requirements to ascertain that corrosion control systems installed on buried or submerged structures are properly designed, operated, and effectively maintained. This book also provides information concerning techniques, equipment, measurements, and test methods used in field application. It deals with inspection of coatings in conjunction with CP for its efficiency on current distribution.

Acknowledgments

I would like to thank the Elsevier editorial and production team, Ms. Katie Hammon, and Ms. Kattie Washington of Gulf Professional Publishing for their editorial assistance.

1 Principle of Electrochemical Corrosion and Cathodic Protection

Cathodic protection (CP) is a technique used to control the corrosion of a metal surface by making it the cathode of an electrochemical cell. The simplest method to apply CP is by connecting the metal to be protected with a piece of another more easily corroded "sacrificial metal" to act as the anode of the electrochemical cell. The sacrificial metal then corrodes instead of the protected metal. For structures in which passive galvanic CP is not adequate, for example, in long pipelines, an external direct current (DC) electrical power source is sometimes used to provide current.

CP systems are used to protect a wide range of metallic structures in various environments. Common applications are steel water or fuel pipelines and storage tanks, such as home water heaters; steel pier piles; ship and boat hulls; offshore oil platforms; onshore oil well casings; and metal reinforcement bars in concrete buildings and structures. Another common application is in galvanized steel, in which a sacrificial coating of zinc on steel parts protects them from rust.

CP can, in some cases, prevent stress corrosion cracking. If two dissimilar metals are touching and an external conducting path exists, corrosion of one of the metals can take place. Moisture or other materials acting as an electrolyte between the metals create an electrochemical cells (similar to that of a battery). Depending on the metals, one will act as a cathode and one will act as an anode of the cell.

Under this arrangement, stray DC currents will flow. In the same way in a normal cell, an electrochemical reaction takes place, and there is a resulting corrosion of the anode.

CP works by converting all anodes that are likely to corrode the cathodes. There are two principal methods of doing this:

1. by attaching a more active metal to form a new anode (making the existing anode the cathode), resulting in the new material (sacrificial anode) being corroded rather than the protected material;
2. by injecting a DC current (impressed current), which uses an anode connected to an external DC source to provide the protection.

CP provides an effective method of mitigating the corrosion damage to metal surfaces exposed to a conducting (corrosive) electrolyte.

- **Types of CP**

A galvanic sacrificial anode attached to the hull of a ship is shown in Fig. 1.1.

Cathodic Corrosion Protection Systems. http://dx.doi.org/10.1016/B978-0-12-800274-2.00001-6

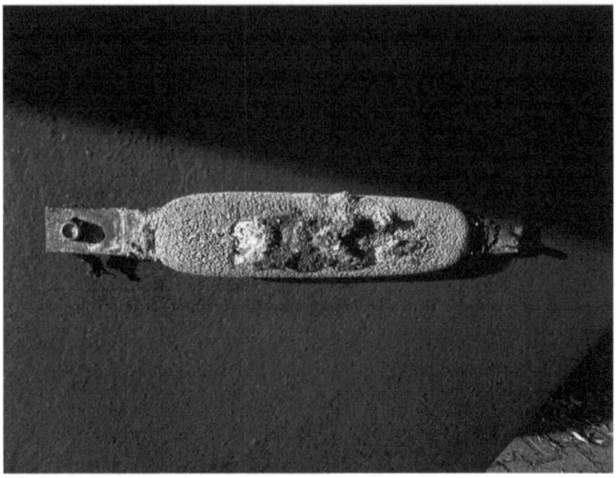

Figure 1.1 Galvanic sacrificial anode attached to the hull of a ship.

- ***Galvanic anode***

In the usual application, a galvanic anode, a piece of a more electrochemically "active" metal, is attached to the vulnerable metal surface where it is exposed to the corrosive liquid. Galvanic anodes are designed and selected to have a more "active" voltage (more negative electrochemical potential) than that of the metal of the target structure (typically steel).

For effective CP, the potential of the steel surface is polarized (pushed) more negatively until the surface has a uniform potential. At that stage, the driving force for the corrosion reaction with the protected surface is removed. The galvanic anode continues to corrode; this consumes the anode material until it must eventually be replaced. Polarization of the target structure is caused by the electron flow from the anode to the cathode, so the two metals must have a good electrically conductive contact. The driving force for the CP current is the difference in the electrochemical potential between the anode and the cathode.

Galvanic or sacrificial anodes are made in various shapes and sizes using alloys of zinc, magnesium, and aluminum. American Society for Testing and Materials International publishes standards on the composition and manufacturing of galvanic anodes.

In order for galvanic CP to work, the anode must possess a lower (i.e., more negative) electrochemical potential than that of the cathode (the target structure to be protected).

- ***Impressed current systems***

In the simple impressed current CP (ICCP) system, a source of DC electric current is used to help drive the protective electrochemical reaction.

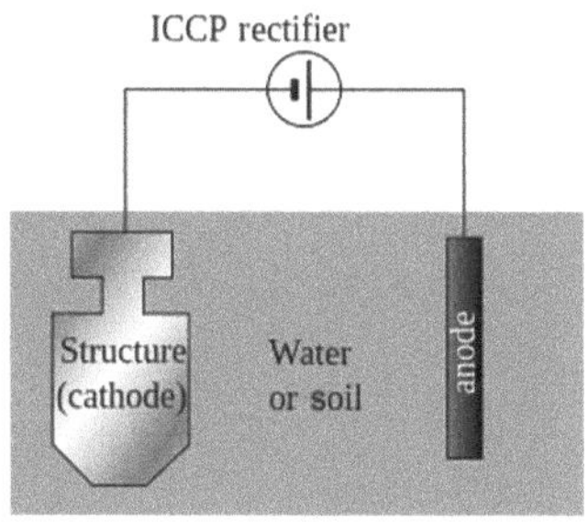

Figure 1.2 Impressed current cathodic protection (ICCP) systems.

For larger structures, galvanic anodes cannot economically deliver enough current to provide complete protection. ICCP systems use anodes connected to a DC power source, often a rectifier from a local alternating current (AC) system (Fig. 1.2). In the absence of an AC supply, alternative power sources, such as solar panels, wind power, or gas powered thermoelectric generators, may be used. For example, all telephone lines are biased to −36 to −60 V compared to the earth, to reduce galvanic corrosion.

Anodes for ICCP systems are available in a variety of shapes and sizes. Common anodes are tubular and solid rod shaped or are continuous ribbons of various materials. These include high silicon, cast iron, graphite, mixed metal oxide, platinum and niobium-coated wires, and other materials.

For pipelines, anodes are arranged in ground beds either distributed or in deep vertical holes depending on several design and field condition factors, including current distribution requirements.

Rectifier units are often custom manufactured and equipped with a variety of features, including oil cooling, automatic output adjustment, various types of electrical enclosures, remote monitoring, remote output adjustment, an AC electrical outlet, selectable AC input setting, and three-phase AC input. The rectifier output DC negative terminal is connected to the structure to be protected by the CP system. The rectifier output DC-positive cable is connected to the auxiliary anodes. The AC power cables are connected to the rectifier input AC cable terminals.

The output of the rectifier is usually determined by a CP expert to optimize the level of protection on the target structure. Many rectifiers are designed with taps on the transformer windings and jumper terminals to vary the voltage output of the rectifier unit. Rectifiers for water tanks and those used in other applications are made with solid-state circuits to automatically adjust the operating voltage to maintain a target current output or structure-to-electrolyte potential. Analog or digital meters are often installed to show the operating voltage (DC and sometimes AC) and current output.

The principle of ICCP forces the structure to be protected to become the cathode by connection to an anode and injection of a DC. The DC power supplies typically vary the current to achieve the required protection potential (Fig. 1.3).

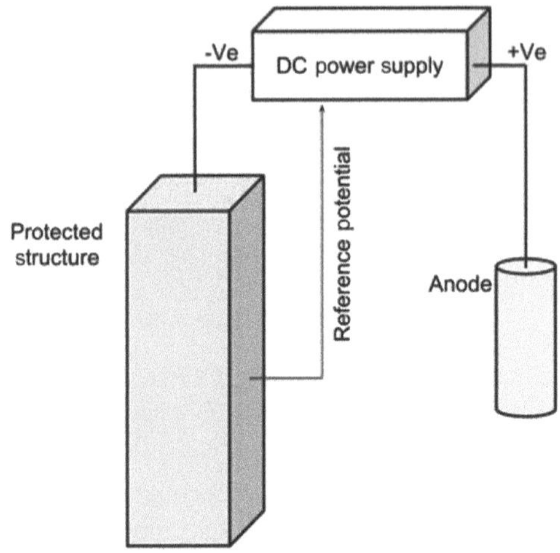

Figure 1.3 Principle of ICCP.

In ICCP systems, anodes can range from low-end consumable metals to more exotic materials that will exhibit little or no corrosion.

- ***Sacrificial Anode***

This is the practice of using a more active metal (sacrificial anode) connected to a structure to be protected, knowing that this metal will be corroded. One example of this would be the use of aluminum sacrificial anodes to protect steel structures in seawater.

Sacrificial anodes need to be electrically connected to the structure being protected.

- ***Cathode and Anodes***

When two metals are connected, determination of which will be the cathode and anode is made by looking at the relative galvanic potentials of each material. Of the two materials, the metal with the lowest potential will be the anode.

When measuring metals to find their galvanic potential, each needs to be measured against a common cathode (hence the term "Anodic Index" is often used). Table 1.1 shows the typical galvanic potentials of several metals as measured using a gold anode.

The amount of potential difference required between metals for corrosion to occur varies and is dependent on the environment. As a rule of thumb, many people take a 0.25-V difference of normal environments, 0.5 V where the humidity (and temperature) is controlled and 0.15 V for harsher industrial environments.

As an example of using the table, we can see that the potential difference between copper and aluminum is of the order 0.6 V, giving a combination that is to be particularly avoided. In practice, special bimetallic connections need to be employed whenever aluminum conductors are to be connected to copper conductors.

Table 1.1 Typical Galvanic Potentials of Several Metals as Measured Using a Gold Anode

Metal	Potential
Gold	0.00 (Most cathodic)
Rhodium	−0.05
Silver	−0.15
Nickel	−0.30
Copper	−0.35
Brass and bronzes	−0.40 to −0.45
Stainless steels	−0.50
Chromium plated	−0.60
Tin	−0.65
Lead	−0.70
Aluminum (wrought)	−0.75 to −0.90
Iron, wrought	−0.85
Aluminum (cast)	−0.95
Zinc	−1.20 to −1.25
Magnesium	−1.75
Beryllium	−1.85 (Most anodic)

1.1 Behavior of Buried or Immersed Metals in the Absence of CP

1.1.1 The Nature of Metallic Corrosion

Metals corrode because we use them in environments where they are chemically unstable. Only copper and the precious metals (gold, silver, platinum, etc.) are found in nature in their metallic state. All other metals, including iron—the metal most commonly used—are processed from minerals or ores to metals that are inherently unstable in their environments.

All metals, with the exception of precious metals, will get oxidized when exposed to oxygen and to an electrolyte (i.e., atmospheric moisture). It is a chemical reaction of the metal surface with the oxygen present in the air that causes some of the metal to corrode (or oxidize) and form the respective metal oxide on the surface. In some metals, such as steel, the corrosion products formed are very visible and loose. Everyone has observed the red color of iron oxide (rust) seen on improperly protected steel products.

The red rust formed is generally scaly and loose and easily falls off. However, metals such as stainless steel (steel with added nickel and chromium) gets oxidized as well. The difference is that nickel and chromium oxides formed are more uniform and tenacious oxide layers that protect the underlying material by "sealing" the surface from further oxidation once formed.

In addition to the surface oxidation that occurs on individual metals, any two dissimilar metals placed in contact with each other in an electrolyte (such as atmospheric moisture or water) will form a corrosion cell. This is the very basis of batteries used in everyday products.

One of the two metals in contact will corrode in preference to the other and form that metal's respective oxide. Metal corrodes is based on what chemists call the electromotive series of metals. The selection of the plated layer to be used is an important since electroplating in its very essence is the process of placing two dissimilar metals in contact with each other.

The plated layer (or layers) can be either an anodic coating (coating corrodes in preference to the substrate) or cathodic coating (substrate corrodes in preference to the plated layer). Whether a coating is an anodic or cathodic coating greatly impacts how the finished system will perform once in service, and there are advantages and disadvantages to each coating.

When a metal corrodes in contact with an electrolyte, natural atoms pass into solution by forming positively charged ions. Excess negative electrons are left behind. For example, in the case of iron,

$$Fe \rightarrow Fe^{++} + 2e^{-} \tag{1.1}$$

Thus, corrosion is accompanied by the flow of an electric current from a metal to an electrolyte due to the movement of positive ions into the electrolyte and of negatively charged electrons into the metal. Any area at which current flows in this direction is referred to as an anodic area.

The metallic ions may react with negative ions in the solution to give insoluble corrosion products (e.g., rust in the case of steel). Such reactions do not materially affect the argument that follows, except in cases when the corrosion product is such that it stifles further attack. The electric circuit is completed by the passage of current from the solution to the metal at other areas known as cathodes. Various reactions occur at cathodes; these do not, in general, cause corrosion.

Because the same number of electrons is related for each atom of the metal going into solution, the current is proportional to the corrosion rate. For example, in the case of iron or steel, two electrons flow for each atom going into solution, as shown in Eqn (1.1), and a corrosion current of 1 A corresponds to a loss of about 9 kg per year.

1.1.2 Polarization

The potential difference between any metal and the surrounding electrolyte varies with the density and direction of any current crossing the interface. This variation is referred to as polarization. The relationship between potential and current may be determined by an arrangement as shown in Fig. 1.4(a). It is not necessarily linear (Fig. 1.4(b)). Section Z–A of the curve corresponds to corrosion, and the more positive the potential the greater the corrosion rate. In practice, it is difficult to draw firm conclusions as to the corrosion rate from measurement of the potential difference between the metal and the solution or soil because the shape of the curve and the potential corresponding

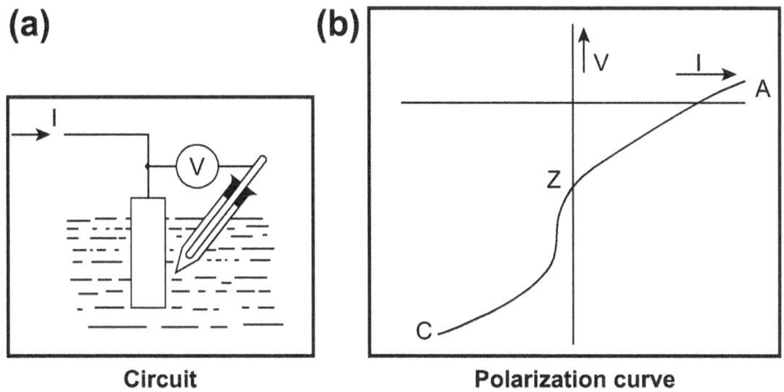

Figure 1.4 Measurement of polarization: (a) circuit and (b) polarization curve.

to zero current flow both vary according to the properties of the surrounding electrolyte. However, it will be seen that any current flow that makes the potential more positive, normally increases the probability that corrosion will occur.

Conversely, changing the potential in the negative direction reduces the corrosion rate and may prevent corrosion entirely. The way in which the potential difference between a metal and the surrounding electrolyte is measured should be specified. If a metal electrode in direct contact with the electrolyte is used, the result will depend, to some extent, on the effect of the electrolyte on the particular metal chosen.

For this reason, a reference electrode, for example, one of several types, should be used and the type of reference electrode should be stated when any results are quoted.

1.1.3 Formation of Cells

Suppose that the potentials of two different metals with respect to a solution are measured with the arrangement shown in Fig. 1.5, in which switch S is open, and the metal marked A is found to be more negative.

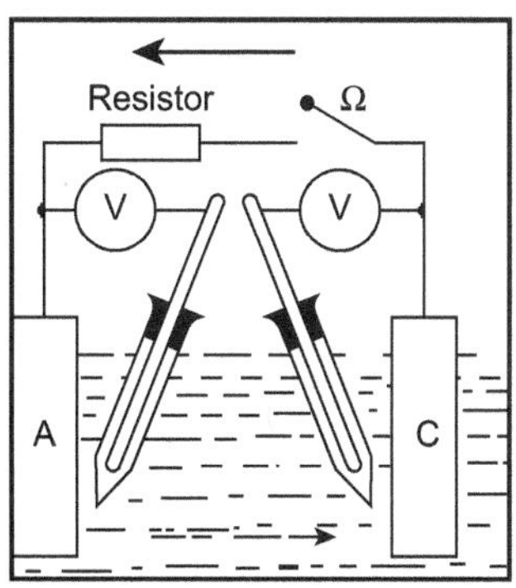

Figure 1.5 A typical cell.

When the switch is closed, current will flow in the direction shown by the arrows. Metal A will therefore be the anode, and will be corroded, while C acts as the cathode. Metals and conducting materials commonly used are listed below in such an order that each normally acts as the anode with respect to all the materials that follow it:

- Magnesium (most electronegative of the materials listed), Zinc;
- Aluminum (certain aluminum alloys may be more electronegative);
- Iron and Steel;
- Lead;
- Brass;
- Copper;
- Graphite, Coke, etc. (most electropositive of the materials listed).

Thus, the connection of magnesium to iron results in a cell in which the magnesium acts as the anode and the iron acts as the cathode. Cells may also arise due to differing properties of the electrolyte in contact with different parts of the same metal surface. Thus, an increased concentration of oxygen tends to make the potential of a metal more positive so that variation of soil density and porosity is a common cause of corrosion cells.

The size of the cells may vary greatly. The anodic area may be small, and the resultant pitting can, however, lead to rapid penetration.

It may be recalled that the anode was the electrode that, with the switch open, had the more negative potential with respect to the solution. If there is resistance in the circuit, this will still be true, even with the switch closed although the difference in potential will be smaller.

This will also be true in cases such as in Fig. 1.6 in which the anode and cathode are parts of the same metal surface in contact with different environments. If the total resistance in the circuit is low, there will be little difference in the metal/electrolyte potential at the anodic and cathodic areas, but corrosion will occur at the former, the potential at the anodic area being more positive than it otherwise would be due to the current flowing in the cell. This illustration has been included to emphasize the difficulty of determining whether corrosion is occurring by measuring the metal/soil potential without having other information.

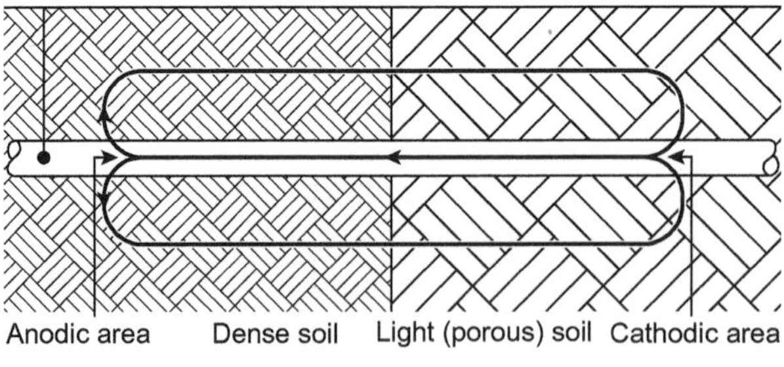

Figure 1.6 Cells due the differential aeration.

1.1.4 Passivity

If the corrosion product forms an adherent film on the surface of the metal, further attack may be prevented. The corrosion resistance of stainless steel, for example, is due to protection by films. The metals titanium and tantalum form highly resistant and adherent films and can therefore withstand strongly positive potentials without corroding.

1.1.5 Reactions at Cathodic Areas

The following are among the most common reactions that occur at cathodes:

$$2H^{+} + 2e^{-} \rightarrow H_2$$

$$\text{Hydrogen ions} + \text{Electrons} \rightarrow \text{Hydrogen gas} \quad (1.2)$$

$$\tfrac{1}{2}O_2 + H_2O + 2e^{-} \rightarrow 2\text{OHOxygen}^{+}$$

$$\text{Water} + \text{Electrons} \rightarrow \text{Hydroxyl ions} \quad (1.3)$$

The first of these reactions is favored by acidity (excess of hydrogen ions) while the second is favored by the presence of dissolved oxygen. Both tend to make the solution near the cathode alkaline (excess of hydroxyl ions over hydrogen ions): In contrast to the anodic reaction (e.g., Eqn (1.1)), cathodic reactions do not involve the passage of metal into solution; hence, in general, corrosion does not occur at cathodic areas.

In practice, the rate of corrosion is often determined by the rate at which the cathodic reaction can be maintained. For example, if the relevant reaction is that given by Eqn (1.3), replenishment of oxygen may be the controlling factor.

In near-neutral anaerobic soils, sulfate-reducing bacteria give rise to a further type of cathodic reaction, and soils of this kind are often particularly aggressive to iron and steel. It is possible, by determining the pH and redox potential, to assess whether conditions are such that sulfate-reducing bacteria are likely to be active.

Although, as previously indicated, the reactions occurring at cathodes do not directly result in corrosion, it should be noted that the environment of the metal is altered, for example, it becomes more alkaline. In the case of aluminum and, occasionally, lead, corrosion may result. Alkalinity may also cause deterioration of paints and other coatings by saponification.

1.2 Cathodic Protection

1.2.1 Basis of CP

Corrosion implies the existence of anodic and cathodic areas (Fig. 1.7).

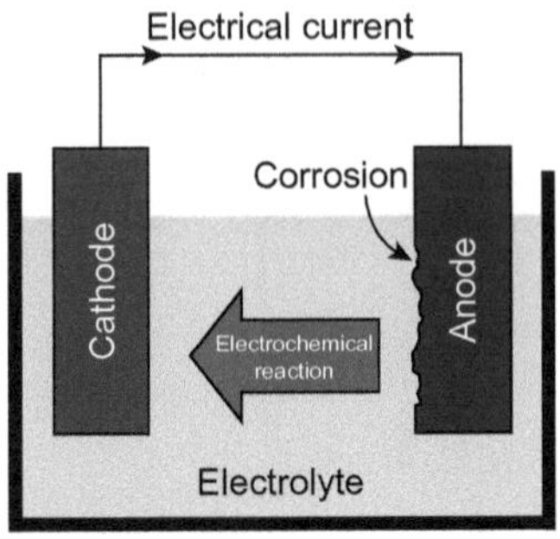

Figure 1.7 Schematic of anodic and cathodic areas in a corrosion process.

In applying CP, a current is superimposed in such a direction that the structure to be protected acts as a cathode.

If the current is sufficient, no part of the structure acts as the anode. This entails the use of an auxiliary anode. If this anode is of a material such as magnesium, the protection current will flow due to the electromotive force (EMF) arising from the cell formed (Galvanic anode system).

Alternatively, the EMF may be derived from a separate DC source, giving a wide choice of materials for the auxiliary anode including some which are not consumed (Impressed current system).

The criterion for CP is that the current flowing in the anode circuit is reduced to zero or reversed. However, anodes and cathodes are often parts of the same metallic surface (as shown in Fig. 1.7), and the individual anodic areas may be small. It is thus impossible, in most cases, to confirm that CP has been achieved by measuring the relevant current. For most of the metals commonly encountered, however, it is possible to state values of the metal/electrolyte potential at which corrosion does not occur in environments such as soil or natural waters.

1.3 Considerations Applicable to Most Types of Structures

1.3.1 Range of Application

CP can, in principle, be applied to any metallic structure or plant that is in contact with a mass of soil or water. The application of CP to metal surfaces that are intermittently immersed, for example, due to the action of tides, may also be beneficial. However, economic considerations sometimes restrict the range of application. It may, for example, be uneconomical to protect certain types of existing structures because the cost of making them suitable for CP is excessive.

The function of the structure or a plant under consideration will determine the benefit to be expected by suppressing corrosion. Thus, a certain amount of attack in the form of pitting may be tolerable on some structural members, while the same severity of attack would cause failure of a pipe. If the consequences of penetration by corrosion are important, for example, hazards due to the leakage of a flammable gas or liquid, or interruption of the operation of a large plant, or the failure of a ship's plate, the need to ensure complete reliability will become overriding and CP would be regarded as economical even under otherwise unfavorable circumstances.

1.3.2 Basis of Design

CP is achieved by causing the current to flow from the surrounding electrolyte into the structure at all points, the criterion being that the structure/electrolyte potential is, at all positions, more negative than the appropriate protective potential given in the literature. Fig. 1.8 represents, in outline, the system required, which consists of a

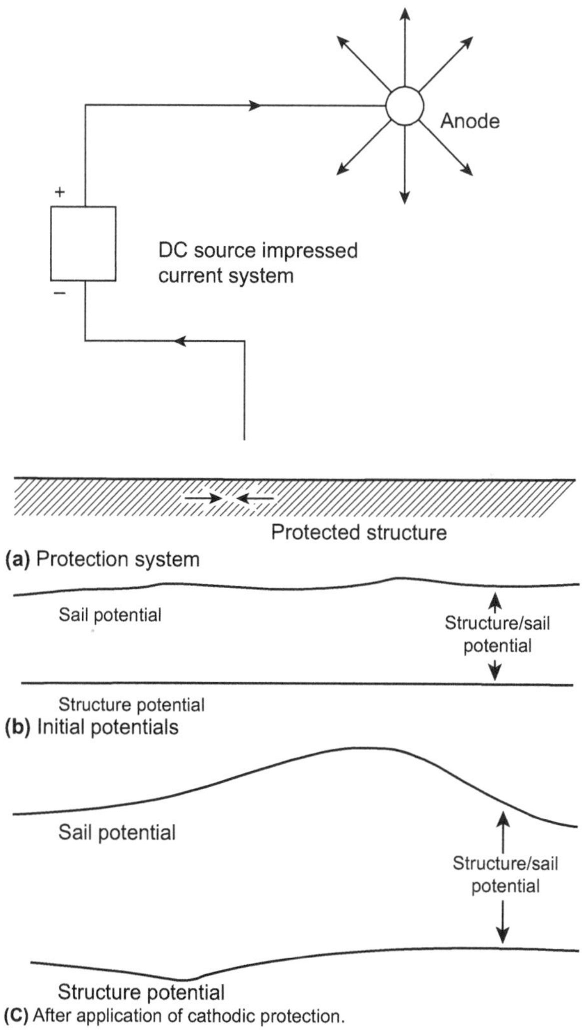

Figure 1.8 Cathodic protection system and distribution of structural/electrolyte potential.

buried or immersed anode, a connection to the structure to be protected, and (in the case of impressed current systems only) a source of EMF. Current flows in the metallic parts of the circuit in the directions indicated by the arrows and returns through the electrolyte (soil or water) to the protected structure. When the potential drop through the electrolyte and/or the structure is appreciable, the potential change due to the CP is nonuniform as shown in the lower parts of Fig. 1.8. The following factors tend to increase the nonuniformity of the CP:

1. Small separation between the anode and the structure (particularly if the electrolyte resistivity is high).
2. High resistivity of soil or water (particularly if the anodes are close to the structure).
3. High current density required to protect the structure (the current density will be governed by the quality of the coating, if any, and the availability of oxygen at the surface of the metal or the activity of anaerobic bacteria).
4. High electrical resistance between different parts of the structure.

The tendency for the current density to be the highest at points nearest the anode may occasionally be an advantage since it is possible to concentrate the effect at a point where it is most needed. For example, when corrosion of iron or steel occurs due to the proximity of a more electropositive metal, the attack is often local; only a small proportion of the surface may require protection.

Normally, however, the whole of the metal surface is to be protected and nonuniformity, as shown in Fig. 1.8, is uneconomical because some parts of the surface receive more current than is required to attain the protection potential. Moreover, since the potential should, generally, not be made too strongly negative for some reasons, it may be impossible to compensate for poor initial design by increasing the current and thereby making the potential more negative. Additional anodes will therefore be needed and, in the case of protection by impressed current on extensive structures, this will also entail the provision of additional sources of EMF.

Thus, if the use of CP is envisaged, the first step is to consider whether the structure or plant can be designed, or modified if it already exists, in such a way as to make the installation of CP more economical. Consideration should also be given to the correct placing of anodes both as regards separation from the structure and their distribution over the surface. When structures such as pipelines are being protected with impressed current, considerations such as the availability of power supplies may, however, have an important bearing on the design.

The characteristic of two systems, that is, galvanic anodes and impressed current, are compared in this chapter. Economical design of structure or plant and its associated CP system entails striking the best possible balance between factors that affect the initial cost (effectiveness of structure coating, electrical conductance between sections of the structure or plant, extent and position of the anode system, number of separate units, etc.) and factors that affect the running cost, notably the power required and frequency of replacement of anodes.

There are, in addition, certain considerations that relate only to particular types of structures. For example, in the CP of ships' hulls or of pumps, consideration should be given to the hydraulic drag arising from the installation of the anodes. In the case of buried structures, possible effects of the DC flowing in the soil on other structures in the vicinity may also have an important effect on the economics of the scheme. There are also some secondary effects of CP that need to be taken into account.

1.3.3 Design or Modification of Structures to be Protected

1.3.3.1 Electrical Continuity

It may be necessary to install continuity bonds between different sections of the structure or plant before CP is applied. The resistance of these bonds should be sufficiently low to ensure that the potential drop due to the passage of the protective current through the structure is small. In the case of impressed current installations, it may be economical to improve the connections between different parts of the structure, even though metallic connections already exist, in order to reduce the total resistance.

It should be noted that if the structure is not metallically continuous, part of the protection current flowing in the electrolyte toward a protected section may pass through the isolated sections of the structure. Corrosion may be accelerated where such currents are discharged from the structure and return to the electrolyte. This accelerated corrosion could be internal where conducting fluids are being conveyed in pipelines.

1.3.3.2 Protective Coating

The function of a coating is to reduce the area of metal exposed to the electrolyte (soil or water). By this means, it is possible to greatly reduce the current density required for CP. As indicated in previous sections, the fact that the current is spread more uniformly may reduce the number of points at which CP needs to be applied.

A coating should, ideally, have a high electrical resistance and be continuous, that is, there should be few holidays. It should be resistant to any chemical or bacterial action to which it might be exposed to, and should withstand all temperature variations to which it may be subjected to; no blisters should exist, and the coating should adhere strongly to the surface to be protected; it should have satisfactory aging characteristics and adequate mechanical strength. The ability to resist abrasion may be important in some applications.

Coatings may take the form of paints, or materials, such as bitumen and coal tar, which are often reinforced with glass fiber or other fibrous material. Plastic sheets or tapes may also be used for certain structures. The most suitable form of coating depends on the type of structure and its environment. In deciding upon the type of coating to be used, the aim should be to achieve overall economy in the combined cost of the protected structure and of the initial and running costs of the protection schemes. Due regard should be given to the life expected of the structure and to the economics of preparing the coating should this become necessary.

In the case of buried structures, a secondary but important function of the coating is to reduce the potential gradients in the surrounding soil and thereby decrease interaction with neighboring buried structures.

The protection current, particularly if strongly negative potentials are used, may produce sufficient alkali to affect the coating adversely. The extent to which coatings are alkali resistant is therefore important in some applications. It is, however, possible to give only general information in respect of coatings. If, in a particular case, the coating performance were critical, it would be desirable to determine the properties by test beforehand. Concrete cannot be considered to be a substitute for an insulating coating, and such a coating should be provided in addition wherever possible.

Metal spraying is not treated as a coating method for the purposes of this chapter; its use in conjunction with CP is unlikely. The adverse effect of a nonadherent coating should not be oversimplified. A nonadherent coating is a barrier that will prevent, from a pipeline, the flow of CP current from the soil. In other words, CP current could not flow to the pipe metal through the soil or water between the nonadherent coating and pipe metal.

However, if the disbonded or nonadherent coating (which acts as a cathodic shield) is sufficiently porous to absorb enough soil moisture to become conductive, the moisture may help pass enough current to protect the pipe metal (which is in contact with soil or water) under the nonadherent or disbonded coating. Such a disbonded coating would not then act as a complete shield or barrier.

This phenomenon has been proved beyond doubt on a number of gas transmission pipeline running through marshy land or terrain with low water levels.

1.3.3.3 Isolation

It often happens that a well-coated structure, to which CP could be applied economically, is connected to an extensive and poorly coated metallic structure, whose protection is not required or would be uneconomical. In such a case, the well-coated structure should be isolated before applying CP to it. In the case of coated pipelines, for example, the inclusion of isolating joints and terminal installations is normally considered to be essential.

A further application is the isolation of a section of a structure to prevent or reduce excessive effects on neighboring structures due to interaction. If the isolated section is so placed that the required continuity of the structure is interrupted, this should be restored using an insulated cable. It may, on occasion, be desirable to shunt an isolating device by means of a resistor. For example, by choosing an appropriate value for the resistor, it might be possible to adjust the current so that it is sufficient to protect the relevant section of the structure but is insufficient to cause unacceptable interaction on nearby structures.

Isolating joints are sometimes a required part of the safety precautions at oil terminal jetties. They should not be installed in above-ground situations where concentrations of flammable gas or vapor occur.

The protection of only part of a structure may accelerate the corrosion of nearby isolated sections of the structure. For this reason, it may be advisable to apply a coating with a particularly high insulation resistance to the protected section of the structure where it is near unprotected equipment or to take other measures to prevent possible damage.

With equipment containing electrolytes, corrosion could similarly occur on the inner surface of the unprotected section. With highly conducting fluids, for example, brine, such corrosion could well be rapid. For this reason, the inclusion of isolating devices, for example, pipelines containing seawater or strong brine is inadvisable.

1.3.4 Comparison of the Various Systems

The advantages and disadvantages of the galvanic anode and impressed current methods are set out in Table 1.2. A further method known as electric drainage is applicable only to structures affected by stray currents flowing in the soil, and may have advantages in suitable cases.

Table 1.2 A Comparison of Galvanic Anode and Impressed Current Systems

Galvanic Anodes	Impressed Current
1. They are independent of any source of electrical power.	**1.** Requires a mains supply or other source of electric power.
2. Their usefulness is generally restricted to the protection of well-coated structures or the provision of local protection, because of the limited current that is economically available.	**2.** Can be applied to a wide range of structures including, if necessary, large, uncoated structures.
3. Their use may be impracticable except with soils or waters with low resistivity.	**3.** Use is less restricted by the resistivity of the soil or water.
4. They are relatively simple to install; additions may be made until the desired effect is obtained.	**4.** Needs careful designing although the ease with which output may be adjusted allows unforeseen changing conditions to be catered for.
5. Inspection involves testing, with portable instruments, at each anode or between adjacent pairs of anodes.	**5.** Needs inspection at relatively few positions; instrumentation at points of supply can generally be placed where it is easily reached.
6. They may be required at a large number of positions. Their life varies with conditions so that replacements may be required at different intervals of time at different parts of a system.	**6.** Generally requires a small total number of anodes.
7. They are less likely to affect any nearby neighboring structures because the output at any one point is low.	**7.** Requires the effects on other structures that are near the ground bed of protected structures to be assessed, but interaction is often easily corrected, if necessary.
8. Their output cannot be controlled, but there is a tendency for their current to be self-adjusting because if conditions change such that the metal to be protected becomes less negative, driving	**8.** Requires relatively simple controls and can be made automatic to maintain potentials within close limits despite wide variations of conditions. Since the EMF used is generally higher than with

(Continued)

Table 1.2 A Comparison of Galvanic Anode and Impressed Current Systems—*cont'd*

Galvanic Anodes	Impressed Current
EMF, and hence current, increases. It is possible, by selection of material, to ensure that the metal cannot reach a potential that is sufficiently negative to damage paint.	galvanic anodes the possible effects of ineffective control or incorrect adjustment, for example, damage to paintwork or coatings, are greater.
9. Their bulkiness may restrict flow and/or cause turbulence and restrict access in circulating water systems. They introduce drag in the case of ships' hulls.	**9.** Allows more compact anodes by the use of suitable materials; drag is negligible.
10. They may be bolted or welded directly to the surface to be protected, thus avoiding the need to perforate the metal of ships' hulls, plants to be protected internally, etc.	**10.** Requires perforation in all cases on ships' hulls, plant to enable an insulated connection to be provided.
11. Their connections are protected cathodically.	**11.** Requires high integrity of insulation on connection to the positive side of the rectifier that is in contact with the soil or water: Otherwise, they will be severely corroded.
12. They cannot be misconnected so that polarity is reversed.	**12.** Requires the polarity to be checked during commissioning because misconnection, so that polarity is reversed, can accelerate corrosion.

1.3.5 Special Considerations

1.3.5.1 Secondary Effects of CP

The application of CP may give rise to secondary effects such as the development of alkalinity or the evolution of hydrogen at the protected surface. The effects that may occur are described in the following paragraphs.

1. Alkalinity may cause the deterioration of paints. The effect can be minimized by avoiding the use of very negative potentials and by using paints that are less susceptible to such damage.
2. Alkalinity produces, in the case of seawater or similar solutions, a white calcareous deposit (chalking). This is beneficial since the current density needed to maintain CP is reduced. If, however, formation of the deposit is excessive, water passages may be obstructed or moving parts may be impeded.
3. Alkaline environments can corrode aluminum, which can therefore be cathodically protected only if the potential is maintained within certain limits. Since Aluminum is an amphoteric metal and is sensitive to Alkali, the CP of aluminum pipes is a special problem. The reaction in a CP circuit generates alkali at the cathodic surface. If too much CP is applied, the alkalinity at the surface of an aluminum pipe may become

strong enough to consume the aluminum chemically. The danger is that a buried aluminum pipeline under strong CP actually may corrode faster than it would if not cathodically protected at all.

4. An alkaline environment can exceptionally corrode lead when protected cathodically. (e.g., cables installed in asbestos–cement pipes).
5. Hydrogen evolved at strongly negative potentials may create an explosion hazard in enclosed spaces.
6. Hydrogen embrittlement of high tensile steel poses a possible danger.
7. Hydrogen produced at the flaws in a coating may progressively detach the coating from the surface of the metal.
8. Rust and scale sometimes detach from a surface during the initial period of operation of CP and may block water passages or cause other difficulties during a short period. If iron or steel has been seriously corroded, removal of rust that is plugging holes may cause a number of leaks to become apparent during this period.
9. Chlorine may evolve at the anodes of an ICCP installation if the electrolyte contains chloride. This may cause a nuisance or create a hazard.

1.3.5.2 Effects of Stray Currents from Protection Installations

Where a protected structure, or the anode(s) or ground bed(s), lies near other buried or immersed metallic structures that are not fully insulated from the earth, the latter (secondary) structures may, at certain points, pick up a proportion of the protective current due to potential gradients in the soil or water and return it to the earth at others. The secondary structures may corrode at these latter points.

1.3.5.3 The Avoidance of Damage or Hazard due to Overvoltage

Overvoltages due to faults on power equipment or to lightning may cause serious damage to equipment installed to provide CP. If isolating joints have been inserted in a protected structure, there is a risk of flashover and explosion if the structure contains a low flash-point material. The following recommendations should be read in conjunction with any other relevant standards or Regulations.

1.3.5.3.1 Damage to CP Equipment by Overvoltages

The ground bed of a CP system will often be the best available connection to earth in a particular locality, and this may result in the associated equipment being subjected to overvoltages or excessive current that originate from either faults on power equipment or lightning as follows:

1. Faults on power equipment via the protective earthing of equipment

In high-resistance areas, where it is difficult to obtain a good connection to the earth, a system of protective earthing is often employed. This involves the bonding together of all the earth and/or neutral terminals of plant and equipment so that they are at the same potential, although this potential may be appreciably higher than the true earth potential.

2. Lightning

Any currents due to strikes to the protected or associated structures are liable to flow to the earth via the ground bed. This could damage the meters of the transformer/rectifier equipment and may also damage the rectifier stack. In either case, over-voltages can arise across the terminals of the equipment, and a suitable surge diverter or protective spark gap should be installed across the output terminals of all transformer/rectifier equipment. Further advice on lightning considerations is given in BS 6651 (1985).

1.3.5.4 Isolation of Buried Structures that are Associated with a Lightning Protection System

Care is needed if isolating joints are to be installed in buried structures where lightning protection has been installed in accordance with BS 6651 (1985). That standard requires that metal cable sheaths, metal pipes, and the like entering a building or similar installation be bonded as directly as possible to the earth termination of the lightning protection system, at the point of entry to the building. This bonding is necessary to avoid breakdown through the soil as a result of a lightning discharge with a consequent risk of damage to the pipes, cables, etc.

The installation of isolating joints for CP purposes where buried structures approach terminal or other installations clearly runs counter to these requirements since the deliberate electrical separation of the metallic services from other earthed components, including the earth termination of the lighting protection system, could, in the event of a lightning storm, result in a breakdown through the soil or flashover of the isolating joint, with a consequent risk of damage or explosion.

To satisfy the opposing requirements, the isolating joints should be bridged by discharge gaps to effect adequate connection between the two earthed systems during the discharge of lighting current. The impulse breakdown voltage of these gaps should lie below that of the isolating joints.

The gaps should be capable of discharging lightning currents without sustaining damage and should be encapsulated to provide complete protection from moisture.

1.3.5.5 Buried Structures in the Vicinity of a Lightning Protection System

Where the structures to be cathodically protected pass close to, but are not already incorporated in, a lightning protection system installed to protect some other structure or installation, the question may arise as to the minimum distance between a lighting protection earth and other buried metalwork, for example, the ground bed of a CP system.

This distance, S, can be estimated from the relationship $S = IR/E$ where I is the crest value of the lightning current discharged through an earth termination of resistance, R, and E is the impulse breakdown strength of the soil.

Although no systematic tests have been carried out, tests on a variety of soil specimens have indicated values of E to be from 0.2 kV/mm to 0.5 kV/mm. Assuming the lower of these, together with a current of 200 kA (an exceptionally severe lightning current), separation, S, (in meters) is given by S = R, where R is the resistance of the earth electrodes before any bonding to other structures has been carried out. BS 6651 (1985) requires that the resistance to earth of the whole lightning protection system be not >10 ohms.

1.3.5.6 Factors Affecting Design

The following factors affect the application of the principles outlined previously in this Section.

Variations of conditions affecting the CP of buried structures are generally slow. Manually adjusted control equipment is therefore usually sufficient. Automatic control may, however, be required if the structure to be protected is affected by stray currents from electric traction systems; the control system has to be quick acting. Suitable equipment is available. The position of the sensing electrode should be carefully chosen.

The nature of the coating and the method of application will determine the most negative potential that can be applied without the likelihood of coating damage. Conventional limits of structure/soil potential for coal tar and asphaltic pipeline enamels are −2.0 V (off) with an absolute minimum of −2.5 V (off) (copper/copper sulfate reference electrode). Other coatings may be more susceptible to over-protection, and the structure/soil potential may need to be limited to less negative values.

For steel structures, the usual criterion of protection is −0.85 V (without allowance for IR drop error) relative to a copper/copper sulfate reference electrode. Where anaerobic conditions occur and sulfate-reducing bacteria may be present, a more negative value of −0.95 V (without allowance for IR drop error) should be adopted.

Special backfills can be used to assist in obtaining a low resistance between anodes and the soil. Special care is needed to avoid accelerating the corrosion of other buried structures by interaction.

For buried structures near electric traction systems, electric drainage can be used. The application of CP to a buried or immersed structure (referred to as the primary structure) causes DC to flow in the earth or water in its vicinity. Part of the protection current traverses nearby buried or immersed pipes, or cables, jetties or similar structures, or ships alongside (termed secondary structures), which may be unprotected and the corrosion rate on these structures may, therefore, increase at points where the current leaves them to return to the primary structure. This effect is described as "corrosion interaction."

1.3.5.7 Electric Drainage

In DC traction systems, the negative of the DC supply is usually connected to the running rails that are in electrical contact with the soil. Thus, the soil will provide an additional path, parallel to the track, for current flowing from the traction unit toward the point of supply particularly in the case of extensive structures such as pipelines or cables, part of the current flowing in the soil may be picked up in one area and discharged in another, leading to accelerated corrosion.

This can be prevented by bonding the pipeline or cable sheath and armoring to a return rail at the most negative portions of the track, that is, near substations or where negative feeders are connected to the rails. The bonding cable will then carry most of the return traction current back to the point of supply, and will thus ensure that the structure receives partial, or sometimes complete, CP.

This form of CP is known as "electric draining" or "drainage." Under no circumstances should any connection or bonding be made to railway running rails or structures without consultation with, and subsequent written permission from, the railway authorities.

The electric drainage method can be applied in all circumstances because of the likelihood of current reversals in the drainage bond; a rectifier (or other unidirectional device) is therefore usually provided as illustrated in Fig. 1.9(a). This is referred to as "polarized electric drainage."

The track voltage may attain relatively high values, and it may be necessary to protect the rectifier and bond against excessive currents by means of suitable series resistors and/or inductors and overload circuit breakers or fuses and by providing more than one drainage connection (Fig. 1.9(a)).

For railway signaling purposes, a relay and power supply are usually connected between the two rails of a railway track so as to provide remote indication that a train is in the section. This arrangement is known as a track circuit and there are

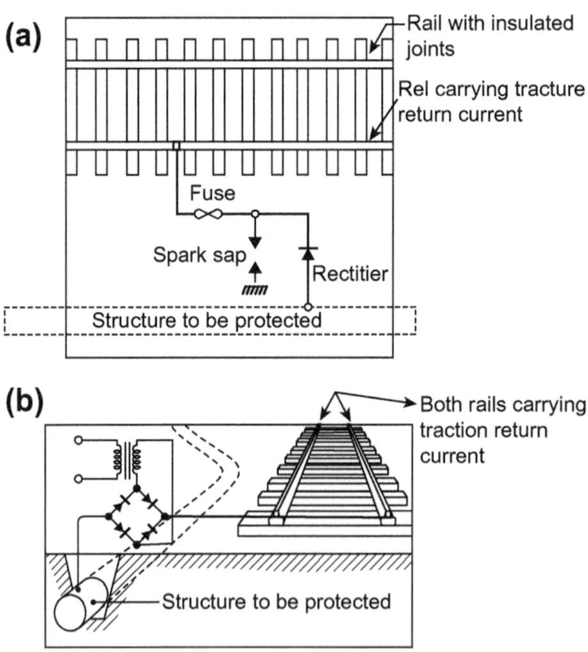

Figure 1.9 Typical electric drainage systems: (a) polarized drainage without a resistor (on an electrified line with single-rail track circuits). (b) forced drainage on an electrified line without track circuits (drainage supplemented by rectified AC).

"single-rail" track circuits and "double-rail" track circuits. In the first type, there are insulated joints in one rail at each end of each signaling section, and the traction return current is confined to the other rail. In the second type, there is an impedance bond at each end of each track circuit, and the traction return current flows in both rails as it does also when there are no track circuits.

Irrespective of the type of drainage system used, the connection to the rail is made to both rails where there are no track circuits, to the traction return rail only where there are single-rail track circuits, and to the midpoint of an impedance bond where there are double-rail track circuits.

Figure 1.9(a) shows the case with single-rail track circuits, and Fig. 1.9(b) shows the case with no track circuits. The amount of CP applied to nearby buried structures by means of drainage bonds may be increased by the use of "forced electric drainage," which entails the insertion in the drainage connection of an independent mains-operated CP rectifier as shown in Fig. 1.9(b). It may be necessary to limit the output from the rectifier by means of saturable reactors or transformers or similar devices.

1.3.6 Measures to Safeguard Neighboring Structures

The application of CP to a buried or immersed structure (referred to as the primary structure) causes a DC to flow in the earth or water in its vicinity. Part of the protection current traverses nearby buried or immersed pipes; cables, jetties, or similar structures; or ships alongside (termed secondary structures), which may be unprotected and the corrosion rate on these structures may, therefore, increase at points where the current leaves them to return to the primary structure. This effect is described as "corrosion interaction."

Corrosion interaction can be minimized by taking care during the design stage; as discussed, it can be assessed by interaction testing and criteria for corrosion interaction and can be corrected, if necessary, by measures methods. One method is to bond the secondary structure to the primary structure so that the former is also cathodically protected. When this method is proposed, consideration should be given to safety aspects.

Corrosion interaction affecting neighboring structures is unlikely to occur as a result of applying CP to the plant internally because appreciable currents flow only through and inside the protected plant.

1.3.6.1 Criteria for Corrosion Interaction

Any current flow that makes the potential of a metal surface with respect to its surroundings more positive is liable to accelerate corrosion. The structure/electrolyte potential is therefore used as the basis for assessment. Positive structure/electrolyte potential changes are more important. Steel surrounded by concrete needs special consideration. Occasionally, negative changes have to be limited.

1.3.6.2 Limit of Positive Structure/Electrolyte Potential Changes

The maximum positive potential change at any part of a secondary structure, resulting from interaction, should not be $>$20 mV. The adoption of a single criterion for all types of structures, irrespective of the value of the structure/electrolyte potential, is oversimplification. It is, however, believed to be reasonable on the basis of evidence available. Where, however, there is a definite reason to suppose that the secondary structure is already corroding at an appreciable rate, even a small potential change that will reduce the life of the structure/electrolyte potential should be permitted.

1.3.6.3 Positive Structure/Electrolyte Potential Changes Steel in Concrete

The foregoing criterion is inapplicable to steel that is completely covered by concrete. Under such conditions, steel becomes passive so that corrosion is prevented. The governing consideration may, therefore, be the disruptive effect of the evolution of oxygen that occurs when the steel is more positive than about +0.5 V (copper/copper sulfate reference electrode). However, the behavior of the steel may be affected by the presence of chlorides (whether introduced initially or due to a saline environment), which may prevent passivation, so that it is impossible to make firm recommendations.

Another complication is that it is not a simple matter to evaluate the structure/electrolyte potential or to measure changes in it across the steel/concrete interface. Changes in the steel/soil potential measured simply by placing the reference electrode in the soil close to the concrete (as distinct from close to the steel) may need to be referred to a criterion other than the 20 mV. However, until another criterion more appropriate to these circumstances is approved, it may be convenient to use the 20-mV criterion as a basis for decision as to whether or not corrective measures should be undertaken.

These considerations apply only to steel fully enclosed in sound concrete. If the steel is only partially encased, the provisions may apply to any area of the surface in direct contact with the soil. It may be noted that in these conditions a cell may be formed in which the steel in contact with soil acts as an anode. The structure/soil potential is likely, therefore, to be more positive at positions near the concrete, and there may be corrosion quite apart from any effect of interaction.

1.3.6.4 Negative Changes of Structure/Electrolyte Potential

If the recommendations made in the previous sections are followed, excessive negative changes of structure/electrolyte potential on the secondary structure will normally be avoided. Large negative changes may, however, occur if the ground bed of an ICCP scheme is unduly close to a secondary structure. Except in the case of aluminum (and, exceptionally, lead in an alkaline environment), corrosion is unlikely to result from making the structure/electrolyte potential more negative. The considerations are, therefore, the secondary effects, particularly the disruption of coatings.

In the absence of any special considerations, structure/electrolyte potentials more negative than −2.5 V (copper/copper sulfate reference electrode) should be avoided on buried structures. In the case of immersed structures, in areas with potentials more negative than −0.9 V (silver/silver chloride/seawater electrode) high-duty coatings must be used. These are based on epoxy resin, chlorinated rubber, vinyl, or other alkali-resistant materials.

Economic considerations will determine whether these should be applied overall or only to area near anodes. Some paints, for example, coal tar epoxy, can withstand potentials more negative than −1.1 V (silver/silver chloride/seawater electrode). Success depends on an adequately prepared surface that ideally should be freshly blast cleaned and free from weathered or unsuitable shop primer.

Where the potentials foreseen are more negative than can be withstood by a paint coating, an insulating shield must be applied near the anode. The boot-top area should be coated with a high duty coating, such as chlorinated rubber or epoxy paint, in preference to oleoresinous types.

1.3.6.5 Insulating Shields for Impressed Current Systems

The high current densities at which impressed current anodes may be required to operate result in very negative potentials immediately adjacent to the anodes. As most hull paints are unable to withstand these potentials, it is important that the surface around each anode be covered by a robust protective shield, extending well beyond the anode mount itself.

The shape and size of the anode shields will be determined by the shape of, and maximum current anticipated from, the anodes.

A disk-shaped anode, for example, will require a circular shield, whereas a long-strip anode will require a rectangular shield of a smaller width but a greater total area.

The dimensions of a shield should be large enough to ensure that the structure/electrolyte potential around its edge is unlikely to cause the breakdown of the adjacent hull paint.

The potential, E, at a distance, r (in meters), from the center of a disk-shaped anode may be calculated approximately from the formula:

$$\mathrm{E} = \mathrm{E}_0 - \frac{\rho I}{2\pi r} \tag{1.4}$$

where E_0 is the general hull potential when protected (volts); ρ is the water resistivity (ohms meter); and $\boldsymbol{I}$ is the current (amperes):

$$\mathrm{E} = \mathrm{E}_0 - \frac{\rho I}{\pi L}\left(\ln\frac{2L}{d} - 1\right), \tag{1.5}$$

where E_0, ρ, and L are as given above, and ln is the natural logarithm.

The value of E_0 is normally about −0.80 V (silver/silver chloride/seawater reference electrode).

In the case of a linear anode, the most negative potential occurs on either side of the center of the anode and the corresponding approximate formula for this potential, at distance d from an anode of length L (meters), is equation 1.5. L is length of anode (m).

The size of the shield should be such that the potential E at its edge is not more negative than can be withstood by the paint that will be applied to the surrounding area of the hull. The shield may consist of either a high-duty coating, for example, epoxy or glass-reinforced polyester resin of an about 1.0-mm thickness built up on the hull steel, or a prefabricated shield usually made as an integral part of the anode mount.

The latter form is more durable and resistant to mechanical damage, but its size may be limited by difficulties of mounting and cost of fabrication. Where necessary, therefore, this type should be supplemented by an surrounding area of a suitable high duty coating.

Care should be taken to see that the hull plating beneath performed shields is adequately protected and, in those cases where it is applicable, the shields are firmly bonded to the plating. It is also recommended that such shields be secured at their outer edges by welded steel fillets or by other suitable means to prevent lifting or striping. Examples of typical anodes and shields are shown in Fig. 1.10(a)–(c).

As a rule, antifouling paint should be applied to the anode shield, but it is imperative that no paint be applied to the working surface of the anode.

1.3.7 Design of CP Installations to Minimize Corrosion Interaction

It is impossible to precisely estimate the amount of corrosion interaction likely to be caused by a CP scheme. The magnitude of any positive changes of structure/electrolyte potential on neighboring secondary structures will depend mainly on the following:

1. Quality of the coating on the primary structure

The better the coating, the smaller will be the current required for protection and the lesser will be the interaction effects.

2. Quality of the coating on the secondary structure

A coating on the secondary structure tends to increase the measured positive changes of structure/electrolyte potential. The greater part of the increased change of the potential difference occurs across the coating and may be regarded as a measurement error that arises because it is impracticable to place the measuring electrode sufficiently close to the surface of the metal. However, high positive changes of structure/soil potential across a resistive coating indicate a possibility of enhanced corrosion, should local coating defects exist or develop later.

3. Magnitude of the structure/electrolyte potential change on the primary structure in the vicinity of the secondary structure

Because interaction effects are roughly proportional to this structure/electrolyte potential change, it should be kept to the minimum required level to provide protection at positions remote from the point of application. A larger structure/electrolyte potential change is necessary at the points of application if the length of the structure protected from any one point is increased.

Thus, interaction can be reduced by applying protection at a larger number of points so that the structure/electrolyte potential change on the primary structure is more uniform and by ensuring, as far as other considerations allow, that the points of application and associated larger structure/electrolyte potential changes are remote from other structures.

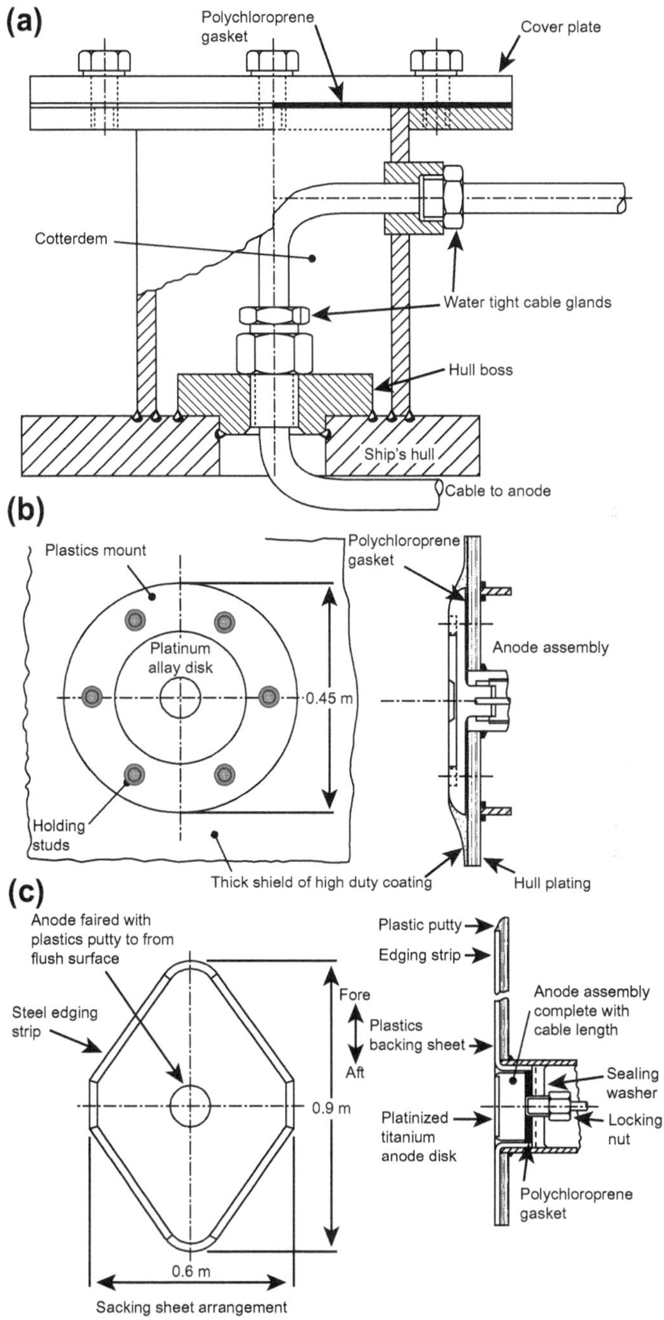

Figure 1.10 (a)–(c), Hull anodes for impressed current: (a) Example of hull penetration gland. (b) Platinum alloy disk anode mounted on a ship's hull. (c) Platinized titanium disk anode recessed into a ship's hull.

4. Spacing between primary and secondary structures

Interaction will be greatest at a crossing point or other proximity. The greater the separation of the structures, the less the effect will be.

5. Distance between the ground beds or anodes and the secondary structure

Structures close to the anode system may be affected by the potential gradient around the anode. Anodes or ground beds should not, therefore, be placed close to other structures.

6. Soil or water resistivity

The potential gradient at any point in the soil is the product of current density and resistivity. Thus, in general, interaction is minimized by siting ground beds in low resistivity areas.

1.3.7.1 Galvanic Anodes

The current available from a single galvanic anode of typical size in most soils is generally of the order of tens of milliamperes compared with impressed current installation where tens of amperes may be produced. If the total current is <100 mA, interaction testing may be omitted. Even if the current is >100 mA, corrosion interaction is unlikely, particularly if anodes are placed at least 2 m away from any secondary buried structure so that the secondary structure does not lie between the anode and the primary structure.

If anode outputs in excess of 100 mA are used, or groups of anodes installed together are used, or if anodes are sited so that another underground metallic structure lies between the anode and the primary structure, interaction testing may be required.

It may be important to reach an agreement at an early stage as to whether testing is necessary for a particular anode system, as connecting links to facilitate disconnection for testing purposes may be necessary. Links may, of course, be required for testing the output of anodes, whether interaction testing is considered necessary or not.

1.3.7.2 Impressed Current Installations

The following precautions should be taken:

1. Structure/electrolyte potentials on the primary structure kept to the minimum, consistent with the required level of protection being obtained.
2. High-quality coatings provided to minimize protection current on a new buried or immersed structure that is to be protected cathodically.
3. The new structure sited as far from neighboring structures as is practicable and the spacing at all crossing points ascertained as being the maximum that conditions permit.
4. The longitudinal resistance of the structure to be cathodically protected made as low as is practicable by means of continuity bonds, welded joints, or other means.
5. The ground bed sited as far from neighboring structures as practicable.
6. Consideration given to installing anodes at a considerable depth, for example, 15–30 m.
7. The total current to be applied distributed from a sufficient number of units to ensure a reasonably uniform distribution of structure/electrolyte potential on the primary structure.

1.3.8 Measures to Reduce Corrosion Interaction

1.3.8.1 Choice of Method

In addition to reconsidering the precautions taken during the installation of the CP scheme, and to ensuring that the current is the minimum necessary to provide an acceptable level of protection, one or more of the following methods should be considered by the parties concerned as a means of reducing corrosion interaction at the points on the secondary structure where positive changes in excess of the recommended maximum have been measured.

The method adopted should aim at restoring the structure/soil potential of the secondary structure to the original value, or preferably making it more negative than the original value. Bonding between structures may be precluded by safety considerations. For example, the bonding together of electric transmission towers and pipelines containing flammable liquids or gases is generally to be avoided. In such cases, the other measures for eliminating the effects of interaction, and below in items 3–5, are to be preferred.

The following techniques are available:

1. A joint cathodic protection scheme should be used so that full protection is given to both structures.
2. Connection of the two structures by means of one or more remedial bonds should be made, which may include suitable resistors to limit the current to the minimum necessary to correct the interaction. This is one of the most effective methods of reducing possible corrosion interaction. A remedial bond should preferably be connected to the secondary structure at or near the point where the maximum positive structure/electrolyte potential change was measured, but if the structures are some distance apart at this point, and it is more convenient, the installation of a remedial bond at a point not too distant, where the structures are closer together, may be satisfactory.
   For reasons, it is essential that any structure to be so bonded be electrically continuous. If it is suspected that the secondary structure may be discontinuous, joints should be tested and continuity bonds should be installed as necessary. The bond, the connections to the two structures, and any resistors, should be constructed to specifications satisfactory to both parties. The bond should be an insulated copper conductor and of an adequate size to carry any fault current that may flow but should not be $<16\text{-mm}^2$ cross-sectional area.
   The bond should be installed inside a suitable housing or building where it can be regularly inspected. When a buried bond is required, mechanical protection against digging operations may be necessary; it is desirable that suitable identification markings be provided. In circumstances where it is clear that corrosion interaction will occur, much testing time will be saved if bonds are installed at suitable positions before interaction tests are conducted. It is always necessary to measure the current flowing in the bond, but if a regular check of the current is required, for example, where resistance bonds have been provided to control the current, a suitable enclosure and a removable link or other bolted connection may be installed at a point accessible to both parties.
3. A galvanic anode should be connected to the secondary structure if the positive changes are small and localized, for example, a meter or so at each side of a point where two structures are close to each other.

4. Resistance should be increased between the two structures, at a point where a positive change is measured, by applying locally an additional good quality coating or warping to the primary and/or secondary structure. If the secondary structure is coated, it is essential that it be free from holidays.
5. In the case of a pipeline, a section of pipe adjacent to secondary structures should be isolated by means of isolating joints. The isolated section can then be bridged by means of an insulated cable of an appropriate size to maintain continuity along the main section of the pipeline. The isolated section can be generally protected by means of galvanic anodes.
   Alternatively, one of the isolating joints can be shunted by a resistor of such a value that the pipe between the isolating joints is protected without causing excessive interaction. This method is particularly applicable at parts of the route near ground beds where the primary structure/electrolyte potentials are strongly negative, creating possible severe interaction on secondary structures. The method can be applied at road crossings where there may be a number of secondary structures, or at railway crossing, to limit interaction with signaling and other equipment. A section of the pipe can also be isolated at crossings with electrified railways to reduce the effects of stray traction of the pipeline.
   It is very important to ensure that methods that entail extending the CP to further structures are not used in such a way that consequent negative changes on such structures, in turn, cause corrosion interaction on a third structure.

1.3.8.2 Temporary Operation of an Installation

If it is urgently required to operate a CP installation that is completed except for arrangements necessary to reduce structure/electrolyte potential changes in neighboring plants. It is recommended that, wherever it seems unlikely that material damage will be caused to the secondary structure, the parties concerned should agree to a limited period of operation without remedial measures. This is to enable the protection to be maintained while the necessary additional work, which should be put in hand as soon as possible, is carried out.

The period during which such a temporary operation may be permitted depends on the following: the amount by which the measured structure/electrolyte potential changes exceed the agreed limit, any available evidence as to the existing state of the two structures, and the period, if any, during which the protection has already been operated prior to testing. If there is no history of corrosion on the secondary structure, it is suggested that, where it is expected that the remedial measures will eventually eliminate the adverse changes entirely, temporary operation may be permitted for periods not exceeding three months with structure/electrolyte potential changes of up to +50 mV.

1.3.9 CP Systems Installed Adjacent to Telecommunication Services

Electrical interference: If a telecommunication cable is bonded to, or is close to, a CP system energized from a source of AC through rectifiers, or if the ground bed of the CP system is sited close to the earth electrodes of a telecommunication system, there may be interference to telecommunication circuits.

This is due to the harmonic currents of the fundamental frequency of the AC supply feeding the rectifier unit inducing unwanted noise voltages into the telecommunication circuits. With rectifier output currents of the order of $\leq$5 A, interference is unlikely. With greater currents, or if interference occurs, consideration should be given to the provision of smoothing for the rectifier.

1.3.10 CP Systems Adjacent to Railway Signals and Protection Circuits

It is imperative that the use of CP on, or adjacent to, railway property be so planned and operated as to ensure that it cannot cause false operation of railway signaling plants.

Railway authorities should be advised of any preliminary planning tests proposed to be carried out in the vicinity of railway running lines. Possible causes of such false operation include the following:

1. Current from the CP system interfering with the operation of the track relays of railway track circuits of the DC type.
2. AC components in the rectified current of the CP system interfering with the operation of the track relays of railway track circuits of the AC type.
3. Induction in cable or line wires from the harmonics in the rectified current of the CP system interfering with remote control transmission associated with signaling installations.

If the railway authorities agree to the bonding of the protected structure to the running rails, or to structures that are themselves connected to the running rails, special precautions may be required to safeguard the railway signaling equipment.

1.3.11 Interaction at Discontinuities in Cathodically Protected Structures

As already discussed, if CP is applied to a pipeline containing a conducting electrolyte, interaction across any discontinuities can cause corrosion of the internal surfaces. Testing to ensure that the pipeline is electrically continuous, and bonding between sections, if necessary, is, therefore, particularly important. Similarly, if isolating joints are inserted in pipes containing weak electrolytes, remedial measures should be taken. Insertion of isolating joints in pipes containing highly conducting electrolytes is inadvisable.

1.3.12 Jetties and Ships: Corrosion Interaction at Sea and River Terminals

Corrosion interaction may arise due to CP of a ship if it is moored alongside an unprotected vessel or jetty, or due to CP on jetties that may cause interaction while unprotected ships are alongside.

In either case, the interaction may be reduced by bonding the unprotected ship to the protected ship or jetty. This will be effective only if the resistance of the bond is low compared with that of the current path through the water. Bonding is sometimes dispensed with in the case of small vessels, notably tugs, that are alongside a protected jetty for short periods. Fortuitous contact may provide some alleviation.

Bonding is sometimes required for other purposes. Interaction can be minimized in designing CP for jetties, by locating anodes on the landward side or within the peripheral boundary of the outer piling. The bonding of ships moored alongside a cathodically protected jetty will increase the current required for protection, and the installation should be designed accordingly.

1.4 Safety Aspects

1.4.1 Danger of Electric Shock

The current normally used for ICCP is rectified AC, taken from a step-down transformer and a rectifier. The transformer should comply with the requirements of relevant standards. The core of the transformer, unless double insulated, and all the exposed metalwork should be effectively earthed.

It is not generally practicable to earth the output circuit of the equipment other than by means of the combined effect of the ground bed or immersed anode and the natural earthing of the structure to be protected. Therefore, only if the transformer is of a double-insulated isolating type or is of a type having an earthed metallic screen between the windings, or if it has windings on separate limbs of the core, should steps be taken to ensure that the combined resistance to the earth of the ground bed and the protected structures is low enough to permit operation of the protective fuses or switch gear in the event of a fault between the input or high-voltage windings of the transformer (i.e., mains voltage) and the DC output of the associated rectifier.

1.4.1.1 Installations on Buried Structures

In the case of buried ground beds, attention should be paid to the danger of possible harm to persons and cattle due to the voltage gradient at the surface of the soil. This will depend on the depth and geometry of the ground bed, the soil resistivity, and the current flowing into the soil, and should at no point be such that the voltage occurring between the feet of people or animals having access to the site could be dangerous. The voltage gradient on the surface of the soil will not, however, be dangerous to people and animals in the following circumstances:

1. The DC output voltage of the CP rectifier does not exceed 50 V DC.
2. The anodes and the upper surface of the backfill surrounding the anodes are buried at least 300 mm below the ground level, thus ensuring that only a proportion of the output voltage appears across the surface of the earth.
3. The leads are fully insulated and protected against mechanical damage between the anode connections and a point well above the surface of the ground.

The danger might be enhanced if, for example, a wire fence on wooden supports passed close to the ground bed. If the wire were earthed at some distance from the ground bed, the voltage between the wire and the soil near the ground bed would be a substantial proportion of the total voltage drop through the soil. These points should be considered at the design stage, and any necessary confirmatory tests should be carried out during commissioning. A similar danger would occur if a wire fence on wooden supports were altered in order to install a ground bed, and a wooden support replaced by a metal support buried close to the ground bed.

1.4.1.2 Installations on Immersed Structures

When ground beds are placed in open water, possible dangers to bathers or fish should be considered. The risk is greater in fresh water than it is in seawater. There is a possibility of danger from electric shock to divers if they are within 1–2 m of impressed current anodes that are in operation: this is particularly important when the underwater scrubbing of a ship's hull is in operation. The impressed current systems of ships should be switched off before divers commence work of any kind in submerged areas.

1.4.1.3 Installations for the Internal Protection of Plant

Voltages >50 V DC. are seldom used for CP; thus, the danger of electric shock would appear to be small, but safety procedures should be adopted that make it impossible for personnel to enter tanks or pipes that normally contain water, while the supply to the anodes is switched on.

1.4.2 Fault Conditions in Electricity Power Systems in Relation to Remedial and/or Unintentional Bonds

There is a possible risk in bonding a CP system to any metalwork associated with the earthing system of an electricity supply network, whether done intentionally or not. This is particularly important in the vicinity of high-voltage substations.

Bonds between metalwork associated with an electricity power system (e.g., cable sheaths) and cathodically protected structures, can contribute an element of danger when abnormal conditions occur on the power network. Principal danger arises from the possibility of current flow, through the bonds, to the protected structure, due to either earth–fault conditions or out-of-balance load currents from the system neutral.

The current, together with the associated voltage rise, may result in an electric shock, explosion, fire or overheating, and also a risk of electrical breakdown of coatings on buried structures. Such hazards should be recognized by the parties who install the bond, and any necessary precautions should be taken to minimize the possible consequences. The rise in temperature of conductors is proportional to i^2t, where i is the fault current and t is its duration. Conductors, joints, and terminations should be sufficiently robust, and of such a construction so as to withstand, without deterioration, the highest value of i^2t expected under fault conditions. For extreme conditions, duplicate bonding is recommended.

Precautions should also be taken against danger arising from the high electro-mechanical forces that may accompany short-circuit currents (see BS 6651, last edition). It is difficult to ensure that current-limiting resistances comply with the foregoing requirements; their insertion in bonds through which heavy fault current might flow should therefore be avoided as far as possible. If they are used, it is essential that they be carefully designed for the expected conditions. Bonds and any associated connections should be adequately protected from damage or deterioration.

1.4.3 Hydrogen Evolution

In impressed current systems, and sometimes with magnesium anodes, excessive polarization can cause evolution of hydrogen on the protected structure. Thus, in situations such as closed tanks where hydrogen can collect, an explosion hazard can arise.

Where hydrogen evolution could produce an explosion hazard, the structure/electrolyte potential should be carefully monitored: hydrogen evolution is not significant at structure/electrolyte potentials less negative than -1.0 V with reference to silver/silver chloride for steel in seawater.

1.4.3.1 Special Precautions for Ships

Hydrogen gas forms an explosive mixture with air, and for this reason, all protected tanks that contain ballast water or have just been deballasted cannot be regarded as gas-free spaces until tested and found safe. It should be borne in mind that the highest concentrations of hydrogen in a tank will be in the upper part of the tank, that is, immediately below the deck head or within the hatch coating. It is essential, therefore, that an escape route to the atmosphere for this gas be ensured at all times.

No dangerous accumulation of gas is likely if the tank hatch lids are in the raised position, but if, for any reason, they have to be lowered and fastened, it is essential that the gas be able to find its way through suitable venting pipe. In the event of such a pipe being fitted with a pressure/vacuum valve, this should be set in the "open" position, giving completely free access to the atmosphere.

During dry docking, it may be necessary, for trim or other reasons, to ballast or partly ballast one or more tanks that are cathodically protected by magnesium. To ensure that hydrogen gas is readily dispersed, the tank lids should be secured in the open position throughout the period in which the vessel is in dry dock.

1.4.4 Installation in Hazardous Atmospheres

1.4.4.1 Explosion Hazards

CP can introduce danger in areas in which a flammable mixture of gas, vapor, or dust (i.e., a hazardous atmosphere) may be present, which could be ignited by an electric arc or spark.

Typical examples of such installations are tanks, pipelines, manifolds, jetty piles, floating craft, etc. Incendive sparking might arise, due to CP, from the following:

1. Intentional or unintentional disconnection of bonds across pipeline joints or any other associated equipment under protection or fortuitously bonded to protected equipment.
2. Intentional or unintentional short circuiting of isolating devices, for example, by tools or breakdown due to voltage surges on the pipeline induced by lightning or by electrical power faults.
3. Unintentional short circuiting by fortuitous bridging of points of different potential, for example, by metal scraps, odd lengths of wire, and mobile plant.
4. Connection or disconnection of loading lines to tankers, barges, and rail car gantry structures and associated pipelines.

5. Disconnection or breakage of cables carrying CP current.
6. Unintentional short circuiting of impressed current anodes when the liquid level is lowered in plants under internal CP.
7. Connection or disconnection of instruments employed for measuring and testing of CP systems.

In locations where any of the above hazards may arise, operating personnel should be suitably instructed and durable warning notices should be authoritatively displayed as appropriate.

It should be noted that the likelihood of incendive sparking may be greater in the case of impressed current systems than with systems using galvanic anodes. However, there is a danger if a suspended or supported galvanic anode, or portion of an anode, becomes detached and falls on to a steel member beneath; the risk, however, is not present with zinc anodes.

1.4.4.2 Measures to Avoid the Explosion Hazard

CP systems that are to operate where flammable concentrations of gas or vapor occur should conform to the statutory and other safety regulations applicable to the particular installation and industry concerned, for example, the Ship Classification Societies have laid down requirements governing the use and inspection of anodes within the tanks of ships classified by them, and approval should be obtained in each individual case as appropriate.

The following safety measures should be adopted where applicable:

1. The enclosure (see BS 4683: part 2 and CP 1003) of transformer/rectifiers or other apparatus should be flame proofed when it is impossible to site it outside the area of risk.
2. Enclosures (see BS 4683: part 2 and CP 1003) of resistive bonds should be flame proofed.
3. A temporary continuity bond across any intended break should be provided, before any break is made in protected pipelines or in other structures or equipment included in the CP scheme.
   It is essential that these bonds be securely clamped to each side of the intended break and remain connected until the work is completed and normal continuity is restored.
4. Any isolating devices in above-ground pipelines should be sited outside the area of risk. Where this is not practicable, measures to avoid arcing or sparking, due to the reasons given in should be adopted. These would include the use of resistive bonds or zinc earth electrodes connected to each side of the insulating device.
5. An isolating device and, if necessary, an encapsulated spark gap should be inserted in each of the loading lines, at oil terminal jetties.
   This is to ensure that the line is electrically discontinuous as a precaution against the dangers associated with incendive sparking. An independent ship/shore bonding cable does not eliminate the hazard. It is essential that any other cable brought on board is connected and disconnected outside the area of risk, or equivalent measures are taken to avoid incendive sparking upon connection and disconnection, for example, by the use of appropriate flameproof techniques (see BS 4683: part 2 and CP 1003).
   Where the loading line is wholly flexible, the isolating device (flange) should be fitted to the jetty manifold. Where the line is partly flexible and partly a metal loading boom, the insulating device should be inserted between the flexible hose and the loading boom.

An isolating device should be incorporated in an all-metal flow boom, and care should always be taken to ensure that the flow booms are not fortuitously earthed to the ship by tools or loads suspended from the ship's gear. Similar considerations apply to ship-to-ship transfers if either or both vessels are cathodically protected and certain low flash-point cargoes are being handled. An isolating device should be fitted at the manifold of one vessel, and the line should be securely earthed to the manifold of the other vessel.

Alternatively, electrically discontinuous hoses may be used, that is, hoses specially made with the bonding wire omitted. For tankers at submarine line berths, at least two hoses that are electrically discontinuous should be inserted into the string of flexible hoses, at the end of the rigid line. These should preferably be the second and third hoses from the tanker manifold. These precautions are not normally considered necessary in the case of single-buoy moorings.

6. Protection devices (surge diverters) should be installed to safeguard the rectifier and associated equipment, for example, instruments, against overvoltages due to lightning or to other external cause.

   This applies particularly to rectifiers that are supplied from overhead lines.
7. Double-pole switches should be provided in each DC circuit entering an area in which flammable concentrations of gas or vapor might occur, to ensure that both poles are isolated during maintenance, etc.

1.4.4.3 Chlorine Evolution

For an ICCP installation in a marine environment, the anode reactions result in the electrolytic formation of chlorine. Seawater is normally slightly alkaline, and the chlorine will form sodium hypochlorite; other side reactions, such as the oxidation of hypochlorite to chlorate and the formation of bromine from bromides, are possible.

Under stagnant conditions, the chlorine will be evolved as a gas and will present a hazard to inspection and maintenance personnel. It is sometimes impossible to use internal protection of enclosed vessels or plants to completely drain a vessel before entering it for maintenance. If it happens that anodes remain energized and immersed, the chlorine level in the remaining water will increase. Disturbing the water, for example, by walking through it, will release enough chlorine in the restricted air space to cause acute discomfort; CP should, therefore, always be switched off before a vessel is entered.

The formation of hypochlorite and gaseous chlorine will be minimized by the incorporation of a system of two levels or automatic control into the CP installation.

2 Application of Cathodic Protection

Cathodic protection can be effectively applied to most steel structures that are in consistent contact with a corrosive electrolyte. Commonly protected structures include the following:

- **Underground Pipelines** are the primary market for cathodic protection. Both sacrificial and impressed current systems are used. Federal and state regulations require cathodic protection for most petroleum or gas pipeline systems.
- **Underground storage tanks** used for fuel are now required by the Environmental Protection Agency to either have functional cathodic protection systems or to be of a noncorrosive material. Both types of systems are widely used.
- **Above-ground storage tank** bottoms can be protected from soil-side corrosion with cathodic protection. Most major tank operators include cathodic protection in their corrosion control program. Unique problems involved with tank applications include the difficulty of distributing current uniformly over the tank bottom and monitoring the effectiveness of the systems.
- **Production well casings** usually require impressed current systems due to higher current requirements. The economics of cathodic protection are excellent until production volumes decline and field near the end of their effective life. This application of cathodic protection is common, but it tends to be concentrated in established fields with a known corrosion history.
- **Internal surfaces** of tanks and vessels are commonly protected by cathodic protection systems. With some exceptions, most of these utilize sacrificial anodes. Possible applications range from heater treaters, heat exchangers, water storage tanks, and hot water heaters.
- **Offshore structures** such as production platforms, docks, and pipelines are almost always protected with cathodic protection systems. Sacrificial anode systems with aluminum anodes are the most common applications.

Cathodic protection should be applied to metal structures where they are in contact with corrosive soils or water, whenever economically justified. Materials other than steel or iron should be considered as separate cases. Metal structures should be coated whenever practical for maximum cathodic protection efficiency.

Impressed current cathodic protection should be applicable to but not necessarily limited to the following:

1. Buried land pipes.
2. Submarine pipelines within the area of influence of shore or offshore platforms where alternating current (ac) power is available.
3. Offshore structures (where power is available).
4. Piers.
5. Storage tank bottoms (where exposed to soil).
6. Water tank interiors.
7. Ship hulls (unless galvanic protection is used).
8. Well casings.

Cathodic Corrosion Protection Systems. http://dx.doi.org/10.1016/B978-0-12-800274-2.00002-8

9. Buried plant piping.
10. Seawater intake systems.
11. Desalination plants.

Galvanic cathodic protection should be applied to but not necessarily limited to the following:

1. Submarine pipelines (beyond the influence of impressed current schemes).
2. Short sections of buried land pipes in areas where soil resistivities are <5000 ohm-cm and ac power is not available or in special cases such as road crossings.
3. Offshore structures and piers where power sources are not available also for all offshore structures for temporary protection during construction and for achieving rapid polarization and maintaining protection in conjunction with impressed current systems.
4. Special electrical grounding facilities on land pipelines.
5. Heat exchanger water boxes.
6. Water tank interiors.
7. Ship hulls.
8. Small boat piers where impressed current systems are not economical.
9. Process vessel interiors.

All buried land pipes and sea pipelines should be provided with an anticorrosive, dielectric coating.

Cathodic protection systems should be so located to prevent the possibility of disturbance or damage to other structures.

2.1 Criteria for Cathodic Protection

When dissimilar metals are in electrical or physical contact (the former through an electrolyte), galvanic corrosion can take place. The process is akin to a simple direct current (dc) cell in which the more active metal becomes the anode and is corroded, whereas the less active metal becomes the cathode and is protected. The electromotive force series shown in Table 2.1 can be used to predict the metal that will corrode in contact with another metal, based on whether it is cathodic or anodic with respect to the other.

Cathodic Protection is an electrochemical means of corrosion control in which the oxidation reaction in a galvanic cell is concentrated at the anode and suppresses corrosion of the cathode in the same cell. The steel pipeline is cathodically protected by its connection to a sacrificial magnesium anode buried in the same soil electrolyte.

Cathodic protection was first developed by Sir Humphrey Davy in 1824 as a means of controlling corrosion on British naval ships. Virtually all modern pipelines are coated with an organic protective coating that is supplemented by cathodic protection systems sized to prevent corrosion at holidays in the protective coating. This combination of protective coating and cathodic protection is used on virtually all immersed or buried carbon steel structures, with the exception of offshore petroleum production platforms and reinforced concrete structures.

Criteria for cathodic protection are in general agreement with the current revision of NACE RP-01-69-83 and in specific agreement with the facts that state that "the selection of a particular criterion for achieving the objective of using cathodic protection is to control the corrosion of metallic surfaces in contact with electrolytes, depends, in part, upon past experience with similar structures and environments wherein the criterion has been used successfully."

Because the conditions and environments make it impractical to comply with Paragraph 6.2.4 of NACE RP-01-69, which states that "the voltage measurements on all buried structures are to be made with the reference electrode positioned as close as feasible to the structure surface being investigated" and "the corrosion engineer should consider voltage (IR) drops other than those across the structure–electrolyte boundary, (IR) drops the presence of dissimilar metals, and the influence of other structures for valid interpretation of his voltage measurements," specific standard criterion incorporates allowances for the intrinsic inaccuracies of structure-to-electrolyte potential measurements, the standard criterion for steel structures in different environments is stated in the following paragraphs.

2.2 Buried Pipes

The criterion most widely used on buried pipes is based on the measurement of potential differences between the pipe and its environment.

Coated buried pipelines should have a minimum pipe-to-soil polarized potential of −1.0 (off) Volts (see Clause 3.2 of DIN 30676, 1985) with reference to the copper/copper sulfate half cell. The potential should be measured immediately after interrupting all the cathodic protection systems influencing the pipeline structure, where it is not practical to measure the interrupted or polarized potential on a pipe, the general guideline for the protection criteria should be as follows:

- For buried pipelines in low-resistivity soils (<2000 ohm-cm), the protection criteria should be −1.3 (on) volts measured at the 1-km test stations or −1.1 (on) volts measured at close interval spacing (15 m or less) with the cathodic protection currents applied.
- For buried pipelines on high resistivity soils (>2000 ohm-cm), the protection criteria should be −1.5 (on) volts measured at the 1-km test stations or −1.2 (on) volts measured at a close interval spacing (15 m or less) with the cathodic protection currents applied.
- Buried facilities in plant areas should be considered to be protected at a minimum of −0.85 (on) volt, measured at test holes over the metallic structure with the current flowing. Another criterion in use is the potential change from the natural value to the value obtained after the application of cathodic protection. In this method, potential change is used as a criterion rather than as an absolute value (say −0.85 V to copper/copper sulfate electrode). In this system, the intent is to change the pipe potential by 0.25 or 0.3 V in the negative direction when the cathodic protection is applied.

Note: Special conditions (e.g., backfill, pipe coating characteristics, or other parameters may require further interpretation of voltage readings) (see NACE RP 01-69-83 Paragraph 6.5).

Table 2.1 Minimum and Maximum Potentials for Cathodic Protection of Bare Metals (volts, without IR drop)

		Reference Electrode							
		Copper/Copper Sulfate		**Silver/Silver Chloride SeaWater (silver/silver chloride/seawater (salinity 32–38%) in brackish water (salinity <32%), the potential reading from an open electrolyte reference electrode must be corrected for the lower chloride concentration. For measurements in such water, a closed electrolyte reference electrode is advantageous)**		**Silver/Silver Chloride Saturated kCl**		**Zinc/(clean) Seawater**	
Metal or Alloy	**Condition**	**Min**	**Max**	**Min**	**Max**	**Min**	**Max**	**Min**	**Max**
Unalloyed and low alloy ferrous materials	At temperatures <40 °C	−0.85	Not applicable	−0.8	Not applicable	−0.75	Not applicable	+0.25	Not applicable
	At temperatures >60 °C	−0.95	Not applicable	−0.9	Not applicable	−0.85	Not applicable	+0.15	Not applicable
	In anaerobic media with a high activity of sulfate-reducing bacteria and sulfate	−0.95	Not applicable	−0.9	Not applicable	−0.85	Not applicable	+0.15	Not applicable

	In aerobic and in anaerobic media with low activity of sulfate-reducing bacteria and sulfates	−0.85	Not applicable	−0.8	Not applicable	−0.75	Not applicable	+0.25	Not applicable
	In sandy soils with resistivities >50,000 ohm-cm	−0.75	Not applicable	−0.7	Not applicable	−0.65	Not applicable	+0.35	Not applicable
Stainless steels with a chromium content of at least 16% by weight, for use in soil and fresh water	At temperatures Below 40 °C	−0.1	Not applicable	−0.5	Not applicable	0.00	Not applicable	+1	Not applicable
	At temperatures >60 °C	−0.3	Not applicable	−0.25	Not applicable	−0.2	Not applicable	+0.8	Not applicable
Stainless steels with a chromium content of at least 16% by weight, for use in salt water	–	−0.3	Not applicable	−0.25	Not applicable	−0.2	Not applicable	+0.8	Not applicable
Copper; copper/nickel alloys	–	−0.2	Not applicable	−0.15	Not applicable	−0.1	Not applicable	0.9	Not applicable
Lead	–	−0.65	−1.7	−0.6	−1.65	−0.55	−1.6	+0.45	−0.6
Aluminum	In fresh water	−0.8	−1.1	−0.75	−1.05	−0.7	−1	+0.3	0
Aluminum	In salt water	−0.9	−1.1	−0.85	−1.05	−0.8	−1	+0.2	0
Aluminum	In soil	−0.95	−1.2	−0.9	−1.15	−0.85	−1.1	+0.15	−0.1
Steel in contact with concrete	–	−0.75	−1.3	−0.7	−1.25	−0.65	−1.2	+0.35	−0.2
Galvanized steel	–	−1.2	Not applicable	−1.15	Not applicable	−1.1	Not applicable	+0.1	Not applicable

2.3 Tanks Exteriors

Tank undersides may be considered partially coated due to contact with sand asphalt padding.

However, contact with the soil will vary with flexing of the base. The potential recorded at the periphery of the underside should be −1.10 to 1.20 (on) volt with reference to a copper/copper sulfate half cell; where permanent reference electrodes have been installed under the tank bottom, a minimum potential of +0.25 (on) volt zinc to steel indicates adequate protection.

2.4 Submerged Pipelines

Offshore submerged pipelines should have a minimum Pipe-to-water potential of −0.90 (on) volt with reference to a silver/silver chloride half cell.

2.5 Offshore Structures and Ship Hulls

Steel structures, other than pipelines, should have a minimum structure-to-water potential of −0.90 (on) volt with reference to a silver/silver chloride half cell.

2.6 Tank, Pipe, and Water Box Interiors

Structures storing or transporting conductive waters or other conductive liquids should have a minimum electrolyte to internal surface potential of −0.90 (on) volt with reference to a silver/silver chloride reference cell or +0.15 (on) volt with reference to an internal zinc reference electrode.

2.7 Well Casings

In general, onshore well casings should be considered to be adequately protected when a polarized casing-to-soil potential of −1.0 (off) volt to a copper/copper sulfate reference cell is measured with the cell located close to the casing and the cathodic current is momentarily interrupted. Alternatively, the potential measured with the cathodic current should be −1.2 (on) volts to the copper/copper sulfate reference cell with the cell located remotely a minimum of 75 m from the well and preferably 180 away from the anode bed.

Where it is impractical to obtain valid casing-to-soil potential measurements, current requirement and polarization test data may be used in interpreting the protected status of well casings. Offshore well casings should be considered protected when the casing-to-water potential is −0.90 (on) volt to a silver chloride reference cell placed close to the casing. Table 2.1 lists the observed protection potentials, that is, potential without allowances for (IR) drop error for full protection of various metals, measured against difference standard electrodes.

Below are some technical notes:

1. All potentials have been rounded to the nearest multiple of 0.05 V. The figures for electrodes in which seawater is the electrolyte are valid only if the seawater is clean, undiluted, and aerated.
2. Aluminum. It is not at present possible to make firm recommendations for the protection of aluminum since this metal may corrode if made too strongly negative. There are indications that corrosion can be prevented if the potential is maintained between the limits shown in the table. Alternatively, it has been recommended in the case of pipelines to make the metal electrolyte potential more negative than its original value by 0.15 V.
3. Lead. In alkaline environments, lead may occasionally be corroded at strongly negative potentials.
4. Stainless steels. In many environments, stainless steels will not require any form of protection. In some cases, anodic protection is used. Stainless steels are however often susceptible to crevice corrosion. A crevice may be encountered between two metals, for example, at a riveted or bolted seam, or between a metal and nonmetal or at a gasketed joint. Crevice attack is a particular from of a differential aeration corrosion and is most often encountered in a marine environment.

   It has been found that cathodic protection will significantly reduce the incidence and severity of this form of corrosion; polarization to potentials given in Table 2.1 is necessary. Difficulty can, however, arise if the crevice can seal itself off from the environment; the protective current cannot flow to the seat of the attack, which may proceed unabated. Polarization of stainless steels to excessively negative potentials may result in hydrogen evolution, which can cause blistering and loss of mechanical strength. Experience has shown that random pitting of stainless steel may not be influenced by cathodic protection, despite the evidence from certain laboratory studies.
5. Steel in concrete. If steel, whether buried or immersed, is only partially enclosed in concrete, the protection potential is determined by the exposed metal and is as indicated in Table 2.1.

Iron or steel fully enclosed in sound concrete free from chlorides would not normally require cathodic protection because of the alkaline environment. For circumstances where cathodic protection needs to be applied, for example, because there is doubt as to the adequacy of the concrete cover or to provide very high reliability, it has been suggested that potentials less negative than are normally required for the protection of steel may be suitable.

2.8 Types of Cathodic Protection Systems

2.8.1 Impressed Current Systems

In this type of system, an external source of dc power, usually an ac/dc transformer rectifier, is used to provide the driving voltage between the anode and the structure to be protected. The negative terminal of the dc source is connected to the structure by a suitable cable, and the positive terminal similarly is connected to the anode. The dc power source should be adjustable so that variations in cathodic protection current and voltage are available.

Rectifiers are especially applicable where electric power is available and current requirements are large or soil resistivity is too high for sacrificial anodes. They are a very flexible source since practically any combination of current and voltage ratings is available and the voltage of a rectifier is normally adjustable over a wide range.

2.8.2 Galvanic Anode Systems

In a galvanic anode system, the driving voltage between the structure to be protected and the anodes is provided directly by the potential difference between the materials involved, If the galvanic anode is subject to possible wetting by oil, Aluminum alloy or magnesium anodes should be used.

Galvanic anodes, either singly or in groups or ribbon, are connected directly to the structure to be protected and are consumed at a rate proportional to the current output. Because of the limited driving voltage, their use is generally in soil or water conditions of electrical resistivity <5000 ohm-cm. This system is applicable where the soil resistivity and current requirements are low and also where a power supply does not exist or can be provided only at uneconomical costs.

The use of magnesium anodes is usually not feasible where the soil resistivity is >5000 ohm-cm (although magnesium ribbon may be used in soils of a higher resistivity). A general rule of thumb suggests that zinc anodes are better used in the lower soil resistivities (<1500 ohm-cm).

2.8.3 Galvanic versus Impressed Current System

In cases where either a galvanic anode or Impressed Current System can be used successfully, the choice is dictated by economics. A cost study is required to determine which type of system is more economical.

2.9 Equipment and Facilities for Impressed Current Systems

2.9.1 Cathodic Protection Transformer Rectifiers

A transformer rectifier in conjunction with an anode system offers the most reliable and effective means for supplying cathodic protection when an ac power supply is available. Transformer rectifiers for cathodic protection use are generally rated in the range of 10–400 A and from 10 to 100 V.

Transformer rectifier construction may use either air cooling or oil immersion. In the latter, all current carrying circuits are immersed in oil-filled tanks. Oil immersed units are preferred for installation in severe corrosive environments. However, air cooled units may be acceptable in less corrosive atmosphere or in nonclassified areas. The air-cooled units are used at an ambient temperature that is $<40\,^{\circ}C$.

2.9.2 Alternative Cathodic Protection Power Sources

Where there is no access to ac supplies, any of the other sources of power as listed below, which technically and economically are justified, may be considered:

1. Diesel–alternator–rectifier set.
2. Gas engine–alternator–rectifier set.
3. Thermoelectric generator (gas operated).
4. Closed circuit vapor turbine (gas or kerosene operated).
5. Solar-powered generator with battery storage.
6. Wind generators with battery storage.
7. Alternating current turbine operating from bypass pipeline flow.
8. Fuel cells consisting of two conducting plates and electrolyte.
9. Petrol engine alternator.

In the case of a Gas trunkline, gas from the pipeline is used to power the gas engine. If the trunkline carries a petroleum product that is suitable for engine fuel, it may be taken directly from the line as well. Otherwise, fuels such as gas, oil, or petrol, depending on the type of engine selected, must be brought to the generator station periodically to fill the fuel storage tank at the site.

2.9.3 Impressed Current Anodes

- Ferro Silicon (an iron alloy with a silicon content of $\sim$14 W% and a carbon content of $\sim$1 W%), graphite, scrap iron, and magnetite (Fe_3O_4 plus additives) are to be considered suitable materials for anodes in contact with soil.
- Ferro silicon anode with 4 W% chromium is recommended for use in soils containing chloride ions. For anodes in contact with salt water, inert anodes of platinized titanium, tantalum, or niobium or lead/silver anodes are commonly used.
- Polymeric anodes are used for cathodic protection of reinforcing steel in salt-contaminated concrete and can also be used for the cathodic protection of in-plant facilities. The system consists of a mesh of wirelike anodes, which are made of a conductive polymer electrode material coated onto copper conductors. The conductive polymer not only serves as an active anode material but also shields the conductors from chemical attack. These mesh anodes are designed and optimally spaced to provide long-term uniform protection at low current densities. In the case of protecting steel in concrete, the anode mesh is placed on the surface of a reinforced concrete structure, covered with an overlay of Portland cement or polymer-modified concrete and then connected to a low voltage dc power source.
- Ceramic anodes are the newest materials available for cathodic protection anodes. These anodes are supplied either as oxide coatings on transition metal substrates or as bulk ceramics.

Their use as cathodic protection anodes for protecting reinforcing steel in concrete is new, but oxide-coated transition metal anodes have been used since the late 1960s in the chloralkali industry as anodes for chlorine production and seawater electrolysis, and cathodic protection of water tanks and bund steel structures, among other industrial processes. These anodes are produced in sheet, mesh, and wire form.

- The following considerations should be considered when selecting the anode for an impressed current system:
  - The relative importance of cost and consumption rate varies according to the conditions. Thus, if the electrolyte resistivity is high, an extensive anode installation will be needed.

    Cheap material is therefore normally chosen, and it will not be excessively costly to provide an additional amount to allow for its higher consumption. For anodes in a low-resistivity environment, such as in seawater, a large mass of material is not generally required so that a compact anode of material capable of working at a high current density may be more economical. Compactness may often be essential, for example, to minimize interference with water flow. In such applications, the mechanical properties of the material and ability to withstand abrasion by suspended matter may also be important.
  - Graphite used as an anode material can be operated at a current density of 2.5–10 A m^2. It is normally impregnated with wax or a synthetic resin to fill the pores and the consumption rate can then be appreciably lower than that of iron and steel. Chlorine that can be generated electrolytically in saline waters is particularly aggressive toward graphite anodes.
  - High silicon cast iron will operate at current densities in the range 5–50 A/m^2 with consumption rates in the range 0.2–1 kg/A year, according to the anode shape and conditions. It is suitable for buried or immersed conditions.
  - Magnetite (Fe_3O_4) is a natural mineral that can be made into a cathodic protection anode by proprietary methods. Such anodes are available in the form of cast hollow cylinders with an internal metallic lining to which a cable is connected and sealed. Magnetite has a low consumption rate of <5 g/A year and may be used in soils, fresh water, and seawater. This material is brittle and requires careful handling.
  - Lead alloys of various compositions are used in seawater applications but are unsuitable for installations where deposits may form. These anodes rely on the formation of a lead dioxide (PbO_2) surface film that constitutes a conducting medium that does not deteriorate rapidly and is self-repairing. For example, lead alloyed with antimony and silver can operate at current densities of up to 300 A/m^2. In another type, the lead is alloyed with silver and tellurium. The performance may be affected adversely in deep water or waters of a low oxygen content.

    A thin platinum wire may be inserted into the lead alloy surface to form a bielectrode that can be beneficial in the initial formation of a film of lead dioxide. This mechanism operates at a current density >250 A/m^2. Such anodes can be operated at current densities up to 2000 A/m^2. Other lead alloys are available, and the operating conditions may be affected by the type of alloy.
  - Platinum and platinum alloys, for example, platinum/iridium are too expensive for use as anodes except in special applications. They are, however, frequently used in the form of a thin layer of platinum or platinum alloy that is about 0.0025 mm thick on a titanium, niobium, or tantalum substrate.

These metals are used as the substrate because they are protected under anodic conditions by an adherent, inert, nonconducting surface oxide film, and therefore resist corrosion at any gap in the platinum. The application of platinum or platinum alloy (platinizing) can be confined to those areas where current transfer is required.

Anodes using these materials can operate satisfactorily at current densities up to 1000 A/m^2 of platinized surface, but operating voltages dictate that a lower figure is used as a basis for design. In electrolytes containing chlorides, the oxide film on titanium may break down if the voltage is >8 V; with niobium and tantalum, 40 V is permissible. Higher voltages may be permissible with fully platinized anodes or in nonsaline environments. The life of the platinum film may be affected by the magnitude and frequency of the ripple present in the dc supply. Ripple frequencies <100 Hz should be avoided. Therefore, anode/electrolyte potential needs to be suitably limited and full-wave, rather than half-wave, rectification is preferable. The life of the platinum film may also be affected by the electrolyte resistivity, and the consumption rate increases with resistivity.

Recent developments have included the use of oxides of precious metals (e.g., ruthenium and iridium), mixed with other oxides, on a titanium substrate. These have operating conditions similar to those of the platinized anodes described above and are said to withstand current reversal. Another material available is a conducting polymer extruded onto a copper conductor.

Characteristics of the principal anode materials are summarized in Table 2.2. These are arranged in the order of cost consumption rate and are given only as examples.

2.9.4 Cables

Cable conductors should be sized such that the conductor is capable of carrying maximum designed current for the circuit without excessive voltage drops except that bonding cables should be a minimum of 16 mm^2. Anode cable insulation should be compatible with the anode environment, which will include generated chlorine gas.

2.9.5 Insulating Devices

The structure to be protected should be isolated from other structures or facilities by means of applicable insulating devices such as insulating flanges, insulating joints, and insulating spools. In the case of pipelines, these include, for example, pressure raising or reducing systems, transfer systems, and domestic service installations.

In areas where there is a risk of explosion, insulating devices should be fitted with explosion-proof spark gaps and should be equipped to prevent sparking such as may occur by bridging caused, for example, by contact with tools.

Insulating flange sets should not be buried. Insulating joints and spools may be installed below grade and buried, provided that they have a suitable bond station. They should be properly identified by marker posts.

2.9.6 Anode Beds

To minimize the anode to earth resistance, anode beds should be constructed with a number of anodes connected in parallel to a common positive dc source.

Table 2.2 Characteristics of Principal Anode Materials

Material	**Density, g/cm^3**	**Maximum Operating Current Density, A/m^2**	**Consumption Rate (mass a year)**	**Maximum Operating Voltage, V**	**Probable Utilization Factor, %**	**Remarks**
Steel	7.85	5	9	50	30–50	Used with carbonaceous backfill
Impregnated graphite	1.1	10	0.5	50	50	
High-silicon iron	7	50	0.2–1	50	50–90	Higher consumption occurs in chloride Containing environment unless Chromium or molybdenum is added
Carbonaceous backfill	0.7–1.1	5	1–2 kg	50	50	Only used in conjunction with other anode materials
Magnetite	5.8	100	<5 g	50	60	
Lead alloys	11.3	300	25 g	25	80	
Platinum on titanium, niobium or tantalum	21.5	1000	10 mg **(platinum on titanium)**	8 **(platinum on titanium)** **40 (platinum on niobium and platinum on tantalum)**	90	Cannot withstand current reversal Operating current density under oxygen Evolution conditions should not exceed 200 A/m^2. Wear rates of platinized (and platinum/iridium) anodes increase with decreasing chloride levels

The construction of anode beds should be of the following forms:

1. Horizontal trench, with carbonaceous backfill and multiple inert anodes laid horizontally in the center of the backfill.
2. Vertical augered or excavated holes each with a single anode surrounded by a carbonaceous backfill.
3. Deep anode beds, constructed vertically to depths that may approach 150 m. These beds may or may not contain a carbonaceous backfill and casing along with the selected anode material.

2.9.7 Electrical Bonding Station

Where one system may affect another, the structures involved should be bonded together. A minimum 16-mm^2-sized cable should be used. Resistors if required should be of the fixed (soldered) tap type or Nickel–chrome wire.

Bonding stations should also be installed across each insulating device in easily accessible locations as required to mitigate interference. Bonding wires from both sides of an insulating device should be terminated in a common bond box.

2.9.8 Test Stations

Stations providing facilities for cathodic protection measurements should be installed. Bonding stations that include bond current regulating resistors and ammeter shunts should be connected between main and secondary cathodically protected structures so that the currents may be regulated and measured.

2.9.9 Connections

All negative drain or bonding cable connections to steel should be thermit welded, or brazed.

2.9.10 Electrical Continuity

It may be necessary to install continuity bonds between different sections of the structure or plant before cathodic protection is applied.

2.9.11 Protective Coatings

The function of a coating is to reduce the area of metal exposed to the electrolyte.

2.9.12 Insulation

It often happens that a well-coated structure, to which cathodic protection could be applied economically, is connected to an extensive and poorly coated metallic structure the protection of which is not required or would be uneconomical. In such a case, the well-coated structure should be isolated before applying cathodic protection to it.

2.9.13 Safety

Sparking hazards may occur when opening or connecting in-service cathodically protected pipelines.

2.9.14 Site Survey

A corrosion survey of the location of the proposed facility to be cathodically protected should be made. Information gathered should include ac power sources, electrolyte resistivities, possible anode bed and rectifier locations, and layout of existing facilities that may be interfered with by the new installation.

2.9.15 Provision for Testing

A cathodic protection installation is unlikely to remain effective unless it is adequately maintained. It is important that the necessary facilities be considered at the design stage so that testing can be carried out conveniently from the time of commissioning the installation. Recommendations relating to the provision of testing facilities for different types of structure or plant are given in each section.

2.10 Cathodic Protection of Buried Steel Pipes

Buried steel pipes are defined as crosscountry pipelines and distribution networks.

2.10.1 Application

This section defines the design requirements for Cathodic Protection of buried steel pipes such as gas or water pipelines and distribution lines.

- Buried steel pipes should be cathodically protected with impressed current systems except in special cases based on available standards.
- All buried steel pipes should be coated prior to installation.
- Short buried lengths of above-ground pipelines, such as road crossings, should be coated and insulated from the rest of the pipeline and protected by galvanic anodes, if such crossings are not an integral part of an impressed current system. An exclusive galvanic anode system is deemed necessary for road crossings only when the soil resistivities are $<5000\ \Omega$-cm and the pipelines do not have an impressed current system. In cases where the road crossings are located in soil resistivities of ≥ 5000 ohm-cm and the impressed current system is not economical, galvanic anodes should be used as an insurance against corrosion.
- As far as possible, the design should utilize existing equipment and available power.
- A site survey should be made to determine soil resistivities and suitable locations for anode beds unless information is available from existing systems or previous surveys made by the owner of the corrosion control department.
- The level of protection for buried pipes should be in accordance with that given in relevant standards. In general, the pipe-to-soil potential should be limited to values specified in Table 2.3, without the cathodic protection current flowing (off).

Table 2.3 Potential Limits for Cathodic Protection of Coated Buried Pipes

Coating System	Off Potentials, Volt (Ref. $Cu/CuSO_4$)
Epoxy powder fusion bonded	−1.5
Asphalt and coal tar enamel	−2
Tape wrap (laminated tape system)	−1.5
Epoxy coal tar	−1.5
Polyethylene (two layers)	−1
Polyethylene (three layers)	−1.5

- In cases where coated buried pipes with high-quality coatings, such as epoxy, are electrically continuous with buried pipes with lower quality coatings, such as plastic tape, the cathodic protection design current capacity for the worst case should be used.

The real amount of current required to attain protection is determined by the current requirement test. Results of a soil resistivity test and a current requirement test will provide the information necessary to design a ground bed. Soil resistivity tests are also necessary to locate the ground bed in an area of low soil resistivity and also to determine the type of anode bed.

- Cathodically protected pipes should be electrically continuous. They should be insulated from other buried metallic structures, except where stray current interference bonds are required.
- Facilities for pipe-to-soil potential measurements should be installed at intervals not >1-km markers for pipelines and at 250–300 m for distribution lines.
- On Pipelines >19 km, in-line current measurement facilities should be available at intervals of not >10 km. In addition, these facilities should be installed at all branches or intersections with other pipe work. The current measurement facilities should consist of two pairs of cables, each pair connected to the pipeline 60 m apart.
- All equipment, including the anodes and power source, should be designed for the life of the pipeline, or 25 years whichever is lesser.
- There should be direct access to rectifiers from the Gas-Oil-Separator-Plant (GOSP) or plant areas. If located within an electrical substation, the rectifier should be separately fenced, with a separate entry from the plant area or a separate gate if in a remote area.
- All pipeline anchors and all pipeline supports except above-grade concrete pads should be electrically (metallically) separated from the pipeline.
- At a plant or GOSP, the rectifier negative connection should be made to the pipeline side of the insulating device. The insulating device should be installed immediately after the pipe becomes exposed.
- All buried pipes in a corridor will have an integrated cathodic protection system irrespective of the type of coating on the new and existing pipes.
- The pipe should, whenever possible, be surrounded by a backfill that is free from stones and carbonaceous material such as coke. If this is impracticable, an additional wrapping of heavy bituminized felt or other suitable material, such as rock shield, should be provided. Where pipes cross streams or rivers, additional protection, in the form of concrete encasement over the pipe coating, may be required.

As the coating deteriorates, an appreciably higher current density is likely to be required to maintain protection, and this should be allowed for in the design of the cathodic protection system. Factors contributing to coating deterioration include soil bacteria, soil stress, pipe movement, effects of pipeline product temperature, water absorption, etc.

2.10.2 Types of Cathodic Protection Systems

2.10.2.1 Impressed Current System

An impressed current system should consist of one or more stations comprising a dc power source, anode bed, and connecting cables. The location of these stations for pipelines will be influenced by the availability of ac power and the attenuation of potential. The attenuation of protection along a pipeline from a single cathodic protection installation will depend on both the linear resistance of the pipeline and the coating conductance. It may be calculated using the formulas given in Chapter 3. The attenuation formulas are not applicable to distribution lines, and the current requirements and availability of land for station will have an influence on the location of the station in this case.

2.10.2.1.1 Surface Anode Beds

Horizontal or vertical anode beds should be constructed. The choice between horizontal or vertical anode beds should be made after field survey results. Anode beds should be rated for the maximum output of the power source. The minimum anode bed distance from the buried pipe or adjacent structure should be 100 m, unless field data show that interference is not a problem. Preferred criteria for this minimum distance are 50 m for a 30-A rated output, 100 m for a 50-A output, 200 m for a 100-A output, and 300 m for 150-A output. Multiple rectifier anode bed installations in the same general area should have anode beds separated by a minimum distance of 300 m. Multiple anode beds are to be separated by a minimum distance of 100 m. The anode bed (ground bed) should be installed at right angles or parallel to the pipeline.

To achieve the desired objective, careful consideration should be given to selection of the installation site for the anode bed. The most important consideration is the determination of effective soil resistivity, which is a prerequisite for the site selection; this consideration must also be taken into account when selecting a site for the anode bed. One point to be considered in this respect is to ensure that no other underground metallic structure exists within the area of influence surrounding the anode bed, since such a metallic structure will pick up the current from the anode bed and create a stray current interference problem that will require corrective measures. It is essential that interaction effects be taken fully into account and suitable precautions adopted.

Before the final decision is made on the selection of the site for the anode bed, coverage of each bed and optimum spacing between the beds are other points of consideration in selecting the site for the anode bed.

2.10.2.1.2 Deep Anode Beds

Deep anode beds are generally employed where top soil strata have a high resistivity and/or where the horizontal or vertical anode bed is impractical as in gas distribution systems. Resistivity generally decreases considerably with depth, especially within the water table.

If more than one deep anode is to be connected to a single rectifier, they should be spaced at a minimum distance of 30 m from each other unless closer distances are justified. Deep anode bed installation should be in accordance with standard drawings.

2.10.2.1.3 Cables

Conductor sizes should be sufficient to carry the maximum designed current load of the circuit without any excessive (5%) voltage drop.

- **Connections**

All positive anode cables from rectifiers to anode beds should be run in continuous lengths to a maximum length of 150 m. An above-grade test box (bond box) should be used for every 150 m cable run exceeding this length. Only in exceptional cases should buried splices be used in epoxy-filled splice case. Generally, buried splices should be restricted to header to anode cable connection(s) and to negative cable splices. Negative drain or bonding cable connections to steel should be termite welded or brazed.

- **Anode Materials**

Type of anode materials that may be considered for use in impressed current. Ground beds.

2.10.3 Galvanic Anode Systems

In cases where galvanic protection is required, that is, where relatively small increments of current are required in areas with a low soil resistivity (listed below), it should consist of block or prepackaged anodes installed along the buried pipe, or a ribbon anode installed in the same trench along the pipeline according to the installation of anodes:

1. On bare or very poorly coated systems, in which complete cathodic protection may not be feasible from a cost standpoint.
2. On well-coated pipes having overall impressed current cathodic protection systems, where additional small amounts of current are needed.
3. On short increments of well-coated pipes such as in distribution and service lines.
4. Galvanic anodes are also placed at specific points on a pipeline (often termed "hot spots") and may be expected to protect only a few feet of the pipe.
5. Galvanic anodes may be used to provide temporary protection of sections of buried pipes where particularly corrosive condition exist, for example, at river estuary crossings, pending commissioning of an overall impressed current system.

2.10.3.1 Magnesium

High potential magnesium 1.70 V ($Cu/CuSO_4$) will be used as a galvanic anode material as per material specification. A magnesium anode may be used in soil resistivities <5000 ohm-cm. For soil resistivities of <700 ohm-cm, zinc anodes can be used.

2.10.4 Isolation of Buried Pipes

In the design of the ground bed for gas or oil transmission pipeline, it is a common practice to locate ground beds at compressor or pumping stations, where a power source is available. This is a sound practice from the technical and economical standpoint provided that the following consequences are taken into consideration and the relevant remedial actions are implemented (see also NACE Standard RP 0288-86 item No. 53060):

- If station pipings, equipment, and machineries are not insulated from the mainline, most of the protective current for the mainline will be drained from poorly coated buried pipes and valves. This would inevitably reduce the protection span of the mainline.
- Lack of isolation of the mainline from station piping and facilities would cause protective current to flow preferentially through low resistance copper grounding of the station; the apparent result is reduction in the length of the mainline being protected.
- To obviate the above-mentioned undesirable consequences, it is essential to isolate station facilities and piping from the mainline using insulating joints or flanges. This isolation could in turn create stray current problems on station-buried piping and accelerate corrosion.

Resistor bonding and test for stray current interference should be made to ensure that the effect of stray current is reduced to the acceptable limit before a separate cathodic protection system is installed for station piping.

This risk of corrosion acceleration can be reduced by paying special attention to the coating of the pipeline near such equipment and by locating the ground bed as distant as possible from such equipment. It is essential that the decision be based on an assessment of the relative costs of providing complete or partial protection and an assessment of the likelihood and consequences of corrosion of the various parts of the isolated equipment.

Once the cathodic protection is commissioned, procedures described in standard enable any effects of interaction to be assessed and overcome. If the pipe contains an electrolyte, there is a danger of corrosion being stimulated at the inner surface of the pipe. With weak electrolytes, for example, potable waters, it may suffice to coat the pipe internally for at least two diameters on either side of the isolating joint. Replacement of a length of pipe by a length of nonmetallic piping, if practicable, may be a more economical method.

In the case of highly conducting electrolytes, such as seawater or brine, it will probably not be feasible to coat or replace a sufficient length of pipe to eliminate the danger of corrosion occurring internally near the flange. The tests to determine the length of pipe that should be treated would be complicated. Where the provision of an isolating joint is considered to be essential and conditions allow, it may be considered best to install an extra heavy walled pipe on either side of the joint and to accept the need for periodic inspection and replacement.

2.10.4.1 Insulating Devices

Cathodically protected pipelines should be electrically isolated from plant piping, wellheads, etc., by the installation of insulating devices. Insulating devices should also be installed at transitions between cathodically protected buried pipes and non-cathodically protected above-ground pipes, where economically justified, and also between submarine pipelines and land pipelines where current control and distribution justify their use. Pipes of different metals should be insulated from one another.

Insulating flange sets should not be installed in buried portions of pipes. Insulating joints may be installed below grade and buried. Insulating joints that are buried should be provided with a protective coating that is equal to or better than that applied to the pipeline.

Insulating devices should not be used to electrically isolate gas-operated valves and new and existing pipelines installed in a pipeline corridor.

Bounding stations should be installed across each insulating device in easily accessible locations as required to mitigate interference.

Bond wires from both sides of an insulating device should be terminated in a common bond box in a way so as to allow the electric continuity of the pipelines to be opened when necessary.

Distribution networks require insulating devices to be fitted to all domestic service pipes and sometimes also to individual sections since it may be convenient to subdivide the area to be protected into smaller zones.

Distribution networks also require insulating devices to limit the zone of influence of the impressed current stations. They should also be installed at pipeline nodes so as to prevent signals emitted anywhere on the network from returning to their send-out point.

Isolating joints for pipelines need to be adequate for the maximum pressure and temperature conditions encountered on the particular installation. The usual method of insulating flanged joints is to employ insulating sleeves on the studs or bolts and insulating washers under the nuts and/or the heads of the bolts, bright mild steel washers being provided to prevent damage to continuity within the cathodic protection system. Insulating sleeves and washers are usually made of synthetic resin-bonded fabric; insulating gaskets may be made of compressed nongraphited asbestos or other suitable materials. Wire-reinforced gaskets or electrically conducting gaskets are considered unsuitable. The internal bore and centralizing of the insulating gasket should ensure that no recess is formed internally, between flanges, in which material that would bridge the insulating gap might accumulate.

Under dry conditions and before connection to other structures on either side, an electrical resistance >1 M should be attainable for pipes of 500 mm diameter or less. Care should be taken to ensure that the testing voltage is appropriate for the intended service conditions. The gap between the flanges should be filled with the compound, and the outer faces of the flange should be taped to minimize the possibility of accidental short circuit. Exposed isolating devices should be painted a distinctive color for identification.

It is preferable to assemble isolating joints and test them both hydrostatically and electrically before they are installed in the pipeline; the isolating joint are supplied with short lengths of pipe so that the joint can be welded into the pipeline without being dismantled. If insulation is being introduced at a pair of flanges already provided in the pipeline, care should be taken to ensure that it is not damaged. In either case, the final testing of the isolating device after assembly will be required. This can usually most conveniently be carried out on commissioning the cathodic protection system by measuring the change of potential across the flange.

2.10.4.2 Pipe Supports

Cathodically protected pipelines installed above ground and supported, with steel supports, should be insulated from the supports by means of a minimum 3-mm thickness of insulating material (not wood) placed between the cradle and the pipe. Insulating material should extend at least 25 mm beyond all support edges. Buried pipe supports have diminished the effectiveness of cathodic protection systems and have caused pipeline leaks. Pipe supports should not be buried at the support pipeline interface. Insulating devices should be considered at the termination points of above-grade piping run when electrical insulation of large quantities of pipeline supports becomes economically impractical. The buried sections of pipe should then be appropriately bonded together if necessary to maintain continuity within the cathodic protection system.

2.10.4.3 Electrical Ground Isolation

For maximum effective electrical isolation, insulating flange kits, spools, or joints should be located on the pipeline side of motor operated valves, instrumentation (electrical or pneumatic), and corrosion monitoring devices. Grounding cables, electrical neutrals, feed back lines, etc., act as low resistance shunt paths, and thereby "shorten" the insulating flange kits and joints. Where isolation is not practicable, the equipment should be electrically isolated from the pipeline or alternatively separate grounding for the equipment should be installed and isolated from the general grounding system for the power supply. Polarization cells can also be used to isolate pipelines from grounding systems. Cathodic protection junction boxes and conduit runs should have a separate ground rod and not be tied into the general grounding system of a plant.

2.10.5 Test and Bonding Stations

2.10.5.1 Potential Measurement

Test stations for potential measurement should be installed at intervals not >1-km markers for pipelines and at 300–500 m of distribution lines and when required at valve boxes, bonding stations, road crossings, and other appropriate locations dictated by operational needs. The cable conductor size should be 16 mm^2 for pipelines and 10 mm^2 for distribution lines.

2.10.5.2 Bonding Stations

Minimum 16-mm^2 cables should be connected to the structures and brought into the bonding stations. Bonding and test stations may be combined where feasible.

Fixed tap resistances or nichrome wire is required, and a shunt is included to enable current measurements.

2.10.5.3 Line Current Measurement

Two test leads separated by a fixed distance (usually 30–60 m) are required for the measurement of current flow in the line. The current flow is calculated from the voltage drop measured in the span of the pipe between the two test leads. In an impressed current system, as a minimum, facilities for measurement of the current flow should be installed in both the direction of each drain point and the midpoints between drain points.

2.10.6 Cased Crossing Test Station

Casing pipe used at road and railroad crossings must have test stations installed to allow a periodic check for shorts between the carrier pipe and the casing or the vent pipe (Fig. 2.1).

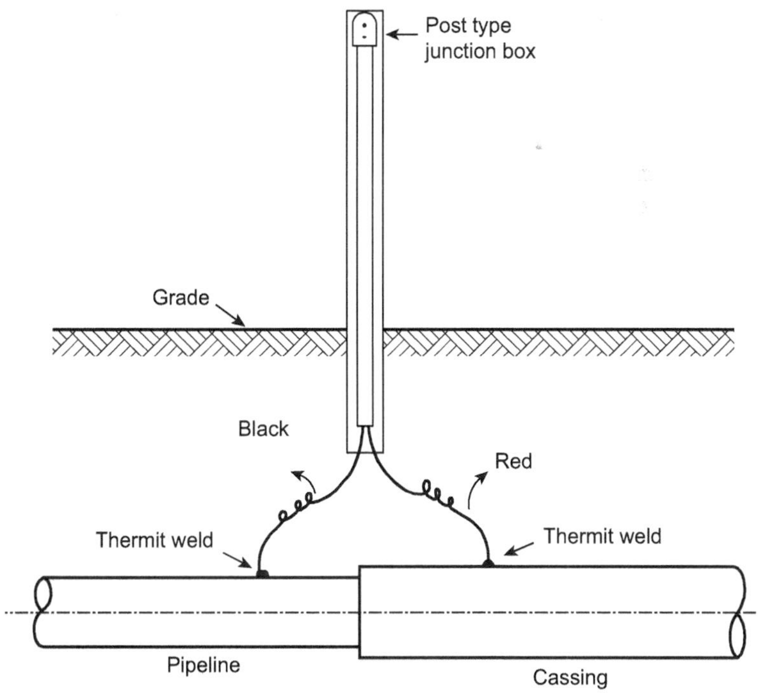

Figure 2.1 Typical cased crossing station.

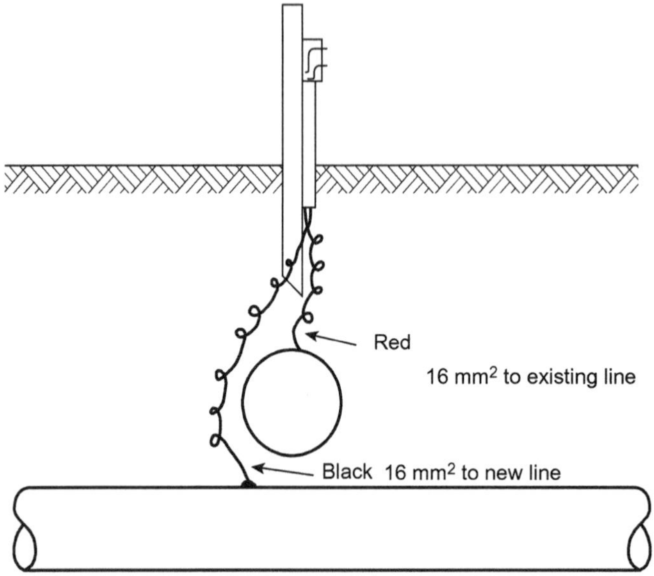

Figure 2.2 Typical line crossing test station.

2.10.7 Line Crossing Test Station

The pipeline crossing test station should be installed at locations where the possibility of interference problems may exist (Fig. 2.2). If stray current is found, a resistance bond wire can be installed.

2.10.8 Insulated Fitting Test Station

All underground insulated joints must have test stations installed to allow a periodic check of the effectiveness of the insulator (Fig. 2.3).

2.10.9 System Design

The design principles of cathodic protection for buried pipelines are given in next sections.

2.10.10 Particular Considerations

2.10.10.1 Casings

Pipe sleeves (casings) may act as a shield preventing the flow of the cathodic protection current to the carrier pipes inside them; their use should therefore be

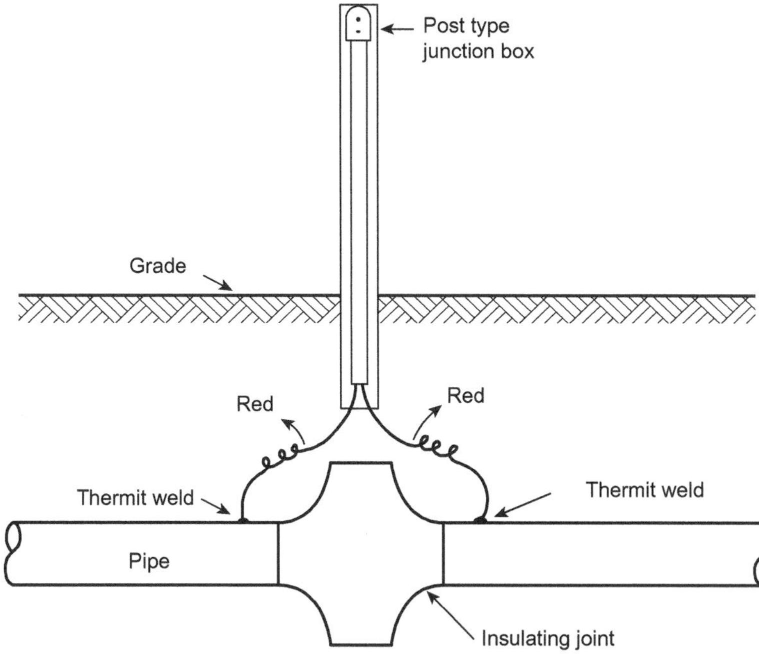

Figure 2.3 Typical insulating joint test station.

avoided where possible. If, however, their use is mandatory, measures as indicated below should be taken:

1. Protection of carrier pipe

Any section of the pipe installed inside concrete or steel pipe sleeves should be coated to at least the same standard as the remainder of the pipeline. If it is feasible to keep the pipe inside the sleeve dry by means of suitable end seals; this will ensure freedom from corrosion. If, however, it is impossible to exclude water, consideration may be given to installing galvanic anodes in the form of a magnesium or zinc strip attached to the carrier pipe. The carrier pipe should be supported by insulating centralizers to prevent coating damage during installation and, where steel sleeves are used, to prevent electrical contact between pipe and sleeve. Such contact is to be avoided since, unless protection has been provided by galvanic anodes inside the sleeve, the carrier pipe will be left without protection because of the shielding effect of the sleeve. If the sleeve is poorly coated, it will also cause a marked increase in the current required for cathodic protection. If insulation is provided, it should be checked before backfilling.

2. Protection of sleeve casing

If a steel sleeve is used and is insulated from the carrier pipe, corrosion will be accelerated, particularly at the inner surface if moisture has not been excluded from the space between the sleeve and the pipe. If the sleeve is to be cathodically protected, it should be coated to the same standard as the pipeline. The sleeve can be cathodically protected by connecting it to the pipeline through a surface connection box or by means of galvanic anodes. If the sleeve is cathodically protected, it is particularly desirable to safeguard the carrier pipe inside the sleeve either by complete exclusion of moisture or by use of galvanic anodes.

It may be possible to seal the pipe sleeve to the pipe by welding at each end and filling the annular space with an inert gas, for example, Nitrogen. The sleeve should then be coated externally to reduce the cathodic protection current demand, but it would be unnecessary to coat the pipe inside the sleeve.

2.10.10.2 Concrete Encased Pipe

The pipe to be encased in concrete, as at a river crossing, valve boxes, etc., should first be coated to the same standard as the rest of the pipeline.

2.10.10.3 Protecting Existing Pipelines

Cathodic protection, usually by an impressed current, may often be applied to old pipelines, even though they are bare or have badly deteriorated coatings, to prolong their life. It is rarely economical to lift and recoat such pipelines except possibly at selected points where interaction may be expected. High current densities will therefore be necessary. The current initially required can be determined by means of a drainage test carried out after isolating the pipeline as far as possible from other structures in contact with the soil. The final current requirement will usually be much less due to long-term polarization effects.

2.10.10.4 Scale Detachment

After a cathodic protection system has been put into operation, it may be found that there is a temporary increase in the number of leaks due to scale detachment at heavily corroded areas where the pipe wall has been seriously weakened or perforated. The majority of such leaks become manifest in the first one or two years of operation of the cathodic protection system.

2.11 Cathodic Protection In-Plant Facilities

2.11.1 Application

- This Section of engineering standards applies to in-plant facilities such as tank bottoms, metallic pipelines, and other metal structures buried or in contact with soil or water.
- The design of new plants should provide an integrated cathodic protection system for all buried metallic pipelines and where required for electrical grounding grids if they are electrically connected to the buried metallic pipelines or other protected facilities.
- The design of additional facilities in existing plants should provide an integrated cathodic protection system. If economically justified, the new cathodic protection system should incorporate protection for the existing plant-buried facilities where required to minimize interference or ensure operation of the system.

- If steel Ground bed is selected for temporary protection, the consumption of steel at the design output rate should be taken into consideration. One ampere of direct current discharging into the usual soil can remove approximately 9 kg (20 pounds) of steel in one year.
- Cathodic protection of the in-plant facilities may be provided by impressed current or galvanic systems or a combination of both.
- The choice between impressed current and galvanic systems should be based on the economic analysis of the availability of electric power and the effectiveness of either system.
- As far as possible, the design should utilize existing equipment and power available.
- A site survey should be made to determine soil resistivities and suitable locations for anodes unless the right information is available from existing systems or previous surveys made by company corrosion control department or other departments.
- The design current density for plant cathodic protection system should be determined by the current requirement test.
- The cathodic protected facilities should be electrically continuous.
- All equipment, including the anodes and power source(s), should be designed for the life of the in-plant facilities or 25 years whichever is lesser.
- There should be direct access across to the transformer/rectifier(s) from the plant areas. If located within an electrical substation, the transformer/rectifier should be separately fenced, with a separate entry from the plant area or a separate gate if in a remote area.
- All above-grade facilities, such as transformer/rectifier(s), junction boxes, ac step-down transformers, should be located in the lowest electrically classified area.

2.11.2 Types of Cathodic Protection Systems

2.11.2.1 Integral Impressed Current Systems

The following design guidelines determine the outline and definition of an integral impressed current system:

All metallic buried sections within the plot limits should be electrically bonded together by means of accessible cables. The cable connections should be made by means of over-ground test boxes. No significant section should be left out to minimize the rise of interference. The primary reason for bonding is electrical safety during ground fault conditions.

Distributed anode configurations and/or polymeric anodes are preferred for new construction sites and where demolition of concrete and asphalt is minimal. Deep well anode designs may be used where justified by structure complexity and installation economics.

Insulating flanges, joints, or spools should be specified only where needed to limit protective current flow between the on-plot and off-plot pipeline sections. Bonding stations should be provided across insulating flanges, joints, and spools. All installations that have limited accessibility for accurate potential readings, such as tank bottoms, should be provided with permanent packaged reference electrodes. Current density to be applied should be selected from the results of cathodic protection tests.

Tank undersides may be considered to be partially coated due to contact with sand asphalt padding. However, contact with soil will vary with flexing of the base. The potential recorded at the periphery of the underside should be −1.10 V minimum with reference to a copper/copper sulfate half cell. Where permanent reference electrodes have been installed under the tank bottom, a potential of −0.85 V versus that of a copper/copper sulfate half cell indicates adequate protection.

2.11.2.2 Galvanic Anode System

Galvanic anode systems may be used in conjunction with segregated impressed current systems to protect small buried metal structures such as road crossings. Where conditions indicate that cathodic protection cannot be fully achieved with packaged magnesium anodes, a helically wrapped magnesium ribbon and plastic envelope system should be used at road crossings.

When the ac power for permanent impressed current will be delayed after plant construction, temporary galvanic anodes should be installed in soil <5000 ohm-cm.

2.11.3 Impressed Current System Details

An impressed current system should consist of one or more stations comprising ac power source, anodes, connecting cables, and necessary test facilities. The location of the transformer/rectifier(s) will be influenced by the availability of ac power and anode locations.

2.11.3.1 Anode Beds

Horizontal or vertical anode beds should be installed in accordance with relevant standards. The choice between horizontal and vertical anode beds should be made after field survey results and evaluation of local conditions. Anode beds should be rated for the maximum output of the power source. The design of the ground bed is based mainly on the following points:

- Current requirement for the section of the pipeline to be protected by the particular ground bed.
- Soil resistivity of the site selected for the ground bed.
- Type of anode material and its dimensions.
- Type of anode installation (vertical, horizontal, spacings between anodes, and backfill material surrounding anodes).
- Life expectancy of the ground bed based on the consumption of anode and backfill resulting from current discharge.
- Cables used in the cathodic protection circuit (material and size).

2.11.3.2 Cables

Conductor sizes should be sufficient to carry the maximum designed current rate of the circuit. Power loss in cables should not be >8% of the total dc power. Cable run design should use loops or multiple returns to minimize (IR) drop and increase the security of the anode system.

2.11.3.3 Connections

Splices should be made in above-ground boxes except for splices of the header cable to anode cable connections. Underground splices should be made with split bolt connectors (line taps). All underground splices should be of the waterproof type and insulated with epoxy kits and standard rubber tape or vinyl tape.

2.11.3.4 Anode Materials

Various types of standard anodes may be considered for use in impressed current ground beds.

2.11.4 Galvanic Anode System Details

In special cases where sacrificial anode protection is required, it should consist of anodes distributed around in-plant facilities. The weight of the anodes required and the current output can be calculated. Magnesium may be used as the sacrificial anode material in soil resistivities of <5000 ohm cm. Zinc anodes may be used in soil resistivity <700 ohm-cm.

2.11.5 Test and Bonding Stations

2.11.5.1 Potential Test Stations

To allow for a systematic inspection scheme, a number of test points with marker plates should be installed at predetermined locations. The marker plate should identify the protected buried structure and include the cathode protection test station number. The cable from this station should be connected to the buried structure in cemented and asphalted areas.

2.11.5.2 Reference Electrodes

Where testing with a portable test electrode is inaccurate, that is, under tank bottoms and foundations, permanent reference electrodes should be used. These electrodes should be installed prior to the construction of the protected structure. Reference electrodes used for this purpose are prepackaged high purity zinc electrodes or permanent $Cu/CuSO_4$ electrodes.

2.11.5.3 Current Monitoring Stations

Four point test stations may be required in the case that the designer can foresee that a high electrical current may occur in one of the multiple legs of the system.

2.11.5.4 Bonding Stations

All bonding stations should provide for measuring the current by means of a shunt and space to install a resistor if required.

2.11.6 Hazardous Locations

Transformer rectifier specifications must follow the rules and regulations concerning area classification for electrical equipment. Other electrical equipment such as above-ground splice boxes, bond boxes, or test stations are generally nonexplosion proof and require a hot work permit for opening and for test work.

2.11.7 Oil Storage Tank Bottoms

Large oil storage tanks are often erected on earth foundation mounds. The top of the foundation mound may be provided with a bit-sand carpet, or some form of coating may be applied to the underside of the tank bottom. During operation, the bottom of the tank is subject to flexure and settlement. The bit-sand carpet is therefore liable to be damaged, and this thus allows parts of the tank bottom to come into contact with the soil of the foundation mound. Any coating applied to the tank bottom plates is also likely to be partially destroyed by the heat of welding of the lap joints, with the same result.

Hence, if the environment is corrosive, cathodic protection may be applied to supplement the protection provided by carpeting or tank bottom coating. The above general considerations also apply to similarly constructed tankage for use with liquids other than with petroleum or its products.

2.11.7.1 Structure Preparation

The tank foundation mound should as far as possible be constructed so that it will distribute protection current uniformly to the whole of the underside of the tank. This means that the use of rubble, rock fill, etc., should be avoided and the mound should consist of fine-grained and well-compacted material.

Oil storage tank bottoms are constructed by lap welding individual plates and are therefore electrically continuous. Where groups of tanks are to be cathodically protected, provision will need to be made for bounding between individual tanks. If it is desired to confine the protection current to the tanks, isolating joints will be required in all pipelines connected to the tanks. If flammable liquids are being stored, such joints should be located outside the tank bund.

2.11.7.2 Application to Oil Storage Tank Bottom

Current density figures of the order of 10 mA/m^2 are normally required for uncoated tank bottoms resting on bit-sand carpets and, owing to the large areas involved, impressed current is usually preferred. Galvanic anodes may, however, be worth considering for small isolated tanks on soil with low resistivity.

Ground beds should, as far as possible, be symmetrically arranged around the tank or group of tanks, and a distance of not less than one tank diameter from the tank periphery is desirable. Ground beds should not be located inside tank bunds if flammable liquids are being stored. Where space is restricted, consideration may be given to installing anodes at considerable depths.

2.11.7.3 Particular Considerations

Structure soil potentials vary across the underside of a cathodically protected tank bottom and rim potentials usually need to be more negative than −0.85 V to ensure protection at the center. The main factors concerned are the insulation provided by the tank bottom coating in relation to soil resistivity, and variations in soil resistivity with depths at the tank site.

On new constructions, it may be possible to install two or three permanent reference electrodes in the foundation mound prior to tank erection so that the least negative rim potential necessary to secure full protection of the tank bottom can be assessed and suitable mechanical protection will be required.

Where measurement is impracticable, an arbitrary value of the rim potential has to be adopted. Potentials in the range of −1.10 to −1.20 V (copper/copper) sulfate reference electrode have been found to be satisfactory. Tanks that remain empty for prolonged periods are therefore unlikely to receive complete protection because upward flexure of the tank bottom when the tank is emptied leads to a loss of contact with the foundation.

Cases may arise where an old tank on a granular bed, for example, granite chips, is lifted for rebottoming and the subsequent application of cathodic protection. Under these circumstances, the spaces between the granite chippings should be filled with suitable fine-grained material to enable protection current to flow to the whole of the underside surface of the new bottom.

2.12 Cathodic Protection of Vessel and Tank Internals

This section applies to all types of water storage tanks, and considers water boxes, salt water strainers, produced oil field brine handling and treating vessels, and other vessels in which water may be in contact with the interior surfaces. Due to the varied nature of vessels and tank designs, this section contains specific designs for only the most common types and is intended to be a general guide for all vessel and tank design used.

The need for providing cathodic protection to tanks containing high resistivity (low dissolved solid content) waters should be dependent on the severity of existing or anticipated corrosion and also the end use of the water itself. Internal cathodic protection may be provided by galvanic or impressed current systems. Internal coatings should be used in conjunction with cathodic protection to protect internals of vessels containing moderately corrosive waters, for example, aquifer or process raw water.

For vessels containing severely corrosive waters, such as produced oil field brines, all corrosion control methods should be considered including chemical inhibition, internal coatings, and cathodic protection. For waters whose end use precludes the possibility of contamination due to cathodic protection anode material, such as demineralized or treated boiler feed water, inert platinized anodes should be used, if cathodic protection is required. Potentials of coated vessels should be maintained as near the minimum protective potential as possible to prevent coating disbondment and excessive scale build-up.

The minimum design current density requirement is dependent on the oxygen content, temperature, resistivity, and velocity of electrolyte. In general, a minimum criterion of 100 mA/m^2 of bare steel surface should be used for relatively clean saline or raw waters. Vessels handling water containing depolarizers, such as hydrogen sulfide and oxygen or operating at high flow rate, usually require higher current densities to maintain protective potentials.

Internal coatings decrease the area of bare steel in contact with water and thus reduce the current required for protection. The design life of a period galvanic anode system should be for the testing and inspection period. Impressed current systems should be designed for the life of the vessel or for 25 years.

2.12.1 Types of Cathodic Protection Systems

2.12.1.1 Impressed Current Systems

Impressed current systems are generally used in applications with high power demands, such as for large water storage tanks or where current requirements may change.

Automatic potential control rectifiers can be considered for use in applications where varying conditions combine to create a changing current density requirement to maintain protective potentials. These decrease the likelihood of underprotection or coating disbondment due to overprotection.

High silicon-cast iron or platinized titanium anodes are the most common types of impressed current anodes used. Installation details for impressed current systems for water storage tank internals should be in accordance with applicable Standard Drawings. Separate transformer rectifier for impressed current systems should be installed for each tank or vessel to be protected. Where cathodic protection is applied internally to a plant, corrosion interaction affecting neighboring structures is unlikely and appreciable current flows only through and inside the protection plant.

2.12.1.2 Galvanic Anode Systems

Galvanic anodes are considered for use in applications where water resistivity and power requirements are low, electrical power is not available, if their use is economically feasible, or in cases where sufficient impressed current anodes cannot be provided or placed for proper current distribution, as would be in the case of a compartmented vessel.

Zinc, magnesium, and aluminum alloys are the common galvanic anode materials used. Zinc anodes are commonly used in low-resistivity brines due to their longer life. Zinc may show a decrease in the driving potential and a possible polarity reversal $>65\ ^{\circ}C$ and passivation in the presence of phosphates, chromates, and sulfides. The use of zinc anodes should not be specified for electrolytes in which these conditions may exist.

Magnesium anodes are generally specified for high resistivity waters. In low-resistivity brines, current output should be controlled by resistors in the external circuits or by partially coating anodes to control exposure of active metal.

Aluminum alloy anodes have a lower driving potential compared with magnesium, and provide a longer life in low-resistivity electrolyte. Like zinc, they are self-regulating and do not need resistors for external current control.

Installation details for galvanic anodes in water tanks and also for other types of vessels or tanks, anodes may be internally supported or mounted horizontally through the vessel sides. Care should be taken in anode placement design to ensure proper current distribution to all internal surfaces in contact with the electrolyte. Vessel redesign or internal modification should be considered, if necessary, for providing effective cathodic protection throughout.

Where large stand of anodes are used for the protection of tanks, the resistance should be determined using the following formula:

$$R = \frac{\rho}{2\pi L}\left(\ln\frac{8L}{D} - 1\right), \tag{2.1}$$

where R is the anode-to-electrolyte resistance of a single vertical anode to remote reference (ohms); ρ is the electrolyte resistivity in (ohm centimeters); L is the length of the anode in centimeters (including backfill, if used); and D is the diameter of the anode in centimeters (including backfill, if used).

Where flat plate anodes are used, their resistance is to be determined from the following formula:

$$R = \frac{\rho}{2S}. \tag{2.2}$$

However, if flat plate anodes are close to the structure or are painted on the lower face, then the resistance is to be determined using the following:

$$R = \frac{\rho}{4S}, \tag{2.3}$$

where R is the anode-to-electrolyte resistance of a single flat plate anode; ρ is the resistivity of electrolyte in ohm centimeters; S is the mean length of the anode sides, in centimeters.

2.12.2 Current Density

Uncoated tanks used as permanent water ballast are to have a minimum current density of 110 mA/m^2, but this may have to be increased to at least 130 mA/m^2 if hot oil is stored on the opposite side of the bulkhead. Uncoated tanks used for the storage of crude oil at an ambient temperature alternating with water ballast are to have a minimum current density of 90 mA/m^2; however, this should be increased for higher temperatures. Unless shown otherwise, the resistivity of the water should be assumed to be 25 ohm-cm.

2.12.3 Anode Distribution

Having determined the number and the size of anodes, they are to be distributed as follows:

1. Ballast-only tanks: evenly over all the uncoated steel work.
2. Crude oil/ballast tanks: evenly but with some emphasis on horizontal surfaces in proportion to the area of these surfaces.

Anodes should be arranged in the tank so that protection can be provided to all surfaces without exceeding potentials (in the vicinity of the anodes) that will be detrimental to coating.

Where freezing occurs, provisions should be made for periodic replacement or for an anode installation unaffected by either freezing or falling ice. Alternatively, the tank may be operated so that ice accumulation does not occur on the anodes or in a solid mass across the tank, which could damage the anodes while collapsing.

2.12.4 Reference Electrode Position

In vertical cylindrical vessels equipped with roof hatches, electrodes may be inserted through these hatches. If this is anticipated, the anodes should be at the maximum distance from these hatches. In compartmented vessels, reference electrodes should be installed as far as possible from the anodes to obtain potential representative of the steel surface. Zinc reference electrical should not be installed in tanks containing high-temperature brine (65 °C and above).

2.12.5 Protection of Specific Installations

2.12.5.1 Storage Tanks

Impressed current cathodic protection is the method normally used for large tanks, such as overhead water towers, oil refinery seawater storage tanks, boiler feed water tanks, and river water tanks in power stations. In the case of open topped tanks, anodes of silicon iron or platinized titanium, or lead anodes, if possible toxicity is not a consideration, may be suspended from adequate supports.

Alternatively, it may be preferable to drill the tank and insert anodes and reference electrodes through the wall. Anodes should be so distributed that they are in the best position to ensure even current distribution, that is, at points roughly equidistant from the bottom and the sides, and so placed as to ensure adequate penetration of current into the corners.

When the water resistivity is exceptionally high, as, for example, in boiler feedwater tanks, copper-cored platinized titanium continuous anodes may be used; the high resistivity of the water necessitates considerable anode lengths being used to achieve adequate current distribution. Where galvanic anodes are used, they are normally attached directly to the metal surface to be protected.

Corrosion is most probable near the waterline where differential aeration effects are the greatest; anodes will normally need to be distributed up to about 600 mm below the minimum water level. For domestic water tanks, sufficient current may be obtained from a single magnesium anode.

- **Tanks for storage of chemicals**

Among the types of plant that have been protected are brine storage tanks, acid storage facilities, and electrolytic resistance tanks. The methods of applying cathodic protection to tanks containing chemicals are similar to those for water storage tanks.

Care needs to be taken in selecting the anode material to ensure that it does not react with the chemicals involved. Galvanic anodes are seldom suitable, and it is normal to use impressed current with platinized titanium anodes.

An alternative system, used to protect certain types of chemical plant, is that of anodic protection.

2.12.5.2 Water Circulating Systems

Water circulating systems are often constructed from cast iron[1] or mild steel, and where the metal is not subject to onerous conditions, such as mechanical erosion, high temperature, or chemically aggressive coolants, a moderate increase of thickness will generally ensure a reasonable life. With cast iron, mechanical considerations in casting and handing often result in an adequate wall thickness. There is, therefore, generally little economic incentive to protect, for example, runs of iron or steel pipework particularly as this requires anodes to be mounted about every four pipe diameters along the length of the run, and there has been insufficient experience of such applications for detailed guidance to be given.

The corrosion rate may be locally high due to the galvanic action between the various combinations of metal and alloys in the system. Generally, the ferrous component acts as an anode and thus corrodes. This process is more important if the oxygen content of the water and the water velocity are high; in such circumstances, the protection of large heat exchangers or coolers becomes economical.

Difficulties of application have prevented the benefits of cathodic protection to be extended to the inside of other items of plant, even where the combination of materials can give rise to corrosion at appreciable rates. The basic principles of sound design practice have not, therefore, emerged. Thus, in the case of pumps, even with the impressed current system, it is difficult to position sufficient anode material to provide the relatively large protection current densities that are needed, particularly at points where galvanic action may occur, without interfering with water flow.

Similarly with valves, care must be taken to position the anodes, according to the proportions of the service life during which the valve will be in the open and closed positions, in such a way as to ensure that the current reaches those parts of the exposed surface that are subject to severe corrosion. If the water is stagnant for long periods, it may be necessary to take measures to prevent the accumulation of chlorine developed at the anodes on parts of the system.

- **Heat exchangers (tube and shell)**

A typical plant has cast iron water boxes with a tube plate and tubes of copper-based alloys. Where this is the case, the cast iron corrodes, and the effect decreases with increasing distance from the components in copper-based alloys. Severe local attack with deep penetration may thus occur on cast iron adjacent to tube plates.

[1] With cast iron, graphitic corrosion occurs in which metal is removed, leaving the graphite. This does not materially alter the application of cathodic protection except that, the graphite being electropositive, the current density required may be increased.

Nonferrous tube plates and tube ends may also be attacked, principally due to water turbulence and impingement, and cathodic protection may help to reduce the rate of attack. It has been observed that the coating of a ferrous water box or its replacement by a nonmetallic type has been accompanied by accelerated tube end attack due to the removal of the cathodic protection previously afforded by the corroding water box. In such a case, provision should be made for alternative means of protecting the tube plate and tube ends, for example, by installing galvanic anodes.

In water circulating systems, the flow velocity is usually much higher than is encountered with immersed structures and this, together with the presence of metals that are more electropositive than iron, increases the current density required for protection. A rough guide to the amount of current required for cathodic protection of condensers and coolers has been found by experience to be 2.5 A/M^2, estimated on the area of the tube plate without deducting the cross-sections of the tubes. The figure relates to power station condensers using seawater as the coolant and without internal coating of the water box or tube sheet.

A protective current may be provided by either galvanic anodes or impressed current, the factors determining the methods including current requirements, power availability, possible hazards, and the accessibility of the components for maintenance. Impressed current has advantages in that the current may be manually or automatically varied to maintain full protection at all times irrespective of the changes of flow rate or composition of the cooling water.

Galvanic anodes cannot normally be so adjusted and also require comparatively frequent renewal. On large power station condensers, galvanic anode renewals may not only be costly but they may also involve considerable effort. Galvanic anodes are generally more suitable for smaller units, particularly if periodic shut downs provide regular opportunities for inspection and anode renewal.

They may also be of advantage in plants where hazardous atmospheres exist since the use of flameproof electrical equipment, which would be necessary with an impressed current system, can be avoided. Cathodic protection is not usually applied to condensers or coolers that are about $<$500 mm in diameter.

- **Water box coatings**

The use of coatings in conjunction with cathodic protection does not generally lead to economy because, even with a well-coated box, the current that will flow to the heat exchanger surface, which cannot be coated, will still be considerable. Suitable coatings include one or more coats of epoxy resin paint (minimum thickness: 0.25 mm) or of epoxy resin mastic, applied after abrasive grit blasting to a high standard. Coatings should not be used that are adversely affected by the alkaline environment produced near anodes; coatings that may come off as sheets, causing tube blockage, should be avoided.

- **Galvanic anode systems**

Anodes are usually made of either zinc alloy or magnesium. Zinc anodes ares generally more suitable where there is cooling water; magnesium anodes, are generally more suitable because their higher driving potential enables them to provide a greater current output than would be the case with zinc anodes. In certain cases, for example, where the water box is very well coated or is of nonferrous material, soft iron anodes may be fitted to provide cathodic protection of the tube plate and tube ends. Iron corrosion products are beneficial in assisting film formation on nonferrous metals; ferrous sulfate is sometimes injected to produce inhibiting films.

In any particular installation, anode wastage rates will be mainly determined by the mean temperature, flow velocity, and resistivity of the cooling water. In view of these variable factors, it is often advisable to determine anode wastage rates by trial. If cathodic protection is being applied to an existing design of a water box, the maximum amount of anode material that can be installed will be limited by the physical dimensions of the water box; anodes should not cause excessive water velocity or turbulence.

In many industrial plants, regular opportunities for opening up condensers for anode inspection and renewal occur during plant shutdowns. In the case of oil refinery units, this may be every 2–3 years, and it may be possible to install sufficient anode material to provide continuous protection over this interval. In some cases, this will entail modification to the water box, for example, dished floating head covers can be made deeper to allow space for the anodes and to maintain sufficient crossover area to avoid undue turbulence.

For zinc alloy anodes in seawater-cooled condensers, the minimum anode weight to be installed on the cover of a coated water box for a five-year life in a temperate region can be estimated from the formula:

$$W = 80D^2, \tag{2.4}$$

where W is the weight of zinc anodes in kilograms and D is the nominal diameter of the wetted part of the tube plate in meters.

In tropical regions, the anode life will be about half of that stated above.

- **Impressed current**

High-quality coatings are not normally used in conjunction with impressed current. Cantilever or continuous platinized titanium anodes are generally installed. In the latter case, the anode should be located at approximately one-third of the depth of the water box from the tube plate up to a maximum separation of 300 mm. Where water boxes have depths that are appreciably >1 m, additional anodes may be needed to protect remote parts of the water box.

For long lengths of continuous anode, it may be preferable to use copper-cored platinized titanium to reduce voltage drop along the length of the anode. The electrical input to the anode can be at any technically suitable and convenient point along it, or it can be fed from two points to reduce voltage drop.

If cantilever anodes are used, they need to be distributed about the water box to ensure adequate distribution of current. Reference electrodes will be required so that the effectiveness of the cathodic protection can be measured, and these should be mounted near the point of maximum galvanic influence and remote from anodes. If adequate levels of protection are reached at such points, it can be assumed that all parts of the box are adequately protected.

It is not general practice to protect coolers that are <500 mm in diameter. If, however, this needs to be done for special reasons, the factors relating to large condensers also apply, except for the location of anodes, which can be of the cantilever type mounted centrally in the end covers.

- **Box coolers**

"Coil in box" coolers comprise a series of pipe coils of a 75- to 150-mm diameter in a steel box through which a flow of cooling water is maintained. The coils, which usually carry hot oil, are generally arranged in vertical banks, and the necessary anodes can be installed only in the restricted space between these banks of tubes.

To obtain an adequate spread of protection current both along the tubes and vertically down the bank of tubes, either a series of vertical or horizontal anodes are required, or a combination of both is required.

Depending on the conditions, impressed current with either lead alloy or platinized titanium anodes can be used. It is usually impossible to accommodate sufficient galvanic anode material to maintain protection over a useful life.

It should be noted that a significant temperature gradient may exist from the top to the bottom of the box that will cause an appreciably higher output current density from anodes located near the water surface. Under certain circumstances, this can lead to overprotection near the water surface at the expense of underprotection in the bottom corners of the box. This tendency should be taken into account in determining the configuration of the anode system.

Because of the higher output of anodes near the water surface, vertical anodes that are suspended from above the water level are subject to wastage in the form of necking. This problem may be overcome by a suitable anode shape, for example, a tapered anode, or by sleeving the anode at the waterline.

2.13 Cathodic Protection of Marine Structures

This section on Standards specifies the general design requirements for cathodic protection of the submerged zones of marine structures and the buried parts of integral onshore/offshore structures.

The specifications are applicable to all types of marine structures and pipelines. Although detailed requirements are given only for the most common types of structures, the system elements are basic to the protection of all types of structures.

Seawater is an excellent medium for the application of cathodic protection. Because of its homogeneity and uniform low electrical resistivity, an even current distribution over the protected surface of the structure is readily obtained. Bare steel in seawater is polarized without difficulty if an adequate current density is maintained.

A galvanic anode may be placed very close to or in contact with the structure to be protected. Impressed current anodes, however, must be well insulated from the protected structure if they are mounted on it. Where it is necessary to place impressed current anodes very close to the structure, dielectric shielding between the anode and the structure should also be provided.

Impressed current cathodic protection is preferred for all installations where power is available. Silicon cast iron, lead alloy, and graphite anodes may be used for the buried sections of integral onshore/offshore structures. Lead silver alloy anodes and platinized titanium anodes may be used for the submerged zones of marine structures and the buried parts of the integral offshore structure.

Provisions should be made for readily accessible dc measurements using shunts or clamp-on ammeters. Galvanic anode systems should be designed for the life of the facility or for 25 years, whichever is lesser. Impressed current anode systems should be designed for 10 years life on offshore structures.

The level of protection should indicate the minimum potential required. The size and design of the cathodic protection scheme are based on the current density necessary to bring the structure up to the level of protection required; seawater velocity and oxygen content affect the current density requirements and should be considered in the design.

For remote offshore bare steel structures, an initial 216-mA/m^2 current density seems to be sufficient to achieve rapid polarization. In order to maintain the required cathodic protections level, 54–86 mA/m^2 is usually sufficient.

2.13.1 Current Density

Current densities to be used for bare and coated steel work (buried or in seawater) with galvanic or impressed current systems are given in Tables 2.4 and 2.5.

Current densities for offshore structures are discussed in the following:

Table 2.4 Current Densities in Milliamperes per Square Meter for Offshore (platforms, gosps, etc.) and Piers

Cathodic Protection System	Required Current (mA/m^2)
Below mud line (Fig. 2.4)	22
In water velocity up to 0.60 m per second	110
In water velocity >0.60 m per second	220

Table 2.5 Current Densities in Milliamperes per Square Meter for Shore-Side Structures (wharves, sheet piling)

Cathodic Protection System	Required Current (mA/m^2)	
Steel environment	Bare	Coated
Backfilled or below mud line (Fig. 2.4)	22	11
In water	54	27

Note: For more details of the current density, refer to NACE PR-01-76 (1988 revision) corrosion control of steel fixed offshore platforms associated with petroleum production.

2.13.1.1 Current Densities for Shore-Side Structures

For all coated pipelines, the current density should be 2.5 mA/m^2 of submerged surface. Each rectifier and/or generator should be provided with its own anode system. Negative drain or bonding cable connections to steel should be thermit welded or brazed.

Drain or bonding cable connections should be made outside classified areas wherever possible. Where cable-to-steel connections must be made in classified areas, two parallel cables should be installed with separate connections to the steel. All equipment, including connection boxes, rectifiers, test points, etc, should be suitable for area classification.

The minimum size of bonding cables should be 16-mm^2 copper. Permanent test lead connections should be insulated copper conductor, of a minimum size of 10 mm^2 and colored coded as needed. Resistor bonding station should be installed outside classified areas wherever possible. Electrical components, such as anodes, drain cables, anode cables, and exothermal weld connections, for impressed current circuits should be rated for the maximum loading possible on each component of the circuit. Conductors should be selected and sized so that the voltage drop does not exceed 5%. All bonding stations, test cable terminations, etc. should be clearly identified.

In certain cases, it may be advantageous to provide a flexible bond connection to the vessel that lies alongside a cathodically protected structure for long periods in order to protect their hulls and to prevent interaction effects. Account should be taken of the additional current required when designing the cathodic protection installation. It should be borne in mind that alterations in berthing will affect the current requirement, and care should be taken so that the removal of a vessel does not lead to overprotection and consequent damage to paint protection of the piling.

2.13.2 Marine Structural Zones

Corrosion on offshore structures may be divided into four zones of attack. These zones, which are shown in the generalized diagram of Fig. 2.4, overlap somewhat, and some differences in the corrosion rate may be expected within the same zone. The four zones are as follows:

1. The atmospheric (spray zone) where the metal appears to be dry most of the time.
2. The splash zone, above the waterline, where the wave action usually keeps the metal wet (sometimes subject to scour from ice).
3. Tidal zone, the portion of the structure between mean high tide and mean low tide; it is alternately immersed in seawater and exposed to a marine atmosphere.
4. The submerged or underwater zone, where the metal is always covered by water (low oxygen <8 m, high oxygen >8 m).
5. The mud zone, or subsoil zone, the portion of the structure lying below the sea bed (below the mud line), where the structure has been driven into the ocean bottom.

2.13.3 Zone Protection

All structural steel work is to be suitably protected against loss of integrity due to the effects of corrosion. Suitable protective systems may include coatings, metallic claddings, cathodic protection, corrosion allowances, or other approved methods.

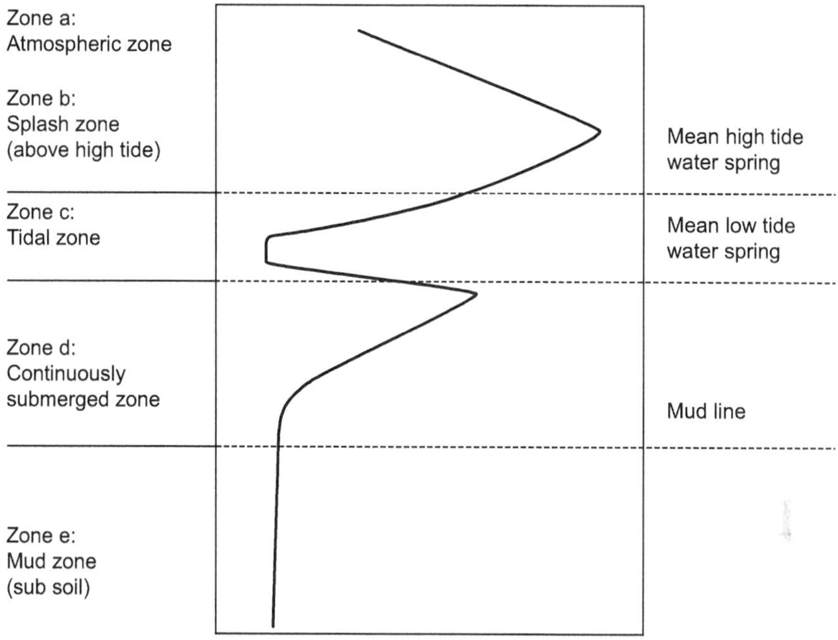

Figure 2.4 Zone of Corrosion for Steel Piling in SeaWater and Relative Loss of Metal Thickness in each Zone.

A combination of methods may be used. Consideration should be given to the design life and maintainability of the surfaces in the design of the protective systems. Special consideration will be given to the corrosion protection of floating fixed installations.

2.13.3.1 Atmospheric Zone

Steel work in the atmospheric zone is to be protected by suitable coatings.

2.13.3.2 Splash Zone

Steel work in the splash zone is to be protected by one or more of the following:

- **Wear plates**
  - Wear plates should compensate for the anticipated corrosion and wear during the life of the platform. Wear plates are usually 13–19 mm thick.
  - Wear plates are needed because damage can occur to coatings or sheathings used in the splash zone. Repair of coatings of or near the waterline is extremely difficult and costly.
  - In addition to providing a corrosion allowance, wear plates add stiffness and strength, thereby providing greater impact resistance.
  - Wear plates are usually coated or sheathed for additional corrosion protection.
    - Alternative corrosion control measures for steel in the splash zone are as follows:
    - A 70/30 nickel–copper alloy (UNS NO4400[1]) or 90/10 copper–nickel alloy (UNS C70600) sheathing.

- This sheating {UNS N04400, typically 1–5 mm (18–4 AWG gage) thick, UNS C70600, typically 4–5 mm thick} is attached to tubular members in the splash zone either by banding or welding. Welding is preferred because the annulus between the steel member and the sheathing is sealed.
- The metal sheathing should be protected from impact damage. Alloy clad steel plate may aid in preventing damage.

- **Vulcanized chloroprene**
  - Vulcanized chloroprene is typically applied at a thickness of 6–13 mm.
  - Since this coating cannot be applied in the fabrication yard, it is normally restricted to straight runs of tubular members.
  - A minimum of 51 mm should be left uncoated at each end of a tubular member to prevent damage to the chloroprene during the welding operations.
- **High-build organic coatings**
  - These coatings are usually filled with Silica glass-flake, or fiberglass.
  - These coatings are typically applied to a thickness of 1–5 mm over an abrasive blast cleaned surface.
  - Since these coatings may be applied in the fabrication yard after assembly, joints may be protected as well as straight tubular runs.

2.13.3.3 Submerged Zone

The submerged zone should be considered as extending from the base of the structure, including all steel below the mud line, up to and including the tidal zone, hereafter. Steel in the submerged zone is to be protected by an approved means of cathodic protection using sacrificial anodes with or without an impressed current system. High duty coatings may be used in conjunction with the cathodic protection system. For design purposes, the upper limit of the submerged zones should be Mean High Tide Water Spring (MHWS) Level.

2.13.3.4 Calcareous Coating

In the case of cathodic protection of steel in seawater, the creation of alkaline environment adjacent to steel structure (cathode) has an especially beneficial effect. When the pH of an electrolyte, such as seawater containing calcium and magnesium salts is increased, there will be a gradual precipitation of these salts.

Therefore, when the pH of seawater adjacent to a metal surface increases as a result of applying cathodic protection, there will be precipitations on the surface of the metal. Precipitations are calcareous coatings that consist of calcium and magnesium hydroxides and carbonates. The composition of these coatings will depend on the composition of the seawater, the temperature of the water, and the current density on the surface of the metal. The denser coatings are formed initially at relatively high current densities. When such coatings have been formed, the current requirements to maintain adequate cathodic protection on the surface of the bare metal will be drastically decreased. In the light of this, after the formation of calcareous films it is quite possible to cut down the number of rectifiers in operation.

2.13.4 Protection of Specific Installations Including Vessels

2.13.4.1 Marine Sheet Piling and Wharves

Marine sheet steel piling and wharves should be cathodically protected on both sides, including all anchor piles, anchor ties, bracing, etc. the joints between anchor piles, anchor ties, and the walking beam should be bonded to ensure good electrical continuity. Bonds should be insulated copper conductors that are not $<35\ mm^2$ and should be brazed or thermit welded to the steel. The overall electrical continuity of sheet piling must be ensured by a continuous walking beam or some other means, sized to pass the maximum cathodic protection current output of the system without excessive voltage drop.

In general, two separate anode systems, one offshore and one onshore, should be installed to deal with the land-side and water-side faces (where only one transformer rectifier is utilized, adjustments to the current distribution will be accomplished by installing variable resistors in the anode circuits).

The ideal anode location is remote and behind the wharf structure with the cable routes above the high water mark on the shore. Where this is not practicable, a remote sea-bed installation may be used for the water-side face of the wharf.

The lateral spread of current along the wharf from distributed anodes will be improved by increasing the distance between the wharf and anodes. Provisions should be made for boosting the polarization on the water-side face.

2.13.4.2 Piers

Small piers and jetties may require separate offshore and onshore anode systems. The selection of a remote distributed anode systems or remote sea bed installation depends upon the shape and continuity of the structure, the number and type of anodes required, and the availability of a location for remote anode systems, safe from mechanical damage. More uniform current spread will be obtained from anodes located approximately 30 m from the structure.

Large piers should in general be protected by multiple rectifier impressed current systems. Each rectifier should have a separate negative connection to the structure and a separate anode system so that each anode system is energized by only one rectifier. Cathodic connections should thus be distributed over the structure. Platinized titanium or other inert anode located and mounted in such a way as to be free from the risk of mechanical or wave damage are preferred. Systems that are either distributed and rigidly fixed below low water level or a sea-bed installation remote from the structure are also preferred. Electrical continuity between separate structural parts of large piers and between individual piles should be provided by means of bonding cables.

2.13.4.3 Offshore Wellhead and Tie-In Platforms

Each platform should be considered as a separate unit for cathodic protection. All pipeline risers should be provided with insulation flanges at platform elevation. Test stations should be installed on all insulating flanges in a position accessible for inspection and testing.

Where power is available, impressed current cathodic protection should be applied generally as for piers. In all other cases, platform structures should be protected by means of galvanic anodes mounted in positions where they will not be subject to damage by platform operation procedures.

Pipe supports on the seaward side of all insulating flanges should be electrically insulated from the pipe. In the case of wellhead platforms, for design purposes, an allowance of 20-A-per-well casing should be added when calculating current requirements of the platform structure.

2.13.4.4 Offshore Gosps and Sea Islands

Multiple interconnected structures, such as offshore GOSPS or sea islands, should in general be protected by multiple rectifier impressed current systems. Each rectifier should have a separate anode system and separate negative connections distributed over the structures. Platinized titanium or lead silver alloy anodes are preferred. They should be rigidly mounted below the water level and distributed on the structures where they are not likely to be subjected to mechanical damage.

Alternatively, one or more separate remote anode systems could be constructed and connected to the rectifier by means of submarine cables. Where structures are interconnected by walkways, bridges, or by other steel structures, these should not be used to provide the continuity of the whole structure. Bonds should be provided at all junctions between these adjacent structures, bridges, or walkaways to avoid any possible sparking hazard. These bonds should be sufficiently flexible to allow relative movement between the sections and must be sized to the maximum output of the cathodic protection system.

Insulating flanges should be provided at deck level for all pipe risers. Bonding stations should be provided on all insulating flanges. Pipe supports should also be electrically insulated from the pipe.

2.13.4.5 Submarine Pipelines

Submarine pipelines should include all lines installed on the sea bed. This includes lines laid from land to land, offshore structure to land, and offshore structure to offshore structure. These pipelines will generally be coated with a dielectric coating and concrete weighted. For design purposes, the conductance of the coating should be 5300 micromhos/m^2 (500 micromhos/ft^2).

Many of the considerations outlined in previous section are applicable to submarine pipelines. All submarine pipelines should be cathodically protected. Where power is available, coated submarine pipelines should be protected by impressed current systems up to a point where attenuation reduces the potential to an unacceptable level. Where power is unavailable or beyond the point where the potential is unacceptable, a galvanic anode system should be installed. Submarine pipelines should be isolated at their extremities by means of insulating flanges with bonding stations.

Submarine trunklines and tielines should be isolated at the shore end by means of insulating devices. The galvanic anode system to be used will be aluminum alloy bracelet anode system or zinc bracelet anode system (Fig. 2.5).

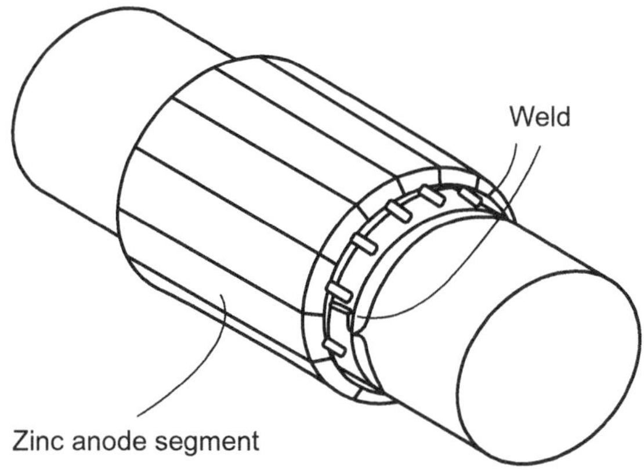

Figure 2.5 Typical pipeline bracelet anodes.

2.13.4.6 Vessels

Steel hulls of tankers, barges, launches, and other vessels can be protected from external corrosion by means of cathodic protection. Both impressed current and galvanic anode systems have been successfully employed. Graphite or high-silicon cast iron anodes are usually used for impressed-current systems, although platinum anodes have been used. Magnesium is normally used for galvanic anode systems. Special anodes of the various materials are available for ship bottom protection, and special anode mounting methods are required for either type of system. With either type of system, the anodes must be distributed over the bottom to provide a relatively uniform current distribution on the submerged portion of the hull.

For the design of a ship-bottom cathodic protection system, the following factors should be taken into account:

1. Variation of current requirement with the speed of the ship.
2. Variation of circuit resistance if the vessel moves from the seawater to brackish or fresh water.
3. Susceptibility of some paint coatings to damage by cathodic protection.
4. Possibility of damage to and removal of the anodes by striking objects in the water.
5. Possibility of losing protection by electrical connection of a protected vessel to an unprotected vessel or pier.

2.13.5 Anodes for Impressed Current Systems

2.13.5.1 High Silicon Iron

High silicon cast iron is a material that corrodes relatively slowly. In seawater, where chlorine is produced at the anode surface, the addition of chromium to the iron improves the performance.

The resistance of a single anode in free flowing seawater is less than that of the anode in silt, mud, or soil. It is common to install groups of anodes, sometimes mounted on wood or a concrete frame work to reduce silting or covering.

The effect of anything rubbing or scraping the active anode surface is to accelerate dissolution at that position. Suspension or stabilizing cables, therefore, must not be attached around the active anode surface. Suspension by the anode cable is prohibited.

The number of anodes is determined by the allowable current density, the life, and the circuit resistance of the system. Design life should be 10 years. For onshore installations, where the anode is placed directly in the ground that is not permanently salt water saturated, the anode surface actually in contact with the soil is inevitably something less than the whole area. Furthermore, as the anodic reaction proceeds, the oxygen or chlorine gas evolved must be able to escape; otherwise, these two factors reduce the active anode area so that the recommended operating current density is exceeded, and rapid but localized dissolution occurs. For this reason, it is usual to install a low resistance carbonaceous (coke) backfill around the anodes in accordance with applicable Standard Drawing.

Anode life is influenced by the current density of operation and the total current emitted. In free flowing seawater, a maximum output of 10 A/m^2 is allowed, and the total anode weight requirement should be based on a consumption rate of 0.4 kg ampere per year. This rate reduces to 0.2 kg per ampere per year when the anode is placed in coke backfill. For anodes buried directly in the soil or which become silted over on the sea bed, the applicable figures are 10 A/m^2 maximum output and 0.8 kg per ampere per year.

2.13.5.2 Platinized Anodes

Platinized anodes are generally constructed from solid or copper cored or titanium cored rod of an approximately 12-mm diameter and approximately 1500 -mm length onto which is deposited, a layer of platinum of a 0.005-mm thickness. Other shapes and configurations are also available. Platinized anodes are limited in use to free flowing seawater in marine designs.

The anodes may be mounted on supports cantilevered off the structures but electrically insulated from them. Alternatively, they may be mounted remote from the structure or supported on immersed nonmetallic frames raised over the sea bed above the level of silt deposition.

Anode-to-cable connections may be made to one or to both ends of the anode rod. The connection must be completely encapsulated in a suitably sealed plastic assembly filled with epoxy resin. When securing the anode to its mounting, contact with the platinized section of the anode rod should be avoided and nothing should touch or rub the platinized active surface.

Platinized titanium anodes should not be used under the following conditions:

1. Where solids such as silt may accumulate on the anode rod.
2. Where electrolyte resistivities, or the dc power source voltage, is of such a value that the voltage between anode and electrolyte is >7 V exclusive of anode and cathode reaction voltages. (The design should thus ensure a terminal voltage that is not >7 V.)
3. Where the frequency of the ac ripple content of the dc power source currents is <100 Hz (three-phase rectification must therefore be used).

4. Where the dc power source is intermittent and does not provide a constant anodic reaction at the anode. (Inert anodes installed in an electrolyte, particularly prior to commissioning, are possibly subject to the loss of anode material and to deposit by marine growths that could detach the platinum.)
5. Where the anode is not totally immersed at all states of the tide.
6. Where the electrolyte is stagnant and does not allow diffusion of anode-produced chlorine gas away from anode assembly.

- **Location of anodes on structures**

Multiple anode assemblies should be mounted on the structure under cathodic protection. It will be necessary to distribute the anodes over the members of complex structures, such as piers, to provide cathodic protection to all the immersed structural members.

Platinized titanium anodes should be designed for operational current densities that are not >700 A/m^2 of platinized area. The 0.005-mm-thick platinizing dissolves at the rate of about 10 mg per ampere year. Although abrasion by suspended solids reduces platinum life, a 0.005-mm platinized surface may be regarded as providing a permanent inert anode with a life in the order of 15 years.

In determining the length of the anode rod to be installed, consideration must be given to the effect of linear voltage drop resulting in nonlinear current dissipation over longer anode lengths. Copper-cored anodes may be used to overcome this problem. The physical flexing of long rods in flowing seawater may result in fatigue fractures so that a larger section is to be preferred if anode movement is likely.

Platinized niobium anodes are also available in various shapes and configurations. One type of these anodes is available under the trade name "Morganode." These anodes may be used in higher or low-resistivity waters and driven at up to 100 V dc.

2.13.5.3 *Lead Silver Alloy*

Lead will normally dissolve under anodic polarization. It becomes a stable anode material only when used in conditions in which the anodic reaction produces a lead peroxide film over the anode surface. The alloying of lead with 2% silver is conductive to the production and retention of the desired lead peroxide film. Lead–silver anodes are suitable for use in seawater. They should be operated at about 215 A/m^2 ($20 A/ft^2$). If operated at this current output, they are regarded as permanent inert anodes also with a life in the order of 20 years. These anodes are of the pile mounting type and are the least expensive of the precious metal anodes. They may be used in a skid-mounted configuration on the sea floor. This anode is available under the trade name "Morganode."

2.13.5.4 *Graphite*

Graphite is a material that corrodes relatively slowly. Graphite anodes are not suitable for freely suspended installation. They may be installed onshore, preferably in carbonaceous backfill (coke).

Graphite anodes buried in carbonaceous (coke) backfill or directly in salt water saturated low-resistivity soils may be operated at a current density of 4 amps per square meter. The consumption rate of anode material is 0.135 kg per ampere year at this current density.

2.13.6 Anodes for Galvanic Anode Systems

Galvanic anodes may be alloys of active metals such as magnesium, zinc, or aluminum. These alloys can be obtained in a variety of shapes and sizes to deliver protective current to a specific offshore platform with optimum current distribution. The method used to attach the anodes to the platform will be governed by their type and application, but low resistance electrical contact must be maintained throughout the operating life of the anodes. The performance of galvanic anodes in seawater depends critically on the composition of the alloy, particularly in the cases of zinc and aluminum (see Appendix B of NACE Standard, RP-01-76, 1983 Revision).

2.13.6.1 Magnesium Anode

Magnesium anodes are very active and furnish high current outputs. For this reason, they can provide cathodic protection with a minimum number of anodes. Because of their properties, they usually are not used for long-life designs.

For marine structures, the use of magnesium is confined to providing boosted polarization current. For offshore structures, the design life of this material is three months.

For the protection of small structures or short buried pipeline sections onshore, magnesium anodes are buried in the soil adjacent to the structure and directly connected. These anodes require moist soils with resistivities <5000 ohm-cm for satisfactory operation. In cases where the soil is occasionally dry, moisture can be retained around the anode by providing a chemical backfill. Prepackaged anodes are installed in a cotton bag surrounded with gypsum bentonite–sodium sulfate backfill. This anode is suitable for land installation.

Standard bare magnesium anodes may be used for immersed, silted, or onshore burial installations. Magnesium is available in a variety of purities with different performances. Efficiency is 50% for all grades, and the consumption rate is 1230 A h/kg. The design potential is normally −1.70 V in reference to that of the Ag/AgCl electrode.

2.13.6.2 Zinc Anode

Zinc anodes are mainly used in seawater for the protection of marine craft. Due to their lower current capacity and higher cost, zinc anodes are not normally recommended for submarine pipeline or offshore structures.

2.13.6.3 Aluminum Alloy Anodes

This type of anodes are generally suitable for use on submarine pipelines as well as on offshore structures and continues to give satisfactory performance in sea-bottom environments. Its chief use will be as bracelet anodes on submarine pipelines, where its use represents an initial cost saving of up to 60% over zinc bracelet anodes due to its higher current capacity.

2.13.7 Boosted Polarization

As cathodic protection commences to operate, a layer of alkaline material is produced on the protected surface by the cathodic electrode reaction. This material persists unless removed by unusual physical conditions, and its presence reduces the amount of current necessary to maintain the cathodic protection.

For onshore-based installations, the difference in cost between providing permanent capacity to polarize rapidly and capacity to maintain the cathodic protection may be significant. This is particularly so in the case of large surface area structures such as sheet piled walls, etc., and in these cases, temporary current sources are provided for the initial polarization. This boosted polarization may be provided by using welding generators connected with the negative output to the structure and the positive output to temporary anode systems.

For offshore structures provided with galvanic anode systems, boosted polarization should be by 9 × 18-mm (3/8-in × 3/4-in) magnesium ribbon connected to the structure in addition to the designed cathodic protection.

2.13.8 Cathodic Protection Design for Marine Structures

2.13.8.1 Plans and Information

The following plans and information are to be submitted:

1. A surface area breakdown for all areas to be protected (seawater immersed and buried) including secondary steel work and details of temporary appurtenances.
2. The resistivity of the seawater.
3. All Current densities used for design purposes.
4. The type and location of any reference electrodes, their methods of attachment, and the method for transmitting the potential data.
5. Full details of any coatings used and the areas to which they are to be applied.
6. Details of any electrical bonding. Piles-to-piles sleeve jacket legs, or insulation.

2.13.8.2 Galvanic Anode Systems

In addition to the information given above, the following plans and information are to be submitted:

1. the design life of the system, in years;
2. anode material and minimum design capacity of the anode material in Ampere hours per kilogram;
3. the dimensions of anodes, including details of the insert and its location;
4. the net and gross weight of the anodes in kilograms;
5. the means of attachment;
6. the location plans of the anodes;

7. calculation of anodic resistance, as installed and when consumed to their design utilization factor, in ohms;
8. closed circuit potential of the anode materials, in volts;
9. details of any computer modeling;
10. the anode design utilization factor.

2.13.8.3 Impressed Current Systems

In addition to the information required in previous paragraphs, the following plans and information are to be submitted:

1. the anode composition and, where applicable, the thickness of the plated surface, consumption, and life data;
2. anode resistance, limiting potential, and current output;
3. details of construction and attachment of anodes and reference electrodes;
4. size, shape, and composition of any dielectric shields;
5. diagram of the wiring system used for impressed current and monitoring systems, including details of cable sizes, underwater joints, type of insulation and normal working current in the circuits, the capacity, and type and make of protective devices;
6. details of glands and size of steel conduits;
7. plans showing the location of anodes and reference electrodes;
8. if the system is to be used in association with a coating system, then a statement is to be supplied by the coating manufacturer that the coating is compatible with the impressed current cathodic protection system.

2.13.9 Fixed Potential Monitoring Systems

Cathodic protection systems should be monitored. For a permanent monitoring, fixed electrodes are to be utilized to measure steel/seawater potential and possibly anode current output and/or cathode current density.

These reference electrodes may be high purity zinc or zinc alloy and silver/silver chloride electrode. The signals from these fixed reference points should be transmitted to the surface indicator, and preferably to the recorder by either hard wire or an acoustic link. The intervals between measurements should accommodate all conditions of weather, tide, and operating conditions and should indicate trends in the cathodic protection performance.

2.13.10 Potential Survey

Potential surveys are to be carried out at agreed intervals. Should the results of any potential survey measured with respect to a silver/silver chloride reference electrode indicate values more positive than −0.8 V for aerobic conditions or −0.9 V for anaerobic conditions, then remedial action is to be carried out at the earliest opportunity.

2.13.11 Retrofits

Where bracelet anodes are proposed for retrofits, then the tightens of the units does not to rely on the anode material being in direct contact with the platform bracing or pipe work. Due to the possible effects of hydrogen, high tensile steel belting materials should be avoided and the hardness for all such bolting materials should be limited to a maximum Vickers Diamond Pyramid Number of 300.

Where it is necessary to weld anodes to the structure, only approved welding procedures are to be used. The welding procedure is to be qualified under fully representative conditions.

2.14 Anodic Protection

This section is included in this electrochemical protection standard to provide a background on the principles of anodic protection and to suggest possible applications that should be considered for this method of corrosion control. It is hoped that the information provided in this Section will assist the users in evaluating the current literature on the subject and in recognizing the possible applications of anodic protection.

2.14.1 Principles of Anodic Protection

The phenomenon of anodic passivity can be illustrated by the simplified experiment shown in Fig. 2.6. The equipment consists of an electrolyte in which are immersed the

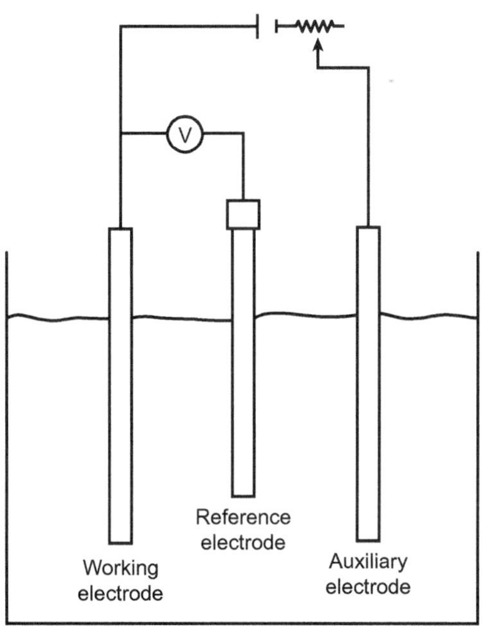

Figure 2.6 The phenomenon of anodic passivity.

metal sample to be protected (the anode), the inert cathode, and the reference electrode. The anode, cathode, and reference electrode are connected to a potential controller ("potentiostat") that is capable of providing a wide range of current while holding the potential of the anode within very narrow limits.

Schematic of experimental apparatus used for anodic polarization studies. The current flow between the working electrode and the auxiliary electrode forces a shift in the potential between the working electrode and the reference electrode (V, voltmeter).

In this example, the anode to be protected is assumed to be iron. The inert cathode is usually platinum, and the reference electrode is a calomel half cell. The potential of the anode with respect to the reference electrode is first measured to ensure that the metal surface is in the active condition and that no films or corrosive products that might inhibit the passivation reaction are present. Current is then applied to the anode. Polarization curves are obtained by plotting the potential between the anode and the reference electrode corresponding to each value of the applied current.

A typical polarization curve obtained in this manner is shown in Fig. 2.7. Point "Ecorr," obtained before the current is applied, represents the initial condition of the system typified by the reaction of the corroding anode:

$$Fe \rightarrow Fe^{2+} + 2e$$

Schematic of the anodic polarization curve. Metal environment systems that have this type of anodic polarization behavior are termed active passive and can be anodically protected. As current is applied to the anode (and returned to the external circuit

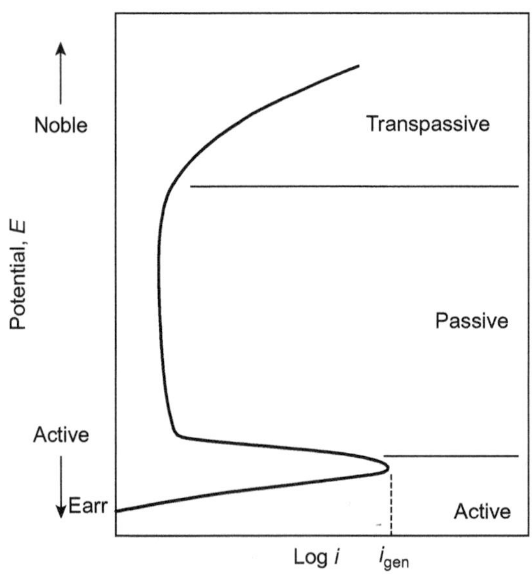

Figure 2.7 A typical polarization curve.

through the cathode), the potential shifts in the direction of greater passivity. However, the loss of metal at the anode also increases as a result of the greater current discharged by the anode to the solution.

The current is further increased until there is an abrupt change in the plotted potential–current relationship. This point, designated as "Ec," is shown as the Flade potential. Further increase in the potential applied by the controller results in a decrease in the current flow. As the potential and current approach point "Ep," the system is in the passive state. The anode reaction has become

$$Fe \rightarrow Fe^{3+} + 3e \tag{2.5}$$

To retain the system in the passive state, the potential of the anode with respect to the reference electrode must be controlled closely within the narrow region shown by the nearly vertical portion of the curve near point "Ep." If the potential is allowed to increase beyond point "EA," the current will again increase and will indicate a corresponding increase in the rate of corrosion. The section of the curve beyond point "EA" is known as the transpassive region.

It is generally agreed that the following conditions must be true for the anodic passivity of iron and sulfuric acid to occur:

1. A protective film must be at least 100 A (angstrom) thick.
2. Not >10–2 coulombs/cm^2 must be required for the film to form.
3. The film must be an electronic conductor.
4. The E–VS–I characteristics defined by Fig. 2.7 must be satisfied.
5. The film must be formed very rapidly.
6. The passivated metal must be oxidized.
7. The film must be hydrophobic.
8. The passive film must be relatively acid insoluble.

The exact nature of the passive film is not completely understood. Three possible theories have been suggested:

- **Oxide film theory:** The iron dissolves in the electrolyte and reacts with oxygen to form protective precipitates of iron oxide. These precipitates form a film that presents a chemical and physical barrier to the corrosive medium.
- **Absorption theory:** Oxygen is absorbed on the surface to form a barrier layer.
- **Sequence model theory:** Oxygen is initially absorbed on the surface and is followed by a slow formation of an amorphous iron–oxygen structure. This occurs by the diffusion of iron from the base metal into the absorbed oxygen film.

The sequence Model theory has received considerable support in recent articles.

2.14.2 Contrast with Cathodic Protection

It is important to distinguish clearly between the phenomenon of anodic protection described in this section and the well-established practice of cathodic protection described previously in this chapter. In cathodic protection, the metal to be protected against corrosion is made the "cathode" of an electrolytic cell. This condition is achieved by impressing current into the cell by means of an anode. Cathodic protection is much simpler to achieve and is less subject to operating difficulties because the potential control is less critical. However, in contrast to anodic protection, cathodic protection is not effective in strong acid solutions.

2.14.3 Effects of Variable Factors on Anodic Protection

Variable factors affecting the tendency of a system to become passive by the application of anodic protection are as follows:

2.14.3.1 Solution Concentration

Electrolyte concentration affects the current density required to establish passivity in a given metal electrolyte system. In the case of iron and sulfuric acid, the current density required for passivity decreases as the concentration of sulfuric acid increases. This is in contrast to corrosivity of sulfuric acid, which is at its maximum at 55% concentration and decreases toward zero at 0% and 100% acid concentrations.

2.14.3.2 Surface Area of the Anode

The current density required to establish passivation is constant for any given combination of metal and electrolyte under a given set of conditions; this value is usually determined by laboratory experiment. Therefore, as the surface area increases, the total current required will increase, but the voltage requirement will remain the same. Hence, the experimental results for the current requirements can be extrapolated to cover process situations under similar conditions.

2.14.3.3 Different Metals

Passivation studies[1] have been made on a number of materials in 67 weight percent sulfuric acid. Those materials that were successfully passivated are shown in Table 2.6, with the current densities required to maintain passivity and with some examples of the very much higher currents required to establish passivity.

These variations in current density requirements should be noted. Carpenter 20 has the lowest current density rating and, therefore, requires only a small current density to retain passivity under these conditions. Variations in solution pH, temperature, concentration, or other physical test conditions will cause a shift in the values given above.

For this reason, it is important to conduct careful laboratory analyses under the precise conditions to be encountered by the metal on process plants to determine the current density required both to initially obtain passivity and to hold the system in the passive state.

From the studies that have been conducted, it appears that any ferrous metal can be passivated successfully, although the current density requirements to retain passivity will vary depending on the metal and the electrolyte. Experimental studies have shown that it is not possible to passivate copper-based alloys, such as brasses and

Table 2.6 Current Density Required for Anodic Protection of Various Metals

	Establish Passivity		**Maintain Passivity**	
Metal	**amp/cm²**	**amp/ft²**	**amp/cm²**	**amp/ft²**
Type 302	512×10^{-6}	0.475	2.2×10^{-6}	2.04×10^{-3}
Type 304			3.8×10^{-6}	3.53×10^{-3}
Type 310			0.5×10^{-6}	4.7×10^{-3}
Type 316	580×10^{-6}	0.539	0.1×10^{-6}	0.093×10^{-3}
Type 405			2.7×10^{-6}	2.51×10^{-3}
Type 410			10.9×10^{-6}	10.2×10^{-3}
Type 446			0.7×10^{-6}	0.65×10^{-3}
Mild steel (SAE1020)			15×10^{-6}	13.9×10^{-3}
Titanium			0.08×10^{-6}	0.075×10^{-3}
Hastelloy			0.5×10^{-6}	0.47×10^{-3}
Carpenter 20			0.03×10^{-6}	0.027×10^{-3}

bronzes, some of the rare metals, such as titanium, hafnium, zirconium, and niobium, which are used to promote corrosion resistance in special alloy systems can be successfully passivated.

2.14.3.4 *Temperature Effect*

As the temperature of the solution is increased, the current density required to maintain passivity increases. For example, the current density required to maintain the passivity on 304 SS and 67 weight percent sulfuric acid is more than three times as great at 85 °C as that at 52 °C. The cause of this radical increase is probably related to the increased solubility of the passive layer at an elevated temperature.

2.14.3.5 *Chemical Environment*

On the basis of developmental work which has been conducted, it appears that passivity can be established in a broad range of oxidizing chemical environments including sulfuric acid, phosphoric acid, nitric acid, sodium hydroxide, lithium hydroxide, ammonium nitrate, and aluminum sulfate.

It is not possible to attain passivity in reducing systems such as hydrochloric acid, ferrous chloride, stannous chloride, hydroiodic acid, and sodium chloride. Although these reducing systems cannot be passivated, it is possible to passivate oxidizing systems that contain small amounts of these reducing agents. However, the addition of these agents will increase the current density required to attain passivity and will also increase the current density required to maintain the system in the passive state.

Agitation of the electrolyte does not adversely affect the ability to passivate the system. In some systems that have been tested, agitation has increased the rate at which the passive state was formed as well as decreased the current density required for passivity.

2.14.4 Applications to Process Plants

Any proposed application of anodic protection should be evaluated by users' experienced technical personnel. The following are some of the factors that should be considered in determining whether anodic protection may have promise for a specific application:

2.14.4.1 Services

Services that have been investigated to determine their suitability for anodic protection include the following:

Sulfuric acid: The use of stainless steel in the storage of sulfuric acid up to concentrations of 85% is practical when anodic protection is applied. For concentrations >85% sulfuric acid, it is probably more economical to use mild steel. The storage of 100% sulfuric acid (oleum) in carbon steel tanks under anodic protection will give complete protection with no contamination of the oleum through reaction with the iron.

Phosphoric acid: Stainless steel is attacked by phosphoric acid over a side range of concentrations. The anodic protection is feasible for pure phosphoric acid. The current density requirements for initiation and maintenance are comparatively

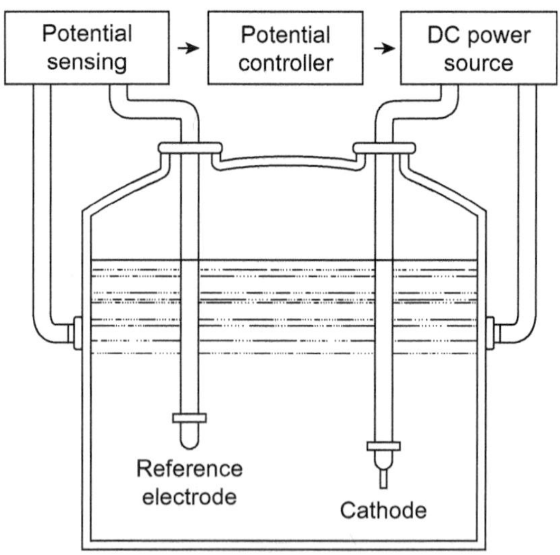

Figure 2.8 Schematic of an anodic protection system for a storage vessel. One or more cathodes, a reference electrode, a potential sensing and controlling circuit, and a dc power supply are required for each anodic protection system. The vessel wall is made the anode of the circuit by current forced between the cathode and the tank wall. The currents are controlled so that the potential of the wall with respect to the reference electrode is shifted and maintained in the passive region (Fig. 2.7).

low, so that the current required to passivate an average process vessel will not be unduly large.

Alkalis: Passivation in alkaline solutions results in the formation of a film that is different from the one associated with oxidation systems. The notable features of the passivation of alkaline solutions are the length of time necessary for passivation to occur and the higher current densities required. The maximum pH limit for passivity of 304 stainless steel in lithium hydroxide is about 9.5.

Salt system: There would be little difficulty in applying anodic protection to systems containing 304 stainless steel in solutions of aluminum sulfate and of ammonium nitrate.

Nitric acid: The systems containing nitric acid are easily passivated. For a combination of halogen acid and nitric acid, the maximum allowable halogen content is about 3% HC1, with 304 stainless steel. This halogen acid–nitric acid system does not seem to be feasible for anodic protection.

Fig. 2.8 shows the schematic of an anodic protection system for a storage vessel.

3 Design Considerations on Cathodic Protection for Buried Pipelines and Marine Structures

3.1 Principles of Cathodic Protection

Corrosion in aqueous solutions proceeds by an electrochemical process, and anodic and cathodic electrochemical reactions must occur simultaneously. No net overall charge builds up on the metal as a result of corrosion since the rates of the anodic and cathodic reactions are equal. Anodic reactions involve the oxidation of metal to its ions, for example, for steel, the following reaction occurs:

$$Fe \rightarrow Fe^{2+} + 2e$$

The cathodic process involves reduction, and several reactions are possible. In acidic water, where hydrogen ions (H^+) are plentiful, the following reaction occurs:

$$2H^+ + 2e \rightarrow H_2$$

In alkaline solutions, where hydrogen ions are rare, the reduction of water will occur to yield alkali and hydrogen:

$$2H_2O + 2e \rightarrow H_2 + 2OH^-$$

However, unless the water is deaerated, reduction of oxygen is the most likely process, again producing alkali at the surface of the metal:

$$O_2 + 2H_2O + 4e \rightarrow 4OH^-$$

Reactions (1) and (2) are shown schematically in Fig. 3.1 where anodic and cathodic sites are nearby on the surface of a piece of metal. We can change the rates of these two reactions by withdrawing electrons or by supplying additional electrons to the piece of metal. It is an established principle that if a change occurs in one of the factors under which a system is in equilibrium, the system will tend to adjust itself so as to annul, as far as possible, the effect of that change.

Thus, if we withdraw electrons from the piece of metal, the rate of reaction (1) will increase to attempt to offset our action and the dissolution of iron will increase,

Cathodic Corrosion Protection Systems. http://dx.doi.org/10.1016/B978-0-12-800274-2.00003-X

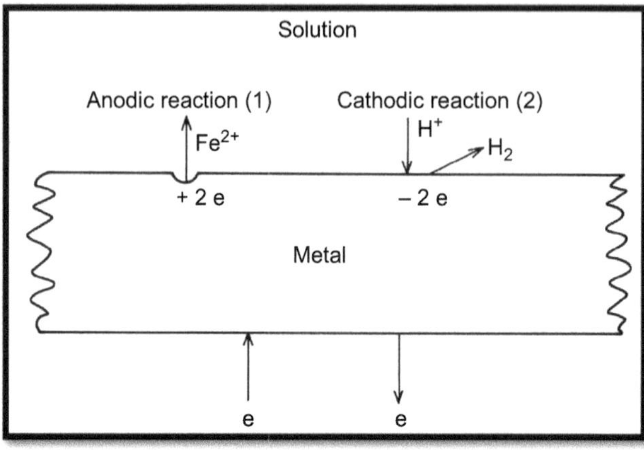

Figure 3.1 Anodic and cathodic reactions on a metal surface.

whereas reaction (2) will decrease. Conversely, if we supply additional electrons from an external source to the piece of metal, reaction (1) will decrease to give reduced corrosion and reaction (2) will increase. The latter case will apply to cathodic protection. Thus, to prevent corrosion, we have to continue to supply electrons to the steel from an external source to satisfy the requirements of the cathodic reaction. Note that the anodic and cathodic processes are inseparable. Reducing the rate of the anodic process will allow the rate of the cathodic process to increase.

These principles may be expressed in a more quantitative manner by plotting the potential of the metal against the logarithm of the anodic and cathodic reaction rates expressed as current densities. Typical anodic and cathodic curves are illustrated in Fig. 3.2.

The corrosion current, Icorr, and the corrosion potential, Ecorr, occur at the point of intersection of the anodic and cathodic curves, that is, where anodic and cathodic reactions rates are equal. If electrons are "pumped" into the metal to make it more negative, the anodic dissolution of iron is decreased to a negligible rate at a potential E_I, whereas the rate of the cathodic current is increased to I_1.

Hence, a current I_1 must be supplied from an external source to maintain the potential at E_1 where the rate of dissolution of the iron is at a low value. If the potential is reduced to E_2 (Fig. 3.2), the current required from the external source will increase to I_2. Further protection of the metal is insignificant, however, and the larger current supplied from the external source is wasted. The metal is then said to be overprotected.

In aerated neutral or alkaline solutions, the cathodic corrosion process is usually the reduction of oxygen. The kinetics of this cathodic process are controlled by the rate at which oxygen can diffuse to the surface of the metal, which is slower than the rate of consumption of oxygen by the cathodic reaction. Thus, the rate of this reaction does not increase as the potential of the metal is made more negative but remains constant unless the rate of supply of oxygen to the surface of the metal is increased by,

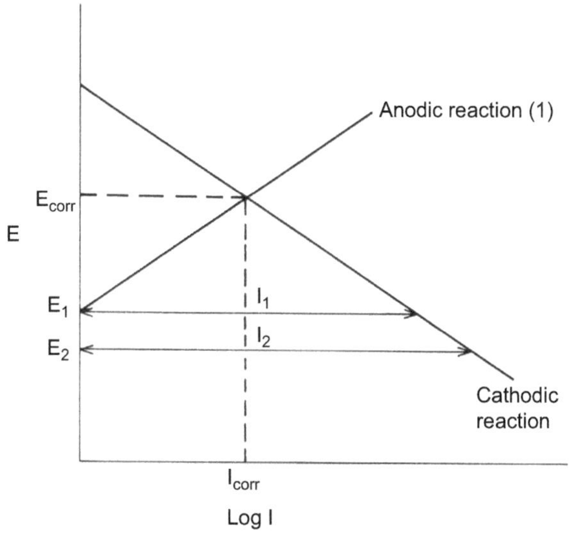

Figure 3.2 Kinetics of anodic and cathodic reactions.

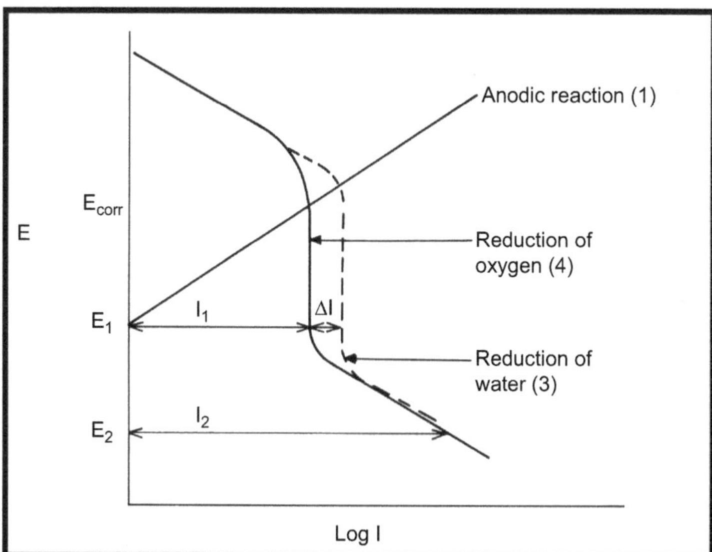

Figure 3.3 Diffusion control of the reduction of oxygen.

for example, increased fluid flow rate. The influence of flow velocity on cathodic-protection parameters is illustrated in Fig. 3.3.

A current of I_1 is initially required to maintain the metal at the protection potential E_1. However, if the flow rate is increased, the limiting current for the reduction of oxygen is increased (dotted line), and the current required to maintain the metal at the protection potential is increased by ΔI Thus, the current density required to maintain the correct protection potential will vary with service conditions. Clearly, cathodic current density is not a good guide as to whether a structure is cathodically protected. The correct protection potential must be maintained if corrosion is to be prevented.

If the structure is overprotected and the potential is reduced to a potential region where the reduction of water (reaction 3) can take place, further current will be required from the external source and current will be wasted. In Fig. 3.3, reducing the potential from E_1 to E_2 will increase the current required from the external source from I_1 to I_2 as a result of an increased rate of reduction of water.

Excessive negative potentials can cause accelerated corrosion of lead and aluminum because of the alkaline environments created at the cathode. These alkaline conditions may also be detrimental to certain paint systems, and may cause a loss of the paint film. Hydrogen evolution at the cathode surface may, on high-strength steels, result in hydrogen embrittlement of the steel, with subsequent loss of strength. It may also cause disbanding of any insulating coating: the coating would then act as an insulating shield to the cathodic-protection currents.

3.2 Methods of Applying Cathodic Protection

Cathodic protection may be achieved in either of two ways: By the use of an impressed current from an electrical source or by the use of sacrificial anodes (galvanic action).

3.2.1 *Impressed Current*

The arrangement for protecting a buried pipeline is illustrated in Fig. 3.4. The buried pipe receives current from a direct current (DC) power source via an auxiliary inert electrode buried in the ground. The pipe becomes the cathode, and the auxiliary electrode is the anode. The auxiliary electrode sometimes consists of scrap iron. In this case, the iron will dissolve from the anode by reaction (1) and the electrode is described as a consumable anode.

If the anode is a noble metal or an electrochemically inert material, the surrounding environment will be oxidized and in water reaction we have:

$$2H_2O \rightarrow O_2 + 4H^+ + 4e$$

In saline solutions, however, chlorine may be produced at the anode. This may present problems in confined spaces. A range of materials have been used as nonconsumable anodes for impressed-current systems. The properties required by these anodes are as follows:

- good electrical conduction,
- low rate of corrosion,

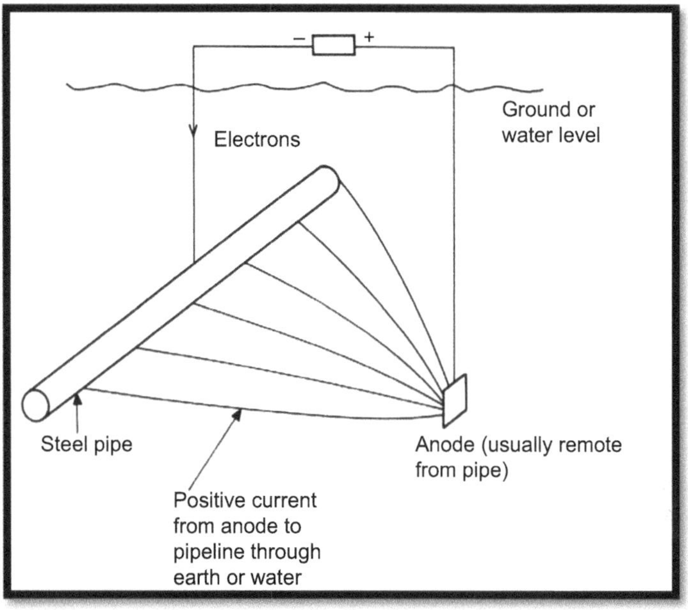

Figure 3.4 Application of cathodic protection by an impressed current.

- good mechanical properties, that is, ability to withstand the stresses that they may be subjected to during installation and in service,
- ability to be readily fabricated into a variety of shapes,
- low cost,
- ability to withstand high current densities at their surfaces without forming resistive barrier oxide layers, etc.

The following materials have been used as anodes: magnetite, carbonaceous materials (graphite), high silicon iron (14–18% Si), lead/lead oxide, lead alloys, platinized materials (e.g., tantalum, niobium, and titanium). Platinum, with its high resistance to corrosion, would be an ideal anode material but it has the major disadvantage of very high cost.

In practice, voltages up to 100 V and high current densities are possible on impressed-current anodes (Table 3.1). Thus, large areas of a structure can be protected

Table 3.1 Properties of Impressed-Current Anodes

Anode Material	Max Volts (Dependent on the Environment)	Typical Anode Current Density A/m^2
Platinum/niobium	100	250–1500
Lead/silver/antimony	100	250–1000
High silicon iron	100	10–100
Graphite	–	200

from a single anode and, because of the high driving voltage, the anode can be placed remote from the structure.

3.2.2 Sacrificial Anodes

To understand the action of sacrificial anodes for cathodic protection, it is necessary to have in mind the galvanic series of metals. The galvanic series for a few selected metals in seawater is shown in Table 3.2. When the tendency for a metal to go into solution as metal ions increases (leaving an excess of electrons on the metal surface), that is,

$$M \rightarrow M^{+} + e$$

the metal becomes more electronegative. Thus, since zinc, aluminum, and magnesium are more electronegative than steel, they are increasingly able to supply electrons to the more electropositive steel when in electrical contact in water, and will effect the cathodic protection of the steel surface. Clearly, if steel was coupled to copper in seawater, steel would supply electrons to copper, which would become cathodically protected, and the corrosion of the steel would be enhanced.

The cathodic protection of a steel pipe with sacrificial anodes is illustrated in Fig. 3.5. Electrons are supplied to the steel pipe, via the electrical connection, and a corresponding amount of anode material goes into solution as metal ions, according to the laws of electrolysis. Some anode material is lost by self-corrosion, and the anodes are not converted to electrical energy with a 100% efficiency. Zinc, aluminum, and magnesium are the metals commonly used for sacrificial cathodic protection. Some anode properties are shown in Table 3.3.

The driving voltage of sacrificial anodes is now compared with impressed-current anodes, and sacrificial anodes must be located close to the structure being protected.

Table 3.2 Galvanic Series of Some Metals in Seawater

Electropositive
Platinum
Titanium
Stainless steel
Monel
Copper
Lead
Iron, cast iron, or steel
Cadmium
Zinc
Aluminum
Magnesium
Electronegative

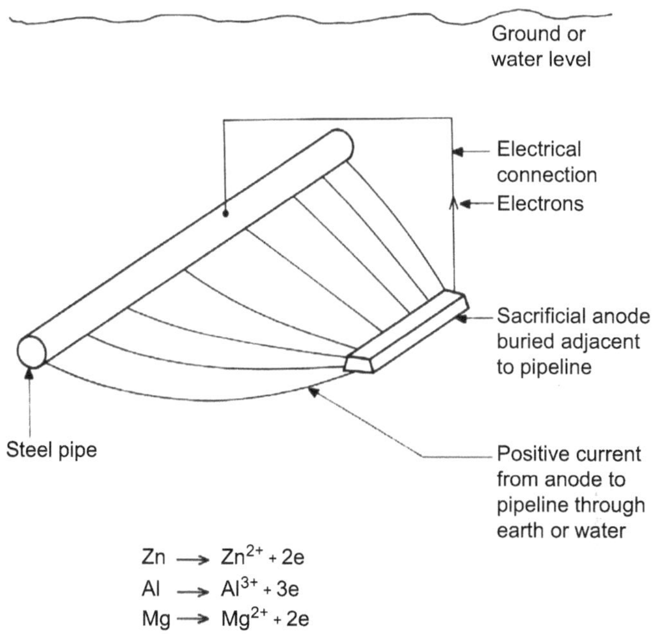

Figure 3.5 Application of cathodic protection by the sacrificial anode.

Table 3.3 Properties of Sacrificial Anodes

Anode Material	Density g/cm^3	Potential Volts Cu/$CuSO_4$	Ah/kg	Typical Anode Current Density A/m^2
Zn	7.1	−1.1	780	0.5–2
Al	2.7	−1.15	2700	0.6–2.5
Mg	1.7	−1.55	1230	1.5–5.6

Although almost any piece of zinc, etc. could provide cathodic protection over a short period of time, cathodic-protection schemes are usually required to operate over periods of several years. Anodes can lose their activity and become passivated, and develop a nonconducting film on their surfaces so that they no longer are able to supply current.

This can be avoided by careful control of the concentrations of trace impurities in the anode materials, and by alloying. For zinc anodes, the level of iron, for example, must be kept below 0.005% for satisfactory long-term operation of the anodes. To prevent passivation of aluminum anodes, alloying with indium, for example, has been found to be successful. The previously successful alloy with mercury is now disliked on environmental grounds.

3.3 Protection Potentials

In practice, the structure-to-electrolyte potentials are measured using a standard reference electrode based on copper/copper sulfate, silver/silver chloride, or pure zinc. The reference electrode should be very close to the surface whose potential is being measured. For steel in an aerobic electrolyte of a nearly neutral pH, a commonly accepted protection potential is −850 mV; when exposed to sulfate-reducing bacteria, a potential of −950 mV would be required. Both values are referred to a copper/copper sulfate electrode.

Some potential values for the protection of other metals are shown in Tables 3 4. Values for lead and aluminum must be carefully controlled to avoid damage by excess alkali that could build up at the surface of the metals if the protection potentials are too negative.

3.4 Current Density

The current density required to maintain the protection potential is very dependent on local conditions. Increased availability of oxygen at the surface of the metal will directly increase the current density as illustrated previously in Fig. 3.3. Increased availability of oxygen may occur because of an increased concentration of oxygen in the environment, increased water flow, or turbulence. Thus, current densities to structures in seawater, rivers, etc. are likely to vary continuously. The pH of the environment will also be important. The presence of coatings, marine fouling, and calcareous deposits will have a profound effect on the current density. Some typical values of current density for steel are shown in Table 3.5.

Table 3.4 Potential Required for Cathodic Protection

Metal	Potential ($Cu/CuSO_4$) (mV)
Steel	−850
Steel (sulfate-reducing bacteria)	−950
Copper alloys	−500 to −650
Lead	−600
Aluminum	−950 to −1200

Table 3.5 Current Densities Required to Protect Steel

Environment	Current Density (A/m^2)
Acidic solutions	350–500
Saline solutions	0.3–10
Seawater	0.05–0.15
Saline mud	0.025–0.05

Having decided on the appropriate current density, the total anode current can be determined from the area of the structure. The size of the anodes can then be determined from data in Table 3.3 if a sacrificial anode scheme is to be employed, taking into account the working life of the protected structure or the period required between refits.

3.5 Coatings

The provision of an insulating coating to the structure will greatly reduce the current demand for cathodic protection. When first applied, coatings will often contain flaws, and in service, further defects will develop over a period of time. The conjoint use of coatings and cathodic protection takes advantage of the most attractive features of each method of corrosion control. Thus, the bulk of the protection is provided by the coating, and cathodic protection provides protection to flaws in the coating. As the coating degrades with time, the activity of the cathodic-protection system develops to protect the deficiencies in the coating. A combination of coating and cathodic protection will normally result in the most economic protection system.

3.6 Calcareous Scales

In seawater, cathodic protection of bare steel is economic because of the formation of calcareous deposits. The alkali formed at the surface of a protected structure reacts with bicarbonate ions present in seawater to form carbonate ions that, in turn, precipitate as insoluble calcium carbonate on the surface of the metal:

$$OH^- + HCO_3^- \rightarrow CO_3^{2-} + H_2O$$

$$CO_3^{2-} + Ca^{2+} \rightarrow CaCO_3$$

3.7 Choice of a Cathodic-Protection System

In the design of a cathodic-protection scheme, a decision must be made as to whether the scheme should be a sacrificial anode or an impressed-current system or a mixture of the two systems.

Sacrificial anode systems have the advantage of being

1. simple to install,
2. independent of any source of electric power,
3. suitable for localized protection,
4. less liable to cause interaction on neighboring structures.

It is difficult to overprotect the structure and moderately easy to obtain a uniform electrode potential across the structure. The most severe limitation of the sacrificial anode is the small driving force that restricts its use to conductive environments or well-coated systems. To protect a large structure, such as a pipeline, with sacrificial anodes, a large number of them would need to be distributed along it, involving a multiplicity of electrical connections and considerable installation work.

The advantages of the impressed-current system include the following:

1. The large driving force available can protect a large, even uncoated, structure in high-resistivity environments,
2. Comparatively few anodes are needed,
3. The voltage may be adjusted to allow for environmental and coating changes.

The impressed-current system may allow considerable overprotection, however, and considerable variation of potential over the structure is difficult to avoid.

Generally, sacrificial anode schemes have found favor for small well-coated low current demand structures or for localized protection, with impressed-current schemes being utilized for large complex structures that may be bare or poorly coated. However, in North Sea offshore work, it has been found economic to provide galvanic protection to large uncoated platforms and similar structures where the cost of coating is prohibitive. In addition, galvanic anodes offer easily installed robust systems that, being independent of a power source, provide protection immediately on "floatout" of the structure.

3.8 Anode Resistance

One of the most important parameters in the design of cathodic-protection systems is the electrical resistivity of the environment. Resistivities encountered for pipeline environments vary from 1 ohm-cm for brackish river water to >500,000 ohm cm in nonporous granite. Measurement of the resistivity of the environment and calculation of the electrical resistance between the anodes and the structure due to the electrolyte must be made at an early stage in the design of the scheme to ensure that adequate current output will be obtained from the anodes over the design life of the structure (by application of Ohm's Law).

Established methods for determining the resistivity of soils and water are available and will not be dealt within these notes. Exact calculation of the electrical resistances between anodes and structure due to the electrolyte is rarely possible. Ground beds, as cathodic-protection electrodes set in the earth are referred to, can be considered to lie in a semiinfinite electrolyte, and the resistances of electrodes to infinite earth or seawater have been calculated for a number of anode shapes. Thus, if the anodes are remote from the structure, these anode resistances can be used, together with the difference between the required protection potential and the anode potential, to determine the current output of anodes using Ohm's Law. If, however, the anodes are positioned close to the structure, then some correction to the resistance is required.

When current flows from a small anode to a large metal structure, the current density is at a maximum near the surface of the anode. Hence, a major portion of the potential drop between the anode and the structure occurs in the vicinity of the anode. This factor probably enables values of anode resistance to infinite earth to be used reasonably successfully even when the anode and the structure are not well separated. A further consequence of this factor is that surrounding an anode with a backfill of conducting material greatly reduces the anode resistance in resistant soils. In practice, the distribution of current to a structure is difficult to control. If we consider the protection of a pipe by an anode, it should be clear that there will be a higher current density at the point on the pipeline nearest the anode than elsewhere.

The effect is illustrated in Fig. 3.6. Clearly, to ensure that the ends of the pipe are protected, the center of the pipe, nearest the anode, must be overprotected to some degree. The effect can be minimized by using several anodes spaced along the pipe, but this will greatly increase installation costs.

Consideration must also be given to spark hazards created by the introduction of electric currents into structures situated in a hazardous atmosphere.

Any secondary structure residing in the same electrolyte may receive and discharge the cathodic-protection DC by acting as an alternative low-resistance path. Corrosion will be accelerated on the secondary structure at any point where current is discharged to the electrolyte. This phenomenon is called "interaction."

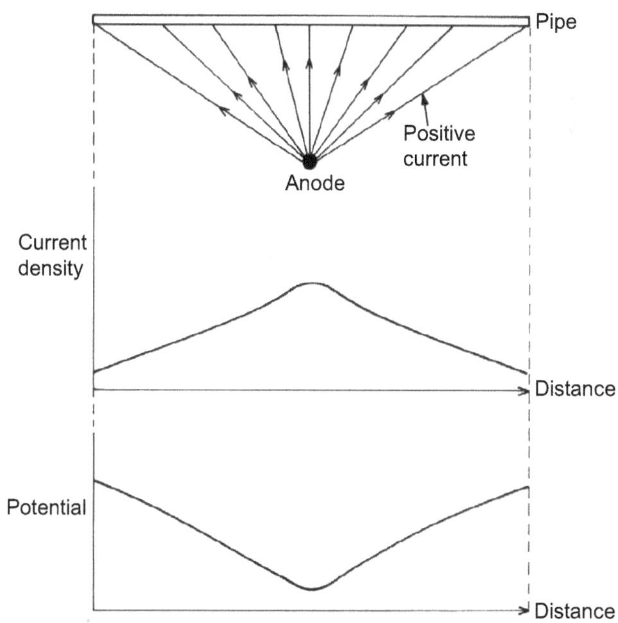

Figure 3.6 Current and potential distribution along a pipe.

Interaction may occur, for example, on a ship that is moored alongside a cathodically protected jetty, or on a pipeline or metal-sheathed cable that crosses a cathodically protected pipeline.

Interaction may be minimized by careful design of the cathodic-protection system, in particular, by design of a scheme to operate at the lowest possible current density and by maintaining the greater separation between the protected structure and the secondary structure, and between the ground beds or anodes and the secondary structure. It is an advantage of sacrificial anode schemes that they are not prone to creating severe interaction problems and therefore are popular for protection in congested and complex locations.

Methods and procedures are available for overcoming interaction, and testing must be carried out in the presence of all interested parties, so that the choice of remedial measures may be agreed upon if and when the acceptable limit of interaction is exceeded.

3.9 Impressed-Current System

3.9.1 Current–Voltage Relationships

3.9.1.1 Attenuation

When current is drained from a buried pipeline, the potential of the line to its surroundings is changed, and current is picked up from the soil and flows in the pipe toward the drain point. The current flow in the line and the potential change are maximum at the drain point. At any point other than the drain point, the current flow in the pipe is less because of the current picked up between that point and the drain point is less. Likewise, the voltage change is less because of the voltage drop caused by the flow of current in the pipe between that point and the drain point is less.

This decrease in voltage and current flow with distance from the drain point is called attenuation. The rate, of attenuation is influenced by pipe resistance, coating conductivity, polarization, length of line, and method of termination (insulating device or metallic connection to other facilities). In the case of bare or poorly coated lines, soil resistivity also has a large effect.

3.9.1.2 Attenuation Formulas

Formulas have been derived to express the relationship between current and potential along a uniform pipeline when current is drained from a single point. The most commonly used forms of these attenuation formulas are given below. These formulas apply specifically to coated lines; they are based on the assumption that the pipe is of a uniform section and resistivity, and that coating conductivity is uniform. The effect of soil resistivity and polarization is neglected. The voltage as used in the attenuation formulas for coated lines is the voltage change across the coated line; this is the voltage change between the pipe and adjacent soil and is the significant voltage change in cathodic protection.

Attenuation formulas are sometimes applied to bare lines. In this case, soil resistivity is assumed to be uniform, and the resistance per unit length of the line to remote earth is usually substituted for the coating resistance. The voltage in the formulas then becomes the voltage change with respect to the remote earth. This voltage does not necessarily indicate the adequacy of cathodic protection. In addition, polarization has a profound effect on attenuation in the case of bare lines, and wide variations in soil resistivity further tend to invalidate calculated results. For these reasons, extreme caution should be exercised in applying attenuation formulas to bare lines.

3.9.1.3 *Calculations*

The calculations in this section are intended to be a guide only. They should not be considered as absolute or the only method of calculation. Cathodic-protection design procedures are empirical at best or based on empirically modified theory. Table 3.6 provides symbols for equations.

- **Formulas**

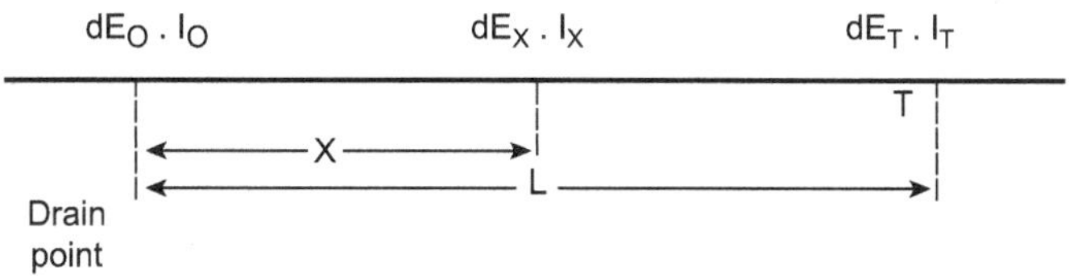

Table 3.6 Symbols for Equations

Symbol	Quantity	Units
R_s	Linear pipe resistance	ohm-m
R_L	Leakage (coating resistance)	Ohm/m
a	Attenuation constant	per kilometer
R_O	Characteristic resistance	Ohm
ρ	Soil resistivity	micro-ohm-cm
ρ'	Steel resistivity	micro-ohm-cm
g	Coating conductivity	micro-mhos/m^2
D	Pipe diameter	mm
L	Pipe length	km
t	Pipe wall thickness	mm
a	Pipe cross-sectional area	cm^2
A	Surface area of the pipe	m^2/m
W	Weight of the pipe	kg/m
dE_O	Change in potential at drain point	Volts
dE_x	Change in potential at point X	Volts
dE_T	Change in potential at the end of the line	Volts
I_O	Current at drain point	A
I_x	Current at point X on the line	A
I_T	Current at the end of the line	A

- **General case**

$$dE_x = dE_O \cdot \cosh \alpha X - I_A \cdot R_O \cdot \sinh \alpha X \quad \text{Volts}$$

$$I_X = I_O \cdot \cosh \alpha X - \frac{dE_O}{R_O} \cdot \sinh \alpha X \quad \text{Amperes}$$

or

$$dE_O = dE_X \cdot \cosh \alpha X + R_O \cdot I_X \cdot \sinh \alpha X \quad \text{Volts}$$

$$I_A = I_X \cdot \cosh \alpha X - \frac{dE_X}{R_O} \cdot \sinh \alpha X \quad \text{Amperes}$$

- **Finite lines**

$$dE_O = dE_T \cosh \alpha L \quad \text{Volts}$$

$$I_O = \frac{dE_T}{R_O} \cdot \sinh \alpha L \quad \text{Amperes}$$

$$dE_X = dE_T \cosh \alpha (L - X) \quad \text{Volts}$$

$$I_X = I_O \frac{\sinh \alpha (L - X)}{\sinh \alpha L} \quad \text{Amperes}$$

- **Infinite lines (in almost all cathodic-protection systems, there are no infinite lines).**

$$dE_X = dE_O \, e^{-aX} \quad \text{Volts}$$

$$I_X = I_O \, e^{-aX} \quad \text{Amperes}$$

$$I_O = \frac{dE_O}{\alpha R_L} = \frac{dE_O}{R_O} \quad \text{Amperes}$$

$$\ln dE_O = \ln dE_X + \alpha X$$

$$\ln I_O = \ln I_X + \alpha x$$

- **Linear pipe resistance**

$$R_S = \frac{\rho' L}{a}$$

It is to be noted that steel resistivity varies from 16 to 23 micro-ohm cm. An average value is 18 micro-ohm cm.

- **Leakage (coating resistance)**

$$R_L = \frac{1}{gAL} \quad (\text{ohm/km})$$

- **Characteristic resistance**

$$R_O = \sqrt{R_S \cdot R_L} \quad \text{ohm}$$

- **Attenuation constant**

$$a = \frac{R_S}{R_L} \quad \text{per km}$$

$$\alpha = \frac{R_O}{R_L}$$

$$\alpha = \frac{R_S}{R_O} = \frac{R_S I}{2E_A I}$$

For design purposes, the coating conductivity of buried, non-fusion-bonded epoxy-coated landlines may be taken as 2000 micro-mho/m^2. Fusion-bonded epoxy-coated lines should be considered to have a coating conductivity of 400 micro-mho/m^2.

3.9.2 Single Drain Point

Figure 3.7 shows the schematic layout of a pipeline with the current drained at a single point at one end, together with typical attenuation curves. If the line is infinitely long, the current flow and voltage change vary exponentially with distance along the line as shown by the curve labeled, "E and I—Infinite Line." If the line terminates in an insulated flange or dead ends, it is called a finite line.

Attenuation curves for finite lines of three different lengths, with the voltage at the drain point the same as in the case of the infinite line, are also shown in Fig. 3.7. The

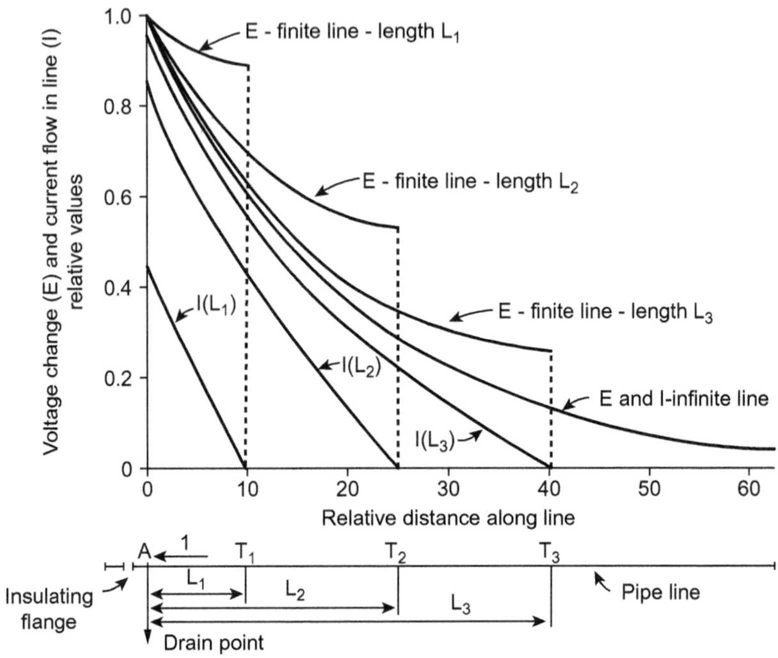

Figure 3.7 Typical attenuation curves—single drain point.

current flowing at the far end of a finite line is, of course, zero, and the potential is substantially higher than the potential at the same distance on an infinite line. Length L_1 corresponds to a short finite line; in this case, the current drained is substantially less than that for the infinite line. Length L_3 corresponds to a moderately long finite line; in this case, the current drained is nearly as great as the current drained from the infinite line, but the potential at the far end is nearly twice as great as the potential at the corresponding distance on the infinite line.

In the above examples, the current drained at the drain point is equal to the current flowing in the line at the drain point because current flows to the drain point from one direction only. In the usual case, the line extends in both directions from the drain point, and the current flows to the drain point from both sides. Under these conditions, assuming a symmetrical, uniform system, the total current drained is twice the current flowing in the line at the drain point from either direction.

3.9.3 Multiple Drain Points

When a long pipeline is placed under cathodic protection, it is usually necessary to drain current at numerous points. A portion of a long uniform pipeline with current drained at two points is illustrated in Fig. 3.8. The attenuation curves for drain points 1 and 2, considering each drain point independently, are shown as dashed curves. These correspond to the infinite line curve of Fig. 3.7.

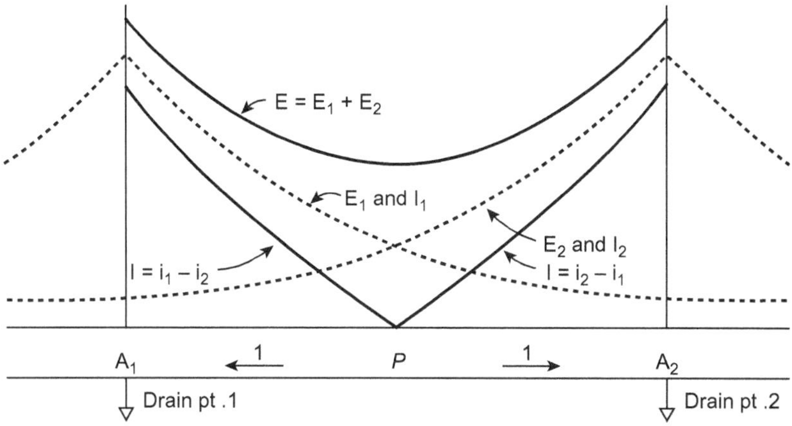

Figure 3.8 Typical attenuation curves—multiple drain points.

The potential change at any point between the two drain points, with current drained at both drain points, can be obtained by adding the potential changes produced by the individual current drains; this is shown as the upper solid curve in Fig. 3.8. Since the currents induced by the two current drains flow in opposite directions between the two drain points, the resultant current flowing in the line at any point between the two drain points is equal to the difference in the currents resulting from the two current drains. At midpoint P, the two currents are equal and opposite, and the resultant current is zero. Since there is no resultant current flow in the line at this point, the current and potential distribution will not be disturbed if the line is cut at this point, and an insulating flange is inserted. From this, it follows that the resultant current and voltage distribution between a drain point and the electrical midpoint between two drain points is the same as for a finite line of the same length. This can be seen by comparing the curves in Fig. 3.8 with those in Fig. 3.7.

3.9.4 Effect of Coating

The coating on the pipeline has a great effect on the current required to protect the line and the rate of attenuation of current and voltage along the line. With well-coated lines, the current requirements are modest, and attenuation is slight, and drain points may be spaced at wide intervals. With bare or poorly coated lines, current requirements are large, attenuation is rapid, and drain points must be spaced at close intervals.

3.9.5 Point of Minimum Protection

Since voltage varies with distance along a pipeline due to attenuation, the point of minimum protection (least negative potential) must be adequately protected to enure complete protection of a pipeline. With a uniformly coated line protected from a single drain point, the point of minimum protection will be the end-most point remote from the drain point. With a multiple-drain-point system, points of minimum protection will exist between drain points. One of the main problems encountered in field tests on pipelines is to determine the locations of these points of minimum protection. Once these points have been located, it is normally a relatively simple matter to adjust the system (assuming that it is adequately designed) to achieve adequate protection at these points and thus complete protection of the line.

3.9.6 Bare and Poorly Coated Lines

The above discussion is strictly applicable only to uniformly coated pipelines. This is not always the case in practice; bare lines, lines partly wrapped and partly bare, or coated lines with the quality of the coating varying widely from point to point are frequently encountered. In the case of a coated line with a variable coating quality, the line will pick up relatively more current from the soil in the areas of poor coating. This will result in larger potential drops in the soil adjacent to the poorly coated sections, and the pipe-to-soil potentials in these areas will be lower (the line will be less negative with respect to adjacent soil) than in nearby areas of good coating.

Where lines are partly coated and partly bare, this effect is even more pronounced; the pipe-to-soil potential at a bare area will normally be much lower than that at an adjacent coated area, even though the coated area may be more remote from the drain point.

Where a bare pipeline is buried in soil of uniform resistivity, the potential along the line will vary in the same manner as for a uniformly coated line. However, if a bare line passes through soil where the resistivity varies widely from point to point, current pick-up will be largely concentrated in the areas of low soil resistivity with resulting large potential drops in the soil and low pipe-to-soil potentials in these areas.

3.9.7 Voltage Limitations

3.9.7.1 Coating Deterioration

An excessive potential applied to a coated line has a detrimental effect on the coating; the bond between the coating and the pipe may be destroyed and the coating conductance will increase many times. The rate of coating deterioration increases rapidly as the potential is increased.

For this reason, the maximum potential applied to a coated pipeline should not exceed approximately 2.0 V (off) to copper/copper sulfate electrode, or the potential change should not exceed about 1.0–1.5 V (off) with respect to the initial potential. The lower values should be observed whenever possible. The minimum potential change should be 0.3 V with respect to the initial potential.

3.9.7.2 Maximum Rectifier Spacing

The potential and current at the drain point required to protect a finite line (or the section between minimum points in the case of a multiple-drain-point system) increase rapidly as the length is increased. Since the potential change at the drain point should not exceed approximately 1.0–1.5 V, this limits the length of the coated line that can be protected from a single drain point and thus determines the maximum spacing between units in a multiple-drain-point system.

Spacings between drain points (or rectifiers) depend on pipeline coating and the diameter of pipeline as well as on soil resistivities of the terrain. It is possible to protect 75 km of well-coated large diameter crosscountry pipeline from one station, whereas it is not practical to protect such a distance of a small diameter pipe from a single station.

3.9.7.3 Depolarization Effect

Any depolarizing parameter will result in increasing current requirement for cathodic protection. The most effective mechanism for depolarization and removal of hydrogen polarization film is sulfate-reducing bacteria. The bacteria are active when conditions are anaerobic. Under this condition, the active sulfate-reducing bacteria reduces sulfates to sulfide and removes hydrogen polarization films by oxidation and water formation.

By maintaining the minimum potential value at −0.95 (with reference to copper/copper sulfate electrode), soil pH will be kept around 9 and sulfate-reducing bacteria would be rendered inactive. Therefore, a minimum potential value of −0.95 is essential to eliminate the depolarizing effect of sulfate-reducing bacteria in soils in which these bacteria are present.

3.9.7.4 Bare Lines

The maximum voltage that can be applied to a bare line is limited only by economics. An excessive potential will not damage bare steel pipe, but it will result in an excessive current pick-up and inefficient utilization of the applied power.

3.9.8 Current Requirements

For pipeline design, requirements should be based on the following minimum current density value in Table 3.7 for the total coated and uncoated pipeline surface:

Two 32-pound magnesium anodes should be installed on all buried valves, on the pipeline at gas reservoir, and in pipeline anchor locations.

An additional current capacity of 50 A should be provided at plant terminations where insulating devices are bonded to control interference.

Table 3.7 Current Density for Some Pipeline Surfaces

Pipeline Surface	Current Density (mA/m^2)
Uncoated	20
Tape or coal tar/bitumen	1.25
Fusion-bonded epoxy or three-layer polyethylene	0.25

3.9.9 Measurements

3.9.9.1 Coated Lines

On existing pipelines, it is desirable to perform field tests to determine the current required for cathodic protection. For coated pipelines, the usual procedure is to conduct preliminary tests to determine the coating conductivity of representative sections of the line. From these tests, it is possible to calculate or to determine from the charts the approximate current required for protection, as discussed in the subsection on "calculations." After this preliminary determination of current requirements is made, the type of system to be installed impressed current or galvanic anode can be selected. If a galvanic anode system is to be installed, no further tests are required until installation of the anodes is commenced.

If an impressed-current system is to be installed, the maximum distance between drain points (limited by voltage across the coating) can be determined once the coating conductivity is known. This is also covered in the subsection on calculations. The actual drain point spacing will be influenced by several other factors; it will normally be considerably less than the theoretical maximum spacing.

3.9.9.2 Bare Lines

In the case of bare lines, the initial drain tests will normally consist of determining, by direct field measurement, the distance that can be protected with a given current drain. In these tests, it should be remembered that, assuming uniform conditions along the pipeline, the potential change at the midpoint between two rectifiers draining equal currents will be twice the voltage change produced by one rectifier alone. Consequently, if tests are conducted using a single drain point, the distance each rectifier can protect if multiple units are installed will be the distance at which one-half of the desired final voltage can be obtained during drain tests with a single drain point.

3.9.10 Sizing of Rectifiers

After rectifier sites have been selected and the anodes and test leads have been installed, it is usually desirable to conduct drain tests to determine by direct measurement the current and voltage required at each drain point. These data determine the output rating of the rectifier at each location. The tests are conducted with batteries or welding generators as sources of DC, utilizing the permanent anodes and permanent test leads.

3.9.11 Calculations

If the approximate coating conductivity is known, the current required to protect a coated line and the maximum distance between drain points may be estimated using the "Attenuation" formulas.

3.9.11.1 *Circuit Resistance*

The total resistance in the cathodic-protection circuit is as follows:

$$R_T = R_C + R_G + R_L + R_S$$

where:

R_T is the total circuit resistance at rectifier terminal.
R_C is the cable resistance, for both positive and negative cables.
R_G is the resistance of the anode bed to the earth.
R_L is the resistance of the coating.
R_S is the resistance of the structure to be protected (may be ignored).

3.9.11.2 *Cable Resistance*

Cable resistance is the resistance from the pipeline via the power source to the first anode of the anode bed, plus effective resistance of the header cable along the line of the anodes. It is practical to use the resistance of one-half of the ground bed header cable as the effective header cable resistance.

Table 3.8 includes data on the resistance of copper conductors in the sizes commonly used in pipeline corrosion engineering work. Table 3.9 gives the correction factors for conductor resistance at temperatures other than 25 °C. Table 3.10 provides some design data on cables.

Table 3.8 Resistance of Copper Conductors

	Conductor Size		
General Use	**mm^2**	**(AWG)**	**Resistance of Stranded Copper Conductors in ohms per meter ×10^{-3} at 25 °C**
Impressed-current ground bed	107.20	400	0.16695
	85.03	3	0.21058
	67.43	2	0.26601
	53.48	1	0.33456
	42.41	1	0.42312
	33.63	2	0.53136
	21.15	4	0.84952
	13.30	6	1.35136
Galvanic anode installations	21.15	4	0.84952
	16.77	5	1.0714
Pipeline test points	16.77	5	1.0714
	10.55	7	1.70361
Instrument test leads	1.31	16	13.71
	0.823	18	21.84
		20	34.77
		22	55.76

AWG - American Wire Gauge

Table 3.9 Correction Factors for Other Temperatures

Temperature		
°C	**°F**	**Multiply Resistance at 25 °C by**
−10	14	0.862
−5	23	0.882
0	32	0.901
5	41	0.921
10	50	0.941
15	59	0.961
20	68	0.980
30	86	1.02
35	95	1.04
40	104	1.059

3.9.11.3 Anode-to-Electrolyte Resistance Calculations

- **Single cylindrical anode**

The following formulas give the anode-to-earth resistance of a single cylindrical anode:

$$R = \frac{\rho}{2\pi L}\left(\ln\frac{8L}{d} - 1\right)$$

Multiple horizontal anodes

$$R_h = \left(\frac{0.00521\rho}{NL}\right)\left[\ln\left(\frac{4L^2 + 4L\sqrt{S^2 + L^2}}{ds}\right) + \frac{S}{L} - \frac{\sqrt{S^2 + L^2}}{L} - 1\right]$$

Where:

R_h = Resistance to the earth of a single horizontal anode (ohms)
L = Length of the anode including the backfill (centimeters)
d = Diameter of the anode including the backfill (centimeters)
ρ = Electrolyte resistivity (ohms centimeters)
S = Twice the depth of the anode (centimeters)
N = Number of anodes.

For a single vertical anode, the following simplified expression can be used:

$$R_V = \frac{\rho K}{L/d}$$

Where

R_V = Resistance to the earth of a single vertical anode (ohms).
ρ = Effective soil resistivity (ohms centimeters).

Table 3.10 Cable Design Data

Nominal Cross-Section (mm^2) of the Copper Conductor	Number of Strands	Actual Cross-Section (mm^2)	Allowable Current (A)
2.5	1	2.49	18
	1	3.31	20
4	7	3.972	24
	1	5.26	30
6	7	5.95	31
	1	8.37	40
10	7	10.02	42
	7	13.3	55
16	7	15.89	56
	7	21.14	70
25	7	25.18	73
	7	26.66	80
	7	33.66	95
35	19	34.93	90
	19	42.45	110
50	19	47.28	145
	19	53.43	125
	19	67.45	145
70	19	68.34	185
	19	85.07	165
95	19	94.75	230
	19	107.76	195
120	37	119.75	260
	37	126.58	270
150	37	147.11	355
	37	151.86	300
	37	177.5	325
185	37	184.54	405
	37	202.78	360
240	61	242.54	480
	37	253.15	405
300	61	304.2	560
	61	304.2	455
	61	354.5	490
	61	380.1	500
400	61	389.1	680
	61	405.2	515
	61	456.3	555
500	61	490.6	600
	61	506.4	585
630	127	633.4	910

Note: Typical current ratings are given as a guide only—always check the manufacturer's rating for the particular site operating conditions, including all relevant deaerating factors.

Table 3.11 The Shape Function

L/D	K
5	0.014
6	0.015
7	0.0158
8	0.0165
9	0.0171
10	0.0177
12	0.0186
14	0.0194
16	0.0201
18	0.0207
20	0.0213
25	0.0224
30	0.0234
35	0.0242
40	0.0249
45	0.0255
50	0.0261
55	0.0266
60	0.0270

L = Length of the anode (centimeters).
K = Shape function, representing the ratio of the anode Length/Anode diameter (from Table 3.11).
L/d = Ratio of the length to diameter of the anode.

- **Single row vertical anode group**

The total anode-to-earth resistance of a group of vertical anodes in parallel and equally spaced in one row is given by the following:

$$R_n = \frac{1}{n} R_V + \frac{\rho}{S} \rho'$$

where

R_n = Total anode-to-earth resistance of vertical anodes in parallel (ohms).
R_V = Anode-to-earth resistance of a single vertical anode (ohms).
n = Number of vertical anodes in parallel.
ρ = Soil resistivity, measured with a pin spacing equal to S (ohms centimeters).
ρ' = Paralleling factor, from Table 3.12.
S = Spacing between adjacent anodes (centimeters).

- **More than one row vertical anode group**

An anode group composed of two or more rows of vertical anodes, separated by a distance substantially larger than that between the anodes within a single row, has a

Table 3.12 Paralleling Factor

n	ρ′ (Paralleling Factor)
2	0.00261
3	0.00289
4	0.00283
5	0.00268
6	0.00252
7	0.00237
8	0.00224
9	0.00212
10	0.00201
12	0.00182
14	0.00168
16	0.00155
18	0.00145
20	0.00135
22	0.00128
24	0.00121
26	0.00114
28	0.00109
30	0.00104

total resistance approximately equal to the total parallel resistance of all the rows. The usual formula for paralleling resistance is used.

- **Vertical and horizontal anode groups—simplified**

If anodes are assumed to be installed vertically (3- to 5-cm diameter and 3-m length) or horizontally (0.1 m^2, 3-m length, and 1.8-m depth), the following two simplified formulas may be used:

$$\frac{1}{R} = \frac{1}{R_1} + \frac{1}{R_2} + \frac{1}{R_3} + \ldots$$

- **Vertical and horizontal anodes group—simplified formulas**

If anodes are assumed to be installed vertically (3- to 5-cm diameter and 3-m length) or horizontally (0.1 m^2, 3-m length, and 1.8-m depth), the following two simplified formulas may be used:

R_V = Anode-to-earth resistance, any number of vertical anodes (ohms).
R_h = Anode-to-earth resistance, any number of horizontal anodes (ohms).
ρ = Effective soil resistivity (ohms centimeter).
F = Adjusting factor for group of anodes, from Table 3.13.

$$R_V = \frac{\rho F}{537}$$

Table 3.13 Adjusting Factors for Parallel Anodes

No. of Anodes in Parallel	Adjusting Factors		Anode Spacing in Meters (ft)		
n	1.5 m (5′)	3 m (10′)	4.5 m (15′)	6 m (20′)	7.5 m (25′)
2	0.652	0.576	0.551	0.538	0.53
3	0.586	0.46	0.418	0.397	0.384
4	0.52	0.385	0.34	0.318	0.304
5	0.466	0.333	0.289	0.267	0.253
6	0.423	0.295	0.252	0.231	0.218
7	0.387	0.265	0.224	0.204	0.192
8	0.361	0.243	0.204	0.184	0.172
9	0.332	0.222	0.185	0.166	0.155
10	0.311	0.205	0.17	0.153	0.142
11	0.292	0.192	0.158	0.141	0.131
12	0.276	0.18	0.143	0.132	0.122
13	0.262	0.169	0.139	0.123	0.114
14	0.249	0.16	0.131	0.116	0.107
15	0.238	0.152	0.124	0.109	0.101
16	0.226	0.144	0.117	0.103	0.095
17	0.218	0.138	0.112	0.099	0.091
18	0.209	0.132	0.107	0.094	0.086
19	0.202	0.127	0.102	0.09	0.082
20	0.194	0.122	0.098	0.086	0.079
22	0.182	0.114	0.091	0.079	0.073
24	0.171	0.106	0.085	0.074	0.067
26	0.161	0.1	0.079	0.069	0.063
28	0.152	0.094	0.075	0.065	0.059
30	0.145	0.089	0.070	0.061	0.056

$$R_h = \frac{\rho F}{483}$$

Note: For n = 1, F = 1.0.

3.9.11.4 Transformer Rectifier Choice

Having determined the current requirement and circuit resistance, the necessary driving voltage can be calculated using Ohm's law, and the proper rectifier size will then be chosen.

3.10 Galvanic Anode System

3.10.1 Symbols

The symbols for equations of Galvanic Anode System are shown in Table 3.14.

Table 3.14 Symbols for Equations of Galvanic Anode System

Symbols	Quantity	Unit
I_a	Anode current output	mA
R_a	Anode Resistance to Earth	ohms
R_c	Cathode resistance to Earth	ohms
E_a	Open-Circuit Potential between anode and Reference Electrode	Volts
E_c	Open-Circuit Potential between cathode and Reference Electrode	Volts
E_o	Open-circuit Potential between anode and cathode = $E_a - E_c$	Volts
E_{cu}	Closed Circuit Potential between cathode and Reference Electrode	Volts
E_d	Driving voltage between anode and cathode for Protected system = $E_a - E_{cu}$	Volts
W	Weight per anode	kg
n	Number of anodes	–
Q	Weight of anode material consumed per unit current	kg/Ayear
$I_{c,d}$	Design current density for protection	mA/m^2
Y	Design life of the system	years
L	Length of the anode	cm
D	Diameter of the anode	cm
W_m	Weight of anode material required	kg
A	Surface area of the structure to be protected	m^2
R	Soil resistivity	ohm-cm

3.10.2 Current Output

$$I_a = \frac{E_O}{R_a + R_C} = \frac{E_a - E_C}{R_a + R_c} \text{ milliamperes}$$

For Negligible Cathode resistance,

$$I_a = \frac{E_d}{R_a} = \frac{E_a - E_{Cu}}{R_a} \text{ milliamperes}$$

where

$$R_a = \frac{\rho}{2\pi L}\left(\ln\frac{8L}{d} - 1\right)$$

Anode life (assumes that the anode will be replaced when it is 85% dissolved).

For zinc anodes,

$$\text{Life} = \frac{71.4\ W_m}{I_a} \text{ years}$$

For magnesium anodes,

$$\text{Life} = \frac{108.7\ W_m}{I_a}\ \text{years}$$

Weight of the anode material required is as follows:

$$W_m = \frac{A \cdot I_{cd}}{1000} \cdot Q \cdot Y,\ \text{kg}$$

where:

I_{cd} = required current density (milliamperes per square meter).
Y = design life in years.

n = Number of anode requires.

3.11 Design Principle of Cathodic Protection for Marine Structures

3.11.1 Design Calculations

The calculations in this section are intended to be a guide for the design engineer. They should not be considered as absolute or the only method of calculation. Cathodic-Protection design procedures are empirical, or at best, based upon empirically modified theory.

3.11.1.1 Impressed-Current Systems

Impressed-current cathodic-protection circuits may be represented by Fig. 3.9, which shows a simplified equivalent circuit.

To size the system, it is first necessary to know the current to be impressed on the structure to achieve cathodic protection. The requirements are detailed in Section 13.2.10 for different environments. In determining the total current required, all metal surfaces submerged in the electrolyte at mean high water should be included in the calculation:

$$\text{Total current (Amperes)} = \frac{\text{Surface area}\ (m^2) \times \text{milli amperes per square meter}}{1000}$$

Having determined the total current requirement, the individual elements in the total load resistance are then adjusted to match the total DC output of the available power

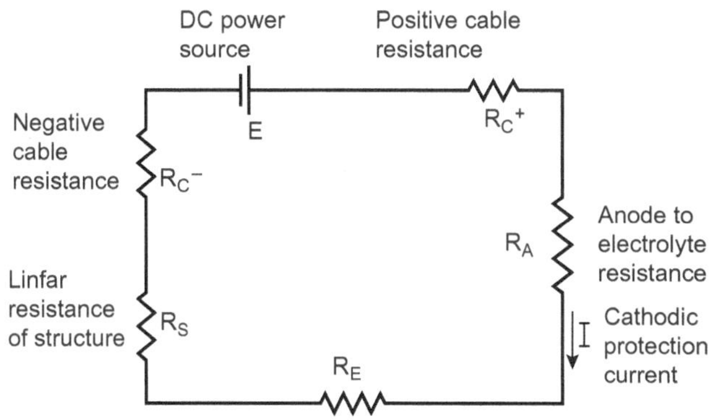

Figure 3.9 Simplified equivalent impressed-current cathodic-protection circuits.

source. From the above diagram, it will be seen that to establish the total circuit resistance, five resistances in series have to be considered as follows:

$R_{C+} + P_{C-}$: The resistance of the positive and negative cables will be dependent on the length and cross-sectional area of the conductor.

R_E: The cathode-to-electrolyte resistance may be calculated using Ohm's law:

$$R_E = \frac{E}{I}$$

where E is the change of the structure-to-electrolyte potential to achieve cathodic protection (usually 1/3 to 1 V) and I is the total current requirement in Amperes. R_a, the anode-to-electrolyte resistance, will be dependent on the shape, number, and spacing of the anodes used, and on the electrolyte resistivity. Design calculations are given below. R_S, the resistance of structures such as platforms, may be ignored. In the case of pipelines and piers, where length is extensive in comparison to the cross-section or where continuity bonds are installed, this factor should be evaluated.

Where a structure such as sheet piling or pier-approach piling has been made electrically continuous by special bonding, the resistance of these bonds must be included in the resistance of the structure. The following formulas are offered to assist in estimating the anode-to-electrolyte resistance.

3.11.2 Single Cylindrical Vertical Anode

The anode-to-electrolyte resistance, R_a, of a single vertical anode, mounted at the surface, is given by the following:

$$R_a = \left(\frac{0.00521\rho}{L}\right)\left[\ln\left(\frac{8L}{D}\right) - 1\right]$$

Where:

R_a = anode-to-electrolyte resistance of a single vertical anode to remote reference (ohms).
ρ = electrolyte resistivity in (ohms centimeters) (For seawater it is 15 ohm-cm)
L = length of the anode in (centimeters) (including backfill, if used).
D = diameter of the anode in centimeters (including the backfill, if used).

3.11.3 Group of Vertical Anodes Equally Spaced in a Straight Line

For a multiple anode system where n anodes are spaced at S centimeters and connected in parallel, the total resistance to the electrolyte of the array, R_N, is given as follows:

$$R_N = \left(\frac{0.00521\rho}{NL}\right)\left[\ln\left(\frac{8L}{D}\right) - 1 + \left(\frac{2L}{S}\right)\ln(0.656N)\right]$$

where:

R_N = Total resistance in ohms.
N = Number of anodes in parallel.
R_a = Resistance of a single vertical anode.
ρ = Electrolyte resistivity in ohms centimeters.
S = Spacing between anodes in centimeters.

The above calculations are generally applicable for anodes in seawater or in soil. For square sectional anodes use equivalent diameter.

- **Vertical anodes surrounded in (coke) carbonaceous backfill**

For design purposes, the diameter and length of the column of backfill, instead of the anode dimensions, apply to the calculations in previous sections.

- **Anodes laid horizontally in free-flowing seawater**

The calculations in previous sections apply, provided the depth of immersion is greater than one anode length. All other cases of horizontally laid anodes are dealt with in this Engineering Standard for cathodic protection of land systems.

3.11.4 Galvanic Anode Systems

- **Equivalent circuit**

A galvanic cathodic-protection circuit may be represented by the following simplified diagram in Fig. 3.10:

Although the diagram shows two voltage cells, one at the anode interface and one at the cathode interface, each are in fact half cells. The same resistance elements will exist as for the impressed-current system; however, galvanic anodes are usually attached close to the cathode with consequent low-resistance connections so that both structure and cable circuit resistance can usually be ignored. Anode-to-electrolyte resistance is usually large in comparison to structure-to-electrolyte resistance and becomes the governing factor in galvanic anode design.

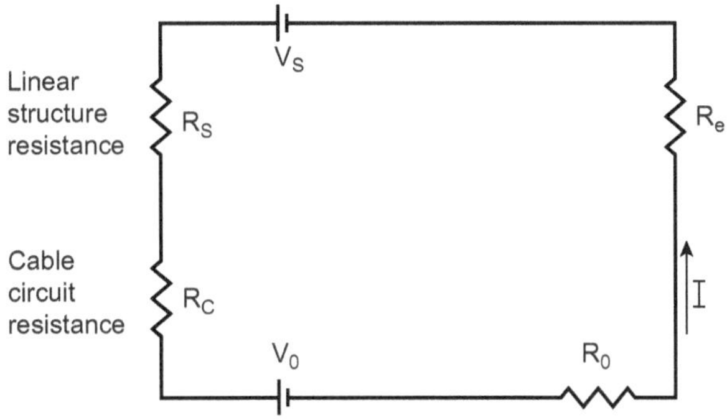

Figure 3.10 A simplified diagram for galvanic cathodic-protection circuit.

The anode-to-electrolyte resistance will be dependent on electrolyte resistivity and the anode size and shape. The cathode-to-electrolyte resistance, however, will increase as the cathodic reaction proceeds, and its potential will tend to approach that of the anode. With this increase in circuit resistance and reduction in potential difference between the cathode and the anode, the current output from the anode will decrease, and subject to its efficiency, its life will be extended. This explains why the number of galvanic anodes can be increased without resulting in overprotection.

The outputs of galvanic anodes should be calculated on the basis that when protected, the structure-to-electrolyte potential is not less than −0.8 V, and approaches −0.9 V referred to Ag/AgCl.

3.11.5 Anode Output Formula

When the steel is polarized to a protective level, Ia is given by the following:

$$I_a = \frac{E_d}{R_a} = \frac{E_a - E_{cu}}{R_a} \text{ amperes}$$

In these formulas, the symbols shown are as follows:

I_a = anode current output (Amperes).
R_a = Anode resistance to the electrolyte (ohms).
E_a = Open-circuit potential, anode to electrolyte (reference electrode) (Volts).
E_{cu} = Structure-to-electrolyte potential when fully protected (Volts)
E_d = Driving voltage between the anode and the protected structure

$$(\text{Volts}) = E_a - E_{cu} = V_S + V_A$$

It is to be noted that E_a, the open-circuit potential between different types of anodes and electrolyte ($Cu/CuSO_4$ reference electrode) is nominally as follows:

Galvalum III—1.05 V
Zinc—1.05 V

Magnesium—1.50 V
Galvomag—1.70 V

It is also to be noted that E_a, the open-circuit potential between different types of anodes and electrolyte ($Cu/CuSO_4$ reference electrode) is nominally as (see British Standard (BS) 7361-1) in Table 3.15.

3.11.5.1 Determination of R_a (anode resistance to electrolyte)

For a single anode freely suspended in seawater and is remote from the cathode, the following formula applies:

$$R_a = \rho\frac{K}{L}\left[\ln\left(\frac{4L}{r}\right) - 1\right]$$

In the case of an anode fastened directly to the structure in seawater,

$$R_a = \frac{1.1}{L + 0.8W + 0.8D}$$

Where:

$K = 0.500/\eta$ or $0.15q$
L, W, and D are the length, width, and depth of the anode in centimeters.

For bracelet anodes, the anode resistance R_a can be calculated from the following formula:

$$R_a = \frac{0.315\rho}{\sqrt{A}}\ \text{ohms}$$

Table 3.15 Open-Circuit Potential between Different Types of Anodes and Electrolyte

Alloy	Common Environment	Open-Circuit Potential (for Reference Electrode Stated) (V)	Capacity (Ah/kg)	Consumption Rate (kg/Ayear)
Magnesium				
1.5% Mn	Soil/fresh water	−1.7 ($Cu/CuSO_4$)	1200	7.5
6% Al, 3% Zn	Soil/fresh water	−1.5 ($Cu/CuSO_4$)	1200	7.5
Zinc				
0.5% Al, 0.1% Cd	Seawater	−1.05 (Ag/AgCl/ seawater)	780	11.25
Aluminum				
0.4% Zn, 0.04% Hg	Seawater	−1.05 (Ag/AgCl/ seawater)	>2800	3.125
5% Zn, 0.04% Hg	Sea-bed mud	−1.05 (Ag/AgCl/ seawater)	1800–2000	4.38–4.86
3–5% Zn	Seawater	−1.1	1800–2000	4.38–4.86

Where:

ρ is the resistivity of seawater in ohms centimeters.
A is the exposed surface area of the anode in square centimeters.
R_a is the anode resistance in ohms.

The formula is from corrosion and protection of offshore steel structures. In brackish water, the resistivity of the water can increase to several 1000 ohm-cm (compared with open seawater resistivities of 15–35 ohm-cm). Anode current outputs in brackish water will therefore be reduced considerably. For plate anodes (Flush-mounted Hull or Bracelet anodes),

$$R_a = \frac{\rho}{2S}$$

$$S = \frac{a+b}{2}$$

(i.e., the mean of the anode sides and b < 2a), where a is the length of the anode in centimeters and b is the width of the anode in centimeters.

Alternatively,

$$R_a = \frac{\rho}{0.58A^{0.727}}$$

where

A = Exposed surface area of the anode (square centimeters).

The number of anodes N required should satisfy both of the following:

1. $N = \frac{I_r}{I_a}$

2. $N = \frac{W_r}{W_a}$

where:

I_r = Total current, required in Amperes.
I_a = Current output of the anode, in Amperes.
W_r = Net weight of the anodes material, in kilograms.
W_a = Net weight of the individual anode, in kilograms.

$$I_r \frac{Y\ 8760}{CU} = W_r$$

here,

Y = Life of the structure, in years;
C = Practical electrochemical capacity of the alloy in Ampere hours per kilogram; and
U = Utilization factor, that is, proportion of the net weight consumed at the end of the anode life. For fully supported tubular inserts, U = 0.9.

To optimize the performance and efficiency of the anodes, the values for (a) and (b) should be similar.

It is to be shown by appropriate calculations that the system is capable of polarizing the structure initially and also when the anodes are consumed to their design utilization factor.

It should be assumed that at the end of the anode life, the anode length has been reduced by 10% and that the remaining material is evenly distributed over the steel insert.

3.11.6 Pipelines and Attenuation of Potential

Calculations to determine the current demand and maximum pipe length protectable from a single impressed-current cathodic-protection station for long pipelines were dealt with in this chapter.

For Cathodic Protection of Ships, see also Section 6 of BS 7361, Part 1 and for Galvanic anode system, see NACE RP0196.

3.11.7 Sacrificial Anode Materials

There are three materials currently used for the sacrificial cathodic protection of ships' internal and external surfaces. These are zinc, aluminum, and magnesium alloyed with small percentages of other elements to ensure adequate performance. Zinc and aluminum alloy materials are used for protection in seawater, and the magnesium is used in fresh water condition, for fresh water tanks, and for other specialized purposes. When designing a cathodic protection system, the procedure in NACE RP0196 should be considered by the corrosion engineer.

3.11.8 The Use of Sacrificial Anodes for Internal Cathodic Protection

In 1962, after fairly extensive trials, the Classification Societies amended their rules to permit up to 10% reduction in scantlings, of bulkhead plating, stiffeners, and certain internal girders and webs, provided an approved system of corrosion control was fitted. A further amendment to the rules in 1964 allowed a 5% reduction in the thickness of the main longitudinal strength members.

An approved system of corrosion control can be the use of coatings, anodes, or a combination of both. Listed below are the most widely used systems:

In cargo, cargo/ballast and ballast-only tanks, all surfaces are to be coated with an approved corrosion control coating.

In cargo/ballast and ballast-only tanks, the steelwork in the ullage space, with a minimum of the top 1.5 m, is to be protected with an approved coating, and the remainder of the tank is to be protected by anodes, made of zinc, aluminum, or magnesium.

A selective system using defined ballasting conditions applies only to ships intended for the transport of crude oil, using a defined ballasting condition, and is as shown in Table 3.16.

The summary of the classification society rules concerning the use of anodes for internal cathodic protection is as follows: In 1971, an IACS working party on tanker

Table 3.16 A Selective System Using Defined Ballasting Conditions Applies Only to Ships Intended for the Transport of Crude Oil

Tank	Coating	Cathodic Protection
Ballast-only tanks	All surfaces	Anodes below normal ballast level plus coating of all surfaces above ballast level
Cargo oil/ballast tanks	All surfaces above normal ballast or cargo level and the upper surfaces of all horizontal members in the remainder of the tank. Also the tank bottom, longitudinal frames, and girders up to a level of the top of the longitudinal	Anodes below normal ballast or cargo level plus coating of all surfaces above nonballast level
Cargo oil only	All surfaces above the normal liquid level	

safety agreed on certain unified interpretations relating to cathodic protection as follows:

1. Impressed-current systems are not permitted in oil cargo tanks.
2. Magnesium or magnesium alloy anodes are not permitted in oil cargo tanks or tanks adjacent to oil cargo tanks.

Aluminum anodes are only permitted in cargo tanks of tankers where the potential energy is not >28 kg m (200 ft lb). The height of the anode is to be measured from the bottom of the tank to the center of the anode, and its weight is to be taken as the weight of the anode as fitted (including the fitting devices).

However, where aluminum anodes are located on horizontal surfaces such as on bulkhead girders and stringers that are not <1 m wide and fitted with an upstanding flange or face flat projecting not <75 mm above the horizontal surface, the height of the anode may be measured from this surface.

Aluminum anodes are not to be located under tank hatches or Butterworth openings in order to avoid any metal parts falling on the fitted anodes. Aluminum anodes containing magnesium should not be used in cargo tanks.

There is no restriction on the positioning of zinc anodes, but it is recommended practice to ensure that the potential energy is not >540 kg m. The anodes should have steel cores, and these should be sufficiently rigid to avoid resonance in the anode support and be designed so that they retain the anode even when it is wasted. The steel inserts are to be attached to the structure by means of a continuous weld of an adequate section. Alternatively, they may be attached to separate supports by bolting, provided a minimum of two bolts with lock nuts are used. However, approved mechanical means of clamping will be accepted. The supports at each end of an anode should not be attached to separate items that are likely to move independently. When anode inserts or supports are welded to the structure, they should be arranged so that the welds are clear of high stress areas.

The application of sacrificial anodes for internal cathodic protection where reduced scantlings are applied. Although the procedure below should be followed in the design of any anode system, the effectiveness of such a system in terms of both anode life and corrosion control can only be ensured by regular inspection.

System life should be for not less than four years, using an estimated ballast factor. Provision should be made for additional anode consumption if it is expected that residual ballast water will be left in the bottom of the tank if ballast voyages are expected to be for less than five days, the recommended current densities should be increased by 20 percent.

The resistivity of the ballast water is assumed to be 25 ohms cm. If brackish water of a higher resistivity is expected to be used, the system should be designed accordingly. The top 1.5 m of the tank must be coated to give optimum protection. In peak and double bottom tanks where tanks can be pressed up, coatings are not required. Table 3.17 shows current densities for some typical tanks.

In large Very Large Crude Carriers (VLCCs) with permanent tank washing machines, the degree of protection given by the residual oil film is virtually nonexistent, and an increase in the current density to 120 mA/m^2 in cargo/clean ballast tanks is recommended.

For the long-term protection of coated surfaces, adequate allowances must be made for coating breakdown. The above minimum recommended figure of 5 mA/m^2 should be increased in proportion to the anticipated coating breakdown.

3.11.9 The Type and Number of Anodes Required

The total current required is as follows:

$$\text{current (amperes)} = \frac{\text{Protected area}\,(\text{m}^2) \times \text{current density}\,(\text{mA/m}^2)}{1000}$$

The total weight of the anode material required is

$$\text{Weight (kg)} = \frac{\text{Current (amps)} \times \text{design life (years)} \times 8760}{\text{Capacity of material (amp hrs/kg)}}$$

Table 3.17 Current Densities for Some Typical Tanks

Cargo/Clean Ballast Tanks	86 mA/m^2
Ballast-only and ballast white oil cargo tanks	108 mA/m^2
Upper wing tanks	120 mA/m^2
Forward and after peak tanks	108 mA/m^2
Coated surfaces	5 mA/m^2
Lower wing tanks	86 mA/m^2
Double bottom tanks, ballast-only	86 mA/m^2
Cargo/dirty ballast tanks	Depend on trade

(8760 = number of hours in one year). The number and type of anodes selected must at least satisfy both the total current and total weight requirements as follows:

$$\text{Number of anodes} = \frac{\text{Current required}}{\text{Individual current output}}$$

$$\text{Number of anodes} = \frac{\text{Weight required}}{\text{Individual net weight}}$$

3.11.10 Anode Location

In ballast-only or cargo ballast tanks of clean oil tankers, anodes should be distributed evenly over all the uncoated structure but with some emphasis on horizontal surfaces. In cargo ballast tanks of crude carriers, those anodes on horizontal surfaces should be distributed in proportion to the areas of these surfaces, that is, on bottom longitudinals, horizontal bulkhead girders, face flats on bottom structural members, and on horizontal stiffeners. The majority of anodes will therefore be located on the bottom structure.

In cases where residual ballast water is expected to be left in the bottom of the tank, a low location of the bottom anodes should be provided, for example, on the webs of bottom longitudinals, and provision should be made for the more continuous anode consumption that this will initial.

The pipelines used in larger ships have given rise to significant corrosion problems. The control of external corrosion on pipelines can be achieved by bonding the pipelines and fitting pipeline anode bracelets.

3.11.11 The Use of Sacrificial Anodes for External Cathodic Protection

Zinc and aluminum anodes are used for this purpose in seawater. Both zinc and aluminum anodes have a normal design life of one, two, three, or four years to suit the owners' requirements.

Hull anodes are usually welded directly to the ship structures, but they can be bolted if required. The applied current density usually varies from 10 to 20 mA/m^2 dependent on the paint system used. In certain instances, this can be increased to 40 mA/m^2 to meet arduous operating conditions.

3.11.12 The Type and Number of Anodes Required (Full Hull Protection)

To obtain the approximate wetted hull area, the formula below may be used:where LBP is the length between perpendiculars; D is the draft; BC is the block coefficient; B is the breadth: (1.8 × LBP × D) + (BC × LBP × B). The typical block coefficients for various vessels are given in Table 3.18:

The total current required is as follows:

$$\text{Current (Amperes)} = \frac{\text{Area}\left(\text{m}^2\right) \times \text{current density}\left(\text{mA/m}^2\right)}{1000}$$

Table 3.18 Typical Block Coefficients for Various Vessels

Passenger Vessels, 0.6	Naval Vessels, 0.55
Cargo vessels, 0.75	Coasters, 0.75
Tankers, 0.8–0.9	Dredgers, 0.8
Tugs, 0.6	Yachts, 0.4–0.5
Trawlers, 0.55	Launches, 0.4

The total weight of the anode material required is as follows:

$$\text{Weight (kg)} = \frac{\text{Current (A)} \times \text{design life (years)} \times 8760}{\text{Capacity of material (A h/kg)}}$$

The number of anodes must fulfill both current and weight requirements as for internal protection.

3.12 Anode Location

1. Full hull protection

Anodes should be located equidistantly around the hull about 6 m apart. Fifteen percent of the anodes should be located around the stern in a manner similar to that for the stern-only protection described below. Anodes situated in the forward part of the vessel should be located to prevent damage or removal by anchor chains. If necessary, anodes can be fitted with cable guards.

2. Stern only protection

Anodes should be positioned on the stern area and rudder adjacent to the propeller, care being taken to minimize disturbance of the water flow to the propeller. Anodes should not be fitted within 300 mm of the line of the propeller tips and should be parallel to the flow lines of the hull. Twenty percent of the anodes required for full hull protection are required for stern-only protection.

Magnesium anodes are used externally for vessels operating in fresh water.

3.12.1 Fitting Out Protection

Magnesium anodes are used to provide protection during fitting out, particularly in estuarine or polluted waters.

3.12.2 Tank Descaling

The high cost of decaling cargo tanks by mechanical methods can be avoided by using magnesium ribbon, which gives an equally clean surface by an electrolytic process and enables decaling to be done during the ballast voyage without disrupting the crew from normal work. In addition, noise, dust, and attack on the metal is avoided.

Extruded flexible magnesium ribbon anodes are usually produced as rectangular sections or ribbed sections and has a continuous steel core. Its particular advantages are that the current output is three to seven times greater than for cast anodes of the same weight. The current is also evenly distributed along the structure.

The method of use is as follows: A current density of 1000 mA/m^2 is normally used in seawater. This is achieved by using 1 m of ribbon for every 3 m^2 of steel area. The ribbon is usually supplied in coils of 150 or 300 m. This should be cut into lengths of about 30 m. The wire should be bared for about 10 cm at each end of every length by means of a hacksaw and a cold chisel or torch. The ribbon is spread around the bulkheads, attached at the top and the bottom so that it stands 1–2 m away from the steel surface. An even spread of ribbon should be used to "cover" all the areas of the tank. The tank is filled with clean seawater and left for four to six days.

Black oil tanks should be pressure washed before descaling. All tanks should be pressure washed after descaling to remove any loose pieces of scale. The white chalk deposit that appears is harmless to food cargoes and should be brushed off before painting. This process releases some hydrogen gas and appropriate precautions should be taken.

3.12.3 Individual Anode Output Determination and Calculation (Sacrificial Anodes)

Anode output (Amperes) is the difference in potential (volts) between the anode material and steel polarized to protection divided by the resistance (ohms) of the anodes in the electrolyte. This is expressed as follows:

$$I = \frac{E}{R}$$

Here, I is in Amperes, E is in volts, and R is in ohms. To obtain a high output, the anodes should have a small cross-section in relation to their length. The resistance of slender-rod anodes (Tank Anodes) in an electrolyte can be obtained from the following formulae:

$$R = \frac{\rho}{2\pi L}\left[\ln\left(\frac{4L}{r}\right) - 1\right]$$

where

R = The resistance of the anode in ohms.
r = The mean effective radius of the anode in centimeters. Here,

$$r = \frac{\text{cross-section area}}{\pi} \times \frac{60^*}{100}$$

Where:

ρ = The resistivity of water in ohms centimeters.
L = The length of the anode in centimeters.

This procedure is traditionally used in calculating anode requirements in the marine industry as being indicative of anode output during its life.

For flat plate anodes (Hull Anodes), the following formula can be used:

$$R = \frac{\rho}{a + b}$$

Where:

a = Length in centimeters.
b = Width in centimeters.

3.12.4 Impressed Current

It is usual to design on the basis of 35 mA/m^2 although this may be varied to suit the following:

1. Coating condition, for example, epoxy or other paint systems.
2. Operating conditions, for example, high-speed operation, low-temperature conditions.

Two types of anodes are usually produced:

1. External type with a current output of 75 and 100 A. These are also supplied in pairs to give 150- and 200-A output.
2. Recessed type, of a circular design, for use in forward installation where mechanical damage from cables is likely to occur. The recessed anode may be produced in one size only with a maximum current output of 50 A.

Anodes are usually located externally to the engine room or suitable void spaces to simplify hull penetrations and cable routes.

For vessels up to 150,000 dwt, it has been found that satisfactory current distribution can be achieved with all anodes sited adjacent to the engine room. Larger vessels normally require additional anodes, of the recessed type, to be mounted forward.

System control and monitoring is by a minimum of two reference electrodes sited forward and after and mounted at least 3 m from the nearest anode. For design calculation, the cross-sectional area of the anode to be used in determining the mean effective radius is that corresponding to the anodes consumed by 40 percent.

3.13 Anodic Protection

In contrast to cathodic protection, anodic protection is relatively new. Edeleanu first demonstrated the feasibility of anodic protection in 1954 and tested it on small-scale stainless steel boilers used for sulfuric acid solutions. This was probably the first industrial application, although other experimental work had been carried out elsewhere.

Anodic protection possesses unique advantages. For example, the applied current is usually equal to the corrosion rate of the protection system. Thus, anodic protection not only protects but also offers a direct means for monitoring the corrosion rate of a system.

Anodic protection can substantially decrease corrosion rate. The primary advantages of anodic protection are its applicability in extremely corrosive environment and its low current requirement. Anodic protection has been most extensively applied to protect equipment used to store and handle sulfuric acid. Anodically protected heat exchangers used to cool sulfuric acid manufacturing plants have represented one of the more successful ventures for this technology. These heat exchangers are sold complete with the anodic protection systems installed and have a commercial advantage in that less costly materials can be used.

This is mainly because of the limitations on metal chemical systems for which anodic protection will reduce corrosion. In addition, it is possible to accelerate corrosion of the equipment if proper controls are not implemented. However, anodic protection has its place in the corrosion control area, provided some important basics are satisfied.

4 Materials and Cathodic Protection Systems

Cathodic Protection (CP) products and systems allow corrosion prevention in metal structures that are partially or completely buried or submerged.

This section consists of materials for the following parts, as listed hereunder:

- Impressed Current Anodes
- Carbonaceous Backfill for Impressed Current Anodes
- Galvanic Anodes for Underground Applications (Magnesium and Zinc)
- Galvanic Anodes for Submerged Applications (Magnesium and Zinc)
- Bracelet-Type Galvanic Anodes for Submarine Pipelines (Aluminum and Zinc)
- Cast Galvanic Anodes for Fixed Offshore Installations (Aluminum)
- CP Cables

4.1 Impressed Current Anodes

As in sacrificial anode systems, impressed current systems depend on a supply of high-energy electrons to stifle anodic reactions on a metal surface (Fig. 4.1). In the case of an impressed current system, these high-energy electrons are supplied by a rectifier.

As shown in Fig. 4.1, low-energy electrons that are picked up at a nonreactive anode bed are given additional energy by the action of a rectifier to be more energetic than the electrons that would be produced in the corrosion reaction.

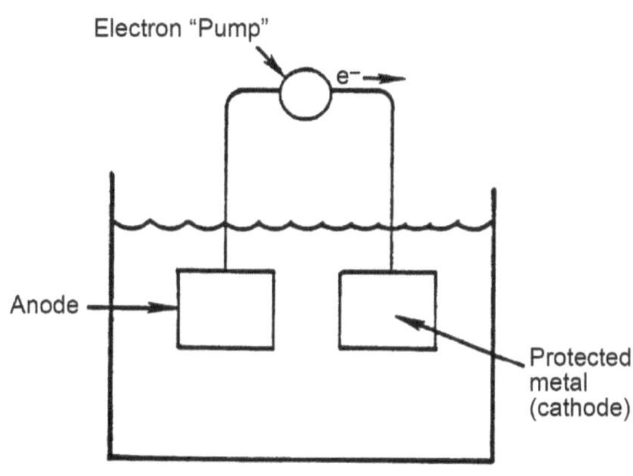

Figure 4.1 Schematic of sacrificial anode systems.

Cathodic Corrosion Protection Systems. http://dx.doi.org/10.1016/B978-0-12-800274-2.00004-1

This section covers the minimum requirements for anodes (High-Silicon–Iron, Graphite, and Magnetite) for use in impressed current CP systems. It specifies the composition, materials, manufacture, properties, inspection, and testing for High-Silicon–Iron, Graphite, and Magnetite anodes.

The technical data should include the following:

- Production procedures that cover all stages of the manufacturing process, from receipt of raw materials to finishing and packing for shipment, including the following:
  - chemical analysis;
  - measurements of weight, dimensions, etc.;
  - inspection and testing procedures and frequencies;
  - anode manufacture;
  - anode connections.

Anode consumption rates at the maximum applicable current density in different environments (e.g., seawater, fresh water, and soil).

Full information about anode lead wire indicating product identification number, construction details, voltage rating of primary insulation, physical properties of the primary insulation, and jacket material including tensile strength and ultimate elongation of finished wire, insulation resistance of primary insulation, notch propagation of the finished wire, abrasion resistance of the finished wire, concentricity tolerance of the finished wire, and continuous service temperature limits of the finished product.

Preservation; packing; and shipping procedures, such that anodes are not damaged or deformed and do not deteriorate during handling; delivery to the fabrication yard; storage, including methods, materials, and any requirement for periodic inspections, are discussed.

4.2 Manufacturing of Materials

Manufacturing of anodes should be carried out using approved production procedures covering each stage of the process from supply of raw materials to finishing and packing for shipment. The procedures should be fully referenced in an inspection and test plan that should also refer to requirements of relevant Standard specification, as applicable.

The manufacturer should ensure that full traceability of all the materials is maintained throughout the manufacturing process.

The quality control at the anode manufacturing plant should include control systems on the following:

- Raw materials, that is, checking of documentation
- Production equipment and process
- Testing during and after production
- Identification of anodes
- Chemical analysis
- Weight and dimensional tolerances
- Surface condition of produced anodes
- Documentation (material certificates)

4.3 Anode Casting

4.3.1 Casting of High-Silicon–Chromium–Iron Anode

Anodes should be made from High-Silicon–Chromium–Iron casting conforming to the American Society for Testing and Materials (ASTM) A 518M-86 (Grade 3) and should have either a solid or "rod-type" configuration with the anode lead attached at one end of the anode or a "tubular type" configuration with the anode lead attached at the center of the anode.

The casting may be produced by any process at the option of the manufacturer, capable of meeting the chemical composition and physical properties specified.

Where solid or "rod-type" anodes are required, the anode should be of a single end with an enlarged head. The anode's enlarged head should be cored to provide cavities for lead wire connection.

Where "tubular-type" anodes are required, the anode should be a hollow, straight-walled tubular anode having walls of a uniform thickness and having an open cylindrical interior for lead wire connection.

The casting should be dense and homogeneous to ensure that the anode properties are uniform throughout their length.

The chemical composition of the anode alloy content should conform to the limits prescribed in Table 4.1.

4.3.2 Casting of Graphite Anode

Anodes should be of the plain graphite type extruded in circular or square shapes. They should be of the "treated type," and should have been impregnated with linseed oil by a vacuum-pressure autoclave process.

The graphitizing ovens should be designed so as to ensure that a uniform internal graphite structure is obtained as well as an overall consistency from one anode to the other is present.

If great care is not taken during the manufacturing process, internal cracks and flaws may develop in the graphite. It is therefore necessary that the manufacturer

Table 4.1 Chemical Composition Limits for High-Silicon–Iron Anodes

Element	Composition, wt%	Method of Analysis
Carbon	0.9–1.0	ASTM A 518M
Manganese	1.5 max.	
Silicon	14.25–14.75	
Chromium	4.3–5.00	
Molybdenum	0.2 max.	
Copper	0.50 max.	
Iron	Determined arithmetically by difference	

takes the necessary steps to maintain a consistent material free from internal and external flaws.

The aggregate and grain should be small and should have similar physical, chemical, and electrical properties. The material should be free of voids and impurities.

The anodes should be sufficiently resistant to permit firm tamping of backfill around them and should be immune to chemical attack.

Anodes should have one end drilled for cable connection.

4.3.3 Casting of Magnetite Anode

Anodes should be made by casting magnetite (Fe_3O_4) with the melting point of about 1540 °C, with an addition of small quantities of alloying elements, into a hollow cylinder closed at one end. The whole inner surface of the anode should be coated with a thin plated copper layer. Hundred millimeters below the anode top, a copper tag should be fixed to the inside copper layer for cable attachment.

The anode body should be filled with a soft plastic core (e.g., polystyrene or polyurethane) to absorb the differential rates of expansion of the resin and the magnetite.

Casting may be produced by any process at the discretion of the manufacturer, capable of meeting the requirements specified here.

The casting should be dense, and homogeneous and should ensure that the anode properties are uniform throughout its length.

4.4 Lead Wire-to-Anode Connection

The anodes should be supplied complete with lead wires connected to the casting.

For "rod-type" High-Silicon–Iron anodes, the method of connecting the lead wire to the anode should either be by leading and tamping (caulked lead connection), or by the use of a tapered pin connection.

For Graphite anodes, the method of connecting the lead wire to the anode should either be by a brass solder or by the use of a compression connection.

For Magnetite anodes, the method of connecting the lead wire to the anode copper-conducting layer should be by the use of a brazed connection.

The anode lead to anode connection should be fully sealed to prevent water intrusion by the application of an epoxy compound. Finally, an internally mastic-lined heat shrink cap that covers the anode head by a minimum of 75 mm and the lead wire by a minimum of 50 mm should be applied and shrunk into place to provide a complete water-tight and leak-free seal capable of withstanding severe service conditions.

A heat shrink anode cap used for sealing the anode lead connection, as will be specified by the Purchaser, should be either of the standard or chlorine-resistant type in accordance with the following:

- **Standard Type:** Heat-shrinkable polyethylene,
- **Chlorine Resistant Type:** Heat-shrinkable polyvinylidene fluoride (PVDF).

For "tubular-type" High-Silicon–Iron anodes, the type and method of connecting and sealing the lead wire to the anode should be approved by the anode manufacturer.

The internal moisture seal must withstand the hydrostatic pressure and resist degradation from oxidizing gases released at the anode.

4.5 Anode Lead Wire

4.5.1 Conductor

Copper conductors should be single core stranded, compressed, soft annealed, uncoated copper in accordance with ASTM specification B8 Class B or IEC Publication 228 Class 2.

Unless otherwise specified by the Purchaser, the lead wire size should be 10 mm^2.

Unless otherwise specified by the Purchaser, the lead wire length should be 3 m.

4.5.2 Lead Wire Insulation

The lead wire insulation, as will be specified by the Purchaser, should be either of the standard or chlorine-resistant type and should conform to one of the following types:

- **Standard type insulation**

The insulation should be in accordance with one of the following alternatives:

A single jacket of High-Molecular-Weight Polyethylene (HMWPE) insulation complying with the physical and electrical requirements of ASTM Specification D 1248, Type I, Class C, Category 5, Grade J3. The average insulation thickness should be 2.8 mm.

A double jacket of HMWPE primary insulation complying with the physical and electrical requirements of ASTM Specification D 1248, Type 3, Class C, Category 5, Grade J3; sheathed with a chemical and abrasion resistant polyvinyl chloride (PVC) extruded jacket conforming to ASTM specification D 1047. The average thickness should be 2.8 mm.

A double jacket of chemically Crosslinked Polyethylene (XLPE) primary insulation conforming to ASTM specification D 2655, sheathed with a chemical and abrasion-resistant PVC extruded jacket conforming to ASTM specification D 1047. The average thickness should be 2.8 mm.

In some cases in which the anode lead wire with the above insulation material may not be available, a double jacket PVC/PVC insulation is also acceptable, subject to the Company's approval.

- **Chlorine-resistant type insulation**

The insulation should be in accordance with one of the following alternatives:

A single jacket of modified PVDF, double-extruded insulation: The minimum insulation thickness should be 0.75 mm.

A double jacket of a radiation Crosslinked Polyvinylidene Fluoride (XLPVDF), primary insulation, sheathed with a chemical and abrasion-resistant HMWPE extruded jacket conforming to ASTM specification D 2308: The minimum thickness should be $0.5 + 1.7 = 2.2$ mm.

Lead wire insulation should have a minimum of 600-V rating.
The minimum thickness at any point should not be <90% of the specified average thickness.
The insulation should provide a continuous coverage, adequate dielectric properties, and have a high resistance to abrasion, stress cracking, notch propagation, and notch sensitivity.
Unless otherwise specified by the Purchaser, the color of the anode lead wire insulation should be fade-resistant black for the standard type and fade-resistant red for the chlorine-resistant type.

4.6 Properties

The finished product should meet the following requirements:

- Workmanship, Finish, and Appearance should be checked.
- Anodes should be free from defects that may impair the performance of the finished component. Anodes should be free of flaws, slags, cracks, blow holes, and excessive porosity consistent with good commercial practice.
- The lead wire insulation should be continuous and have a thickness as uniform as possible and should be completely free of nicks, scratches, or other discontinuities.

Notch propagation refers to the tendency of a cut, nick, or scratch in a material to increase in depth and/or length when the material is stressed.

4.6.1 *Mechanical Resistance Test*

When a uniaxial force equivalent to 100 kg, or five times the mass of the anode metal, whichever is greater, is applied to a cast anode, it should not result in damage to the cable or to its connections.

The Silicone–iron alloy should meet transverse bond test results (according to the test method A518/A518 M and acceptance test criteria).

4.6.2 *Electrical Resistance Test*

The electrical resistance at the connection of a completely assembled anode should not be >0.03 ohm, when tested in accordance with ASTM test method B 539.

Note the following: Connection resistance is defined as total resistance minus the resistance of the anode cable.

4.6.3 *Temperature Dependence*

Anodes should not have a tendency to crack when affected by alternating high and low temperatures.

4.6.4 *Fluoropolymer-Insulated Anode Lead Wire Test*

The Fluoropolymer insulation used on anode lead wires must pass the impulse dielectric test performed in accordance with ASTM D 3032, Section 12, at 600 V (peak) per 0.025 mm or a maximum of 18 kV (peak).

If an outer jacket of fluoropolymer or another material is applied, all finished wire must pass the impulse dielectric test at a voltage of 600 V (peak) per 0.025 mm of primary insulation plus 100 V (peak) per 0.025 mm of the jacket thickness or a maximum of 18 kV (peak).

4.6.5 Radiographic Test

The anode should show no internal defects when tested by means of X-rays in accordance with ASTM Guide E94. The type and degree of discontinuities considered should be judged by reference radiographs ASTM E 802.

The impulse dielectric test is a method of applying voltage to an insulated wire through the use of electric pulses (usually 170–250 pulses per second) to determine the integrity of the wire's insulation.

In addition the "rod-type" High-Silicon–Iron anodes should conform to the requirements of physical properties prescribed in Table 4.2.

4.7 Dimensions and Weights

The dimensions and weights of High-Silicon–Iron "rod-type" anode, as will be specified by the Purchaser, should be in accordance with those given in Table 4.3.

The dimensions and weights of the High-Silicon–Iron "tubular type" anode, as will be specified by the Purchaser, should be in accordance with those in Table 4.4.

Table 4.2 Physical Properties of "Rod-Type" High-Silicon–Iron Anodes

Property	Unit	Requirement	Test Method ASTM
Density	g/cm^3	7–7.05	Approved method
Hardness, Brinell (min.)	–	500	E10
Compressive strength (min.)	MPa	650	A256
Electrical resistivity (max.)	micro-ohm-cm	72	B193
Impact resistance (min.)	J	0.1	A327 (Charpy type)

Table 4.3 Dimensions of High-Silicon–Iron Anodes ("Rod-Type")

Diameter (Nominal), mm	Head Diameter (Nominal), mm	Length (Nominal), mm	Weight (Nominal), kg
51	76	915	14.5
51	76	1220	19
51	76	1525	22.5
76	102	915	29
76	102	1220	38
76	102	1525	49

Table 4.4 Dimensions of High-Silicon–Iron Anodes ("Tubular Type")

Outside Diameter, mm	Inside Diameter, mm	Wall Thickness, mm	Length (Nominal), mm	Weight (Nominal), kg
67	47	10	1067	14
56	36	10	2134	21
67	47	10	2134	29
95	75	10	2134	39
121	100	10	2134	50
121	86	17	2134	79

Table 4.5 Dimensions of Graphite Anodes

Diameter (Nominal), mm	Length (Nominal), mm	Weight (Nominal), kg
76	1525	11
102	2032	28
152	1830	53

The dimensions and weights of the Graphite anode, as will be specified by the Purchaser, should be in accordance with those given in Table 4.5.

The dimensions and weight of the Magnetite anode should be as follows:

Overall length: 800 mm,
Effective length: 740 mm,
Outside diameter: 60 mm,
Inside diameter: 40 mm,
Total weight: 6.0 kg,
Magnetite weight: 4.7 kg.

The actual dimensions should conform to the following:

- Anode mean length should be ±3% of the nominal length or ±25 mm, whichever is smaller.
- The diameter of the anode should be ±5% of the nominal diameter.
- The permissible variation in the weight of each finished anode should be ±5% of the nominal weight.

4.8 Carbonaceous Backfill for Impressed Current Anodes

This Part covers the minimum requirements for the supply of carbonaceous backfill for use in shallow and deep well-type impressed current ground beds.

Carbonaceous backfill material, should be one of the following types:

Type 1: Coal coke breeze,
Type 2: Calcined petroleum coke breeze,
Type 3: Special calcined petroleum coke breeze.

Note: Type 1 and 2 coke breeze are to be used for shallow (horizontal or vertical) ground beds, and type 3 coke breeze is to be used for deep well ground beds.

4.8.1 Requirements

4.8.1.1 Composition

- **Type 1 backfill material**

Backfill should be coal coke breeze of low resistivity and low ash content. The coke breeze may be treated by the addition of 10% (by weight) of commercial grade slaked lime. The proposed material should have the properties given in Table 4.6.

4.8.1.2 Type 2 Backfill Material

Backfill should be petroleum coke calcined (heat treated) to remove all other petroleum products, other than carbon, and should be supplied in granular form. Lime in a proportion of 10% by weight of coke breeze should be added to the product.

The proposed material should have the properties of Table 4.7.

4.8.1.3 Type 3 Backfill Material

Backfill should be petroleum coke calcined (heat treated) to remove all other petroleum products, other than carbon, specially formulated to facilitate pumping, settling, and compaction of carbon lubricants. The product should contain 0.1% wetting agent to enhance the setting of the granular particles and water absorption of the backfill.

Table 4.6 Type 1 Backfill Material (Coal Coke Breeze)

Property	Unit	Requirement	Test Method
Carbon content (min.)	wt%	80	ISO 609 or ISO 625
Moisture content (max.)	wt%	5	ISO 687
Ash content (max.)	wt%	15	ISO 1171
Total sulfur content (max.)	wt%	1	ISO 351 or ISO 334
Volatile matter content (max.)	wt%	5	ISO 562

Table 4.7 Type 2 Backfill Material (Petroleum Coke Breeze)

Property	Unit	Requirement	Test Method
Carbon content (min.)	wt%	90	ISO 609 or ISO 625
Moisture content (max.)	wt%	5	ISO 687
Ash content (max.)	wt%	5	ISO 1171
Total sulfur content (max.)	wt%	5	ISO 351 or ISO 334
Volatile matter content (max.)	wt%	5	ISO 562

Table 4.8 Type 3 Backfill Material (Special Petroleum Coke Breeze)

Property	Unit	Requirement	Test Method
Carbon content (min.)	wt%	95	ISO 609 or ISO 625
Moisture content (max.)	wt%	1	ISO 687
Ash content (max.)	wt%	3	ISO 1171
Total sulfur content (max.)	wt%	1	ISO 351 or ISO 334
Volatile matter content (max.)	wt%	1	ISO 562

Table 4.9 The Bulk Density of Backfill Materials on Dry Packed Basis Conformation Range

Type	Bulk Density (kg/m^3)	Test Method
1	650–800	ISO 1013 or ISO 567
2	700–1100	
3	1050–1200	

It should have a round, uncrushable shape. The proposed material should have the properties as given in Table 4.8.

4.8.2 Bulk Density

The bulk density of backfill materials on a dry packed basis should conform to the following range in Table 4.9.

4.8.3 Resistivity

For Type 1 and 2 backfill materials, the resistivity should be <100 ohm-cm when dry and lightly tamped and prior to the addition of lime. For Type 3 backfill material, the resistivity of the tamped dry backfill should not be >0.1 ohm-cm.

4.8.4 Particle Size

The particle sizes of backfill materials should conform to the following size distribution as given in Table 4.10.

The test method should be in accordance with ISO 2325.

4.9 Galvanic Anodes for Underground Applications (Magnesium and Zinc)

This Part covers the minimum requirements for galvanic (sacrificial) anodes for use in the CP of buried structures against corrosion. It specifies the compositions, materials,

Table 4.10 The Particle Sizes of Backfill Materials and Size Distribution Requirements

Types	Particle Sizes	Size Distribution
Type 1	<10 mm	100%
	>1 mm	90%
Type 2	<3 mm	100%
	>0.5 mm	90%
Type 3	<1.0 mm	100%
	<0.6 mm	80%
	>0.4 mm	90%
	>0.15 mm	99%

manufacture, properties, inspection, and testing for magnesium and zinc anodes. It also includes details of backfill compositions and properties, for magnesium and zinc anodes. This specification covers only the general requirements for the anodes.

4.9.1 Classification

The anodes, as will be specified by the Purchaser, should be one of the following types:

Type I: Cast anodes.
Type II: Extruded rod (or ribbon) anodes.

4.9.2 Materials and Manufacture

Anodes should be manufactured by any suitable casting or extruding process. The anode material should be cast or extruded around a steel insert (core) so designed as to retain the anode material even when it is consumed to its design utilization factor.

Manufacturing of anodes should be carried out using approved production procedures covering each stage of the process from supply of raw materials to finishing and packing for shipment. The procedures should be fully referenced in an Inspection and Test plan that should also refer to the requirements of relevant Standard specification, as applicable.

The manufacturer should ensure that full traceability of all the materials is maintained throughout manufacture and thereafter.

The supplier should submit to the Purchaser his quality assurance and control plan.

The quality control at the anode manufacturing plant should include control systems on the following:

- Raw materials, that is, checking of documentation.
- Production equipment and process.
- Testing during and after production.
- Identification of anodes.
- Chemical analysis.

- Weight and dimensional tolerances.
- Surface condition of produced anodes.
- Documentation (material certificates).

The cast anodes may be supplied bare or packaged in a permeable cloth bag containing a rapid wetting and moisture retaining backfill, which reduces the electrolyte resistivity adjacent to the anode and improves performance, as will be required by the Purchaser.

4.9.3 Anode Composition

4.9.3.1 Magnesium Anodes

Magnesium anodes should be manufactured from virgin materials in conformity with the chemical composition limits given in Table 4.11 as determined by approved spectrographic and analytical procedures.

The anode composition in relevant Standard specification meets the chemical composition requirements of ASTM specification B275 Designation M1C (also called Galvomag).

4.9.3.2 Zinc Anodes

Zinc anodes should be made from special High Grade Zinc conforming to ASTM specification B 6, with suitable alloying additions, with an iron content well below the 0.003% maximum given in ASTM specification B 6. The composition of the anode alloy content should conform to the limits given in Table 4.12.

The anode composition in the relevant Standard specification meets the chemical composition requirements of ASTM specification B 418, Type II.

The following are to be noted:

1. The chemical analysis of the anode samples may be carried out either at the time they are cast, or at the finished or semifinished product stage.
2. The electrochemical properties are highly dependent on the content of alloying elements and impurity elements. The anode manufacturer should thus prove his capability of delivering anodes that satisfy the relevant Standard specification.

Table 4.11 Chemical Composition Limits for Magnesium Alloy Anodes

Element	Remarks	Composition, %
Aluminum	Max.	0.01
Manganese		0.9–1.2
Copper	Max.	0.02
Iron	Max.	0.03
Nickel	Max.	0.001
Other elements, total	Max.	0.3
Magnesium	The magnesium content should be calculated by subtracting from 100.00%.	Remainder

Table 4.12 Chemical Composition Limits for Zinc Alloy Anodes

Element	Composition, % (Max.)
Aluminum	0.005
Cadmium	0.003
Iron	0.0014
Lead	0.003
Copper	0.002
Other elements, total	0.012
Zinc (the zinc content should be calculated by subtracting from 100.00%)	Remainder

3. Small deviations from the specification in the alloy composition may significantly reduce the protective properties. Generally, it is of importance that the fabrication procedure ensures that the melt is not polluted by iron or other impurities. Furnaces, pumps, pipes, etc., should be nonmetallic or properly lined.

4.9.4 Anode Core

The magnesium anode core should be made from commercial hot dipped galvanized steel conforming to ASTM specification A 283, Grade C or equivalent standard.

The zinc anodes should have steel cores conforming to ASTM specification A 36. The steel should be coated with zinc or cadmium to a minimum thickness of 12.5 μm. Hot-dipped or electrodeposition methods may be used. Either coating should be adherent and free from flaking. The coating should adhere tenaciously to the surface of the base metal. When the coating is cut or pried into, such as with a stout knife applied with considerable pressure in a manner tending to remove a portion of the coating, it should only be possible to remove small particles of the coating by paring or whittling, and it should not be possible to peel any portion of the coating so as to expose the steel.

The position of the anode core should be such that the maximum possible mass of anode material is available to provide CP.

Cores should be abraded or otherwise treated to ensure that the requirements can be achieved.

4.9.5 Production Testing

The manufacturer should carry out the following tests:

1. A full spectrographic analysis, or an approved alternative, of each cast—The results should meet the requirements in Table 4.11 (for magnesium anodes) or in Table 4.12 (for zinc anodes).
2. An electrochemical performance test for each cast—The test should be carried out by the manufacturer in accordance with approved procedures designed to demonstrate batch consistency of the as-cast electrochemical properties.

4.9.6 Requirements

4.9.6.1 Performance Properties of Magnesium and Zinc Anodes

Any anode must be capable of a sufficient driving voltage to provide the necessary design current output. The performance properties of anode alloys should be in accordance with that given in Table 4.13.

4.9.7 Workmanship, Finish, and Appearance

The anodes should be free of flash burrs, cracks, blow holes, and surface slag consistent with good commercial practice. The cast anodes should be free of shrinkage cavities that are >6 mm (¼ in) in depth, except that anodes ≥50 mm (2 in) thick should be free of shrinkage cavities that are >9.5 mm (3/8 in) in depth, when measured from a straight edge placed diagonally across the opposite edges of the anode.

Metal core extensions from the anodes should be smooth and free of sharp burrs. The cast-in cores should have metallurgical bonds specified herein, free of air pockets and inclusions consistent with good commercial practice.

The anode lead wire insulation should be continuous and should have a thickness as uniform as possible and should be completely free of nicks, scratches, or other discontinuities.

4.9.8 Mechanical Resistance

When a uniaxial force equivalent to 100 kg, or five times the mass of the anode metal, whichever is the greater, is applied to a cast anode containing a core, no movement of the core or any metal insert should occur. Where a cable is provided, a uniaxial loading of 100 kg should not result in damage to the cable or to its connections.

4.9.9 Electrical Resistance

When determined in accordance with the standard test method, the electrical resistance of the anode to core or, if appropriate, of the anode to the end of the cable, should not be >0.01 for the life of the anode.

Table 4.13 Performance Properties of Galvanic Anode Alloys

Anode Alloy Type	Open Circuit Potential Negative Volts ($Cu/CuSO_4$ Electrode)	Current Efficiency % (Min.)	Energy Capability, amp-h/kg (Min.)	Consumption Rate, kg/Ayear (Max.)
Zinc	1.1	90	780	11.2
Magnesium	1.7 (14-day test, with reference to ASTM test method G 97)	50	1200	7.3

4.9.10 Anode Dimensions and Weights

4.9.10.1 Magnesium Anodes

The cast anode (Type I) dimensions and weights (nominal), as will be specified by the Purchaser, should be as given in Table 4.14.

Core straps should be 20 gage (8 mm) perforated electroplate galvanize steel with 6-mm holes on 12.7-mm (½ in) centers. The core length should be 75% of the anode length.

The Type II anode dimensions and weight (nominal) should be as follows:

Magnesium weight: 370 g/m
Cross-section: 9.5 × 19 mm
Core diameter: 3.2 mm
Coil length: 300 m (unless otherwise specified).

4.9.11 Zinc Anodes

The cast anode (Type I) dimensions and weights (nominal) should be as given in Table 4.15.

The cores should be made of a 6.35-mm (¼-in) diameter electrogalvanized mild steel.

Table 4.14 The Cast Anode (Type I) Dimensions and Weights (Nominal)

	Dimensions, mm			
Weight, kg (lbs)	**Width**	**Height**	**Length**	**Total Packaged Weight (Approx.), kg**
4.1 (9)	90	90	350	10
7.7 (17)	90	90	650	20
14.5 (32)	140	140	540	30
21.8 (48)	140	140	820	45

Table 4.15 The Cast Anode (Type I) Dimensions and Weights (Nominal)

	Dimensions, mm		
Weight, kg (lbs)	**Width**	**Height**	**Length**
2.3 (5)	35	35	230
5.5 (12)	35	35	610
8.2 (18)	35	35	915
13.6 (30)	50	50	760
20.5 (45)	50	50	1140
27.3 (60)	50	50	1525

Table 4.16 The Type II Anode Dimensions and Weights (Nominal)

Weight, kg/m (lb/ft)	Cross-Section, mm	Coil Length, m
1.8 (1.2)	15.8 × 22.2	60
0.9 (0.6)	12.7 × 14.2	150
0.37 (0.25)	8.7 × 11.9	300

The Type II anode dimensions and weights (nominal), as will be specified by the Purchaser, should be as given in Table 4.16.

Dimension and weight tolerances should conform to the following:

- Anode mean length should not deviate >3% of the nominal length.
- Anode mean width should not deviate >3% of the nominal width.
- The straightness of the anode should not deviate >2% of the anode nominal length from the longitudinal axis of the anode.
- The permissible variation in the weight of each finished anode will be ±5% of the nominal weight.
- The total contract weight should not be >2% above and not below the nominal contract weight.

4.9.12 Methods of Attachment of Cable for Cast Anodes

The cast anodes should be supplied complete with cables connected to the anode core. One end of the anode should be recessed to provide a cavity for attachment of the lead wire to the anode core and for subsequent sealing of this connection.

The method of connecting the lead wire to the anode core should be either by the use of a brazed connection or silver solder connection or by any other approved method. The connection should be made so as to bear the weight of the anode.

The lead wire to anode core connection should be sealed to prevent water intrusion by application of a potting compound (e.g., epoxy resin) or mastic-lined polyethylene heat-shrink sleeve to completely seal the connection and to protect the joint mechanically.

For Magnesium anodes, typical methods of attachment are shown in Figs 4.2 and 4.3.

For zinc anodes, a typical method of attachment is shown in Fig. 4.3.

The lead wires should be 6-mm^2 single core, stranded soft annealed copper conductors conforming to IEC Publication 228, Class 2 or ASTM specification B 8 Class B, and 5 m in length.

The lead wire insulation, as will be specified by the Purchaser should be in accordance with one of the following alternatives:

- A single jacket of HMWPE insulation conforming to ASTM specification D 1351, having a minimum thickness of 1.5 mm.
- A single jacket of XLPE insulation conforming to ASTM specification D 2655, having a minimum thickness of 1 mm.

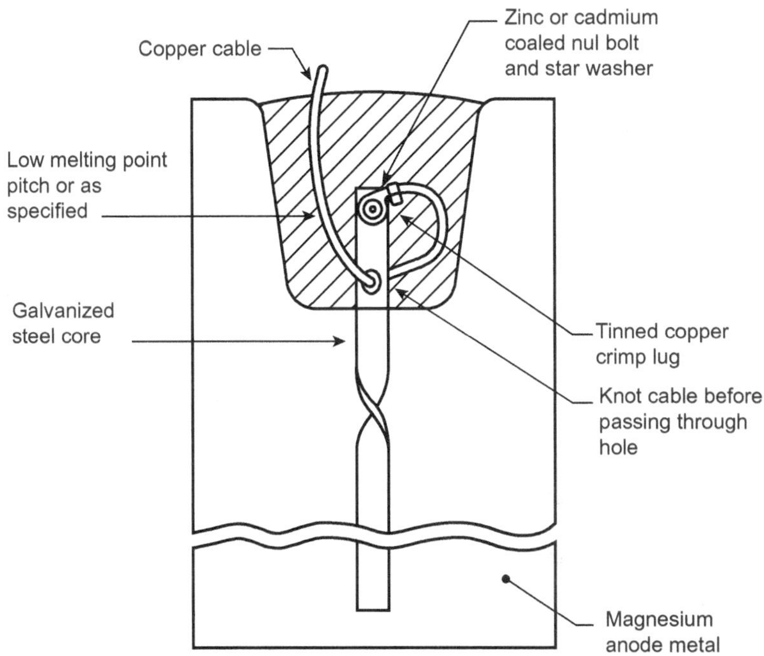

Figure 4.2 Typical method of cable attachment for buried magnesium anodes.

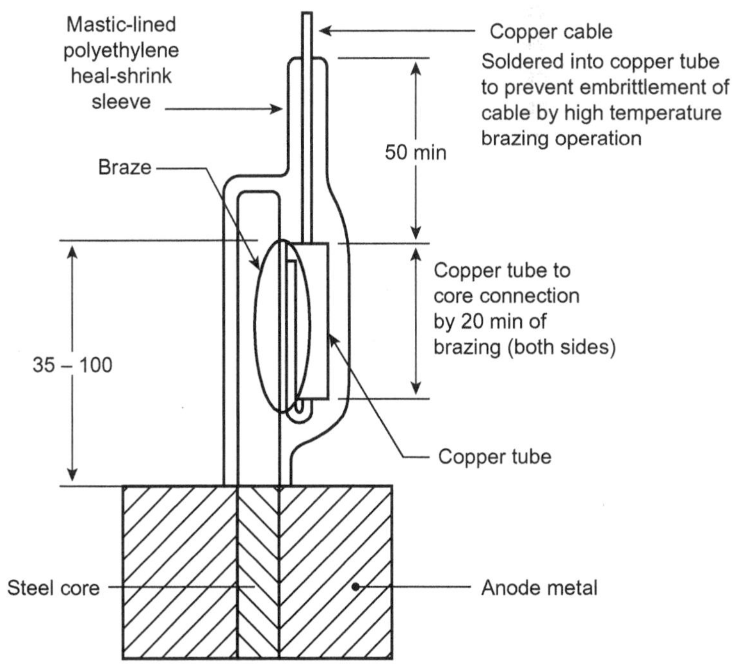

Figure 4.3 Typical method of cable attachment for buried anodes.

- A double jacket of PVC primary insulation conforming to ASTM specification D 2219, sheathed with a PVC jacket conforming to ASTM specification D 1047. The average insulation (of insulation and sheath) should be 2 mm with a 10% tolerance.

Lead wire insulation should have a minimum of 600-V rating.

The color of the insulation of the anode cables should be red, unless otherwise specified.

4.9.13 Identification of Anodes

Each anode should be clearly marked with the type of material (trade name), the cast number, and a piece serial number. The numbers of any rejected anodes should not be used again for replacement anodes.

4.10 Chemical Backfill for Galvanic Anodes

4.10.1 Backfill Composition

When specified by the Purchaser, the cast anodes should be supplied with backfill. The backfill surrounds, as will be specified by the Purchaser with reference to notes to Table 4.17, should be in accordance with one of the nominal compositions given in Table 4.17.

4.10.2 Particle Size

The Backfill should have a grain size so that 100% is capable of passing through a 0.84-mm (20 mesh) screen and 50% should be retained by a 0.15-mm (100 mesh) screen, and should not contain >5% moisture.

4.10.3 Backfill Analysis

The manufacturer should submit to the Purchaser a typical analysis of a random sample of backfill material giving the following information:

- Moisture content
- Gypsum content
- Bentonite content
- Sodium sulfate content (if any).

The manufacturer should perform screening tests on samples of the backfill to ensure that the backfill is in the form of a fine powder.

4.10.4 Anode Packaging

Anodes should be packed in a calico bag, or in some other type of approved porous container. The diameter of the packaging should be at least 50 mm greater

Table 4.17 Backfill Composition

Gypsum ($CaSO_4$)%				
Hydrated	**Molding Plaster (Plaster of Paris)**	**Bentonite Clay %**	**Sodium Sulfate %**	**Approximate Resistivity in ohm-cm**
(A) 25	–	75	–	250
(B) 50	–	50	–	250
(C) –	50	50	–	250
(D) 75	–	20	5	50

Notes:
(1) Backfill mix (A) is useful in low soil moisture areas and utilizes the moisture-holding characteristic of the bentonite clay.
(2) Backfill mix (B) is commonly used with Zinc anodes.
(3) Backfill mix (C) is useful with Zinc or Magnesium in very wet or marshy soils to prevent the rapid migration of backfill from the anode surface.
(4) Backfill mix (D), with low resistivity, is useful in high soil resistivity areas to reduce the anode resistance to the earth.

than the diagonal of the anode. The backfill should be tightly packed in the bag to prevent the movement of the anode and should achieve the following coverage around the anode:

1. A minimum of 25 mm of cover on corners, measured diagonally.
2. A minimum of 50 mm of cover on the top, bottom, and sides.

The backfill should be firmly held around the anode, and the whole should be vibrated during filling to ensure compaction.

The container should be sealed to minimize moisture intrusion during storage and shipment and to ensure against loosening or loss of backfill.

4.11 Method for the Determination of the Anode-to-Core Resistance of Galvanic Anodes

This section describes the method for the determination of the anode-to-core resistance of galvanic (sacrificial) anodes used for CP.

The voltage across the anode metal and anode core is measured when a known current is passed between them. The anode-to-core resistance is then computed by dividing the measured voltage by the known current.

The following apparatus is required:

1. A direct current (DC) supply of a 5-A capacity, with an adjustable current control capable of smooth variation from 5 A to near zero. It should be capable of generating an on-load-voltage sufficient to pass a current of 5 A through a resistance of not <1 ohm.
2. An ammeter capable of reading to at least a 5-A DC with a maximum error of $\pm1\%$ at full-scale deflection. This ammeter may be integral with the DC supply.
3. A millivoltmeter of the electrically protected type capable or reading to a 50-mV DC, with a maximum error of $\pm1\%$.

4. Suitable test leads and clamps or clips to connect Items (a)–(c) to the anode and the core. The clamp system should be of such a quality as to ensure that the connection resistance to each point is <0.05 ohm.

4.11.1 Circuit

The circuit should be arranged as shown in Fig. 4.4(a) or (b), as applicable. The leads connecting the millivoltmeter to the anode and the anode core (or cable) should be independent of the leads from the power supply. All connections should be properly made to minimize contact resistance.

If the millivoltmeter is connected across the anode metal and the terminal of the cable attached to the anode core, the anode-to-core resistance is obtained by subtracting the cable resistance from the total resistance. The total resistance is computed by dividing the measured potential by the known current.

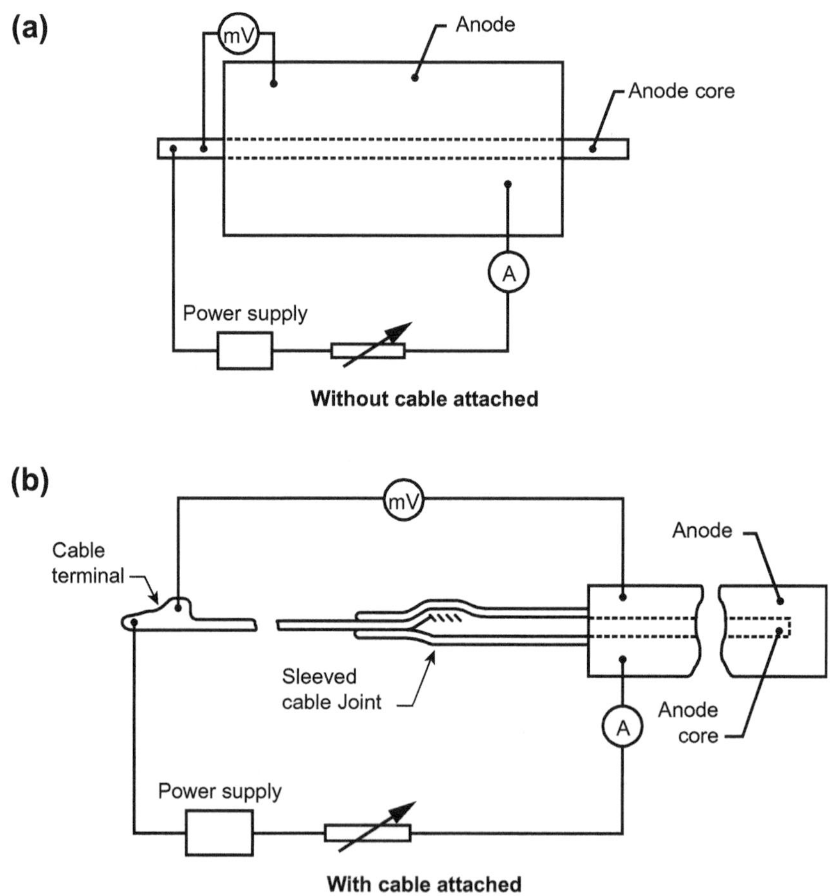

Figure 4.4 (a) or (b) Schematic circuits for measurement of anode-to-core resistance.

4.11.2 Procedure

The procedure should be as follows:

1. Before connecting the power supply, set the variable current control to ensure a minimum current flow when the circuit is switched on.
2. Connect the power supply. Adjust the variable current control until approximately 5 A is registered on the ammeter.
3. Read the voltage difference as indicated by the millivoltmeter.
4. Calculate the anode-to-core resistance (R), in ohms, using the following equation: R = (Millivoltmeter reading in Volts)/(ammeter reading in amperes).
5. Reverse both ammeter supply leads and also the millivoltmeter supply leads, and repeat Steps (a)–(d) above. The results of this second calculation of R should be the same as the initial result.

4.12 Galvanic Anodes for Submerged Applications (Magnesium and Zinc)

This Part covers the minimum requirements for the supply, identification, inspection, and testing of galvanic (sacrificial) anodes made of Magnesium or Zinc.

The magnesium alloy anodes conforming to this Part of the Standard specification are intended for use in corrosion protection in seawater or underwater surfaces of ships, vented ballast tanks of oilers, piers, dock structures, and other underwater (submerged) metallic structures. Galvanized steel structures may also be protected.

The zinc alloy anodes conforming to this Part of the Standard specification are intended for use in the corrosion prevention in seawater of ship hulls, steel and aluminum equipments and structures, sea chests, sonar domes, and the seawater side of condensers and other heat exchangers.

4.12.1 Materials and Manufacture

The electrochemical properties are highly dependent on the content of alloying elements and impurity elements. The anode manufacturers should thus prove their capability of delivering anodes that satisfy this Part of the Standard specification.

Small deviations from the specification in the alloy composition may significantly reduce the protective properties. Generally, it is of importance that the fabrication procedure ensures that the melt is not polluted by iron or other impurities. Furnaces, pumps, pipes, etc., should be nonmetallic or properly lined.

The quality control at the anode manufacturing plant should include control systems on the following:

- Raw materials, that is, checking of documentation
- Production equipment and process
- Testing during and after production
- Identification of anodes
- Chemical analysis

- Weight and dimensional tolerances
- Surface condition of produced anodes
- Documentation (material certificates).

4.12.2 Magnesium Anodes

The Magnesium anodes should meet the requirements of US Military Specification MIL-A-21412A (ships), unless specified otherwise by this Part of the Standard specification.

4.12.3 Zinc Anodes

The zinc anodes should meet the requirements of US Military Specification MIL-A-18,001 H, unless specified otherwise by this Part of the Standard specification.

4.12.4 Production Testing

The manufacturer should carry out the following tests:

A full spectrographic analysis of each cast—The results should meet the requirements of US Military specification MIL-A-21412A (for magnesium anodes) or MIL-A-18,001 H (for zinc anodes).
A short-term voltage test for zinc anodes only—Duration of test: 3 h minimum, current density: 0.75 mA/cm^2, in seawater (natural or synthetic) at a temperature of 20 °C.
Anode potential (active) to be− 1000 mV (saturated calomel) or more negative. Anode potential (at rest) to be −1050 mV (saturated calomel). The test should be carried out for each cast.
A destructive testing of one anode of each anode type/design, where the net weight of the anode is >100 kg, to check the bonding between the anode material and steel core in cross-sectional areas.
The anode material should be bonded to the steel core over a minimum of 90% of the total surface, while no individual discontinuity or void in the bonding should be >25 mm in length.

4.12.5 Requirements

- **Quality of anode castings**

The as-cast anode surface should be free of surface slag or other embedded material.

Cracks are not acceptable, except in the form of micro (hairline) cooling cracks. Maximum width is 1 mm.

Shrinkage cavities should not be >10 mm in depth, as taken from the anode surface. Slag inclusions are not acceptable.

Tolerance on the dimensions of anodes, position of anode inserts, and gross weight of anodes to be plus and minus 2%.

4.12.6 Quality of Steel Inserts

Any type of steel core, insert, or bracket should be fabricated from fully killed low-carbon steel, pipe material to be ASTM A 106 Gr. B or BS 3602-HFS27, and plate/strip to be ASTM A 185 Grade C or BS 1501-151 Gr. 26A, to be substantiated by mill certificates.

Steel cores and inserts should be blast cleaned to Grade Sa 2½ according to ISO 8501-1, and have this finish at the time of casting.

4.12.7 Electrical Resistance

When determined in accordance with the test method, the electrical resistance of the anode to core or, if appropriate, of the anode to the end of the cable, should not be >0.01 for the life of the anode. This should be measured in ohms.

4.12.8 Identification of Anodes

Each anode should be clearly marked with the type of material (trade name), the cast number, and a piece serial number. The numbers of any rejected anodes should not be used again for replacement anodes.

4.13 Method for the Determination of the Anode-to-Core Resistance of Galvanic Anodes

This section describes the method for the determination of the anode-to-core resistance of galvanic (sacrificial) anodes used for CP.

4.13.1 Principle

The voltage across the anode metal and anode core is measured when a known current is passed between them. The anode-to-core resistance is then computed by dividing the measured voltage by the known current.

4.13.2 Apparatus

The following apparatus is required:

1. A DC supply of a 5-A capacity, with an adjustable current control capable of smooth variation from 5 A to near zero. It should be capable of generating an on-load-voltage sufficient to pass a current of 5 A through a resistance of not <1 ohm.
2. An ammeter capable of reading to at least 5 A DC with a maximum error of ±1% at full-scale deflection. This ammeter may be integral with the DC supply.
3. A millivoltmeter of the electrically protected type capable of reading to 50-mV DC, with a maximum error of ±1%.

4. Suitable test leads and clamps or clips to connect Items (a)–(c) to the anode and the core. The clamp system should be of such a quality to ensure that the connection resistance to each point is <0.05 ohm.

4.13.3 Circuit

The circuit should be arranged as shown in Fig. 4.5. The leads connecting the millivoltmeter to the anode and the anode core should be independent of the leads from the power supply. All connections should be properly made to minimize contact resistance.

If the millivoltmeter is connected across the anode metal and the terminal of the cable is attached to the anode core, the anode-to-core resistance is obtained by subtracting the cable resistance from the total resistance. The total resistance is computed by dividing the measured potential by the known current.

4.13.4 Procedure

The procedure should be as follows:

1. Before connecting the power supply, set the variable current control to ensure a minimum current flow when the circuit is switched on.
2. Connect the power supply. Adjust the variable current control until approximately 5 A is registered on the ammeter.
3. Read the voltage difference as indicated by the millivoltmeter.
4. Calculate the anode-to-core resistance (R), in ohms, using the following equation: R = (Millivoltmeter reading in volts)/(ammeter reading in amperes).

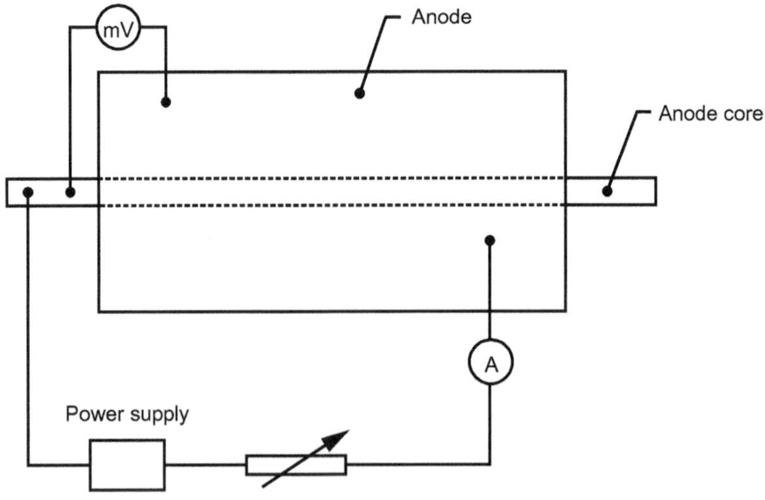

Figure 4.5 Schematic circuits for measurement of anode-to-core resistance.

5. Reverse both ammeter supply leads and also the millivoltmeter supply leads, and repeat Steps (a)–(d) above. The result of this second calculation of R should be the same as the initial result.

4.14 Bracelet-Type Galvanic Anodes for Submarine Pipelines (Aluminum and Zinc)

This Part covers the minimum requirements for the material, manufacture, inspection, and testing, documentation, and supply of half-shell bracelet type galvanic anodes made of Aluminum and/or Zinc, to be used for the CP of submarine pipelines.

4.14.1 Materials and Manufacture

The electrochemical properties are highly dependent on the content of alloying elements and impurity elements. The anode manufacturer should thus prove his capability of delivering anodes that satisfy this Part of Standard specification.

Small deviations from the specification in the alloy composition may significantly reduce the protective properties. Generally, it is of importance that the fabrication procedure ensures that the melt is not polluted by iron or other impurities. Furnaces, pumps, pipes, etc., should be nonmetallic or properly lined.

Manufacturing of anodes should be carried out using approved production procedures covering each stage of the process from the supply of raw materials to finishing and packing for shipment. The procedures should be fully referenced in an Inspection and Test plan that should also refer to requirements of this Part of Standard specification, as applicable.

The manufacturer should ensure that full traceability of all the materials is maintained throughout the manufacturing process.

The manufacturer should submit to the Purchaser his quality assurance and control plan.

The quality control at the anode manufacturing plant should include control systems on the following:

- Raw materials, that is, checking of documentation
- Production equipment and process
- Testing during and after production
- Identification of anodes
- Chemical analysis
- Weight and dimensional tolerances
- Surface condition of produced anodes
- Documentation (material certificates).

4.14.2 Alloy Composition

- **Aluminum anodes**

The aluminum anodes should conform to the chemical composition specified in Table 4.18 as determined by spectrographic and other methods specified herein.

Table 4.18 Chemical Composition Limits for Aluminum Anodes

	Chemical Composition, %	
Element	**Min.**	**Max.**
Silicon	–	0.1
Iron	–	0.1
Copper	–	0.006
Zinc	3	5
Indium	0.02	0.05
Other Elements		
Each	–	0.02
Total	–	0.05
Aluminum	Remainder	

- **Zinc anodes**

The zinc anodes should be made from special High Grade Zinc conforming to meet ASTM specification B 6, with suitable alloying additions. The anodes should conform to the chemical composition given in Table 4.19 in order to reduce the susceptibility to intergranular corrosion at elevated temperature.

The zinc anode composition in this specification meets the chemical composition requirements of ASTM specification B418 (Type I) and US Military specification MIL-A-18,001H.

Certificates of analysis should be submitted for all elements >0.01%, in addition to the elements above listed, and should state that all other elements are below this limit.

Table 4.19 Chemical Composition Limits for Zinc Anodes

	Chemical Composition, %	
Element	**Min.**	**Max.**
Cadmium	0.03	0.06
Aluminum	0.10	0.2
Iron	–	0.002
Lead	–	0.006
Copper	–	0.005
Other Elements		
Each	–	–
Total	–	0.1
Zinc	Remainder	

4.14.3 Production Testing

The manufacturer should carry out the following production tests:

4.14.3.1 Chemical Analysis

No anode casting should commence before confirming by chemical analysis that the alloy constituents fall within the range specified in 6.2.

A full spectrographic analysis, or an approved alternative, should be made of representative samples taken at the beginning and end of each melt to prove that the chemical composition is within the limits of the alloy specification stated in 6.2. If one of these analyses is out of the specified range, the heat will be rejected.

The equipment used for chemical analysis should be calibrated at the start of each shift using reference samples. Calibration samples should be independently certified.

If alloying additions are made in holding furnaces or crucibles, then analysis should be conducted at the beginning and end of pouring from the holding furnace or crucible.

No additions of any kind should be made to the melt following a spectrographic analysis that shows the material to be within specifications and before commencement of pouring.

Each heat should be assigned a unique number that should be transferred to all samples taken from that heat and to all records pertaining to that heat. All samples should be indestructibly marked with the unique heat number.

At least two further samples, taken at the same frequency as those for spectrographic analysis, in the form of sticks of an approximately 30-mm diameter × 200-mm length should be retained from each heat for use in electrochemical testing.

All samples should be retained for a minimum of 2 years beyond the completion of the supply contract and may be subjected to independent analysis and testing.

4.14.3.2 Short-Term Voltage Testing

A short-term voltage test should be carried out on one representative sample from each heat.

The voltage test should be carried out in natural seawater or synthetic seawater (to the composition specified in ASTM D 1141) at 20 °C.

For zinc anodes, the test should be performed at an anodic current density of 0.75 mA/cm^2 and the duration of test should be 3 h min. The anode working potentials should be, respectively, −1000 and −1050 mV vs the saturated calomel electrode (SCE), which is a reference electrode.

4.14.3.3 Short-Term Capacity Testing for Aluminum Anodes Only

A short duration capacity test should be carried out on one representative sample of each heat.

One-fifth of the number of heats should be tested, with a minimum of one test per order.

The test procedure should be in accordance with standard. The anode material capacity should be determined by weight loss. The Anode capacity should not be <2500 Ah/kg.

Anodes failing the above electrochemical test acceptance criteria should be rejected. However, a minimum of two new samples for retests may be cut from two different anodes of the same heat number as the original failed test pieces. In case even one of these two give negative results, the entire heat will be rejected, or all the anodes should be tested individually. In case both of them give acceptable results, the remaining part of the heat is accepted.

4.14.3.4 Destructive and Nondestructive Testing

One anode should be selected for sectioning at random from each of 150 produced anodes, in addition to the first sectioned anode taken at the start of production.

The selected anode should be transverse sectioned by single cuts at 25%, 33%, and 50% of the nominal length and once longitudinally.

The cut faces, when examined visually without magnification, should be free from visible slag/dross inclusions.

Total porosity in any sectioned surface should not be >0.25% of the alloy area, with no individual pores or void >10 mm^2.

The lack of bonding between the anode and the core should not be >5% of the core peripheral length in any sectioned surface.

Should any anode section fail to meet the above criteria, the subsequent anode cast should be also sectioned. If this also fails, the entire production batch (since the last successful section) should be quarantined and a series of backchecks should be made to allow acceptance of part batches.

Alternatively to destructive testing, a nondestructive testing by radiography may be used to check for lack of bond or slag/dross inclusions.

4.15 Anode Core Steel Works

4.15.1 Insert Material

The anode material should be cast around a steel insert so designed as to retain the anode material even when it is consumed to its design utilization factor. The steel inserts should have sufficient strength to withstand all external forces that they may normally encounter.

The anode insert should meet the requirements of EN 10,025 Grade Fe 430C. The maximum carbon content of the steel to be welded should not be >0.23%.

The carbon equivalent of the insert materials should not be >0.45%, as determined using the following formula:

$$\text{Carbon equivalent} = \mathrm{C} + \frac{Mn}{6} + \frac{Cr + Mo + V}{5} + \frac{Ni + Cu}{15}$$

where each element is expressed in weight percent.

Rimming steels should not be used.

All materials should be identified against original mill certificates.

4.15.2 Insert Surface Preparation

Inserts should be stored indoors at a humidity not >85% with steel temperature maintained at least 3 °C above the dew point.

For aluminum anodes, the steel fabrication to be inserted into the cast anode should be prepared by a dry blast cleaning process to a minimum quality complying with ISO 8501-1, Grade SA 2½ or equivalent standard.

At the time the aluminum anodes are cast, “dulling” of the blast-cleaned surface of the insert should be permitted. Rust discoloration and/or visible surface contamination should not be permitted.

For zinc anodes, the steel fabrication to be inserted into the cast anode should be prepared by dry blast-cleaning process to a minimum quality complying with ISO 8501-1 Grade SA 2½ or equivalent standard, galvanizing to BS 729,or Zinc electroplating to BS 1706, or equivalent standards.

At the time the zinc anode is cast, “dulling” of blast-cleaned or zinc-coated steel insert surfaces should be permitted. Rust discoloration and/or visible surface contamination of the blast-cleaned or zinc-coated surface should not be permitted.

Inserts not used within 8 h of blast cleaning, or showing signs of rusting, moisture, grease, oil, or other surface contamination, should be recleaned.

4.15.3 Welding

All fabrication welding and acceptance tests on anode insert should be in accordance with AWS D1.1, latest edition.

Welding procedures and welder qualifications should be in accordance with AWS D1.1, latest edition.

All welds should be visually inspected and be free of deleterious defects. Adequate examination of production welds should be carried out by appropriate nondestructive testing.

4.15.4 Requirements

4.15.4.1 Anode Shape, Weight, and Dimensions

- **Shape**

The shape of anodes, as will be specified by the Purchaser, may be either plain half-shell, or tapered half-shell.

The following are to be noted:

1. The plain half-shell bracelet is used when the pipe is concrete coated and should be manufactured so that the internal diameter fits over the outside diameter of the wrap/composition coat and matches the outside diameter of the concrete.

2. The tapered half-shell bracelet is for pipes without concrete; the purpose of the taper is to enable the anodes more easily to traverse the stringer rolls of the lay-barge.

- **Weight**

The individual casting (half-shell pairs) net weight should be within ±3% of the net weight shown in the drawings.

All half-shell anodes and their cores should be weighed individually to confirm compliance. The individual weights should be recorded. The total supplied net weight should not be>2% above and not below the nominal supplied weight.

- **Dimensions**

The anode dimensions should conform to the tolerances indicated by the manufacturer and/or supplier in the approved drawings.

The position of the insert in the casting should be within 5 mm of the design position, as an average over the length, and not >10 mm from the design position at any point.

Anodes should be free from excessive bowing or twisting. The manufacturer should fabricate a suitable template for the half-shell bracelet anodes. Complete bracelets should be assembled around the template to ensure compliance with the diametrical tolerances specified on the drawings.

Anode insert cross-section dimensions should comply with the appropriate specification for the insert material used. Full dimensional checks should be carried out to ensure compliance with the approved drawing for the insert material used.

4.15.5 Mechanical Resistance

When a uniaxial force equivalent to 100 kg, or five times the mass of the anode metal, whichever is the greater, is applied to a cast anode containing a core, no movement of the core or any metal insert should occur.

4.15.6 Electrical Resistance

When determined in accordance with the test method, the electrical resistance of the anode to core should not be >0.01 ohm for the life of the anode.

4.15.7 Consumption Rate of Aluminum Anodes in Seawater

When determined in accordance with the test method, the consumption rate of an aluminum anode in seawater should not be >3.5 kg/A year.

The following is to be noted:

The test specified above is applicable to aluminum anodes only. Typical consumption rates for zinc anodes in seawater are given in Table 4.20.

4.15.8 Closed Circuit Potential of Aluminum Anodes in Seawater

When determined, the closed-circuit potential of an aluminum anode should be at least −1.05 V with respect to a silver/silver chloride reference electrode.

Table 4.20 Typical Properties of Zinc Anodes

Open Circuit Potential, Ea. Reference Electrode		Typical Anode Consumption Rate in Seawater, kg/Ayear
$Cu/CuSO_4$	Ag/AgCl	
−1.1	−1.05	12

It is to be noted that the test is relevant for aluminum anodes only.

4.15.9 Surface Irregularities in the Anode Casting

Shrinkage depressions in the casting should not be >10% of the nominal thickness of the anode as measured from the uppermost corner of the depression to its bottom.

The anode casting should be free from nonmetallic inclusions. Cold shuts or surface laps should not exceed a depth of 10 mm or extend over a total length of >150 mm.

All protrusion detrimental to the safety of personnel during handling should be removed. No further grinding or mechanical treatment of the anode surface should be permitted.

All anodes should be inspected visually to confirm compliance with the above.

4.15.10 Cracks in Cast Anodes Material

- Within the section of sacrificial anodic material wholly supported by the core, transverse cracks should not be permitted except in the form of micro (hairline) cooling cracks with a maximum width of 2 mm and a length not >75% of the anode bracelet diameter or 200 mm, whichever is smaller. A maximum of two cracks per casting (half shell) should be allowed. Small close cracks should be taken as one crack.
- No cracks should be allowed in anodic material unsupported by the core.
- Cracks that extend around two or more faces of the anode should not be permitted.
- Longitudinal cracks should not be permitted.
- All castings should be visually inspected to ensure compliance with the above.

4.15.11 Anode Connections

- Two anode connections, for bonding to the pipe, should be attached to each bracelet half-shell. Electrical connection to the pipeline should be by means of four 16 mm^2 cables per anode to be thermit welded to the pipeline after anode installation.
- The lead wires should be 16 mm^2 single core, stranded soft annealed copper conductor conforming to IEC 228 Class 2, minimum length 250 mm.
- The lead wire insulation, as will be specified by the Purchaser, should be in accordance with one of the following alternatives:
  - A single jacket of HMWPE conforming to ASTM D 1351, having a minimum thickness of 1.5 mm.

  - A single jacket of XLPE conforming to ASTM specification D 2655, having a minimum thickness of 1 mm.
  - A double jacket of PVC primary insulation as per ASTM D 2219, sheathed with a PVC extruded jacket conforming to ASTM D 1047. The average thickness (of insulation and sheath) should be 2 mm with a 10% tolerance.
- Lead wire insulation should have a minimum 600-V rating.
- Unless otherwise specified, the color of lead wire insulation should be red.
- The insulation should be applied tightly to the conductor without adhering to it and should form a compact and homogeneous body. The lead wire insulation should be completely free of cracks, nicks, scratches, or other discontinuities.

4.16 Method for the Determination of the Anode-to-Core Resistance of Galvanic Anodes

This section describes the method for the determination of the anode-to-core resistance of galvanic (sacrificial) anodes used for CP.

The voltage across the anode metal and anode core is measured when a known current is passed between them. The anode-to-core resistance is then computed by dividing the measured voltage by the known current.

4.16.1 Apparatus

The following apparatus is required:

1. A DC supply of a 5-A capacity, with an adjustable current control capable of smooth variation from 5 A to near zero. It should be capable of generating an on-load–voltage sufficient to pass a current of 5 A through a resistance of not <1 ohm.
2. An ammeter capable of reading to at least 5-A DC with a maximum error of ±1% at full-scale deflection. This ammeter may be integral with the DC supply.
3. A millivoltmeter of the electrically protected type capable of reading to 50 mV DC, with a maximum error of ±1%.
4. Suitable test leads and clamps or clips to connect Items (a)–(c) to the anode and the core. The clamp system should be of such a quality so as to ensure that the connection resistance to each point is <0.05 ohm.

4.16.2 Circuit

The circuit should be arranged as shown in Fig. 4.6 (a) or (b), as applicable. The leads connecting the millivoltmeter to the anode and the anode core (or cable) should be independent of the leads from the power supply. All connections should be properly made to minimize contact resistance.

If the millivoltmeter is connected across the anode metal and the terminal of the cable attached to the anode core, the anode-to-core resistance is obtained by subtracting the cable resistance from the total resistance. The total resistance is computed by dividing the measured potential by the known current.

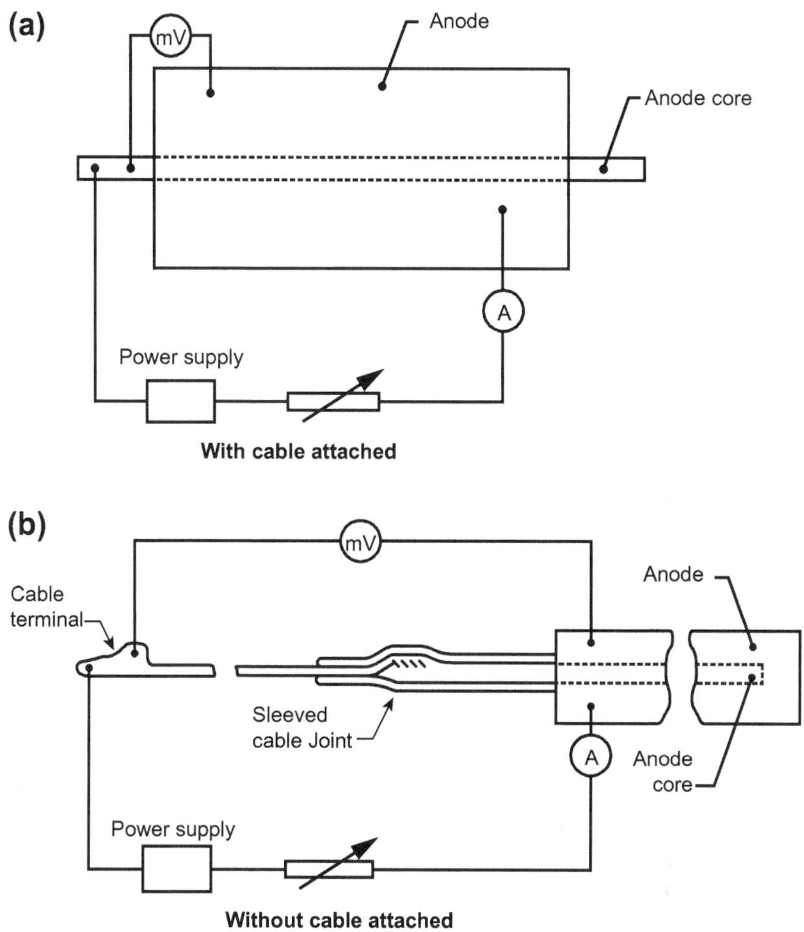

Figure 4.6 Schematic circuits for measurement of anode-to-core resistance. (a) with cable attached, (b) without cable attached.

4.16.3 Procedure

The procedure should be as follows:

1. Before connecting the power supply, set the variable current control to ensure a minimum current flow when the circuit is switched on.
2. Connect the power supply. Adjust the variable current control until approximately 5 A is registered on the ammeter.
3. Read the voltage difference as indicated by the millivoltmeter.
4. Calculate the anode-to-core resistance (R), in ohms, using the following equation: R = (Millivoltmeter reading in Volts)/(ammeter reading in amperes).
5. Reverse both ammeter supply leads and also the millivoltmeter supply leads, and repeat Steps (a)–(d) above. The result of this second calculation of R should be the same as the initial result.

4.17 Method for the Determination of the Consumption Rate of Aluminum Anode Alloys Immersed in Seawater

This section describes the method for the determination of the consumption rate of aluminum anode alloys intended for use in CP of ferrous metals immersed in seawater.

The anode alloy consumption rate, expressed in kilograms per ampere year, is an important factor when the performance of different alloys for the sacrificial protection of ferrous metal structures in seawater is being assessed. The consumption rate may be used to compute the current capacity of an anode system.

The loss of anode mass, which occurs when a known quantity of current flows from the anode to the cathode in seawater, is computed to show the quantity of anode metal consumed in one year by the passage of 1 A.

4.17.1 Apparatus

The following apparatus is required:

1. A constant current DC power supply with an output voltage appropriate to the number of test cells in series, and having sufficient current capacity to maintain the specified anode current density: The output current stability of the power supply should be within ±1%.
2. An ammeter to suit the current requirement, and having a maximum error of ±1%: This ammeter may be integral with the power supply.
3. Aluminum cathodes: one per aluminum anode.
4. Nonmetallic tanks: one per anode, to hold seawater.
5. Clean seawater with a resistivity of not >0.3 ohm m at 20 °C, or synthetic seawater conforming to the requirements of ASTM D 1141.
6. An agitator.

4.17.2 Circuit

The circuit should be arranged as shown in Fig. 4.7.

4.17.3 Preparation of Test Anodes

The as-manufactured anode surface is preferred for test purposes; however, cut and machined surfaces are not excluded for use in this test. All anodes used in the same test series should have the same nominal composition and the same area dimensions exposed to seawater.

Any mounting strap or wiring conductor should be covered to prevent contact with seawater.

The anodes should be cleaned in nitric acid (e20 1420 kg/m^3) at an ambient temperature and then thoroughly rinsed in clean fresh water and dried.

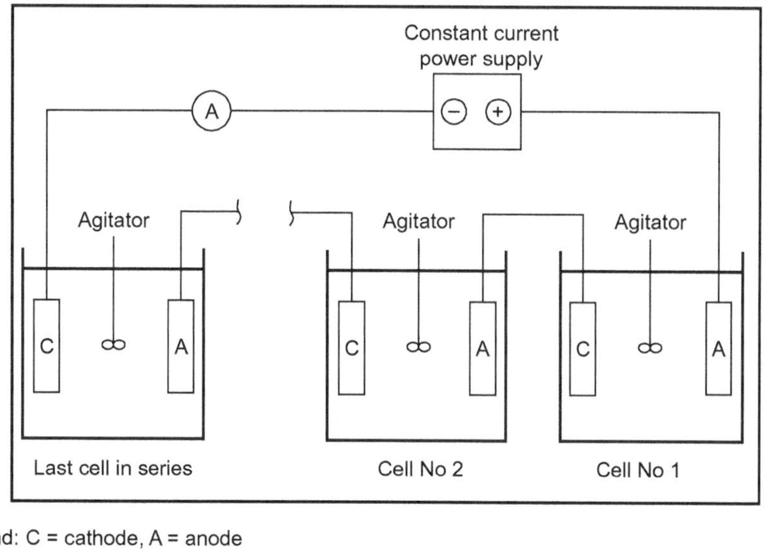

Figure 4.7 Schematic circuit for the determination of alloy consumption rate.

4.17.4 *Procedure*

The anode consumption rate should be determined as follows:

1. Weight each anode and record its mass.
2. Connect the circuit as shown in Fig. 4.7.
3. Adjust the power supply to produce a current output equivalent to a current density of between 0.5 and 0.7 mA/cm^2 of the anode surface area. Note the time of commencement of the test and the set current value.
4. Change the seawater when 0.1 Ah has passed for each 1 L of seawater in a cell.
5. Continue the test for a period of at least 10 days.

Note that the longer the test period, the more reliable will be the result.

6. Remove the test anodes from the cells. Clean, rinse, and dry the test anodes and reweight.
7. Calculate the alloy consumption rate for each anode, in kilograms per ampere year, using the following equation:

$$\text{Alloy consumption rate} = \frac{\text{mass loss} \times 8760}{\text{total current flow}}$$

where mass loss is in kilograms and total current flow is in ampere hours.

Express the result to the nearest 0.1 kg/Ayear.

4.18 Method for the Determination of the Closed-Circuit Potential of Aluminum Anodes Immersed in Seawater

This section describes the method for the determination of the closed-circuit potential of aluminum anodes intended for use in the CP of ferrous metals immersed in seawater.

The closed-circuit potential, as distinct from the open-circuit potential, is a critical factor when assessing the ability of galvanic anodes to protect ferrous metal structures in service.

The closed-circuit potential is measured when current is flowing between an anode and cathode in seawater, by locating a reference cell close to the anode surface.

4.18.1 Apparatus

The following apparatus is required

1. A 12-V DC power source with a variable resistance to maintain the specified anode current density.
2. A suitable ammeter.
3. A high-resistance voltmeter with a minimum input resistance of 1 mega-ohm/V and a resolution of 10 mV or better, and having a maximum error of $\pm 1\%$ at full-scale deflection.
4. A silver/silver chloride or SCE.
5. An aluminum cathode.
6. A nonmetallic tank to hold seawater.
7. Fixed resistors of equal value: The voltage drop across each fixed resistor should be at least 10 V.
8. Clean seawater with a resistivity of not >0.3 ohm m at 20 °C, or synthetic seawater conforming to the requirements of ASTM D 1141.

4.18.2 Circuit

The circuit should be arranged as shown in Fig. 4.8 to enable the testing of one of a series of anodes. All anodes in the circuit should have the same nominal composition and shape and the same area dimensions exposed to the seawater.

Any mounting, wiring conductor, or anode connection should be insulated at the air/water interface to prevent contact with seawater.

4.18.3 Preparation of Test Anode

For test purposes, it is preferred that the surface of the anode is in the as-manufactured condition; however, cut and machined surfaces are also acceptable.

4.18.4 Procedure

The closed-circuit potential should be determined using the following procedure:

1. Arrange the equipment so that the anode-to-cathode distances are approximately the same, and the current flow is not affected by cell geometry or adjacent anodes.

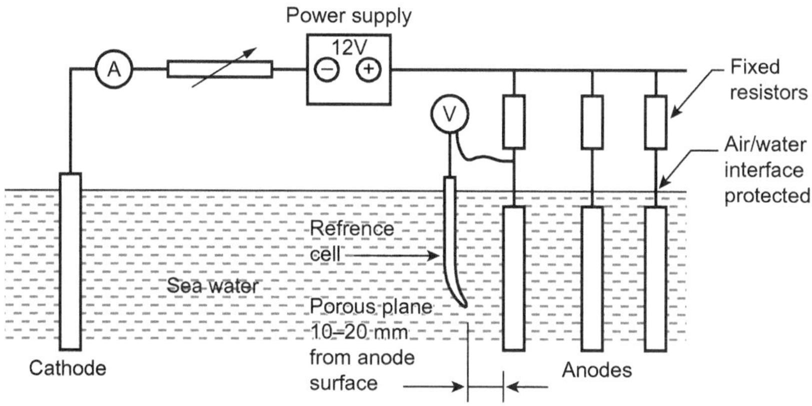

Figure 4.8 Schematic circuit for the measurement of closed-circuit potential.

2. Adjust the test anode current density to a value between 0.5 and 0.7 mA/cm^2 and record the time of commencement of current flow. Change the seawater when 0.1 Ah has passed for each liter of seawater contained in the tank.
3. Record, to the nearest 0.01 V, the closed-circuit potential on the voltmeter. Continue the test until three consecutive stable readings are obtained at time intervals of 1 h.
4. Report the last three stable closed-circuit potential readings obtained with respect to the reference cell, for the anode under test.
5. Repeat Steps (a)–(d) for each anode tested to enable duplication of the results.

4.19 Cast Galvanic Anodes for Fixed Offshore Installations (Aluminum)

This Part covers the minimum requirements for the material, manufacture, inspection, and testing of cast sacrificial anodes made of Aluminum–zinc–indium, intended for use on fixed offshore installations. This specification covers only the general requirements for the aluminum anodes; the specific requirements will be given in pertinent ordering documents.

4.19.1 Materials and Manufacture

The electrochemical properties are highly dependent on the content of alloying elements and impurity elements. The anode manufacturer should thus prove his capability of delivering anodes that satisfy this Part of Standard specification.

Small deviations from the specification in the alloy composition may significantly reduce the protective properties. Generally, it is of importance that the fabrication procedure ensures that the melt is not polluted by iron or other impurities. Furnaces, pumps, pipes, etc., should be nonmetallic or properly lined.

The quality control at the anode manufacturing plant should include control systems on the following:

- Raw materials, that is, checking of documentation.
- Production equipment and process.
- Testing during and after production.
- Identification of anodes.
- Chemical analysis.
- Weight and dimensional tolerances.
- Surface condition of produced anodes.
- Documentation (material certificates).

4.19.2 Type of Anodes

The anodes should be of the "stand-off" type, cast in Aluminum–Zinc–Indium sacrificial alloy, with a minimum stand-off distance of 150 mm, and can have a trapezoid or cylindrical shape.

The alloy should be proven to be free from passivity and intergranular corrosion for the conditions specified in ordering/purchasing documents.

The anodes should be sufficiently rigid to avoid vibration in the anode support.

4.19.3 Chemical Composition

The anodes should conform to the chemical composition specified in Table 4.21 as determined by spectrographic and other methods specified herein.

Table 4.21 Chemical Composition Limits

Element	Composition, % by Weight
Silicon, max	0.1
Iron, max.	0.1
Copper, max.	0.006
Zinc	3–5
Indium	0.02–0.05
Other elements, each	0.02
Other elements, total; other elements—total should be the sum of unspecified metallic elements ≥0.010%, rounded to the second decimal before determining the sum.	0.05
Aluminum	The aluminum content should be calculated by subtracting from 100.00%.

Analysis should regularly be made only for the elements specified in this table. If, however, the presence of other elements is suspected or indicated in amounts greater than the specified limits, further analysis should be conducted to determine that these elements are not present in amounts in excess of the specified limits.

4.19.4 Production Testing

The manufacturer should carry out the following production testing:

4.19.4.1 Chemical Analysis

Two samples from each melt should be taken for chemical analysis. The samples should be taken in the beginning and at the end of casting from the pouring string.

For smaller alloying furnaces (~500 kg), it is sufficient to take one sample per batch. The sample should be taken at the beginning of one batch, at the end of the next, and so on.

Sampling should be in accordance with ASTM Practice E716 or approved alternative standard.

Evaluation should be by spectrographic analysis in accordance with ASTM Test Method E 101 or approved alternative standard; in the case of dispute, the methods of analysis should be agreed upon by the Purchaser and supplier.

If one of these analysis is out of the specified range, the heat will be rejected. Individual anodes may be accepted subject to further analysis, at the option of the Purchaser.

4.19.4.2 Electrochemical Testing for Quality Control

Checking of closed-circuit-potential and current capacity should be done for each 10-ton produced anodes, with a minimum one test per order. The test samples may be separately casted or cut directly from an anode.

4.19.4.3 Destructive and Nondestructive Testing

- At least one anode per delivery or at least 0.5% of the anodes, where the net weight of the anode is >100 kg, should be subject to destructive testing to check the bonding between anode material and steel core in cross-sectional areas.
- All sectioning should be carried out in the presence of the purchaser's inspector(s).
- Anodes should be sectioned transversely, by single cuts, at 25%, 33%, and 50% of nominal cast length.
- Anodes should be sectioned once longitudinally through the center line/pour point.
- Not >5% of the insert peripheral length in any sectioned surface should show a visible lack of bond between the insert and the cast alloy.
- There should be no visible pores and slag/dross inclusions in any sectioned surface.
- Total porosity in any sectioned surface should not be >0.25% of the alloy area, with no individual pores or voids >10 mm^2.
- Should any anode section fail to meet the above criteria, the subsequent anode cast should also be sectioned. If this also fails, the entire production batch (since the last successful section) should be quarantined and a series of backchecks made to allow acceptance of part batches. Acceptance of quarantined anodes should be at the sole discretion of Purchaser's inspector(s). The problem should be investigated, and the cause corrected to the Purchaser's inspector satisfaction, before production continues.
- Alternatively to destructive testing, a nondestructive testing by radiography may be used to check for the lack of bond or slag/dross inclusions.

4.19.5 Anode Insert

4.19.5.1 Insert Material

The platform anode insert should be cut from unspliced weldable seamless low alloy carbon steel line pipe to API 5L, Grade B or approved equivalent standard.

The steel inserts should have sufficient strength to withstand all external forces that they may normally encounter.

The carbon equivalent (Cev) of the insert materials should not be >0.45 percent, determined using the following formula:

$$C_{ev} = C + \frac{Mn}{6} + \frac{Cr + Mo + V}{5} + \frac{Ni + Cu}{15}$$

where each element is expressed in weight percent.

Rimming steels should not be used.

Mill certificates should conform to the requirements of BS EN 10 204, Part 3.1.B, and should be obtained for all steel used in the manufacture of inserts.

For low-temperature applications, consideration should be given to the notch toughness of material, in particular, to the toughness of the material to be welded to the parent structure.

4.19.5.2 Insert Surface Preparation

- All anode inserts should be blast cleaned to ISO 8501-1, Grade SA 2½ prior to anode casting. Sand should not be used for blast cleaning. The surface profile of the cleaned surface should be 50–100 μm.
- Inserts should be stored indoors at a humidity not >85% and such that the steel temperature is maintained at least 3 °C above the dew point.
- Inserts not used within 8 h of blast cleaning, or any that show signs of rusting or surface contamination, or that do not conform to ISO 8501-1 Grade SA 2½ at time of casting should be recleaned in accordance with the requirements of 6.5.2.1.
- At the time the anodes are cast, "dulling" of the blast-cleaned surface of the insert should be permitted. Rust discoloration and/or visible surface contamination should not be permitted.

4.19.5.3 Insert Fabrication

All fabrication welding and acceptance tests on anode inserts should be in accordance with AWS D1.1 latest edition, or approved equivalent standard.

Welding procedures and welder qualifications should be in accordance with AWS D1.1, latest edition, or approved equivalent standard.

All welds should be visually inspected and be free of deleterious defects. Adequate examination of production welds should be carried out by appropriate nondestructive testing.

Tubular inserts should be bent using purpose-built equipment. The manufacturer should confirm that the bend radii given on the contract drawings are suitable. Surface cracking and wrinkling on the inside surface of the bend should not be acceptable.

The manufacturer should ensure that the necessary controls are placed on the heat treatment condition, and other relevant parameters, of the tube material so as to permit satisfactory bending of inserts.

Where inserts are preheated prior to pouring of aluminum alloy, the manufacturer should provide appropriate control and measurement of the insert temperature and should ensure that the specified standard of cleanliness is maintained.

Inserts should be held rigidly during casting.

4.19.5.4 Insert Position

The position of the insert in the platform anode casting should be within 5 mm of the design position as an average over the anode length and not >10 mm from the design position at any point.

4.20 Requirements

4.20.1 Casting Quality

Each anode should be cast in one pour. Intermediate solidification's are not acceptable. Cold shuts or surface laps should not be permitted.

Anodes should be topped up prior to final solidification to present a top cast surface free from shrinkage cavities and depressions. Overpouring after solidification should not be permitted.

No grinding or other surface treatment should be permitted except that which should be carried out to remove mold flashing or other protrusions that might present a safety hazard.

All protrusions detrimental to the safety of personnel during handling should be removed.

The anode surface should be free from cracks that may reduce the performance of the anode.

Any cracks that follow the longitudinal direction of anodes are not acceptable.

Smaller cracks in the transverse direction may be acceptable provided the cracks would not cause any mechanical failure during service of the anode. The combination of cracks and lack of bond to the anode core is detrimental.

For transverse cracks, the following limits should be used:

- Cracks with a length of <50 mm and width <5 mm are normally acceptable.
- Cracks with a length of 50–200 mm should be limited to two per anode face or four per anode.
- Cracks with a length >200 mm or those that are >5 mm in width are not acceptable.

The anode surface should be free from any significant slag/dross inclusions or anything that may be considered detrimental to the satisfactory performance of the anodes. There should be no visible disbondment from the insert surface.

Shrinkage depressions should not be >100 mm in length, 12 mm in width, and 10 mm depth.

Casting surface irregularities should be fully bonded to the bulk anodic material.

Not more than 1% of the total surface of the anode casting should be contaminated with nonmetallic inclusions visible to the naked eye.

Reduction in the cross-section of anodic material adjacent to the emergence of inserts should not exceed 10% of the nominal anode cross-section.

Anode inserts protruding from the anodes should be smooth and free of sharp edges.

4.20.2 Weight and Dimensional Tolerances

The weight tolerance on individual anodes should be within ±3% of the specified net weight.

The total contract weight should not be >2% above and not below the nominal contract weight.

Dimensional tolerances should conform to the following:

- Diameter of cylindrical anodes: ±7.5
- Width: ±5%
- Depth: ±10%
- Length (for anodes >1 m long): ±25 mm (for shorter anodes): ±2.5%

The straightness of the anode should not deviate >2% of the anode nominal length from the longitudinal axis of the anode.

The anode insert location within stand-off anodes should at least be within ±5% of the nominal position in anode width and length and within 10% of the nominal position in anode depth.

Anode insert cross-section dimensions should comply with the appropriate specification for the insert material used.

4.20.3 Performance Requirements

4.20.3.1 Mechanical Resistance

When a uniaxial force equivalent to 100 kg, or five times the mass of the anode metal, whichever is the greater, is applied to a cast anode containing a core, no movement of the core or any metal insert should occur.

4.20.3.2 Electrical Resistance

When determined in accordance with the test method, the electrical resistance of the anode to core should not be >0.01 ohm for the life of the anode.

4.20.3.3 Consumption Rate in Seawater

When determined, the consumption rate of an aluminum anode in seawater should not be >3.5 kg/Ayear.

4.20.3.4 Closed Circuit Potential in Seawater

When determined in accordance with instructions, the closed-circuit potential of an aluminum anode should be at least -1.05 V with respect to a silver/silver chloride reference electrode.

4.20.4 Identification of Anodes

Each anode should be clearly marked with the type of material (trade name), the cast number, and a piece of serial number. The numbers of any rejected anodes should not be used again for replacement anodes.

4.20.5 Anode Protection

Before storage outside, the open ends of the platform anode inserts should be sealed against water ingress by the use of nonmetallic end caps.

The supplier should notify the purchaser of any extra requirements to be carried out to prevent deterioration of the anodes and inserts.

4.21 CP Cables

This part covers the minimum requirements for the materials, construction, dimensions, packaging, and tests of cables, intended for DC services in CP installation for pipelines, well casings, storage tanks, ships, and other buried or water-submerged metallic structures.

4.21.1 Conductor

The conductors should be plain annealed stranded copper. Stranding should be circular (noncompacted).

The material should be copper of such a quality and purity that the finished product should have the properties and characteristics prescribed in IEC Publication 228 Class 2, or ASTM specification B8 Class B Conductor sizes, as will be specified by the Purchaser, should be in accordance with one of the designations listed in Table 4.22 to achieve the resistance value required.

The size, cross-sectional areas, and DC resistance of the conductor in the completed cable should conform to the minimum and maximum values allowed by the standards.

4.21.2 Insulation

Cable insulation should be made from materials that are chemically and physically resistant to the environmental effects to be anticipated in buried or submerged service. It should provide continuous coverage, adequate dielectric properties, and have a high resistance to abrasion, stress cracking, and notch propagation.

Table 4.22 Conductor Sizes

Nominal Cross-Sectional Area, mm^2	Minimum Number of Wires in Conductor	Maximum Resistance of Conductor at 20 °C ohm/km
4	7	4.52
6	7	3.02
10	7	1.79
16	7	1.13
25	7	0.715
35	19	0.514
50 (The actual cross-sectional area is ~47 mm^2).	19	0.379
70	19	0.262

The insulation should be applied tightly to the conductor without adhering to it and should form a compact and homogeneous body.

The sheath (jacketing) should be continuous and have a thickness as uniform as possible and not less than the values specified. It should be possible to remove the sheath of the cable without damaging the insulation of the conductor.

Cable insulation (primary insulation and sheath) should be completely free of cracks, nicks, scratches, or other discontinuities.

Cable insulation should have a voltage rating of 600/1000 V in accordance with the National Electrical Code.

4.21.3 Insulating Materials

The types of insulation and sheath compound covered by this Part of the Standard specification are listed below.

- **Polyvinyl chloride**

Insulation and sheath compound based on PVC or copolymer of vinyl chloride and vinyl acetate conforming to ASTM specifications D 2219 (as Insulation) and D 1047 (as Jacketing).

- **High molecular weight polyethylene**

Insulation and sheath compound based on thermoplastic polyethylene conforming to ASTM specifications D 1351 (as Insulation) and D 2308 (as Jacketing).

The polyethylene before application to the conductor (or cable) should comply with the requirements of ASTM specification D 1248 for type I, Class A, B, or C; category 5; grade E 5 or J 3.

- **Cross-Linked polyethylene**

Insulation compound based on chemically XLPE conforming to ASTM specification D 2655. The base polymer of this insulation consists substantially of polyethylene or a polyethylene copolymer.

- **PVDF**

Insulation compound based on specially chemical-resistant modified PVDF conforming to ASTM specification D 3144.

4.22 Specific Cable Requirements

4.22.1 *Positive Conductor Cable*

Cable for use as a positive conductor should be single core, stranded soft annealed copper.

Cable insulation, as will be specified by the Purchaser, should be either standard or chlorine resistant type and should conform to one of the following alternatives:

- **Standard type cable**

Cable insulation should be in accordance with one of the following alternatives, as will be specified by the Purchaser.

- A single jacket of HMWPE,
- A double jacket of HMWPE primary insulation/PVC sheath,
- A double jacket of XLPE primary insulation/PVC sheath.

The average thickness of cable insulation should be 2.8 mm for conductor sizes up to 35 mm^2 and 3 mm for larger sizes. The minimum thickness at any point should not be $<90\%$ of the specified average thickness. In the case of a double jacket, the thickness of the sheath should not be <1.5 mm.

Unless otherwise specified by the Purchaser, the color of cable insulation should be fade-resistant red.

- **Chlorine resistant type cable**

(To be used as positive cable in saline water and salty soil ground beds).

Cable insulation should be in accordance with one of the following alternatives, as will be specified by the Purchaser.

- A single jacket of PVDF, single or double extruded with the minimum thickness of 0.75 mm.
- A double jacket of a radiation XLPVDF primary insulation/HMWPE sheath (minimum thickness $0.5 + 1.7 = 2.2$ mm).

Unless otherwise specified by the Purchaser, the color of cable insulation should be fade-resistant red.

4.22.2 *Negative, Bond, or Test Conductor Cable*

All cables for use as negative conductor, bond conductor, or test conductor should be single core, stranded soft annealed copper.

Cable insulation should be in accordance with one of the following alternatives:

- A single jacket of HMWPE.
- A single jacket of XLPE.
- A double jacket of PVC primary insulation/PVC sheath.

Minimum thickness of cable insulation should be 1.0 mm for conductor sizes $\leq 6.0\ mm^2$, 1.5 mm for conductor sizes up to 25 mm^2 and 2 mm for larger sizes.

Unless otherwise specified by the Purchaser, the color of cable insulation should be fade-resistant black.

4.23 Quality Assurance Provisions

It is the responsibility of the manufacturer to establish quality assurance by quality control procedures that should ensure that the product will meet the requirements of this Part of Standard specification.

The quality control at the manufacturing plant should include control systems on the following:

- Raw materials, that is, checking of documentation and/or testing for conformity with the specification.
- Production equipment and process.
- Testing during and after production.
- Identification of cables.
- Dimension tolerances.
- Documentation (material certificates).

4.24 Tests

Tests should be carried out to confirm that the materials comply with this Part of the Standard specification and all cables should satisfactorily pass such tests.

The factory acceptance tests should be done on test samples taken from the end of each manufactured length (drum) at a temperature of 20 °C and should cover at least the following:

- Dimensions and construction.
- Dielectric strength of insulation.
- Insulation resistance.
- Electrical resistance of conductor.
- Voltage test.

All tests should be carried out in accordance with relevant specifications.

4.25 Fabrication and Inspection of Monolithic Insulating Joints

This Part covers the minimum requirements for the design, fabrication, and inspection of monolithic insulating joints, for electrical isolation of pipelines from plants and/or other pipelines.

The purpose of insulating joints is to provide electrical isolation between sections of pipelines to prevent detrimental electrochemical interaction between the sections. Insulating joints are used also to ensure effective current distribution for CP systems. The insulating joint is intended to be installed under/above ground or above water. It is intended to be girth welded between two pipeline sections.

The technical bid usually include the following:

1. Comprehensive catalogs, technical data, general arrangement drawings showing outline dimensions, proposed test procedure, etc. of the offered insulating joint(s).
2. Detailed drawings showing all parts with material identification and stress/design calculations.
3. Welding procedure specification.
4. Material specifications for the following items:
   a. Steel forgings
   b. Insulating rings
   c. Seal Gaskets
   d. Epoxy resin filler material
   e. External coating
   f. Internal coating
5. Assembly procedure.
6. Records of tests.
7. Declaration of confirmation with this Part of the Standard specification and/or clear indication of deviations from relevant standard specification.
8. Specification of packing.

4.25.1 Design

All insulating joints should be of monolithic type fabricated by welding and with pups on either side. The insulating joints should be suitable for pigging operations.

The design of the insulating joint(s) should be compatible with the design code adopted for the pipeline in which it is installed, as stated in the data sheet.

ANSI/ASME B 31.4 and B 31.8 are commonly adopted pipeline codes. In some situations, for example, within plant fences, the pipeline sections connected to the insulating joint may be designed to a plant piping code, for example, ANSI/ASME B 31.3.

The insulating joint(s) should be designed so that its bending moment along with projection pressure to be able to generate a longitudinal stress equals to 75% of the adjacent pipe pups.

The insulating joint(s) should be designed following the methodology described in ASME VIII Division 1 Appendix 2 or equivalent, using the design pressure, design temperature and design factor of the pipeline, and the anticipated external loads acting on the insulating joint. Stress due to internal design pressure should not be >50% of specified minimum yield strength of the material.

The insulating joint(s) must be assembled so that its internal components are tightly fixed in the required position. To this purpose, an assembling precompression, adequately calculated, should be applied.

The number of weldings must be as restricted as possible. All joints should be butt welded and have full penetration except for the final closure weld. The final closure weld should be in accordance with ASME Section VIII Div. 1 ULW 17.6.

For sour service conditions (when specified), the materials should comply with the requirements of NACE Standards MR 01 75 and TM 01 87.

Primary and secondary sealing gaskets should be so designed that the joint assembly complies with the mechanical and electrical requirements specified in relevant standard specification. The seal gaskets must oppose the fluid's internal pressure at the same height from the two sides of the rigid insulating ring.

Sealing systems that foresee *o*-rings cannot be accepted.

The gaskets should be made of an age-resisting elastomer of a high-dielectric characteristic. It should resist explosive decompression, and should be suitable for long-term exposure to the transported fluid at the design pressure and temperature conditions.

The insulating materials and filler materials should be suitable for the long-term exposure to the transported fluid at the design pressure and temperature conditions.

The insulating material should be nonhygroscopic and should be of sustaining high compressive stresses. The material should have good long-term stability, excellent dielectric strength, and thermic properties without cracking, distorting, or a loss of insulating properties. The material should be flame resistant, and also resistant to hydrocarbons, sulfides, bacteria, and the climatic conditions.

The internal diameter of the joint should not deviate by $>\pm 2\%$ from the nominal internal diameter of the pipeline.

Construction should not incorporate any flanges, boltings, gaskets, and should be of rigid and rugged design suitable for field installation without a maintenance pit when installed either on a buried or surface laid pipeline.

Tension and compression should be axially transmitted satisfactorily and without appreciable elongation or contraction.

The joint should indicate no unacceptable deformation when under maximum pressure and temperature.

The manufacturer should take into consideration that, during the tie-in to the pipeline, the rings of the rigid insulating material, must not exceed a temperature of 70 °C.

The minimum overall lengths for the joints (as measured between ends) of a given size and pressure class should be as in Table 4.23.

Note: For pipelines with a wall thickness of >25 mm, the manufacturer/supplier should confirm that the length of the joint is suitable for postweld heat treatment of the field welds, that is, the heat generated by the treatment is not detrimental to the isolating and filler materials and the internal coating. If this is not the case, the length of the joint should be increased accordingly.

4.25.2 Materials

Material for the pressure containing parts of the insulating joints should meet the requirements of data sheet.

Table 4.23 The Minimum Overall Lengths for the Joints (as Measured between Ends) of Given Size and Pressure Class

	ANSI Class Rating					
Pipe Size DN	**150**	**300**	**400**	**600**	**900**	**1500**
Up to 350	700 mm	700 mm	1000 mm	1000 mm	1000 mm	1000 mm
400–600	1000 mm	1000 mm	1200 mm	1500 mm	1500 mm	1500 mm
650–900	1500 mm	1500 mm	1800 mm	2000 mm	2000 mm	2000 mm
950–1200	2000 mm	2200 mm	2200 mm	2500 mm	2800 mm	2800 mm
1250–1500	2500 mm	2800 mm	3000 mm	3000 mm	3200 mm	3200 mm

The metallic parts of the insulating joint should be of forged steel conforming to the requirements of MSS-SP-44.

The ladle analysis of each heat of steel used in the manufacture of insulating joint should conform to the material degree that is declared to be used with the restrictions.

Forged, fully killed, normalized, or hardened and tempered materials will have to be used according to the schedule here below in Table 4.24.

The carbon equivalent calculated with the following formula should not be >0.42%:

$CE = C + Mn/6$
CE = Carbon Equivalent
C = Carbon
Mn = Manganese

Materials, quality, and size of the pups should be the same as or better than those of the pipe to be connected.

Insulating materials should be made of age-resisting laminated epoxy resin-reinforced with glass fiber conforming to ASTM specification D 709 Type IV Group G 10/G 11 or equivalent standard.

The primary and secondary sealing gaskets should be made of fluorinated elastomers like polytetrafluoroethylene and/or Butadiene acrylonitrile copolymers such as Nitrile rubber according to ASTM specification D 2000, or equivalent standard.

Table 4.24 Forged, Fully Killed, Normalized, or Hardened and Tempered Materials

Approved Stub Material	Forged Material
API 5L Gr. A	ASTM A 105
API 5L Gr. B	ASTM A 105
API 5L × 42	ASTM A 694 F 42 or F 52
API 5L × 52	ASTM A 694 F 52
API 5L × 60	ASTM A 694 F 60
API 5L × 65	ASTM A 694 F 65

Filling materials should be of an adhesive sealant elastomer having a compressive strength >1500 kg/cm^2.

4.25.3 Manufacture

The manufacturer should be responsible for the integrity of the manufacturing procedures and adherence to this Part of Standard specification.

The assembly of insulating joint(s) should be done in a controlled, clean environment.

The insulating joint(s) should be assembled in such a way that its various components are firmly locked in position and the complete joint is capable of withstanding stresses due to designed operating conditions and field hydrostatic testing.

The surface of the annular space between the pup and the retainer should be abrasive shot blasted to SA 2½ grade in accordance with ISO 8501-1. Assembly of the insulating joint should commence within 2 h after the completion of abrasive shot blasting.

The annular space between the retainer and the pup should be filled with an epoxy resin filler material, to be approved by the Purchaser's representative.

Any depressions in the insulating joint should be filled with an adhesive sealant elastomer to prevent the deposition of sediment at this point.

Insulating joint(s) should not be of harmful defects specified in paragraph 14.5.2 of MSS-SP-75 and the other defects in the surface should be removed according to paragraph 14.5.3 of MSS-SP-75.

All butt welds should be heat treated after all welding has been completed in accordance with the requirements of ASME Boiler and Pressure Vessel Code, Section VIII.

Lamination found at the bevel end parts should be removed wholly. The height of internal weld beads should be <2 mm. Undercutting on weld seam should not be permitted.

The end connections should be beveled for welding in accordance with MSS-SP-75, unless otherwise specified by the Purchaser, to allow welding to the pipeline with only two circumferential butt welds. Welding bevels should be dry and free from grease and other contaminants.

For a distance of 100 mm from the welding ends, the internal diameter should not deviate by > ±1.5 mm from the nominal internal diameter of the pipeline.

Heat treatment for insulating joint(s) should be performed according to ASME VIII, Division I and related ASME specification.

In this case, the temperature in the furnace should be recorded and controlled continuously during processing, and record chart should be submitted to purchaser. When required, before manufacturing, supplier should submit the heat treatment specification described following items:

- Heat-treatment procedure,
- Standard temperature curve chart,
- Temperature control range.

4.26 Approved Welding Processes

Any of the following processes may be used:

1. Shielded Metal Arc Welding (SMAW);
2. Gas Tungsten-Arc Welding (GTAW or TIG);
3. Gas Metal-Arc Welding (GMAW, or MIG);.
4. Submerged Arc Welding (SAW);
5. Combinations of above processes.

All joints should be butt welded and have full penetration welds except for the final closure weld, which should be welded in accordance with ASME VIII ULW 17.6, Figure A.

4.27 Coating

After all tests and inspections required have been carried out, insulating joints should be thoroughly cleaned to remove rust or impurities. Surface preparation should be carried out by shot blasting or sand blasting to SA 2½ grade in accordance with ISO 8501-1. After visual inspection, all insulating joints should be externally and internally coated with an amine-cured epoxy paint to a dry film thickness of 500 μm applied in two coats. The application should be in strict accordance with the coating supplier's specification. Beveled ends for 50 mm approximately should be painted with rust preventive paint.

Note: Epoxy internal coatings may not be suitable for some chemical products. When other coatings need to be specified, advice from a material specialist should be taken. Metallic pigmented coatings should not be used.

The coatings should be subject to holiday detection using a wet sponge tester set at 60 V. The acceptance level should be zero pinholes.

4.28 Quality Assurance Provisions

It is the responsibility of the manufacturer to establish quality assurance by quality control procedures, which should ensure that the product will meet the requirements of this Part of Standard specification.

The quality control at the manufacturing plant should include control systems on the following:

- Raw materials, that is, checking of documentation and/or testing for conformity with the specification.
- Production equipment and process.
- Testing during and after fabrication.
- Checking of dimensions.
- Documentation (material certificates).

As an alternative to the above control system, the manufacturer can have a quality assurance system in accordance with ISO 9002.

4.29 Tests

Tests should be carried out to conform that the materials comply with this Part of Standard specification and all insulating joints should satisfactorily pass such tests (Table 4.25).

4.29.1 Chemical Analysis

Chemical analysis should be performed by the ladle analysis for each heat of steel used and should conform to the requirements of the chemical composition for the respective materials as per relevant standard.

4.29.2 Mechanical Test

Tensile test is performed for each lot after heat treatment.

The same charged, shaped and heat-treated products are considered one lot.

1. Forgings

Tensile tests should be performed according to ASTM A 370 code corresponding to equivalent material and the results should comply with the mechanical requirements in this Part of Standard specification.

2. Pups

Tensile tests should be performed according to Section 3, Mechanical Properties and Tests of API 5L. No leakage or pressure drops should be permitted during the test.

Table 4.25 Items of Test and Inspection

Test Item	Manufacturer	Purchaser Witness	Record
Chemical analysis	Operation & record	Document check	Operation & record
Mechanical test	Operation & record	Document check	Operation & record
Hydraulic test or air tightness test	Operation & record	Operation & record	Operation & record
Radiographic test	Operation & record	Document check	Operation & record
Coating inspection	Operation & record	Document check	Operation & record
Dielectric test and Megger test	Operation & record	Operation & record	Operation & record
Ultrasonic test and magnetic particle test	Operation & record	Document check	Operation & record
Dye penetration test	Operation & record	Document check	Operation & record
Visual inspection	Operation & record	Operation & record	Operation & record
Dimension check	Operation & record	Operation & record	Operation & record
Heat treatment	Operation & record	Document check	Operation & record

4.29.3 Dielectric Test and Megger Test

Upon completion of the assembly and prior to the hydrostatic test, each insulating joint should be megger tested with 1000-V DC. Minimum resistance should be 25 mega-ohm.

Each insulating joint should also be dielectric tested at 5000-V alternating current (50 Hz) for 1 min. There should be no corona effects or breakdown.

The above-stated tests should be repeated after the hydrostatic test. The readings before and after the hydrostatic test should be equal.

4.29.4 Hydrostatic Test

The insulating joint should be hydrostatically tested, in an unrestrained condition, using water with a low surface tension, at 1.5 times the rating (permissible working pressure).

In the test, the insulating joint should be subjected to the full axial load deriving from the hydrostatic pressure unless it has been verified in a type test that the insulating joint is able to resist the full axial load in service.

Unless otherwise specified by the Purchaser, the maximum test pressure should be maintained for at least 3 h. No leaks or unacceptable deformation should occur during the test.

4.29.5 Low Pressure Leakage Test (Air Test)

After the hydrostatic test, a low-pressure leakage test should be carried out on each insulating joint. The joint should be filled with air and will be pressurized to 0.5 MPa and held at that pressure for 10 min. No leakage or pressure drops should be permitted during the test.

4.29.6 Prototype Tests

Prototype tests should be required if the manufacturer cannot provide data of similar tests to the satisfaction of the Purchaser.

Prototype joints, representative of production, selected for the test should be identified as to material, grade, and lot, including heat treatment, hydrostatic cyclic pressure tests, hydrostatic pressure plus bending test, and vacuum test where appropriate.

4.29.7 Radiographic Test

All butt welds should be 100% radiographed by X-rays in accordance with Section V of the ASME Boiler and Pressure Vessel Code Nondestructive Examination.

4.29.8 Magnetic Particle and Ultrasonic Test

The final closure weld should be inspected by ultrasonic or magnetic particle methods in accordance with ASME Section VIII.

All finished bevel ends should be 100% ultrasonically tested for lamination-type defects for a distance of 50 mm from the ends. Lamination should not be acceptable.

The following are to be noted:

1. The ultrasonic examination method should be applied before preparing weld edges, especially for areas of a 25-mm width around the edge.
2. The magnetic particle examination should be made only after the completion of weld edge preparation.

Acceptance criteria should be as per ASME Section VIII Div. I, Appendix 12.

4.29.9 Dye Penetrant Test

Dye penetrant test should be performed on the machined surface according to ASME Section VIII, DIV. I. Appendix 8.

4.29.10 Visual Inspection and Dimensional Check

Visual inspection of all parts prior to assembly and after finishing of the insulating joint should be carried out by the manufacturer. General appearance should show good workmanship.

Dimensions of the insulating joint(s) should be checked against the specifications, purchase order description, and/or approved manufacturer drawings.

The inside and outside thickness of coating should be checked. The coating appearance should be even on the whole surface.

4.30 Flange Insulation Kit

This Part covers the minimum requirements for flange insulation kit, complete, intended for electrical insulation of standard, flat face, raised face, or ring joint type, flanges used for cathodically protected steel pipes and forming part of corrosion protection of the structure.

Each kit should include the following components:

1. Insulating central gasket,
2. Insulating bolt sleeves,
3. Insulating bolt washers,
4. Steel bolt washers.

It is to be noted that the insulating sleeve and washer may be supplied as one-piece.

Gaskets should be manufactured from material having low water absorption and high compressive strength. Preference should be given to materials with low y and m factors. The y factor is a measure of the compressive load required to establish an initial seal, while the m factor is an indication of the additional load required to hold the fluid pressure needed to keep the seal in operation. The smaller these factors are, the less bolt loading is required.

Gasket materials should withstand the expected bolt loading without injurious crushing, and which are suitable for the service conditions. It should not crack or distort under load and should be unaffected by salt water, hydrocarbons, natural gas, sunlight, or atmospheric conditions.

The following are to be noted:

The *m* and *y* factors for various materials are included in the ASME Code, Section VIII and ASME B 16.5.

Insulating bolt sleeves should be manufactured from materials having low water absorption, high dielectric strength, and low cold-flow characteristics. They should be suitable for the service conditions of the particular application.

Insulating washers should be manufactured from materials having high compressive strength, low water absorption, high dielectric strength, and low cold-flow characteristics.

4.30.1 Materials

The central gasket for flat face or raised face flanges should be made of a fabric-reinforced phenolic laminate core material (1) coated on both sides with a chloroprene polymer (i.e., neoprene) sheets.

The central gasket for ring joint flanges should be made from fabric-reinforced phenolic laminate core material (1).

Insulating sleeves should be fabricated from high-density polyethylene material, to which should be added only those antioxidants, ultraviolet (UV) stabilizers, and pigments necessary for the manufacture of sleeves to the specification and to its end use.

The fabric-reinforced phenolic laminated core material should consist of a suitable cotton fabric base or filler properly impregnated and bonded with a phenolic thermosetting resin compound or binder, processed to meet the requirements of this Part of Standard specification, which meet US Military Specification MIL-P-15,035 Type F.B.G or ASTM D 709 Type II Grade CE.

The following are to be noted:

1. The choice of stabilizers and other additives is at the discretion of the manufacturer.
2. Molding compounds of an unknown composition should not be used. Impurities that are occasionally contained in polymers should not be $> 0.1\%$ by mass.

Insulating washers, except when cast integrally with sleeves, should be fabricated from fabric-reinforced phenolic material and should be 3.2 mm thickness (nominal).

Combined insulating sleeve/insulating washer should be fabricated from injection molded acetal copolymer or equivalent in the form of a one-piece casting.

Steel washers should be fabricated from plated mild steel as per BS-3410, a 3.2-mm thickness, and should be treated to prevent corrosion.

4.30.2 Requirements

- **Dimensions**

Unless otherwise specified by the Purchaser, gaskets for flat face or raised face flanges should conform to the dimensions specified in ANSI B 16.21. They should have a nominal thickness of 3.2 mm and protrude into the bore of the pipe by 1.5 mm to prevent electrically conductive bridging over the insulation by debris, etc.

Ring gaskets for ring-joint flanges should be of dimensions established in ANSI B 16.20. Sheet and ring joint gaskets for sizes NPS 10 and smaller should conform to ANSI B 16.5 dimensions and recommendations.

Insulating bolt sleeves are normally designed for standard bolting in standard bolt holes and should be of sufficient length to extend halfway inside the steel washer. Care should be taken to ensure that the dimensions selected will allow the use of the standard size bolt or one size smaller high tensile strength bolt.

Insulating bolt washers should be sized internally so that the bolt sleeve will pass through, and the OD should be sized so that the washer will fit inside the flange spot facing.

Steel washers should be the same size as the insulating washer.

- **Uniformity**

The materials should be uniform in texture, finish, and specified properties.

- **Surface Defects**

The materials should be free from blisters, wrinkles, air marks, or cracks, and reasonably free from other small defects such as scratches, dents, and heat marks.

- **Workmanship**

All materials and workmanship should be in accordance with good commercial practice, and the resulting products should be free from porous areas, air pockets, foreign matter, or other defects affecting serviceability.

4.30.3 Detail Requirements for Insulating Gaskets and Sleeves

- **Gaskets**
- **Type**

The type of insulating gaskets provided for herein that are suitable for use with the flange facing indicated are given in Table 4.26:

- **Sizes**

The gasket nominal size should be the same as used on the standard pipe flanges and flange fittings in accordance with ANSI B 16.21 and/or ANSI B 16.20.

Table 4.26 The Type of Insulating Gaskets Provided for Herein that are Suitable for Use with the Flange Facing Indicated are as Follows

Gasket Type	Flange Facing
Full face	Flat face or raised face
Self-centering flat ring	Raised face
Ring joint (oval or octagonal)	Ring-joint

- **Property values**

Gasket materials should conform to the requirements for physical and electrical properties prescribed in Table 4.27.

- **Machinability**

The material should be such that it can be drilled, tapped, sawed, and machined in all directions in accordance with the manufacturer's recommended technique without cracking, splitting, or otherwise impairing the material for general use.

- **Sleeves**

The insulating sleeves should be made of high-density polyethylene and should meet the requirements of Table 4.28.

Table 4.27 Properties of Insulating Gasket Materials

Characteristic	Unit	Requirement	Test Method ASTM
Thickness (nominal)	mm	3.2 ± 0.15	D 229
Flexural strength, Flatwise (min.)	MPa		D 229
LW		115	
CW		95	
Impact strength (Izod, edgewise) (min.)	J/m of notch		D 229
LW		86.4	
CW		75.6	
Dielectric breakdown parallel to laminations (step-by-step test), (min.)	kV	25	D 229
Dielectric strength perpendicular to laminations (step-by-step test) (min.)	kV/mm	8	D 229
Water absorption (max.)	%	1	D 229

LW: Tested in a lengthwise direction.
CW: Tested in a crosswise direction.

Table 4.28 Properties of Insulating Sleeves

Characteristic	Unit	Requirement	Test Method ASTM
Material	–	Polyethylene	–
Density (min.)	g/cm^3	0.941	D 1505
Thickness (nominal)	mm	0.8	D 374
Dielectric strength (min.)	kV/mm	16	D 149
Water absorption (max.)	%	0.01	D 570
Tensile strength (min.)	MPa	19	D 638

4.30.4 Packaging

The kits purchased according to this Part of the Standard specification should be packaged in suitable containers to ensure acceptance and safe delivery to their destination. Individual kits should be packaged in such a manner as to protect the material against physical and mechanical damage and contamination during shipment, handling, and storage. Each container should contain application instructions. All packaging and packing provisions of ASTM practice D 4088 should apply to this Part of the Standard specification.

4.30.5 Quality Assurance Provisions

It is the responsibility of the manufacturer to establish quality assurance by quality control procedures that should ensure that the product will meet the requirements of this Part of Standard specification.

The quality control at the manufacturing plant should include control systems on the following:

- Raw materials, that is, checking of documentation and/or testing for conformity with the specification.
- Production equipment and process.
- Testing during and after production.
- Identification of materials.
- Checking of dimensions.
- Documentation (material certificates).

4.30.6 Sampling

- **Lot**

The lot or batch should consist of an indefinite number of finished products offered for acceptance, of materials manufactured by a single plant run through the same processing equipment, with no change in ingredient materials.

Unless otherwise specified by the Purchaser, the number of samples for inspection and/or testing should consist of 10 percent of the lot, but in no case should be less than one or more than ten kits.

Sampling should be in accordance with the ASTM test methods for the specific properties to be determined.

The numbers and types of test specimens should be in accordance with the ASTM test method for the specific properties to be determined.

4.31 Polypropylene Rope

This Part covers the minimum requirements for three-strand (hawser laid) rope made of polypropylene for use in dry and water-type deep well ground beds.

The technical bid should include the following:

1. Catalog.
2. Rope size, quantity (net and gross mass), material, and construction.

3. Main characteristics and tolerances.
4. Production procedures that cover all stages of the manufacturing process, from receipt of raw materials to finishing and packing for shipment, including
   a. Measurement of diameter, weight, etc.
   b. Inspection and testing procedure, including acceptance criteria, and frequencies.
   c. Rope manufacture.

4.31.1 Requirements

- **Materials and Manufacture**

Only virgin polymeric material should be used. The polymeric material used in manufacturing the rope should be homopolymers or copolymers of polypropylene processed as continuous multifilament, having a relative density of approximately 0.91 to produce rope having the characteristics specified in Table 4.29. The rope should be stabilized against deterioration due to sunlight by the addition to the polymer of an additive or additives.

Note: Any Ultraviolet (UV)-inhibiting system may be used, such as pigmentation using carbon black, Iron (III) oxide (Fe2O3) or any other coloring product or special UV inhibitor.

It should be resistant to sodium chloride solution when tested in accordance with ASTM method D 543.

- **Flexibility and Uniformity of Strands**

The finished rope should be flexible and should consist of uniform strands, well laid, and as free as practicable from defects in the yarn, strand, and finishing.

- **Linear Density**

The rope should have a linear density (net mass, in grams per meter) given in Table 4.29, measured when the rope is subjected to the tension called the reference tension, within a tolerance of ±10%.

Note: Reference tension to be applied to the rope, conforming to this Part of Standard specification, when measuring linear density and lay should have the nominal value of 8 daN (deca newton) within a tolerance of ±5%.

Table 4.29 Requirements for Polypropylene Rope

Characteristic	Unit	Requirement	Test Method
Nominal diameter	mm	8	ISO 2307
Linear density	ktex	30	ISO 2307
Rope lay (max.)	mm	28	ISO 2307
Breaking force (min.)	decanewtons (daN)	1040	ISO 2307
Color	–	Brown	–

Notes:
(1) The linear density (in kilotex) corresponds to the net mass per meter (in grams per meter) or to the mass of rope (in kilograms) per 1000 m. The linear density (net mass per meter) is measured under tensile loading given in ISO 2307.
(2) The SI unit for force is the newton. A load of 1 kg ≈ 1 daN (daN) or 0.01 kN (kN).

- **Diameter**

The diameter of the rope should not vary by $>\pm4\%$ from that specified in Table 4.29, measured at zero tension.

- **Securing Rope Ends**

The ends of the rope should be secured to prevent unlaying.

- **Finish**

Weighting and loading materials should not be added.

- **Yarn in the Strands**

Each strand should consist of a sufficient and equal number of rope yarns to produce a rope that should comply with the requirements given in Table 4.29. The rope yarns should not contain fibers that have been used or recovered.

- **Structure**

The rope and their strands should be continuous throughout its length and should not contain loose ends, splices or joints.

- **Lay**
- **Direction of lay**

The direction of lay should be Z twist (right hand lay), their strands S twist, and their roping yarns Z twist (Fig. 4.9).

Figure 4.9 Three-strand (plain or hawser laid) "Z" or right-hand lay rope.

- **Length of lay**

The maximum lay of the rope should be as given in Table 4.29, measured with the rope under a specified tension.

4.31.2 Other Characteristics

Other characteristics, for example, the diameter of the circumscribed circle and the elongation of the rope under specific tensile conditions, may be specified, subject to agreement between the manufacturer and the Purchaser, where applicable with submission of a sample.

4.31.3 Quality Assurance Provisions

It is the responsibility of the manufacturer to establish quality assurance by quality control procedures that should ensure that the product will meet the requirements of this Part of the Standard specification.

The quality control at the manufacturing plant should include control systems on the following:

- Raw materials, that is, checking of documentation and/or testing for conformity with the Specification.
- Production equipment and process.
- Testing during and after production.
- Identification of ropes.
- Weight and dimension tolerances.
- Documentation (material certificates).

4.31.4 Splicing Kit

This Part covers the minimum requirements for materials supplied in kits for making cable joints, in the voltage range up to 1000-V DC, for use in coating of buried or submerged electrical connections in CP systems.

4.31.5 Description

Each splicing kit should contain the following components:

- Two-piece snap-together plastic mold.
- Two-pack catalyzed resins (resin and hardener).
- Funnel/funnel cap.
- Proper electrical insulating and sealing tapes.
- **Types**

The splicing kits, as will be specified by the Purchaser, should be one of the following types:

- “Branch” type: Intended for use in shallow type ground beds.
- “Straight-through” (in-line) type: Intended for use in deep-well type ground beds.

4.31.6 Materials and Manufacture

- **Cast mold**

The two-piece snap-together mold should be made of synthetic thermoplastic material (i.e., polycarbonate), cast into a desired shape designed to provide a transparent, self sealing, water proof, and high strength product. The mold size should be such as to form a water-resistant seal over the connection giving a voltage rating of 600/1000 V. It should withstand the weight and exotherm of the compound without undue distortion, such that when supported in accordance with the manufacturer's instructions the design clearances are maintained during setting.

Cast molds should be provided with filling apertures to allow easy filling with resin compound and should be designed such that the filling operation does not entrap air within the joint.

The molds should be designed to permit a minimum of a 10-mm thickness of cured compound over each joint component and cable component involved in the jointing procedure. The mold design should allow the longitudinal seal on each cable sheath to be of a minimum length of 40 mm.

- **Catalyzed resins**

The catalyzed resins (cold pour resin compound) should be an unfilled two-component Epoxy or Polyurethane resin conforming to DIN VDE 0291 Part 2, and this Part of the Standard specification which, after mixing, sets quickly to form a stable, void-free resilient molded parts by casting in molds for low-voltage cables rated up to 1 kV.

The base component not including curing systems should contain no fatty acids, oils, chlorinated-type plasticizers, or polystyrene resins. The resin compounds should contain no toluene diisocyanate and no aromatic substituted isocyanates.

4.31.7 Properties of Catalyzed Resins

The catalyzed resins should comply with the requirements given in Table 4.30.

- **Odor**

The odor of the admixed cast resin, wet or dry, should not be obnoxious.

- **Working properties**

The cold pour resin compound should be fresh and will show no signs of partial cure from having been improperly stored, nor should it possess other detectable defects that adversely affect the ultimate performance of the material.

The fully cured casting resins should not be attacked by oil, petrol, water, salt, and salt solution.

- **Toxicity**

The manufacturer should certify that the materials should have no adverse effect on the health of personnel when used (mixed, applied, and cured).

Table 4.30 Properties of Catalyzed Resins (Cold Pour Compound)

Property	Unit	Requirement	Test Method
Cast resin component open cup flash point (min.) reactant	°C	>100	ISO 2592
Cast resin component open cup flash point (min.)	°C	>55	ISO 2592
Gel time; at 23 °C (max.)	Minutes	30	DIN 16945
Total volume shrinkage (max.)	%	4	DIN 16945
Hardness (min.)	Shore D	70	DIN 53505
Water absorption (max.) in cold water after 24 h at 23 °C	mg	25	ISO 62
Electrolytic corrosion (at least)	Grade	A1.2	DIN 53489
Dielectric strength (min.) 1-min test voltage; at 23 °C	kV	10	ASTM D 149
Electrolytic corrosion (at least) dielectric strength (min.) 1-min test voltage; at 80 °C	kV	10	ASTM D 149
Volume resistivity (min.) at 23 °C	ohm cm	1 × 10^12	ASTM D 257

- **Shelf life**

The two-pack catalyzed resins when stored in the original container at an ambient temperature, between 10 and 45 °C, should have a minimum shelf life of 24 months from the date of delivery and should meet all the requirements specified herein. After such storage, the catalyzed resins should show no instability, hardening of the resin components, settling out of the pigment, and deterioration of the two-compartment plastic mixing bag.

- **Dimensions**

For specified cables and connectors to be joined, the dimensions of mold and the required quantity of resin compound should be submitted by the supplier.

4.31.8 Packaging

Packaging should be sufficient to ensure that deterioration of the components will not occur under covered, dry storage within the stated ambient air temperatures such that the performances of these components when forming part of the total joint are unaffected.

The splicing kits should be packaged, in cardboard boxes, in accordance with the best commercial practice.

The resin and hardener components should be packed in exact doses in a two-chamber transparent plastic bag, with an aluminum resistive layer, sealed against air and damp. The complete package should be designed to protect the item against damage during shipment, handling, and storage. The packaging/packing should also meet the relevant requirements of ASTM Standard Practice D 3951 and D 3892.

All shipments should be marked "store in a cool place out of direct sunlight."

4.31.9 Sampling

Sample splicing kits should be selected from each lot in accordance with inspection Level II of ISO 2859 (Part 1). The acceptable quality level should be 4 percent for all visual, dimensional, and weight defects combined and 2.5 percent for all other type of defects with reference to this Part of the Standard specification. The single sampling plans for normal inspection (Table II A of ISO 2859 Part 1) should be used.

4.32 Split Bolt Connector (Line Tap)

This Part covers the minimum requirements for a mechanical split bolt connector (line tap) for use in making, through or tap, connections of stranded copper conductors, 600/1000 V rating in CP installations (Fig. 4.10).

The technical bid usually includes the following:

1. Catalog (including technical data of the offered connector).
2. Production procedures that cover all stages of the manufacturing process, from receipt of raw materials to finishing and packing for shipment, including the following:
   - **a.** Chemical analysis.
   - **b.** Material specification.
   - **c.** Measurements of weight, dimensions, etc.
   - **d.** Inspection and testing procedures, including acceptance criteria, and frequencies.
   - **e.** Line tap manufacture.
   - **f.** Specification of packaging.

The split bolt connectors (line taps) should be manufactured, inspected, and tested in accordance with ANSI/UL Standard 486 A.

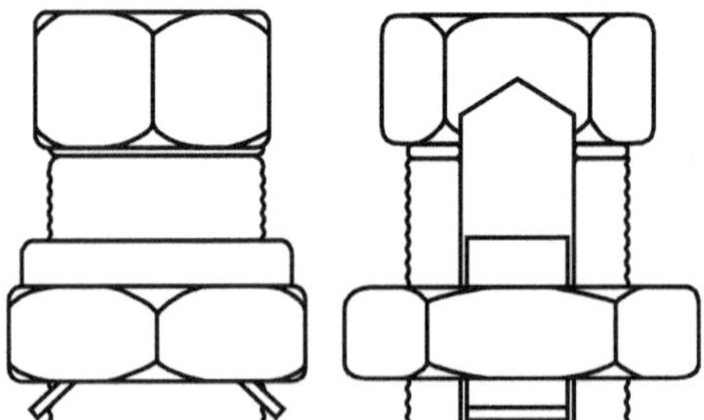

Figure 4.10 Split bolt connector (line tap).

Table 4.31 Chemical Requirements

Element	Composition, %
Copper	55–60
Lead	2.0–3.8
Iron, max.	0.35
Zinc	Remainder

4.32.1 Materials and Manufacture

The split bolt connectors as cast should be manufactured from a copper–zinc–lead alloy (leaded-brass), having a chemical composition conforming to the requirements specified in Table 4.31, tinned all over by electroplating to BS 1872 with a minimum thickness of 5 μm.

The materials should be of such a quality and purity that the finished product should have the properties and characteristics prescribed herein.

The bolt threads should be of a sufficient strength to withstand the tightening required to produce a zero resistance connection.

4.32.2 Dimensions

The split bolt connectors should be of such dimensions as to be capable of joining the specified number of cables of specified dimensions and should be suitable for fittings into the splicing kit. The supplier should submit to the Purchaser, the dimensions and contours, including dimensional tolerances of offered connectors, preferably by means of a drawing.

4.32.3 Workmanship, Finish, and Appearance

Split bolt connectors should have a workmanlike finish free of injurious burrs, seams, laps, irregular surfaces, and other imperfections affecting serviceability.

4.32.4 Quality Assurance Provisions

It is necessary to establish quality assurance by quality control procedures that should ensure that the product will meet the requirements of specification.

The quality control at the manufacturing plant should include control systems on the following:

- Raw materials, that is, checking of documentation.
- Production equipment and process.
- Testing during and after production.
- Identification of line taps.
- Chemical analysis.
- Weight and dimensional tolerances.

- Surface condition of produced line taps.
- Documentation (material certificates).

4.32.5 Splice Coating Materials

This Part covers the minimum requirements for the following coating materials as electrical insulation for use on joints and splices in wires and cables in accordance with the National Electrical Code.

1. Electrical insulating plastic tape.
2. Rubber splicing tape.

4.32.6 Electrical Insulating Plastic Tape

- **Materials and Manufacturer**

The tape should consist of a plastic backing coated on one side with a pressure-sensitive adhesive.

The backing should be either PVC (or copolymer of vinyl chloride and vinyl acetate) or PE (thermoplastic polyethylene) suitably compounded to meet the requirements of this Part of the Standard specification. The backing should be smooth and uniform. The tape edges should be straight and unbroken.

The pressure sensitive adhesive layer should be smooth and uniform and as free from lumps and bare spots as the best commercial practice will permit. "Fish Eyes" should not be cause for rejection. There should be no adhesive transfer when the tape is unwound from the roll.

The tape should provide high electrical resistivity, resistance to corrosive environments, low moisture absorption, and permeability and should form a water-tight bond with itself and with the rubber tape. Table 4.32 shows the requirements for electrical insulation of plastic tapes.

- **Deformation test**

The thickness of the insulation on a splice covered with PE tape should not decrease >40 percent when conditioned, under pressure, in a full-draft circulating-air oven operating at a temperature of $100.0 \pm 1.0\,°C$, and the corresponding decrease in thickness should not be >65 percent for PVC tape.

A bare 3.3-mm^2 (No. 12 AWG) solid copper conductor is to be wrapped with successive layers of tape, with each layer of tape directly over the one below, until a thickness of tape equal to approximately 0.8 mm is in place over the conductor. The thickness of the insulation (tape) is to be measured by means of a dead-weight dial micrometer having no added weight and a presser foot 9.5 ± 0.2 mm in diameter. The specimen is then to be placed in the oven for 1 h of preliminary heating at a temperature of $100.0 \pm 1.0\,°C$ along with the dial micrometer and a weight to cause the foot of the micrometer to press on a specimen placed between the foot and the anvil with 500 gf or 4.90 N. At the end of 1 h, the specimen is to be placed under the foot of the dial micrometer with the weight added and is then to remain in the oven with the oven operating at full-draft at a temperature of $100.0 \pm 1.0\,°C$ for an additional hour.

Table 4.32 Requirements for Electrical Insulating Plastic Tape

Property	Unit	Requirement	Test Method ASTM
Thickness (avg.)	mm	0.02 ± 0.02	D 1000
Breaking strength (min.)	kN/m	2.6	D 1000
Elongation (min.)	%	150	D 1000
Dielectric breakdown (min.) standard condition	kV/mm	40	D 1000
Dielectric breakdown (min.) wet condition **(96/23/96)** conditioning for 96 h in air at $23.0 \pm 1.0\,°C$ and 96 ± 2 percent relative humidity		90% of dry	
Adhesion strength (min.) to steel	N/mm	0.175	D 1000 (method A)
Backing at $-7\,°C$		0.175	
Unwind	N/m		D 1000 (method B)
Minimum		175	
Maximum		1400	
Static elongation	%	15	D 1000
High-humidity insulation resistance (min.)	mega-ohm median	100,000	D 1000

At the end of the second hour, the dial of the micrometer is to be read, without removing the specimen or the dial micrometer from the oven, and the percent decrease in the thickness of the insulation (tape) is to be calculated.

- **Storage test**

A thermoplastic insulating tape should retain its adhesive qualities so as to be capable of being used to insulate splices when an unused roll of tape is conditioned in a full-draft circulating-air oven operating at a temperature of $40.0 \pm 1.0\,°C$ for 60 days or, at the manufacturer's option, a temperature of $65.0 \pm 1.0\,°C$ for 240 h (10 days).

To determine whether a thermoplastic tape complies with the requirements, an unused roll of tape, in its original container, is to be laid flat in a full-draft circulating-air oven operating at the temperature and for the time, after which the tape is to comply with the requirements for adhesion strength.

- **Test for indirect measurement of conductor corrosion**

PVC tape should exhibit a resistance of ≥ 1.0 tera-ohm (1,000,000 mega-ohm) for a 25-mm width of tape when five specimens are tested in accordance with the ASTM D 1000 82a, using copper electrodes. The conditioning temperature of specimens with the copper electrodes should be $23.0 \pm 1.0\,°C$.

- **Roll size**

The roll size should be as follows:

Roll length: 20 m.
Roll width: 19 mm.

- **Color**

The color of plastic backing should be black.

4.33 Rubber Splicing Tape

4.33.1 Materials and Manufacture

A rubber splicing tape should consist of a compound of unvulcanized or partially unvulcanized natural or synthetic rubber, or a blend thereof (Ethylene–propylene copolymer or ethylene–propylene terpolymer), that should contain not >0.5 percent of free sulfur by weight of the original compound when tested in accordance with the ASTM D 297. The tape should be smooth and uniform. The tape edges should be straight and unbroken.

A separator of parchment paper, glazed sheeting, polyester film, or similar material should be interposed between adjacent layers of a roll of tape and should cover the outside of the tape.

4.33.2 Requirements

A rubber splicing tape conforming to this Part of the Standard specification should meet the requirements given in Table 4.33.

When a roll of tape is originally unwound, the separator should not show any undue tendency to stick or to unravel.

4.33.3 Storage Test

A rubber splicing tape should retain its adhesive qualities so as to be capable of being used to insulate splices when an unused roll of tape is conditioned in a full-draft circulating-air oven at a temperature of $40.0 \pm 1.0\,°C$ for 60 days (1440 h).

Table 4.33 Requirements for Rubber Splicing Tape

Property	Unit	Requirement	Test Method ASTM
Thickness	mm	0.75–1.0	D 1000
Breaking strength (min.)	kN/m	1.4	D 1000
Elongation (min.)	%	800	D 1000
Dielectric breakdown (min.)			
Standard condition	V/mm	32,000	D 1000
Wet condition (conditioning for 96 h in air at $23.0 \pm 1.0\,°C$ and 96 ± 2 percent relative humidity)		90% of dry	

To determine whether a rubber tape complies with the requirements in standard, an unused roll of tape, in its original container, is to be laid flat in a full-draft circulating-air oven operating at a temperature of $40.0 \pm 1.0\,^{\circ}C$ for 60 days (1440 h). After this conditioning, the tape is not to become unduly attached to the separator and is to meet the fusion test in the ANSI/ASTM D 119.

4.33.4 Roll Size

The roll size should be as follows:

Roll length: 9 m.
Roll width: 19 mm.

4.33.5 Color

The tape color should be black.

4.33.6 Packaging, and Sampling

- **Packaging**

The individual rolls should be packaged in such a manner as to protect the material against physical and mechanical damage and contamination during transit. The packaged material should be packed in suitable containers to ensure acceptance by common carrier and to provide product protection against loss and damage during multiple shipments, handling, and storage.

- **Sampling**

A Lot or batch should consist of an indefinite number of rolls, offered for acceptance, of tapes manufactured by a single plant run through the same processing equipment with no change in ingredient materials.

Unless otherwise specified by the Purchaser, the number of samples for testing should consist of 10 percent of the lot, but in no case should it be less than one or more than ten rolls. The results of tests on four specimens cut from each sample roll should be averaged for each test specified in this Part of the Standard specification to determine conformance with the specified requirements.

The selection of rolls, sampling, conditioning, and testing should be in accordance with ASTM Test Method D 1000.

4.34 Thermit Weld (Cad Weld) Powder

This Part covers the minimum requirements for the supply of thermit weld (cad weld) powder for connection of CP leads to new or in-service steel pipes, tanks, and structures by the thermit welding process.

The technical bid usually includes the following:

1. Catalog (including technical data of the offered material).
2. Production procedures that cover all stages of the manufacturing process, from receipt of raw materials to finishing and packing for shipment, including the following:
   a. Chemical analysis of alloy (material specification).
   b. Measurements of individual cartridge weight.
   c. Inspection and testing procedures, including acceptance criteria, and frequencies.
   d. Preservation, packing, and shipping procedures, including methods, material and any requirement for periodic inspections.

The thermit weld powder for all welds of copper cable to steel pipe irrespective of cable size should be cartridge No. 15 (15 g).

Each cartridge powder charge should be enclosed in a separate polyethylene tube containing 15 g with starting powder at the bottom.

Individual cartridge tubes along with the required metal discs should be tightly packaged 20 nos. to a plastic box and nos. of plastic boxes should be tightly packaged in a cardboard package.

The material should be of such a quality and purity that when applied produces strong and permanent high conductivity connection.

The material should meet the initial properties after storage for at least 3 years from the date of delivery at normal storage conditions.

The material purchased according to this Part of Standard specification should be packaged in suitable containers, in accordance with the best commercial practice. The complete package should be designed to ensure the item against damage during shipment, handling, and storage.

4.34.1 Sampling

- **Lot**

Unless otherwise specified by the Purchaser, a lot should consist of all thermit weld powders made from the same production run offered for inspection at one time.

- **Sampling for Inspection**

Sample cartridge powders should be selected from each lot in accordance with inspection level II of ISO 2859 (Part 1). The acceptable quality level should be 2.5 percent for all visual, weight, and all other type of defects with reference to this Part of the Standard specification. The single sampling plans for normal inspection (Table II A of ISO 2859 (Part 1)) should be used.

5 Monitoring Cathodic Protection Systems

The purpose of control and maintenance is to ensure that the cathodic protection installation remains effective throughout the design life of the structure. The check procedures must be adapted to the type of structure and to its cathodic protection system. They must be in conformity with the standards in force and be validated once the installation is put into service.

The control of the effectiveness covers at the same time functional controls of the equipment and measurements of the parameters of cathodic protection. It should be noted that the material used for measurements and controls must themselves be regularly checked.

If the installation is not controlled by remote monitoring, the frequency of functional controls must be in conformity with the standards in force.

The filing of data about the operation and maintenance of the installations is essential. Computerized management systems of data are now readily available.

It is necessary to regularly carry out the following potential measurements:

- **Potential without current output (free corrosion potential):**

The corrosion potential of a metal in an electrolytic environment, such as water, soil, or concrete (aqueous porous environments), is the difference in potential between this metal and a reference electrode placed in the vicinity of the surface of the metal.

In natural conditions, when no external current is applied to the metal part, this potential is called potential-free corrosion potential (or null current corrosion potential). It is useful to know it in order to be able to apply the criterion to lower the potential of 100 mV. Its measurement can nevertheless be adversely affected, particularly on large structures, by the circulation of stray currents, telluric currents, or currents induced by power transmission lines.

- **Potential with current switched on (Eon)**

The potential with the current switched on (sometimes called "on") is the difference in potential between metal under cathodic protection and the reference electrode. As indicated above, this potential strongly depends on the position of the reference electrode compared to the protected structure.

If there is no external current at the considered cathodic protection system, the « Eon » measurements are more electronegative than the real potential of the structure. They are optimistic for the checking of the protection level and pessimistic for the one of the overprotection level. For buried structures or structures in concrete, they are only indicative and must be supplemented by Eoff measurements.

Cathodic Corrosion Protection Systems. http://dx.doi.org/10.1016/B978-0-12-800274-2.00005-3

- **Potential with current switched off and depolarization (Eoff)**

The potential measured with the current switched off is measured after the cathodic protection current is switched off. This potential changes with time.

If there is no external current at the considered cathodic protection system, the «Eoff» is close but less electronegative than the real potential of the structure. They are pessimistic for the checking of the protection level and optimistic for overprotection.

- **Checks and measurements at start up**

For cathodic protection by impressed current, it is necessary to measure the output voltage of the rectifier, the protection current, and the potential with known current at various locations.

For buried structures, it should be checked whether there is any interference to adjacent structures.

- **Functional controls in service**

  Functional controls include
  - to check that the transformer-rectifiers work well, including measurements of the voltage and current,
  - for complex structures, to check the connection and/or the insulation of the external structures,
  - to check the mechanical and electrical integrity of safety and protection devices.

- **Control of the protection effectiveness in service**

Generally, the effectiveness of cathodic protection is evaluated by comparing the measured values of certain parameters, with the reference values or with the protection criteria.

The protection criteria concern
- the measurement of the potential with Eon switched on current.
- the measurement, after the current is switched off, Eoff, and eventually of the depolarization
- checking there is no risk of overprotection for the considered structure (types of materials and coatings).

- **Remote monitoring and control**

The cathodic protection current and the potentials of the protected structures can be measured without human intervention, continuously or with definite intervals. The results of these measurements can be recorded in a data collector and then transmitted. The functioning of the installation, with its possible failures, can thus be followed remotely.

The data transmission system, from the protected structure toward the monitoring station, can be supplemented by a device that allows regulation of the protection current of the structure, from the monitoring station.

This chapter provides survey requirements to ascertain that corrosion control systems installed on buried or submerged structures are properly designed, operated, and effectively maintained.

This also provides information concerning techniques, equipment, measurements, and test methods used in field application. This chapter deals with inspection of coatings in conjunction with cathodic protection for its efficiency on current distribution.

Corrosion control system is effective when it is properly designed, constructed, and maintained. Therefore, without a comprehensive inspection program, both during construction for a designed system and operation, the investment for design, material, and installation of protection systems may be wasted.

Inspection of a cathodically protected system is necessary to ensure that protection of the structure is in accordance with applicable criteria, and that each part of the cathodic protection system is operating satisfactorily and if changes are noted, then action is taken to return the system to protection conditions. This inspection is possible by periodic inspection of units and recording their current and voltages, measuring structure to soil (or water) potentials, and most of all interpretation of data obtained.

Records of all information pertinent to effective maintaining of cathodic protection system and other changes and adjustment during the life of system should be kept and be available for further survey and inspection when required.

The conditions of the coating applied to a protected structure will have a considerable effect on structure/electrolyte potentials, then inspection of the coating should be made during construction and commission and when opportunities arise care needs to be taken to minimize damage to coating during such an inspection.

To ensure that overprotection does not cause accelerated disbondment of the coating or other deleterious effects, a potential corrected for voltage gradient error should be measured.

5.1 Criteria for Cathodic Protection

Criteria for cathodic protection is in general agreement with the current revision of National Association of Corrosion Engineers (NACE) PR-0169-2002 and in specific agreement, which state that "the selection of a particular criterion for achieving the objective of using cathodic protection is to control the corrosion of metallic surfaces in contact with electrolytes, depends in part, upon past experience with similar structures and environments where in the criterion has been used successfully."

Because the conditions and environments make it impractical to comply with NACE RP 0169-2002, which states that "the voltage measurements on all buried structures are to be made with reference electrode positioned as close as feasible to the structure surface being investigated" and "the corrosion engineer should consider voltage other than those across the structure-electrolyte boundary (IR drops), the presence of dissimilar metals, and the influence of other structures for valid interpretation of his voltage measurements," specific standard criterion incorporates allowances for the intrinsic inaccuracies of structure to electrolyte potential measurements, the standard criterion for steel structures in different environments is stated in the following paragraphs (see also Clause 3.2 of DIN 30676, 1985).

5.1.1 Buried Pipes

The criterion most widely used on buried pipes is based on the measurement of potential differences between the pipe and its environment.

Coated buried pipelines should have a minimum pipe to soil polarized potential of −0.850 (off) (see Clause 3.2 of DIN 30676, 1985) Volt with reference to copper/copper sulfate half-cell. The potential should be measured immediately after interrupting all the cathodic protection systems influencing the pipeline structure, where it is not practical to measure the interrupted or polarized potential on a pipe, the general guideline for the protection criteria should be as follows:

For buried pipelines in low resistivity soils (less than 2000 ohm-cm), the protection criteria should be −1.1 (on) Volt measured at the 1-km test stations or −1 (on) Volts measured at close interval spacing (15 m or less) with the cathodic protection currents applied.

The presented values for buried pipelines are often based on satisfactory coating application and in soils with low to intermediate resistivity (<2000 ohm-cm).

If the coating is particularly exposed to wear and mechanical damage or deteriorated due to aging and/or the pipe is buried in high-resistivity aerated soils, higher values must be used due to IR drop in (on) potential measurements (IR-drop describes the voltage that develops across a conductor as a result of its electrical resistance. This voltage is proportional to the current that flows though the conductor (V=I.R) and results in a drop in voltage available at the load devices (Vload = Vsupply – Vdrop)).

Buried facilities in plant areas should be considered protected at a minimum of −0.85 (on) Volt, measured at test holes over the metallic structure with the current flowing.

Another criterion in use is potential change from natural value to the value obtained after application of cathodic protection. In this method, potential change is used as criterion rather than an absolute value (say −0.85 V to copper/copper sulfate electrode). In this system, the intent is to change the pipe potential by 0.25 or 0.3 V in the negative direction when the cathodic protection is applied.

Note: Special conditions (e.g., backfill, pipe coating characteristics, or other parameters) may require further interpretation of voltage readings (see NACE RP 01-69-83 Paragraph 6.5).

5.1.2 Tanks Exteriors

Tank undersides may be considered partially coated due to contact with sand asphalt padding. However, contact with the soil will vary with flexing of the base. The potential recorded at the periphery of the underside should be −1.10, −1.20 V (on) with reference to a copper/copper sulfate half-cell; where permanent reference electrodes have been installed under the tank bottom, a minimum potential of +0.25 (on) Volt zinc to steel indicates adequate protection.

5.1.3 Submerged Pipelines

Offshore submerged pipelines should have a minimum pipe to water potential of −0.90 (on) Volt with reference to a silver/silver chloride half-cell.

5.1.4 Offshore Structures and Ship Hulls

Steel structures, other than pipelines, should have a minimum structure to water potential of −0.90 (on) Volt with reference to a silver/silver chloride half-cell.

5.1.5 Tank, Pipe, and Water Box Interiors

Structures storing or transporting conductive waters or other conductive liquids should have a minimum electrolyte to internal surface potential of −0.90 (on) Volt with reference to silver/silver chloride reference cell or +0.15 (on) Volt with reference to an internal zinc reference electrode.

5.1.6 Well Casings

In general, onshore well casings should be considered adequately protected when a polarized casing to soil potential of −1.0 (off) Volt to a copper/copper sulfate reference cell is measured with the cell located close to the casing and the cathodic current momentarily interrupted.

Alternatively, the potential measured with the cathodic current should be −1.2 (on) volts to the copper/copper sulfate reference cell with the cell located remotely, a minimum of 75 m, from the well and preferably 180° away from the anode bed. Where it is impractical to obtain valid casing to soil potential measurements, current requirement and polarization test data may be used in interpreting the protected status of well casings. Offshore well casings should be considered protected when the casing to water potential is −90 (on) Volt to a silver chloride reference cell placed closed to the casing. Table 5.1 lists the observed protection potentials, i.e., potential without allowances for IR drop error for full protection of various metals, measured against difference standard electrodes.

5.1.7 Potential Limits

The potential limits for coated pipes should be as specified in Table 5.2.

5.1.8 Aluminum

It is not at present possible to make firm recommendations for the protection of aluminum since this metal may corrode if made too strongly negative. There are indications that corrosion can be prevented if the potential is maintained between the limits shown in Table 5.1. Alternatively, it has been recommended in the case of pipelines to make the metal electrolyte potential more negative than its original value by 0.15 V.

5.1.9 Lead

In alkaline environments, lead may occasionally be corroded at strongly negative potentials.

Table 5.1 Minimum and Maximum Potentials for Cathodic Protection of Bare Metals (Volt, without IR Drop)

		Reference Electrode							
		Copper/Copper Sulfate		Silver/Silver Chloride[a] Seawater		Silver/Silver Chloride Saturated KCl		Zinc/(Clean) Seawater	
Metal or Alloy	**Condition**	**Minimum**	**Maximum**	**Minimum**	**Maximum**	**Minimum**	**Maximum**	**Minimum**	**Maximum**
Unalloyed and low alloy ferrous materials	At temperature below 40 °C	−0.85	Not applicable	**−0.8**	Not applicable	**−0.75**	Not applicable	**+0.25**	Not applicable
	At temperature higher than 60 °C	−0.95	Not applicable	**−0.9**	Not applicable	**−0.85**	Not applicable	**+0.15**	Not applicable
	In an aerobic media with high activity and sulfates-reducing bacteria and sulfides	−0.95	Not applicable	**−0.9**	Not applicable	**−0.85**	Not applicable	**+0.15**	Not applicable
	In an aerobic and in anaerobic media with low activity of sulfate-reducing bacteria and sulfides	−0.85	Not applicable	**−0.8**	Not applicable	**−0.75**	Not applicable	**+0.25**	Not applicable
	In sandy soils with resistivities greater than 50,000 ohm-cm	**−0.75**	Not applicable	**−0.7**	**Not applicable**	**−0.65**	**Not applicable**	**+0.35**	**Not applicable**

Stainless steel with a chromium of at least 16% by weight, for use in soil and fresh water	At temperature below 40 °C	−0.1	Not applicable	**−0.05**	Not applicable	**0**	**Not applicable**	**+1**	**Not applicable**
	At temperature higher than 60 °C	−0.3	Not applicable	**−0.25**	Not applicable	**−0.2**	**Not applicable**	**+0.8**	**Not applicable**
Stainless steels with a chromium content of at least 16% by weight, for use in salt water		−0.3	Not applicable	**−0.25**	Not applicable	**−0.2**	Not applicable	**+0.8**	Not applicable
Copper; copper/nickel alloys		−0.2	Not applicable	**−0.15**	Not applicable	−0.1	Not applicable	**+0.9**	Not applicable
Lead		−0.65	−1.7	**−0.6**	**−1.65**	**−0.55**	**−1.6**	**+0.45**	**−0.6**
Aluminum in fresh water		−0.8	−1.1	**−0.75**	**−1.05**	**−0.7**	**−1**	**+0.3**	**0.000**
Aluminum in salt water		−0.9	−1.1	**−0.85**	**−1.05**	**−0.8**	**−1**	**+0.2**	**0.0**
Aluminum in soil		−0.95	−1.2	**−0.9**	**−1.15**	−0.85	−1.1	+0.15	**−0.1**
Steel in contact with concrete		−0.75	−1.3	**−0.7**	**−1.25**	−0.65	−1.2	+0.35	

[a]Silver/silver chloride/seawater, (salinity 32–38%), in brackish water (salinity $< 32\%$), the potential reading from an open electrolyte reference electrode must be corrected for the lower chloride concentration. For measurements in such water a closed electrolyte reference electrode is advantageous.

Table 5.2 Potential Limits for Cathodic Protection Buried Pipes Coated with Different Coating Systems

Coating System	Off Potential (Ref. $Cu/CuSO_4$ V)
Epoxy powder fusion-bonded	−1.5
Asphalt and coal tar enamel	−2
Plastic tape (laminate)	−1.5
Epoxy coal tar	−1.5
Polyethylene	−1

All potentials have been rounded to the nearest multiple of 0.05 V. The figures for electrodes in which seawater is the electrolyte are valid only if the seawater is clean, undiluted, and aerated.

5.1.10 Stainless Steels

In many environments, stainless steels will not require any form of protection. In some cases anodic protection is used.

Stainless steels are, however, often susceptible to crevice corrosion. A crevice may be encountered between two metals, e.g., at riveted or bolted seam, or between a metal and nonmetal or at a gasketed joint. Crevice attack is a particular from of a differential aeration corrosion and is most often encountered in a marine environment. It has been found that cathodic protection will significantly reduce the incidence and severity of this form of corrosion; polarization to potentials given in Table 1 is necessary.

Difficulty can, however, arise if the crevice can seal itself off from the environment; the protective current cannot flow to the seat of the attack, which may proceed unabated.

Polarization of stainless steels to excessively negative potentials may result in hydrogen evolution, which can cause blistering and loss of mechanical strength.

Experience has shown that random pitting of stainless steel may not be influenced by cathodic protection, despite the evidence from certain laboratory studies.

5.1.11 Steel in Concrete

If steel, whether buried or immersed, is only partially enclosed in concrete, the protection potential is determined by the exposed metal and is as indicated in Table 5.1. Iron or steel fully enclosed in sound concrete free from chlorides would not normally require cathodic protection because of the alkaline environment. For circumstances where cathodic protection needs to be applied, for example, because there is doubt as to adequacy of the concrete cover or to provide very high reliability, it has been suggested that potentials less negative than are normally required for the protection of steel may be suitable.

5.2 Periodic Inspection

Periodic inspection of cathodically protected structure includes the following survey and measurements.

5.2.1 Potential Survey

The best way to judge the effectiveness of a cathodic protection system is by measurements, at selected points, of the structure/electrolyte potentials, which should be maintained more negative than the minimum values given in cathodic protection criteria.

- **Potential survey intervals**

Cathodic protection potential surveys should be carried out at time intervals determined from a consideration of the system parameters, including the type of cathodic protection system, the nature of the environment, the presence or absence of stray current, and the structure operating conditions.

The frequency at which it is necessary to recheck galvanic anode system depends ultimately upon the history of the structure and similar structures in the same locality. However, toward the end of effective life of galvanic anodes a recheck of protective potentials may be necessary every month, i.e., having regular intervals.

With regard to impressed current system, in addition to complete potential surveys occasional surveys should also be carried out in areas where there is pipeline conjunction with many foreign lines crossings. More frequent checks, e.g., monthly, should be made in areas that are considered particularly from the standpoint of corrosion and are vital to operation. The critical areas include but are not limited to places where interference correction bonds or facilities for stray current correction installed, gaps and crevices on offshore pipelines, platforms and jetties, tidal and splash zone areas, points of minimum protection, and drainage.

Generally following time intervals should be used for potential measurements and associated tests as a minimum. Shorter time intervals may be considered where necessary.

- Pipelines: every 2 months
- Municipal gas network: every 2 months
- Jetties: every month
- Tank farms and plants: every 3 months
- Offshore platforms: every 6 months
- Sea line risers: every month
- Offshore pipelines: every 3 years
- Ships and cargo tankers: every 6 months
- Interior of vessels and tanks: every month
- Cased crossing: every 2 months
- Hot spots protected by galvanic anodes: every month
- Galvanic anode system: every 2 months.

5.2.2 Test Equipment for Potential Survey

Measurement of potential between a structure and its surrounding electrolyte should be made by using suitable equipment. If suitable voltage-measuring equipment are not used (or if the right equipment are not used properly, and care in handling of equipment are not exercised), observed potential value may be misleading or completely meaningless. Therefore, due attention should be given to selection of the test equipment, type, and required accuracy and they should be maintained in good working condition at all times. The following are essential equipment for potential measurements.

- **Reference electrodes**

The potential of cathodically protected structure to be measured with respect to a suitable reference electrode. Normally used electrodes for potential measurements are

- Copper/copper sulfate electrode
- Silver/silver chloride electrode
- Zinc electrode

Approximate comparison of potentials of these electrodes is given in Figure 5.1.

Copper/copper sulfate is the most common type of electrode used in measuring the level of protection of buried structures.

Silver/silver chloride is wildly used in potential measurements of submerged structures such as jetties, offshore platforms, ship hulls, and other marine structures.

Pure zinc electrode may be used as permanent reference electrodes in marine structures and under the bottom of storage tanks.

The reliability of potential measurements can be enhanced by using sophisticated systems such as sonar electrical potential unit.

- **Potentiometer and voltmeter**

Different types of potentiometer or voltmeter may be used depending on requirements. Any conventional voltmeter will take some current and its reading is

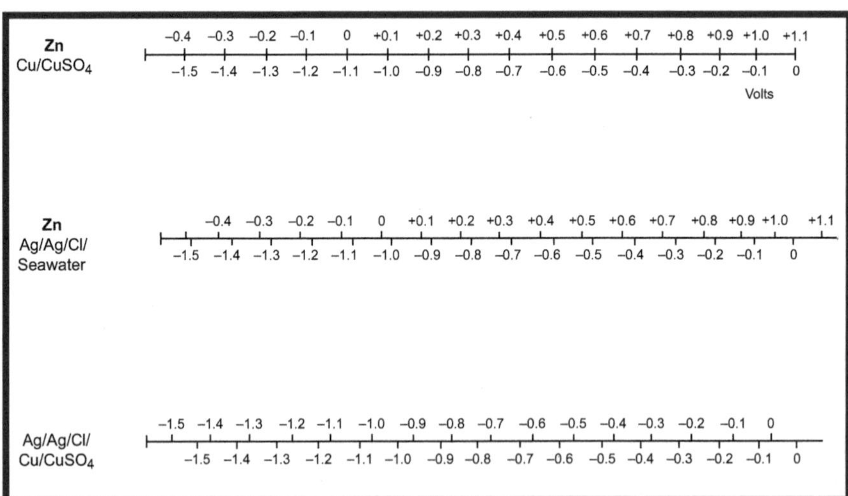

Figure 5.1 Approximate comparison of potentials using zinc, $Cu/CuSO_4$, and Ag/AgCl reference electrodes.

subject to correction. However, if the voltmeter resistance is high in comparison with the external circuit resistance, the correction factor is very low. In view of this, high-resistance voltmeter should be used for cathodic potential measurements.

Potentiometer–voltmeter should be used in measuring potentials in areas having high resistivity (i.e., over 2000 ohm-cm).

Potentiometer is more accurate in potential measurements and can be used as a bias facility to measure potentials by a null current potentiometric method or as a millivolt source for calibration checks on other instruments.

5.2.3 *Procedure to Be Observed*

Following should be observed in taking potential readings:

- Test equipment should be in order or proper working conditions.
- For close potential reading of pipelines the reference electrode should be placed on the ground surface, directly above the pipeline.
- The porous plug of copper–copper sulfate electrode should be in firm contact with moist soil. In dry areas, it is necessary to moisten the earth around electrode with fresh water to obtain good contact and reduce electrode to earth resistance.
- Do not permit grass or weeds (particularly when they are wet) to contact exposed electrode terminals, because they affect the observed figures.
- In high-resistivity soil (more than 2000 ohm/cm) use potentiovoltmeter or potentiometer to obtain more accurate readings.
  Make sure to connect the right wire to the right terminals of the meter, i.e., connect the reference electrode to the positive and test point to the negative terminal of meter.
- Select the most proper instrument range to get the most accurate reading.
- For offshore structures with complex geometry or where shielding can occur at some points such as nodes, the reference electrode may be located by a diver or a remotely operated vehicle at these points, and readings taken. Such readings can then be related to readings taken with a reference electrode placed adjacent to the side of the structure.
- Care should be taken to ensure that the structure component to which the measuring voltmeter is connected is not carrying a substantial cathodic protection current with impressed current systems, in particular, parts of the structure may be carrying a large current and hence cause a significant voltage drop error in the measurement.
- In measuring the potential of a ship or tanker, make sure it is not in contact with jetty structure.
- As corrosion of buried or submerged pipelines is mainly attributable to sulfate-reducing bacteria in an anaerobic environment, the protective potential should be adjusted to at least −0.95 V (off) with respect to $Cu/CuSO_4$ electrode in all soils other than pure sand.
- In municipal gas networks, potential measurements should be done just on provided test points. Measurements on house gas risers are not allowed.

5.3 Potential Survey of Buried Steel Pipes

5.3.1 *Instruments*

Copper/copper sulfate half-cells are the most common type of reference electrodes. Commonly available models operating principle are shown in Figure 5.2.

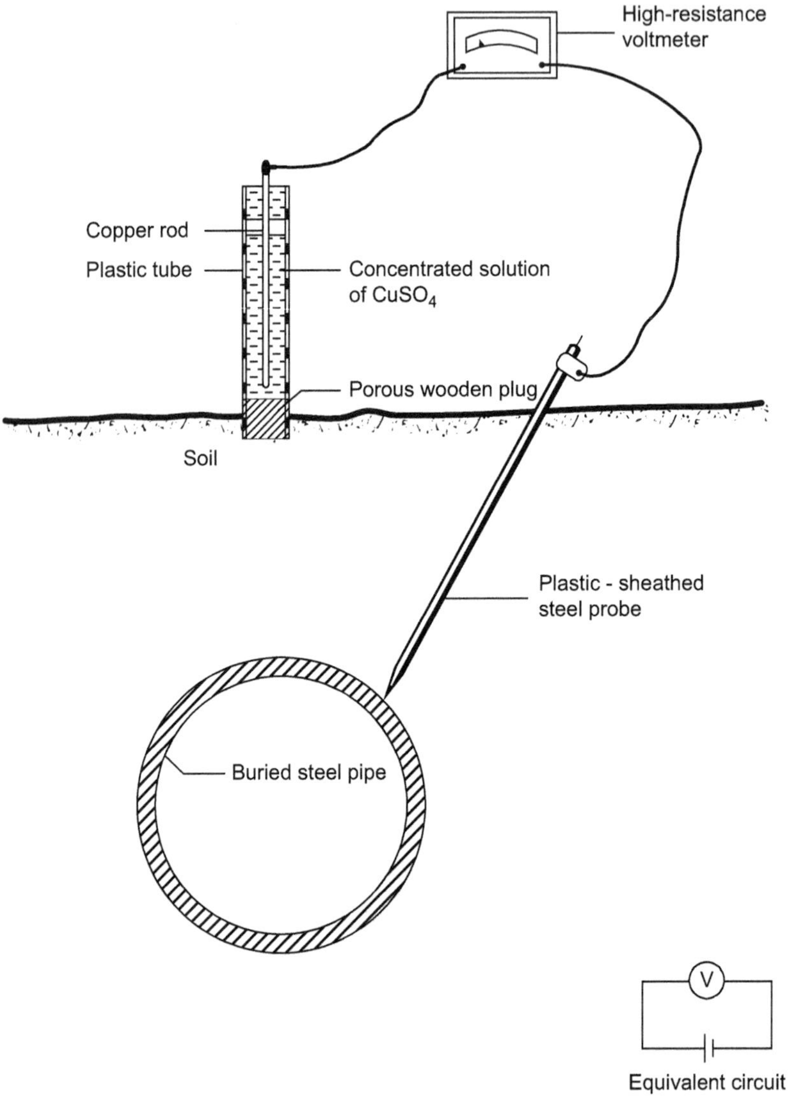

Figure 5.2 Typical measurement of pipe to soil potential.

In measuring the electromotive force of such a cell the current flow should be restricted to the lowest practicable value, or polarization will occur, i.e., the objective is to measure the open-circuit electromotive force.

A convenient type of instrument for field use is a high-resistance voltmeter with a resistance not less than 1 mega-ohm/V; for field use under difficult circumstances robustness should be given consideration. The potentiometer mode should be available. The accuracy of the 0.5–2.5 V range should be high.

5.3.2 Potential Measurements

The method of potential measurement is as follows:

Place the porous plug of the electrode firmly into contact with the soil, moistening the area of contact if the soil is very dry. Electrical contact then exists between the steel pipe, the soil electrolyte, and the porous plug of the reference electrode. The circuit is completed by connecting the copper electrode of the half-cell via a suitable lead to the potentiometer or voltmeter, and connecting the latter to a steel probe, the sharpened point of which is in contact with the pipeline. The probe itself is plastic sheathed except for its point. Alternatively, contact with the pipe may be made via the measuring wires inside test points erected over the pipeline, e.g., in conjunction with the kilometer markers.

The potential measured is the resultant of the potentials of all parts of the pipe surface, which are "scanned" by the electrode and the reading obtained will depend therefore on the location of the electrode with respect to the pipe. The closer the electrode is placed to the pipe, the smaller is the area it will "scan," and the more likely it is that variations in potential from point to point will be detected. It is clearly impracticable to place the electrode successively close to all parts of the pipe surface and normal practice is to place it at intervals on the surface of the ground immediately above the pipe. See Figure 5.2.

Provided that the reference electrode is not more than 0.5 m from above the top of the pipeline, sufficiently accurate results for most purposes will be obtained. Several readings may be made at each contact of the probe with the pipeline by moving the reference electrode to different points on the soil surface along the route. To obtain accurate readings the long cable connecting the reference electrode to the voltmeter should have a low resistance compared with the internal resistance of the voltmeter.

Figure 5.3 shows a typical series of pipe to soil potential measurements together with the potential differences between adjacent points along a pipeline that is not

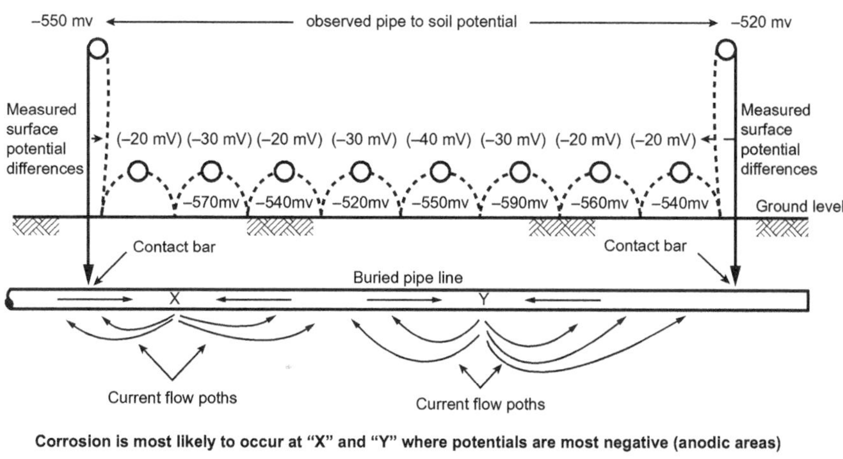

Figure 5.3 Pipe to soil potential survey.

cathodically protected. Typical current-flow paths resulting from these potential differences are also shown. From these it can be seen that the point where corrosion may be expected is at "X" and "Y," where current leaves the pipeline carrying iron ions with it into the soil.

- **Calibration**

Calibration of measuring electrodes should be carried out before and after each potential survey. For calibration, there should be at least two copper sulfate electrodes in the test kit. These should be matched in potential. The matching may be checked before and after each potential survey. The potential difference should not exceed 10 mV when the electrodes are placed side by side in the same soil or water. If the difference is excessive, contamination of one or both is indicated. In this case, they should be cleaned by dipping in 10% solution of nitric acid and rinsing the electrodes.

5.3.3 Compensation for the IR Drop Component in Cathodically Protected Pipelines

Pipe to soil potentials measured against a reference electrode placed on the soil when the protective current is switched on, always incorporate an IR drop caused by the resistance of the ground, coating, holiday, and that in the metallic portions of the circuit. Therefore, the corrosion engineer must consider voltage drops other than those across the structure/electrolyte boundary when interpreting any soil to pipe potential to conclude the true pipe to soil potential. The most frequent techniques used to remove the unwanted IR drops from pipe to soil potentials are as following.

The IR drop component can be eliminated by making I = O if all current to the pipe can be shut off, or by making R = O if the reference electrode is moved extremely close to either a bare pipe or the controlling holiday on a coated pipe (see DIN 30676).

A holiday simulation probe* can be installed to eliminate the stray current IR drops in measured pipe to soil potentials on critical areas of coated pipelines, or all IR drops if the applied cathodic protection current can be interrupted.

Note: A probe is electrically connected to the buried pipeline and then installed close to the pipeline surface to permit the flow of protective and stray currents to the exposed tip. Measurements of IR drop free potentials were then possible, by disconnecting the probe from the pipeline and recording the potential by normal means.

On bare or coated pipelines, the IR drop components can be eliminated, and true pipe to soil potentials can be determined by the stepwise current reduction technique.

Note: The stepwise current reduction technique was developed by Barlo and Fessler. In this technique, current reduces in steps and corresponding IR drops determined with each step. By extrapolating the cumulative IR drops to a value corresponding to zero current, the total IR drop in a potential measurement can be established.

* Note: A measurable gradient of at least 3.3 mV/m measured transverse to the pipe and referenced to a $Cu/CuSO_4$ electrode indicate a pipeline with a poor coating requiring IR drop correction.

The following are some important notes on instant-off potential measurements that corrosion engineers have to consider:

- In most cases, it is not possible to shut off all current to the pipeline as a result of stray current, multiple protection system, galvanic ground beds, and current flowing between more polarized small coating defects and less polarized larger coating defects.
- The measured instant-off potential on a coated pipeline corresponds to that of the largest freely exposed holidays, with smaller holidays having more negative potentials.
- The measured instant-off potential on a coated pipeline having shielded holidays (i.e., detached coating) will have less negative true potentials than a coated pipeline having freely exposed holidays, and the apparent instant-off potentials may not always correspond to the least negative potential of all holidays.
- On a coated pipeline with a transverse voltage gradient of more than 3.3 mV/m (10 mV/ft)* or on a bare pipeline, the IR drop component can cause measured pipe to soil potentials to be more negative than actual.

5.3.4 Overprotection

To ensure that overprotection does not cause accelerated disbondment of the coating or other deleterious effects, a potential corrected for voltage gradient error should be measured.

5.3.5 Potential Survey at Cased Crossings

Complete cathodic protection of pipelines can not be accomplished with short circuit between carrier pipe and casing pipe that is usually used at road or railway crossings. With the short circuit in place, cathodic protection current gathers on the outside surface of the casing (which is commonly uncoated) and flows along the casing to the point of metallic contact between carrier pipe and casing. The protective current then flows along the pipeline back to the rectifier or sacrificial anode. A single bare casing pipe, if in metallic contact with pipe, can absorb as much cathodic protection current as several kilometers of pipeline. Therefore, periodic inspection and testing to ensure the insulation of cased pipe at crossing is essential. This inspection is made by potential measurement and comparing the figures of casing pipe with that of carrier pipe.

- **Potential survey**

Potential survey at cased crossing will consist of casing to electrode potential reading versus carrier pipeline to electrode reading. This test can coincide with the routine survey along pipelines. A difference in the readings is qualitative indication of satisfactory insulation between casing and carrier pipe. A more detailed test of insulation can be carried out by interrupting the cathodic protection current and measuring the potential of casing and pipeline with respect to a reference cell. If pipeline potential changes between on and off conditions while the casing potentials remain steady it indicates effective insulation. If the potentials of both casing and pipeline change similarly as the current changes, a short circuit is indicated.

- **Locating the point of short circuit**

Further test can be carried out to locate the point of short circuit between casing and pipeline.

5.3.6 Potential Tests at Insulating Devices

Insulating equipment consist of insulating flanges, insulating joints/couplings, and insulating unions. Insulating flanges and joints on transmission lines are used to separate the pipelines electrically from refineries, plants, platforms, terminal facilities, pumping station, compression stations, tank farms, and city stations. Insulating joints and flanges may also be used to divide the line into sections or blocks so that failure of cathodic protection facilities or development of contacts with other metallic structures will limit loss of protection to a section. This practice of sectionization is most used in distribution systems and where stray current from underground, domestic networks, subways, and other systems presents problem and reasonably spaced insulating joints may be helpful in controlling stray current pickup and discharge.

Insulating union (monobloc) has wide application on offshore platforms where the possibility of short circuit by water droplets exists if ordinary insulating flanges are used. Also, insulating unions at 2 inches and below are extensively used to separate the city gas network from residential or commercial service line upstream of gas pressure regulator and meter.

- **Insulation test**

Proper functioning of the insulating device is important for effective operation of cathodic protection systems. Test for effectiveness of insulating devices involves interrupting the source of cathodic protection current on one side of the device and measuring the potential on both sides of that. If the insulation is effective, the potential on the protected side will change as the interrupter operates while the potential on the unprotected side will remain constant.

- **Locating of insulation defect and measuring the percentage of "leakage"**

If an insulation flange is found defective, step-by-step test should be made to locate the defect and repair it.

- **Test intervals**

Insulation test should be made during the potential survey along pipeline.

- **Lightning arrestor**

In case the pipelines are connected to oil storage tank farms, the insulating device prevents discharge of lightning current to other structures. This current may damage insulation in the event of lightning storm. To overcome this problem, lightning arrestor is installed across the insulating device. The arrestor bridges the insulating device. By doing so, lightning current is easily discharged without any damage to insulation. During periodic check of the insulating device, the lighting arrestors should be thoroughly inspected to ensure that they are in good shape and have proper connection at either side of the insulating device.

5.4 Potential Survey of Cathodically Protected Reinforced Concrete

Cathodic protection is increasingly being used to prevent corrosion of steel reinforcement (rebar) in concrete structures. Although the use of metal to electrolyte potential survey with respect to reference electrode is the primary technique for ascertaining when protection is achieved, unlike soil and water electrolytes, concrete presents some unique peculiarities when testing with reference electrodes. Measurements are complicated, primarily owing the fact that it is not possible to reach the metal surface when it is buried under 50–100 mm of concrete, and secondly the concrete is usually a fairly high-resistivity electrolyte. Embedded electrodes, corrosion coupons and probes, electrochemical impedance, electrochemical noise, and linear polarization techniques are available for this purpose. The most wildly used in situ test method to investigate the corrosion and protection of steel in concrete is half-cell potential mapping. This technique is covered by ASTM C 876-87 (1987 F).

5.4.1 *Potential Measurement*

To maximize the control and prevent unwanted side effects buildup, the concrete structures under cathodic protection, should frequently be monitored by potential measurement. The frequency of control measure is more critical for impressed current systems than galvanic ones.

Due to the fact that reinforcement bar is embedded in 50–100 mm of concrete, and concrete is high-resistivity electrolyte that its resistivity is not homogeneous due to the simple variation in chemical and moisture content of the aggregate, or climatic conditions from one day to another, care must be taken in measurement to compensate the IR drops. In this regard, the most acceptable criterion is "instantaneous off" potential measurement in impressed current systems. It is defined as the potential obtained not less than 0.1 s and not more than 1.0 s following interruption of direct current (DC). The well-protected structures will have the potential values between −200 and −250 mV more negative than silver/silver chloride half-cell.

Based on another criterion, the concrete society (1989) recommended the following criterion for protection: a minimum of 100 mV potential decay over all representative points, subject to a most negative limit of −1.1 V with respect to a silver/silver chloride half-cell. The decay period is typically 4 h but can vary with the rate of depolarization or potential decay.

For accuracy in measurements note the following:

- Use high internal resistance (i.e., 200 mega-ohm) voltmeter.
- To decrease contact resistance between portable reference electrode and concrete, wet the surface site by using biodegradable detergent in water.

- For reinforced concrete in air, a difference of 200 mV between sites in close proximity may indicate active corrosion.
- Concrete surfaces that have been subject to carbonation indicate potentials of −100 mV more negative than would be ordinarily encountered.
- If concrete is repaired, make sure that half-cell is not placed on nonconductive repair materials.

5.4.2 Electrodes and Probes for Some Concretes

- **Permanent reference electrode**

These electrodes are fitted into the concrete structures during construction of structures. The stability of the available conventional electrodes is questionable over long periods of time in concrete. Therefore, the use of permanent electrodes for accurate steel/concrete potential measurements or for the control of automatic potentiostatic is of doubtful value. The use of relatively stable electrodes (such as graphite) for the short-term measurement of potentials, such as for the 4 h requirement in the 100-mV polarized potential decay criterion, or points of structures that are not accessible, can be considered.

- **Resistance probe (corrosometer type)**

These probes should be fitted within the concrete structures during construction. They measure the extent of corrosion on a tubular steel element connected to the reinforcing in the concrete by accurately measuring the change in resistance of the element resulting from corrosion. The element of the probe essentially becomes part of the reinforcing matrix.

- **Rebar probe**

These probes are used to measure the current or current densities collected at particular locations in the reinforced concrete, to determine the extent of cathodic protection distribution or current flow. Probe comprises a section of reinforcing bar electrically separated from the main reinforcing except for cabled connection, which incorporates a current measurement device.

- **Macro cells**

These cells are similar to rebar probes except that they are deliberately installed in a local surrounding of particularly highly chloride-contaminated concrete. If these macrocell bars corrode, current will flow then into concrete to discharge on other cathodic areas of reinforcing and return through the monitored cable connection. If cathodic protection is fully effective in the area of this most actively corroding bar, it will overcome the corrosion current, and the current measured in the monitored cable connection will change direction.

- **Associated problems with overprotection in concrete**

Overprotection should be controlled because it damages the structures as following:

- Ionic movement in concrete due to passing cathodic protection current resulting in hydrogen gasification at the concrete–steel interface and thereby weakening the mechanical bond.

- Many prestressed and posttensioned structures incorporated the use of high-strength steels (i.e., 100,000 psi yield strength or more) are susceptible to hydrogen embrittlement.
- Acidic ions move toward anode that is embedded in concrete lowering the alkalinity then delimitation of the surface may be initiated.
- Overprotection causes unwanted heat around anode and cracking of concrete.

- **Stray current**

Reinforcement rebar of concrete structures is very susceptible to stray current damage. At discharge point, a current of 1 A can dissolve 9 kg of steel per year, which on thin reinforcing bar can be disastrous and over a relatively short time premature failure can be rapidly induced. Therefore, all reinforcement cages should be bounded together and test should be done and corrective measures should be taken.

5.5 Potential Survey of Offshore Structures

Protective measure of marine structures is greatly dependent on the environmental conditions and furthermore on the geometry and surface condition of the steel surfaces. The decisive environmental parameters are temperature, seawater chemistry, flow conditions, and growth of marine fouling. These parameters vary with the geographical location, depth, and season. The nucleation and growth of a calcareous layer, which retards the diffusion flux of oxygen onto the steel cathode, differ at various locations and depth. Flow velocities of seawater have a detrimental effect on polarization. For deep waters, due to lower seawater current, more current densities require for initial polarization. Therefore, to get correct information about the condition of the cathodic protection system, it is important that the locations for potential measurements are carefully selected and electrodes placed as close as possible to the steel surface to minimize voltage drop in seawater.

5.5.1 Reference Electrodes

In seawater, potential measurements should be performed with reference to silver/silver chloride half-cell. The use of copper/copper sulfate is limited due to the fact it may be contaminated by chloride ions.

Zinc is not an accurate reference electrode since it is unstable in seawater. But due to the long lifetime of zinc electrode, it should only be used to get an estimate of the protective level.

5.5.2 Methods of Measurement

To locate the reference electrode as close as possible to the structure surface, different means may be used:

1. Reference electrode may be carried by divers.
2. Measuring electrode is carried by remotely operated vehicle (ROV) or a submarine.

Where diving or use of ROV is difficult or impossible due to high cost, the reference electrode may be lowered from the surface to a specified series of depth for readings.

Measurements taken in this way are relatively inaccurate as the distance between the reference electrode and steel surface is difficult to control.

5.5.3 Potential Measurements with a Diver-Operated Unit

A measuring electrode, a high-impedance voltammeter ($\geq$10 mega-ohm), and a metal tip for making contact to the steel are built into a single tool that is easy to handle by the diver. When potential measurements are performed, the metal tip is pressed against the steel surface or anode and the potential is read on the unit's digital display. The diver reports the reading orally to the surface.

The metal tip is connected to the positive terminal of the built-in voltmeter while the measuring electrode is connected to the negative. Negative potentials will then be read on the display.

On some diver-operated units, the metal tip is connected to the negative terminal, whereas the measuring electrode is connected to the positive. For such units, the sign of all potential figures given within this section will be opposite.

- **Calibration**

Calibration of measuring electrodes should be carried out before and after each potential survey. In order to become stable, the measuring electrodes should be submerged for at least 1 h before calibration is performed. Saturated calomel electrodes should be used for calibration. These electrodes should also be controlled to select the best electrode. Control of saturated calomel electrodes is only necessary before each potential survey, and thereafter every 24 h.

Three saturated calomel electrodes should be used for calibration. They must be properly marked to distinguish between them. A high-impedance ($\geq$10 mega-ohm) voltmeter should be included in the equipment. Alternatively, if the measuring electrode can be removed from the diver operated unit, the unit's built-in voltmeter may be used. The electrodes should be kept at constant temperature at least 1 h prior to and during the control of the calibration electrodes. The temperature should be the same as for calibration of the measuring electrode.

For simplification, the saturated calomel electrodes are called 1, 2, and 3. The following procedure may be used to select the best electrode.

Remove the rubber cap from the electrode's tip. Saturated calomel electrode No. 1 (SCE 1) is connected to the positive terminal of the voltmeter, while SCE 2 is connected to the negative. The tip of the electrodes is submerged in the seawater, and the potential difference is read on a voltmeter. The value should be logged. Then SCE 2 is replaced with SCE 3. The potential difference between SCE 1 and SCE 3 is measured and logged. At last, SCE 1 is replaced with SCE 2.

The potential difference between SCE 2 and SCE 3 is measured and logged.

Acceptable readings are $\pm$2 mV (from -2 mV to $+2$ mV).

The following principles are suggested for selection of the best electrode:

- If all readings are within the acceptable range, any electrode may be used.
- If one reading is out of range, the electrode not included in this reading should be used.
- If only one reading is within the range, either of the two electrodes used in this reading should be used.

- If all readings are out of range, either of the two electrodes with least potential difference may be used.

However, the electrode used should be labeled and sent to a laboratory for control.

Electrodes that are suspected to give wrong potential readings should be renewed or sent to a laboratory for control.

Various types of diver-operated units exist. Different procedures have to be used to calibrate these types of units. The various types are

- Units with removable measuring electrode.
- Units that can be submerged during calibration using calomel electrodes.
- Units calibrated by use of zinc and saturated calomel electrode.

On some diver-operated units, it is possible to remove the measuring electrode. The following equipment is necessary to perform the potential measurements:

- Diver-operated unit.
- Three saturated calomel electrodes.
- Cable for connection between measuring electrode and diver-operated unit.
- Clamp.
- Saturated potassium chloride solution.

The measuring electrode is pulled out of the connector on the unit. An electric conductor is connected between the electrode's connector and the unit's connector (see Figure 5.4). It should be ensured that the electrical contact is satisfactory. The measuring electrode is to be submerged into seawater. If the electrode's connector or conductors are not watertight, neither of them must be in any contact with seawater (indicated on Figure 5.4). The selected calibration electrode is connected to the unit's metal tip, for instance, by a clamp. The tip of the calibration electrode is submerged, the potential difference is read on the unit's display, and the value is logged. Acceptable readings are -5 mV ± 5 mV (from -10 mV to $+0$ mV). If the potential difference is outside these limits, the results obtained during the potential survey should be corrected.

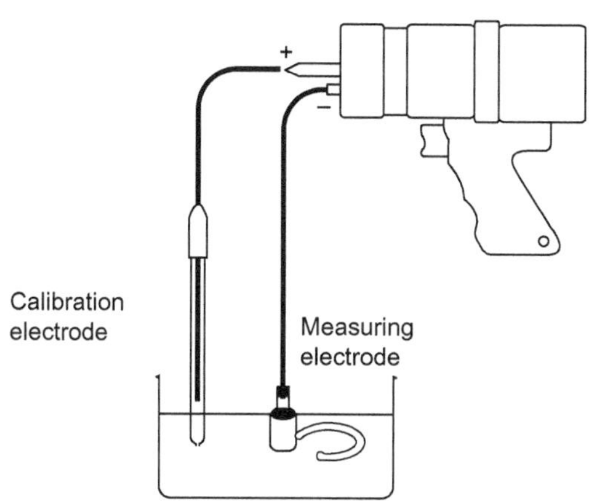

Figure 5.4 Principle for calibration.

An additional voltmeter may be used instead of the unit's built-in voltmeter. The measuring electrode is connected to the negative terminal of the voltmeter, and the saturated calomel electrode is connected to the positive. The procedure is the same as given above.

On some diver-operated units, it is possible to connect the saturated calomel electrode to the metal tip of the unit by an electrical conductor. The conductor is connected to the metal tip by a watertight sealing. The electrical conductor must be water proof. The following equipment is necessary to perform the potential measurements:

- Diver-operated unit.
- Three saturated calomel electrodes.
- Watertight sealing.

Then, diver-operated unit can be totally submerged into seawater while only the tip of the saturated calomel electrode is submerged. The potential difference between saturated calomel electrode and measuring electrode can then be read on the unit's display. Acceptable readings are -5 mV ± 5 mV. If the potential difference is outside these limits, the results obtained during the survey should be corrected.

For several diver-operated instruments, it is not possible to calibrate. For such instruments, an alternative procedure should be used. The following equipment is necessary in order to perform the potential measurements:

- Diver-operated unit.
- Three saturated calomel electrodes.
- High-impedance voltmeter ($>$10 mega-ohm).
- Cable.
- Clamp.
- Piece of zinc, either pure or anode zinc.

The saturated calomel electrodes should be controlled. The selected calibration electrode is connected to the negative terminal of the voltmeter and submerged in seawater. The piece of zinc is connected to the positive terminal by using the clamp and cable. The piece of zinc is then submerged in seawater.

It should be ensured that only the zinc is submerged, not the clamp or cable. The potential difference between zinc and saturated calomel electrode is read on the voltmeter and logged. The reading will be in the range of -1.00 to -1.05 V. The cable and clamp are disconnected from the zinc piece. Then the potential of zinc is measured directly with the diver-held unit, and the result is logged. The potential will normally be in the range -1.00 to -1.05 V. It should be noted that the diver-operated unit should be submerged for at least 1 h before calibration to become stable. It is important that the two measurements are carried out in rapid succession to ensure that the zinc potential does not vary. The potential between the measuring electrode and zinc (EAg/AgCl/seawater/Zn) is subtracted from the potential between saturated calomel electrode and zinc (Ecalomel/seawater/zinc) as follows:

- $E_{Ag/AgCl/seawater/Zn}$
- $E_{calomel/seawater/Zn}$
- $E_{Ag/AgCl/seawater/calomel}$

Acceptable difference is −5 mV ±5 mV (from −10 mV to +0 mV). If the potential difference is outside these limits, the results obtained during potential survey should be corrected.

- **Electrical contact between metal tip and steel or anode material**

Before measurements on steel surfaces are carried out, a light cleaning of marine fouling should be carried out at the point on the steel surface where the measurement is taken. Before the metal tip of the instrument is in contact with the steel surface, the relatively noble potential of the metal tip will be read on the instrument. When electrical contact between steel surface and metal tip is established, the potential reading will immediately become more negative.

At the points on the anodes where the potential measurements are taken, the potentials should immediately become more negative. The potential change will depend on the type of alloy and anode current output. Insufficient electrical contact may cause potential readings in the range −0.60 to −0.90 V vs Ag/AgCl.

- **Check of equipment during potential survey**

The diver should bring a piece of zinc during potential surveys. If the potentials are not within the acceptable ranges, the unit should be checked against the zinc piece. This reading should be reported together with the unacceptable reading. The zinc surface should be thoroughly cleaned by wire brushing. If a value less negative than −1.02 V (Ag/AgCl) is measured on the zinc piece, the equipment should be taken to the surface for calibration with saturated calomel electrode.

- **Check of battery voltage and voltmeter**

The battery voltage should be checked regularly. Minimum allowed battery voltage should be checked and will depend on the type of voltmeter. The voltmeter may be checked by measuring known voltages in the actual range. This may be done by using mercury batteries, which have stable voltages.

- **Check of watertight sealing**

Watertight sealings should be checked visually. Some sealings should be lubricated with silicone grease to be waterproof. Any leak of the sealings in the measuring circuit may cause erroneous readings.

5.5.4 Potential Measurements with Surface Voltmeter, Cable, and Measuring Electrode

The technique consists of using a high-impedance (≥10 mega-ohm) voltmeter, a measuring electrode, and insulated connecting cables. The positive terminal of the voltmeter is grounded to the structure, usually above water level, while the negative terminal is connected to the measuring electrode through a cable. The voltmeter is located above water level (see Figure 5.5(a)). The diver moves the electrode to various parts of the structure. Potentials are read on the voltmeter above water. The reading is taken when the diver is holding the measuring electrode in a stable position at the

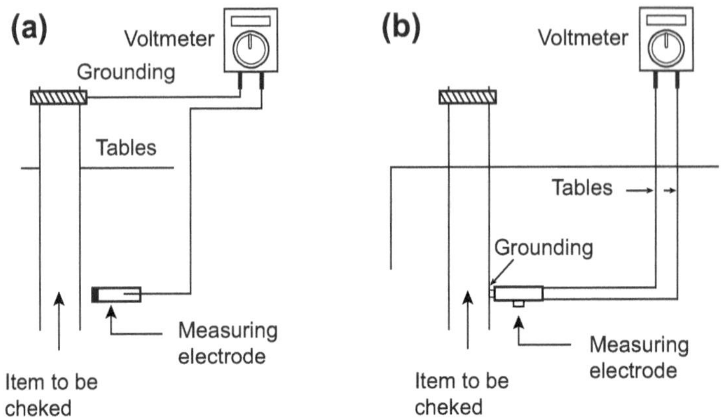

Figure 5.5 Schematic wiring diagram for potential measurements.

measuring point. When the reading is taken, the exact position of the measuring electrode must be known.

Alternatively, the measuring electrode and a metallic tip for making contact to the steel are built into a tool. Two cables or a two-core cable connects the metallic tip and measuring electrode to a surface voltmeter (see Figure 5.5(b)). The diver moves the electrode to various parts of the structure. When potential measurements are performed, the metallic tip is pressed against the measuring point and the potential is read on the surface voltmeter. When the reading is taken, the exact position of the measuring electrode must be known.

The following equipment is necessary to perform the potential measurements:

- Measuring electrode with cable.
- High-impedance voltmeter.
- Two clamps for grounding to structure.
- Grounding cable.
- Three saturated calomel electrodes.
- Saturated potassium chloride solution.

- **Calibration**

Calibration of measuring electrodes should be carried out before and after each potential survey. Saturated calomel electrodes should be used for calibration. However, these electrodes should also be controlled to select the best electrode. Control of calibration electrodes is only necessary before each potential survey, and thereafter every 24 h.

First, the measuring electrode should be submerged in fresh seawater for a minimum of 1 h for stabilization. Then, the measuring electrode is connected to the negative terminal of a high-impedance ($\geq$10 mega-ohm) voltmeter. The selected saturated calomel electrode is connected to the positive terminal of the voltmeter. The tip of the electrode is submerged (see Figure 5.6). The potential difference is read on

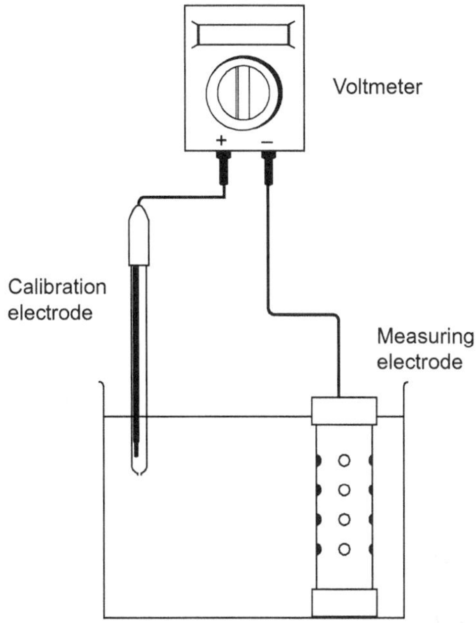

Figure 5.6 Calibration of measuring electrode.

the voltmeter, and the value is logged. Acceptable readings are −5 mV ±5 mV. If the potential difference is outside these limits, the results obtained during the potential survey should be corrected.

An alternative to the use of the silver/silver chloride/seawater electrode as measuring electrode is the use of silver/silver chloride/saturated potassium chloride electrode. The silver chloride plated silver in the Ag/AgCl/Sat.KCl electrode is located in a housing containing a saturated potassium chloride solution. The silver chloride plated silver in the Ag/AgCl/seawater electrode is directly exposed to seawater. The potential of the Ag/AgCl/Sat.KCl is not influenced by the seawater composition. The electrolytical contact between the silver chloride plated silver and seawater is achieved by porous plug.

The same procedure for calibration of the electrode may be used for the Ag/AgCl/Sat.KCl electrode. Acceptable potential difference between the saturated calomel electrode and the Ag/AgCl/Sat.KCl electrode is +40 mV ±5 mV. If the potential difference is outside these limits, the results obtained during the potential survey should be corrected. The positive limit for cathodic protection of steel in aerobic environments will be approximately −0.755 V, and the negative limit will be approximately −1.005 V for this electrode.

- **Check of cable insulation**

Any failure of conductor insulation may cause erroneous potential readings. The magnitude of the error will depend on the size and location of the insulation damage.

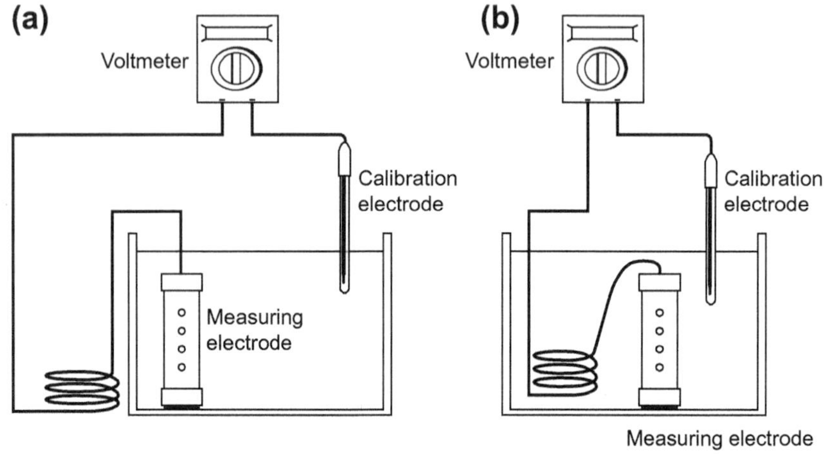

Figure 5.7 Sketches for check of cable insulation. (a) non submerged cable (b) submerged cable.

When checking the cable insulation, the cable with measuring electrode is connected to one of the terminals of a voltmeter. The other terminal is connected to a saturated calomel electrode. The measuring electrode, not the cable, is submerged into seawater. The same is done with the calibration electrode. The potential between the two electrodes is read on the voltmeter (see Figure 5.7(a)). The cable to the measuring electrodes is then also submerged, and the potential between the measuring electrode and saturated calomel electrode is read on the voltmeter (see Figure 5.7(b)). The difference between the two readings should not be more than ± 2 mV. If the difference is more than ± 2 mV, the cable may be damaged, and it should be renewed. If the readings are unstable, it may be caused by the electrodes, which need some time to stabilize (at least 1 h).

- **Check of grounding connection to structure above water**

When potential measurements are carried out, the positive terminal is connected to the structure. This part of the structure should be in electrical contact with the submerged part of the structure. To establish electrical contact to the structure, a clamp is often used. Coating, dirt, etc. should be removed before the clamp is fastened. If the electrical connection is unsatisfactory, the potential readings will be shifted to less negative values.

Special attention should be paid to grounding when the survey is being undertaken from a survey vessel.

The magnitude of the error in potential reading will depend on the impedance of the voltmeter and the resistance across the grounding. The resistance across a coating layer may be several mega-ohms, which may lead to significant errors. The magnitude of the error will decrease with increasing internal impedance of the voltmeter.

The electrical contact between clamp and structure can easily be checked with a multimeter. The clamps should have a metallic tip or needle for easily establishing an electrical contact. The following procedure may be used for checking the electrical contact between clamp and structure.

Coating, rust, and dirt are removed at two locations on the structural part until bright metal appears. Two clamps are fastened at these locations. They are connected to a multimeter with two cables. The resistance between the clamps is measured with the multimeter. The reading should be less than 1 ohm. If not, the electrical connection is unsatisfactory and must be improved. If the electrical connection is found satisfactory, clamp No. 2 is removed, while clamp No. 1 is used further for potential measurements. It is important that clamp No. 1 is not removed since this may change the electrical contact resistance.

5.5.5 Subsea Pipeline Potential Survey

The cathodic potential and electric field gradient of the submarine pipelines may be measured continuously by three electrode systems. The electrodes can be fitted to any pipeline inspection vehicle and the potential profiles and field gradient readings obtained along the pipeline transferred to a digitizer. This system measures five readings per second. This rate gives one reading every 10 cm when the ship travels at 1 knot. Pipeline cathodic potential readings will be plotted on a chart against distance along the pipe.

5.5.6 Time Intervals

When considering a galvanically protected steel structure in seawater, the most critical period during the polarization process is the first 6–12 months. If the structure achieves full polarization during this period and maintenance current density levels are consistent with criteria levels, then it is unlikely that any long-term problems will develop. Thereafter, semiannual potential survey of fixed offshore structures and for subsea installation every 3 years should normally be carried out. For jetties under impressed current systems monthly potential measurement and for ships, interior of water tanks, and cofferdam every 6 month should be done.

5.5.7 The Extent of the Potential Survey

Survey program for offshore structures should cover all major parts of the installation such as nodes, pile guides, pile sleeves, conductors and their supports, and partly closed compartments where underprotection is likely to occur. Also,

- Corrosion seems to be prevented by cathodic protection in crevices filled with stagnant seawater. However, in crevices with flowing seawater, cathodic protection has limited effect. Therefore, in these crevices, the potential outside of them plus the potential drop should be equal to or more negative than the protective level.
- Use high-impedance voltmeter (>10 mega-ohm) to compensate IR drop.
- Before measurements, a light cleaning of marine fouling should be carried out.
- Check the equipment during potential survey.
- Check for grounding connection to structure above water.

- Control of distance from measuring electrode to structure is very important. This distance should be as small as possible (i.e., maximum 50 mm).
- Look for overprotection of highly stressed members.
- Potential measurements may be disturbed by electromagnetic noise, which is pitched up by the cable; this noise may be caused by communication equipment, radar, etc.
- Any defect in reference electrode cable insulation may cause erroneous potential readings. The magnitude of the error will depend on the size and location of the insulation damage.
- Check the battery voltage of voltmeter regularly. This may be done by using mercury batteries, which have stable voltages.

5.6 Inspection of Rectifiers

This section outlines procedures and practices for inspection of rectifiers. Being a source of protection current, rectifiers should be inspected at least on biweekly basis. During the inspection, DC voltage and current outputs should be observed and recorded. Any changes in voltage and current from normal operation figures should be recorded and the reason should be determined.

5.6.1 Records

Readings of rectifier output should be recorded on a form prepared for this purpose and forwarded to designated office where they are compared with prior readings to determine abnormality.

5.6.2 Major Points for Routing Inspection

Inspection of rectifier stations should include the following as a minimum:

- Observing and recording readings of DC voltmeter and ammeter.
- Observing and recording figures of alternating current (AC) kilowatt-hour meter.
- Where seasonal variation causes substantial changes in rectifier current output, rectifier taps may require adjustment during routine inspection to prevent overprotection.
- Inspection for level of transformer oil in the tank.
- Inspection for appearance of silica-gel breather. If the color is changed to purple, replace the cartridge and reactivate the old one by heating in an oven.
- If the system is for protection of internal surfaces of a water tank and the system is not automatically controlled, manual adjustment of current should be done to prevent overprotection when the tank is empty or partially filled. This may cause sever deterioration of coating.

5.6.3 Annual Inspection

In addition to the above-mentioned items, at least once per year (usually at the time of complete annual survey) rectifiers should be inspected systematically for the following:

- Inspection of exterior and interior of units for cleanliness and good shape of paint.
- Inspection of all bolted current carrying connections for tightness and cleanliness.

- Inspection of all ventilating screen in air-cooled units. In case of obstruction of air vent, cleaning the vent.
- Inspection of indicator meters for accuracy and calibration if required.
- Inspection of insulated wires to see if insulation has cracked or been damaged.
- Oil should be clear and almost colorless in oil-immersed units. A cloudy or turbid appearance indicates failing oil. If this is observed, sample of oil should be taken from drain point of the tank. The sample should be tested in accordance with BS 148. Failing oil should be replaced with proper grade of transformer oil suitable for low-tension AC power supply.
- Inspection of all protective devices such as fuses, circuit breakers, lightning arrestors, and earth connections should be done, to ensure they are undamaged and in satisfactory condition.
- If the system is installed in the hazardous atmosphere and the equipment is flameproof, inspect and follow the recommendations in Parts 1 and 2 of BSI CP 1003 and specially inspect earthing system, circuit breaker, and surge diverter.
- Check for performance and efficiency of rectifying stack of unit, especially those located under hot weather condition. This is very important where the rectifier energy supply is from regional supply and the annual cost of power exceeds the replacement cost of stack.

5.7 Inspection of Ground Bed

This section deals with inspection and routines to ensure the continued effective operation of impressed cathodic protection ground bed, both onshore for buried structures and offshore for jetties and ship hulls and submarine pipeline.

5.7.1 On-land Ground Bed

Inspection of ground bed should be made in conjunction with inspection of rectifier. The inspection should include the following as a minimum.

Inspect and ensure that there has been no disturbance of the earth above the header cable and line of anodes. If construction activity is noted in the vicinity of ground bed, the inspector should notify the authorized parties to prevent unintentional damage to the ground bed. If new construction involves installation of underground structures, test should be conducted to determine whether they will be affected by field potential gradient or stray current.

Inspect to ensure that no part of the ground bed is subject to washing (by flooding, etc.) or exposure of cables. If such a failure is observed, the insulation of cables should be checked for damage and damaged points should be marked and recorded for later repair or maintenance.

If the ground bed is the type of replaceable deep anode system, periodic inspection should be carried out for gas blockage, anode dimension, retrofit rope, and cable.

If cathodically protected rebar in concrete structure is located in the ground or above ground, acidic conditions may be generated during cathodic protection current discharge at the anode–concrete interface, which may affect both concrete and anode.

Therefore, visual inspection of ground bed should be done periodically for any apparent changes in concrete surface conditions particularly to detect lamination and cracking due to the temperature rise or acidic condition caused around ground beds in highly resistant concrete.

5.7.2 Submerged Ground Beds

The two most commonly used impressed anodes for offshore structures are platinized titanium and niobium anodes. The anodically formed magnesium dioxide on the surface of these anodes in seawater will adversely affect performance by favoring oxygen evolution over chlorine evolution. The anodic dissolution rate of platinum is approximately 10 times greater when oxygen evolution is the predominant reaction. Silting of platinized anode will shorten the anode life due to acid attack to both the platinum and titanium or niobium substrates.

Also, the formation of scale, biofouling, the presence of certain organic materials, low-frequency AC ripple, and even short periods of current reversal could adversely affect anode performance. Therefore, these anodes should be inspected by divers annually, and in the case of shielding the passage of current by fallen nonconductive sheets, marine growth, or silt, cleaning should be done. Also, divers should thoroughly inspect the route of anode's cable for probable cable mechanical damage by impeller of ships, falling objects, etc.

In harsh environments of sea, it is desirable to inspect impressed anode cables passing thorough the splash zone. If the cables pass through conduits, they should be inspected for mechanical damage.

5.8 Inspection of Offshore Sacrificial Anodes

Offshore platforms, ships, and submarine lines are mostly protected by sacrificial anodes. Although these systems are designed for a specific period of time, known as designed life, due to unknown environmental changes or events of increasing the surface area that has to be protected by anodes, the consumption of anodes will increase and the overall life of the cathodic protection system will be affected. To estimate the remaining life of anodes and for future maintenance planning, it is necessary to inspect at least every 5 years the sacrificial anodes of such structures. This inspection consists of cleaning the anode surface, dimensional measurements, calculating the consumed percentage of anode, and predicting the remaining life of the system. Anodes' potential should be measured during inspection.

Potential readings of sacrificial anodes should be close to the figures in Table 5.3. If the measured potential is less negative than figures in Table 5.3, it means the current output is high and caused the anode to be polarized to relatively positive potentials. Readings more negative than those figures in Table 5.3 may be erroneous. In this case, the measuring equipment should be checked and the reference electrode calibrated.

Table 5.3 Potentials of Some Sacrificial Anode Alloys at Ambient Temperatures

Alloy	Environment	Potential Volt vs Ag/AgCl
Al–Zn–Hg	Seawater	−1.00 to −1.05
Al–Zn–In	Seawater	−1.05 to −1.10
Zn	Seawater	−1.00 to −1.05

5.9 Current Survey

This survey helps to determine the distribution of current along a cathodically protected structure. This test should be conducted on pipeline annually and on plants when and where required.

5.9.1 *Pipelines*

Permanent test leads provided on pipeline are used to determine the direction and amount of current flowing in a particular length and span of pipe. Two methods have been used for current measurement of pipelines:

1. Two-wires method.
2. Four-wires method.

In the first method, the wires bridge a span of known length (usually 30–60 m); therefore, by measuring the potential drop across the span together with the polarity of instrument connections, the direction and amount of current can be determined. The amount of current can be calculated by Ohm's law where the resistance of span of pipe can be extracted from available steel pipe resistance table.

In the four-wire method, at each end of the test span two test wires are connected. The two inner leads serve as current measuring and measuring of external circuit resistance, the two outer leads are used for calibration of the test span. This method permits more accurate pipeline current measurements. The length of pipe span and pipe resistance can be accurately measured in this method by using outside leads and passing a known amount of current between them.

5.9.2 *Offshore Structures*

To ensure the effective current distribution on a cathodically protected offshore structure, many computerized programs are available. Also, in the circuit of each anode of some cathodic protection system a fuse in installed. By using a shunt and voltmeter, it is possible to measure the current passing to each anode. This should be done every 6 months and any differences with respect to the previous measures should be recorded.

If the impressed current cathodic protection system is one of the automatically controlled current type, then the modular units should be inspected and calibrated bimonthly.

5.10 Inspection Following Failure Report

Periodic inspection and survey may indicate failures or unusual conditions in the protection system. Problems revealed during periodical survey are dealt with in this section for further inspection.

5.10.1 Increase in Circuit Resistance

If periodic inspection of rectifier output indicates a drastic increase in circuit resistance of system, further inspection survey is required to locate the nature of the problem.

When an increase in resistance is noted but the circuit is not completely open (i.e., rectifier ammeter indicates some current output, although less than usual), a conductive-type cable locator receiver can be used to follow the AC hum on the cable. If this indicates the cables (negative and positive including ground bed header) are in proper condition, then one or more anodes may have failed. If there is a header cable break along the line of anodes, the signal will drop to essentially zero in the vicinity of the break.

To locate the failed anodes, two reference electrodes should be used to determine over-the-line potential profile (made along the line of anodes with the rectifier energized). Potentials are measured between a remote reference and a copper sulfate electrode located over the line and moved by 60–90 cm increments along the line of anodes. The potential profile will show positive potential peaks at each working anode. Any missing peak, then shows the position of anodes that are no longer working and require repair or replacement.

In the case of deep well ground beds, gas blockage may be the cause for increased resistance. Impressed anodes of submerged systems may be covered by silt, scale, or biofouling and cause the increase in circuit resistance.

5.10.2 Stray Electric Currents

Where the presence of stray electric currents is suspected, e.g., in proximity to DC electric traction systems or where varying structure/electrolyte potentials indicate the possibility of such currents, it is sometimes necessary to determine more accurately the extent of stray current effect on the structure. This can be done by plotting the potential field in the area, using a stationary reference electrode, or a structure, as a reference point. The effect due to the operation of the electric traction system can be found by making a potential survey during the time the system is working and repeating the tests after the system has been shut down.

5.10.3 Tests for Electrical Continuity

Tests should be carried out whenever the continuity of the structure is in doubt, to locate any discontinuities.

5.10.4 Problems Associated with Galvanic Anode Installation

Due to low driving voltages available from galvanic anodes, resistance in the connection can cause a marked decrease in their current outputs. Therefore, if there is a marked decrease in the output of a galvanic anode and there is no reason to believe that it is reaching the end of its life, then inspection should be done for broken header cable or anode lead. Also, all connections between anode lead and pipe cable in test box should visually be inspected for cleanliness and tightness.

5.11 Inspection and Survey for Efficiencies of Coatings

Adequate corrosion protection of buried or submerged metallic pipelines can be achieved by using a combination of two corrosion control methods:

1. Coatings and
2. Cathodic protection.

Also, for structures such as ship's hulls, wharf, bulkheads, and the interior of tanks, the coating and lining must serve as a part of the cathodic protection system. Dielectric materials applied to the metal surface exposed to the corrosive environment isolate the metal from corrodent and provide primary barrier against attack.

Cathodic protection completes the protection program by counteracting the electrochemical dissolution of the metal surface at damaged locations or flaws in the coating. Where the coating is not bonded to the pipe surface, the problem is how to protect beneath the coating. The void or crevice between the pipe surface and the coating may contain corrosion products, deposits, or electrolyte. Unbonding due to poor application of pipe coating and disbonding, which means the bond of the coating to the metal surface has been disrupted, usually by the effects of cathodic protection, have deleterious effect on the efficiency of cathodic protection. The usual cathodic protection inspection and potential measurements do not detect the state of polarization of the pipe surface beneath unbounded coating.

This section describes the different inspection methods and techniques to evaluate the effectiveness of coating. The need for these tests is felt when unusual variations are observed in pipe to soil potential profile.

5.11.1 Coating Resistance Measurement

This is the most practical method used for determining the effectiveness of a coating. This method is based on direct measurement of coating resistance by measuring current and voltages across a section of pipe in field. In this method, useful data on current requirements for cathodic protection can also be taken concurrently.

A current interrupter automatically switches off the circuit of some batteries, as power source, in temporary ground bed and pipeline. Then potential and line current measurements continue to change in accordance with the established on–off cycle. Test data may be taken for a section of 8 km. The average V in millivolts divided by the current collected I in milliamperes will give the resistance to earth, in ohms, of the pipeline section tested.

5.11.2 Attenuation Test Method

Coating effectiveness may be evaluated using the attenuation method, which involves a limited amount of field measurements by interruption of rectifier(s). The measurements lead to a figure that is termed "attention constant" per kilometer.

From this figure, spread of protective current can be evaluated and the approximate distance between drain points can be estimated. Rating tables (see Table 5.4) for attenuation constant can be used for interpretation of taken field data and to describe the relative effectiveness of pipe coating. Also, based on the attenuation constant value, the maximum distance between drain points can be estimated and location for installation of new rectifiers, to supplement ineffective cathodic protection system, may be decided.

5.11.3 Pearson Method

Holidays (defects) in the coating of a buried pipeline may be located by this method. Also, large disbonding in pipe coating may be found. It can be used successfully only if a pipeline has been carefully backfilled and the soil is compact around it. Also, there must be some moisture content in the surrounding soil. The system includes a source of AC transmitter with AC signal being impressed upon the pipe. Most of the current flows between the pipe and earth at discontinuities in the coating, can be detected by means of an amplifier and pick-up headphone at the surface.

5.11.4 Coating Inspection by C-Scan

This system is based on current attenuation along the pipe and is largely automatic. An electrical current applied to a well-wrapped buried metal pipeline will decrease gradually with increasing distance from the current injection point, as the current escapes to earth through the coating. If the coating has as uniform thickness and bonding to the pipe, the strength of the signal current on the pipe will decline logarithmically. But if there is a low-resistance electrical path from the pipe direct to the soil at any point, there will be a substantial local increase in the rate of loss of signal current.

Based on this theory, it will identify sections of pipeline between measurement points where coating defects exist. By using a computer, data can be monitored or stored on disk or tape for further analysis.

5.11.5 Visual Inspection

Where the structure is exposed for any purpose, it should be examined for protection level and if coated, the condition of the coating should be assessed and recorded as

Table 5.4 Spread of Protection Relative to Attenuation Constant

	Spread of Protection		
Rating Level	**Range of Values Attenuation Constant "a" per km**	**Evaluation of Spread of Protection**	**Approximate Distance between Drain Points (see Note) km**
1	Less than 0.025	Excellent (E)	Over 128
2	0.025–0.050	Very good (VG)	64–128
3	0.050–0.10	Good (G)	32–64
4	0.10–0.20	Fair (F)	16–32
5	0.20–0.40	Poor (P)	8–16
6	More than 0.40	Very poor (VP)	Less than 8

Based on assumed values as follows; dVo (to remote cell at each drain point) 1000 mV, dVx (At midpoint between drain points due to each of drain points to give total change at drain point of 200 + 200 = 400 mV) 200 mV, e.g., for "a" = 0.025.

$$x = \frac{\ln 1000 - \ln 200}{0.025} = \frac{6.90 - 5.3016}{0.025} = 64.0$$

Distance between drain points is 2X = 128 km.

part of the history of cathodic protection system. This should also be considered for interiors of storage tanks and vessels, ship's hull, and water box of heat exchanger.

By continuous pipe to soil potential logging technique, localities of pipelines that fail to comply with protective criteria can be identified. Excavation of bell hole to reach the pipe surface and visual examination of coating at this anomaly section reveals the coating condition. Close interval pipe to soil potential survey around the pipe in excavated section indicates the real potential drop in soil above the pipe. Visual inspection also helps to obtain information on bacteria activity around the pipe and if so, cathodic protection level has to be corrected.

Visual inspection also gives evidence of the type of coating deficiency, i.e., unbonding, which is usually due to poor material or application, versus disbonding, which is caused by overprotection. For those offshore structures and ships where weight saving is significant, coating offers special advantages in reducing the total amount of required protection current and anodes. Therefore, diver visual inspections of coating is necessary for providing reliable long-term corrosion protection to offshore structures.

In ship hulls and their ballast tanks, coating prolongs the anode life, then increases the replacement intervals. This in turn brings down the high cost of dry docking. Routine inspection of coating in conjunction with anode's dimensional measurements during deballasting will indicate both efficiency of coating and anode performance.

Since the space in water boxes of exchangers and pressure vessels is limited, it is not possible to install massive sacrificial anodes. Coating will reduce the surface area and required current densities. Routine inspection of these surfaces should be a part of overhaul program in plants to ensure effective corrosion protection.

5.11.6 Over-the-Line Potential Survey

In routine inspection, pipe to soil potential is measured every kilometer along the pipeline or where test points have been installed.

When unusual variations in pipe to soil potentials is observed, an over-the-line potential survey helps to pinpoint locations of low point in potential profile that need close attention. For measuring closely spaced potential readings, i.e., 15 m, two methods can be used.

5.12 Data Recording and Analysis

5.12.1 Data Recording

Records should be used to demonstrate the operational history at any time during the working life of a cathodically protected system. For this reason, it is recommended that records be retained for the life of the structure. Printed forms should be used to record field data. The forms are used for the following purposes:

- To save inspector's time in field, by minimizing the amount of writing.
- To establish a uniform manner of recording and easy comparison of results and recognition of changes.

The corrosion engineers may design forms to suit their particular requirements. However, the following information should be included:

1. Dates of survey and current control procedures applied.
2. For anode replacements, types, location, and date of replacement.
3. Any damage to the structure, nature and extent of repairs carried out.
4. Condition of coating at failure points and remedial action taken.
5. The location of any structure corrosion observed and, if possible, the cause of corrosion.
6. Details of any alterations made to the structure.

5.12.2 Analysis of Data

In evaluation of results from inspection and potential measurements, corrosion specialists should always be consulted when irregular values have been recorded.

Irregular potential readings may be caused by either systematic or random measurement errors.

5.12.2.1 Sacrificial System

5.12.2.1.1 Less Negative Potentials

Possible reasons for potentials less negative than −0.85 V ($Cu/CuSO_4$) in connection with sacrificial anode cathodic protection systems are

- Low design current density
- Consumed anodes
- Loss of anodes
- Passivation of anodes by inferior chemical composition
- Bad distribution of anodes
- Larger coating breakdown than expected
- Larger surface area than expected

5.12.2.1.2 Too Negative Potentials

Too negative potentials seldom occur in connection with sacrificial anode cathodic protection. One reason may be that the potential readings are taken close to anodes.

5.12.2.2 Impressed Current System

- **Less negative potentials**

  When impressed current system is utilized, potentials less negative than the minimum value may be caused by
  - Failure on fixed impressed current anodes.
  - Failure on fixed reference electrodes.
  - Failure on cables.
  - Failure on rectifiers.
  - Unsatisfactory distribution of anodes.
  - Low design current density.
  - Larger coating breakdown than expected.
  - Larger buried submerged steel surface area than expected.

- **More negative potential**
  Potentials more negative than the maximum values may occur in connection with impressed current systems, especially close to anodes. Possible causes may be
  - High anode current output
  - Unsatisfactory distribution of anodes.
  - Failure on fixed reference electrodes for potentiostatiscally controlled impressed current systems.
  - Failure on rectifiers.
  - Failure on potential reading.
  - Failure on anode shields
  - IR drop.

Possible causes for irregular potential reading on sacrificial anodes of submerged structure

- Passivation; may be as high as the potential of −0.7 V (Ag/AgCl).
- Polarization of anode due to large anode current output.

5.12.2.4 Rectifiers Data Analysis

Applied voltage normal, current low but not zero:

- Deterioration of anodes or ground beds.
- Drying out of soil around ground bed, or some anodes no longer immersed.
- Accumulation of electrically produced gas or marine growth around anodes.
- Disconnection of some of the connections to individual anodes of ground bed or anode system.
- Disconnection of part of protected structure.

Applied voltage normal, current zero:

- Severance of anode or cathode cables.
- Complete failure of anodes.

Both current and voltage are low:

- Tapping on transformer set too low.
- Transformer or rectifier failing.
- Electricity supply faulty.

Applied voltage and current correspondingly high:

- Tapping on transformer set too high.

Applied voltage and current normal but structure potential insufficiently negative:

- Break in continuity bond or increased resistance between point of connection and point of test.
- Greatly increased aeration of soil at or near point of test due to drought or increased local drainage.
- Alteration of environment causing rapid depolarization, or increase in oxygen content of water due, for example, to reduced level of pollution.
- Faulty isolation equipment.

- Protected structure shielded or otherwise affected by new structures.
- Failure of cathodic protection system on another part of the same structure or on a secondary structure bonded to it.
- Deterioration of, or damage to, protective coating.

Applied voltage and current normal but potential abnormally negative:

- Break in continuity bonding at position further from the point of application than the point of test.
- Decreased aeration of soil or electrolyte at point of test.
- Reduction in rate of flow of electrolyte.
- Secondary structures have been removed or have been cathodically protected or bond to them broken.

Applied voltage and current normal but structure/electrolyte potential fluctuates:

- Interference from DC source.

Applied voltage and current normal but structure/electrolyte potential is positive:

- Immediate action should be taken by inspector if abnormally positive potentials indicate that connection has, in some way, been reversed.

5.13 Interferences

Interference from cathodic protection arises where a foreign structure intersects the DC path between the anode and cathode. Where the current enters the structure, the effect is cathodic. Where it leaves the structure, the effect is anodic, and the rate of corrosion of that position may be increased.

Galvanic anodes do not usually cause interference to other structures. Interference problems are more probable with impressed current systems because of the electrolyte voltage gradients usually associated with the pipe or anodes. Therefore, interference is discussed under two separate headings as cathode field and anode field interferences.

Testing should be carried out and remedial action taken to correct adverse effect to foreign structures. Follow-up testing is necessary specially when there is an increase in cathodic protection current above that originally applied to the primary structure.

5.13.1 Cathode Field Interferences

When a structure is placed under cathodic protection, the electrolyte in immediate vicinity becomes more negative. This creates a "cathode field" in the surrounding of the protected structure. If a portion of foreign structure is located in this cathode field, current will tend to be discharged from the foreign structure because the soil in the cathode field will be more negative with respect to the foreign structure.

5.13.2 Measurements of Cathode Field Interferences

Test may be conducted by measuring the potential of foreign structure with respect to a half-cell at the area of strongest cathodic field. Readings should be taken with the cathodic protection power source alternately switched "on" and "off." A tendency for interference is indicated if the potential of the unprotected structure increases (because it is less negative) when the power source is turned "on" and current is drained from the protected structure.

5.13.3 Anode Field Interferences

The electrolyte is more positive in the vicinity of an impressed cathodic protection ground bed. If a portion of foreign structure lies within this field, it will tend to pick up current from the surrounding positive electrolyte. This current will flow in the structure toward the positive electrolyte. This current will flow in the structure toward a less positive area (such as at low resistivity area) in order to flow to the protected structure. If the current is discharged over a sufficiently large area, the current density will be very low, and negligible damage to the foreign structure will result, but if it is discharged to the soil in a limited area, rapid corrosion may occur. While anode field interference will not induce corrosion on the foreign structure in the area of current pickup, in case of coated structure, it is quite possible that the induced potential may be high enough to damage the coating.

5.13.4 Measurements of Anode Field Interferences

Where an unprotected structure passes through, or lies within, an anode field, steel to electrolyte potentials of the unprotected structure should be measured near the anode and at several points remote from the anode. Steel to electrolyte potentials will be more negative near the anode and less negative remote from the anode.

5.13.5 Corrective Measurers

If tests indicate any tendency for interference, it should be reported to corrosion specialist for corrective measures.

Bonding the structures through resistor, use of galvanic anodes and extra coating as electrical shield are same customary methods to eliminate interferences.

5.14 Meters and Equipment

5.14.1 Reference Electrode

The potentials of metals given in electrochemical series are related to the so-called standard hydrogen electrode (SHE) of which the potential is set arbitrarily at ± 0 mV. Such an electrode consists of a platinum wire or platinum mesh that is immersed in a solution with hydrogen ion activity of one and flushed with hydrogen gas at atmospheric pressure (760 Torr $= 1.013 \times 105$ Pa). Handling of such an electrode is naturally very awkward. In practice, therefore, it is replaced by so-called reference electrodes. The most important reference electrodes and their potentials

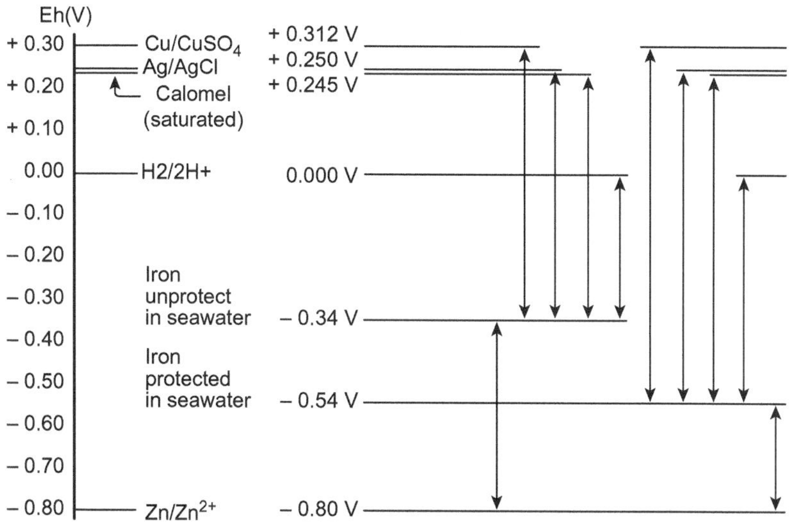

Figure 5.8 Reference electrodes.

against SHE are reproduced in Figure 5.8. The characteristics of these are shown in Table 5.5.

5.14.2 Copper/Copper Sulfate Electrode

The most common type of electrode used for measuring the level of protective potentials of a buried pipeline and other buried structures is the copper/copper sulfate reference electrode. Typical copper/copper sulfate electrodes are shown in Figures. 5.9 and 5.10.

Copper sulfate electrode has a half-cell potential that is reasonably constant under a wide range of soil conditions provided that the electrode is kept clean. There should be at least two copper sulfate electrodes in the test kit. These should be matched in potential. The matching may be checked periodically by measuring the potential

Table 5.5 Characteristics of Reference Electrodes

Reference Electrode	Cell	Potential to Standard Hydrogen Electrode at 25 °C V	dE/dT, V/°C
0.1 N Calomel	**$Hg/Hg_2Cl_2/KCl$ (0.1 N)**	**0.334**	-0.7×10^{-4}
0.1 N Calomel	**$Hg/Hg_2Cl_2/KCl$ (1.0 N)**	**0.280**	-2.4×10^{-4}
Saturated calomel	**$Hg/Hg_2Cl_2/KCl$ (saturated)**	**0.242**	-7.6×10^{-4}
Silver chloride	**Ag/AgCl/KCl (0.1 N)**	**0.288**	-6.5×10^{-4}
Silver chloride	**Ag/AgCl/Seawater**	**0.250**	–
Copper sulfate	**$Cu/CuSO_4$ (saturated)**	**0.316**	9×10^{-4}

0.1 and 1.0 N mean 0.1 and 1.0 "normal solution," respectively.

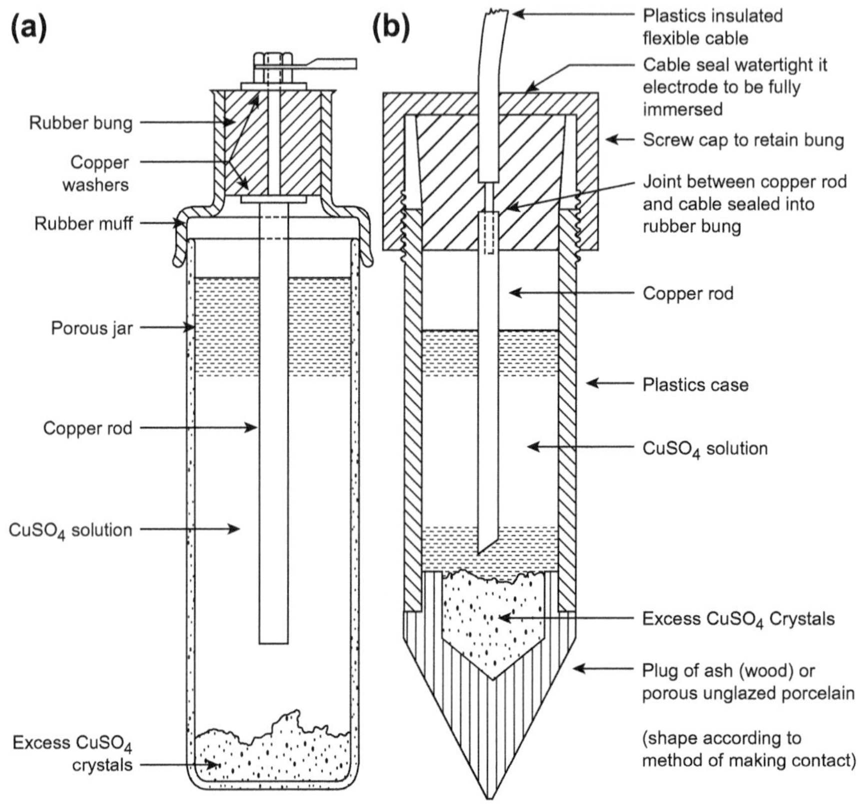

Figure 5.9 Typical copper saturated copper sulfate reference electrodes.

difference between the two when replaced side by side in same soil. For best performance, the potential difference should not exceed about 10 mV. If the difference is excessive, contamination of one or both is indicated. In this case, they should be cleaned and refilled with fresh solution. The copper rod may be cleaned by submerging in a 10% solution of nitric acid for a few minutes. It should be rinsed well with distilled water before reassembling. The electrode should be kept full of saturated copper sulfate solution with a few excess copper sulfate crystals or powder to ensure that the solution will remain saturated. The solution should be replaced periodically to avoid contamination. The porous plug should be kept covered when not in use to avoid drying out. If the electrode does dry out, replace the porous plug.

5.14.3 Silver/Silver Chloride Electrode

This electrode is widely used in potential measurements of submarine pipelines, jetties, wharves, platforms, and interiors of water tanks that are under cathodic

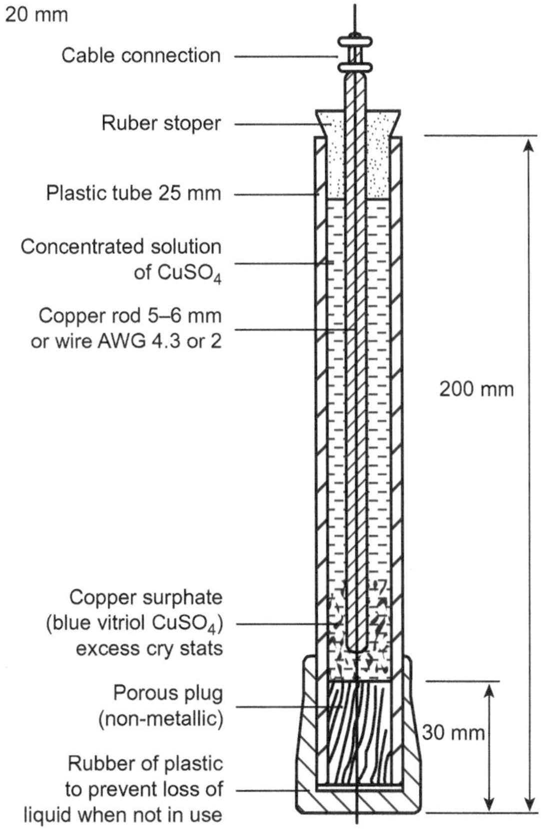

Figure 5.10 Simple Model of Copper/Copper Sulfate Half-Cell.

protection. Copper sulfate may be contaminated by chloride ions; then its use in seawater is limited. Figure 5.11 shows typical silver/silver chloride half-cells.

5.14.4 Pure Zinc Electrode

Zinc should not be considered as an accurate electrode since it's potential is unstable in seawater. It should only be used to get an estimate of the protective level.

Due to the long lifetime of zinc electrodes, they may be used as permanent electrodes. This electrode may be used at points that are not accessible such as center of storage tank's bottom. The reading with respect to zinc depends on the type of zinc and its surface condition. In both seawater and soil variation is in the range of 50 mV. Figure 5.12 shows different types of zinc electrodes.

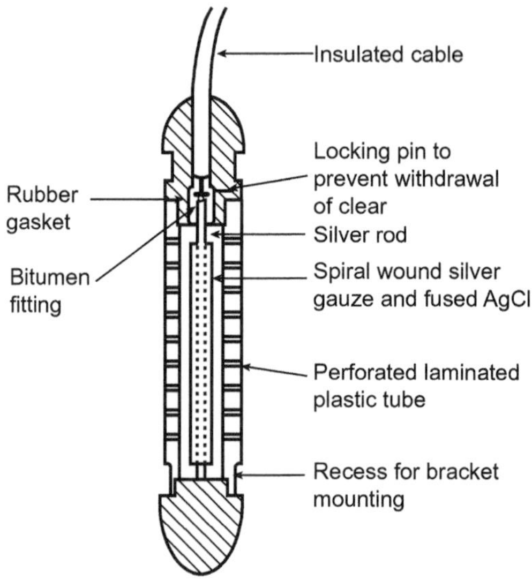

Figure 5.11 Typical silver/silver chloride reference electrodes (admiralty pattern).

5.14.5 Calomel Electrode

The calomel electrode is a half-cell made of mercury/mercurous chloride mixture in contact with a solution either saturated potassium chloride or normal potassium chloride. The solution contacts the electrolyte through a porous plug the same way as copper/copper sulfide electrode does. The stable half-cell potential of calomel electrode makes it suitable for calibration purposes. Laboratory-type electrodes are usually constructed of glass.

5.15 Potentiometer and Voltmeter

5.15.1 Conventional Voltmeter (High Resistance)

In measuring potentials between pipeline and electrolyte, the circuit resistance includes voltmeter resistance and external circuit resistance (the resistance of test leads connecting the voltmeter with pipeline and electrode, the pipe to earth resistance, and the electrode to earth resistance).

In most cases, the major part of the external circuit resistance is electrode to earth resistance; therefore, the electrode should be placed on moist soil during measurement. The electrode in contact with the surface of 1000 ohm-cm soil will have a resistance to earth in the order of 50 ohm. Any conventional voltmeter will take some current and its reading is subject to correction. The correction factor is reversely proportional to voltmeter resistance. If the voltmeter resistance is high compared with the external circuit resistance, the correction factor will be desirably low so that for practical application correction factor is not needed.

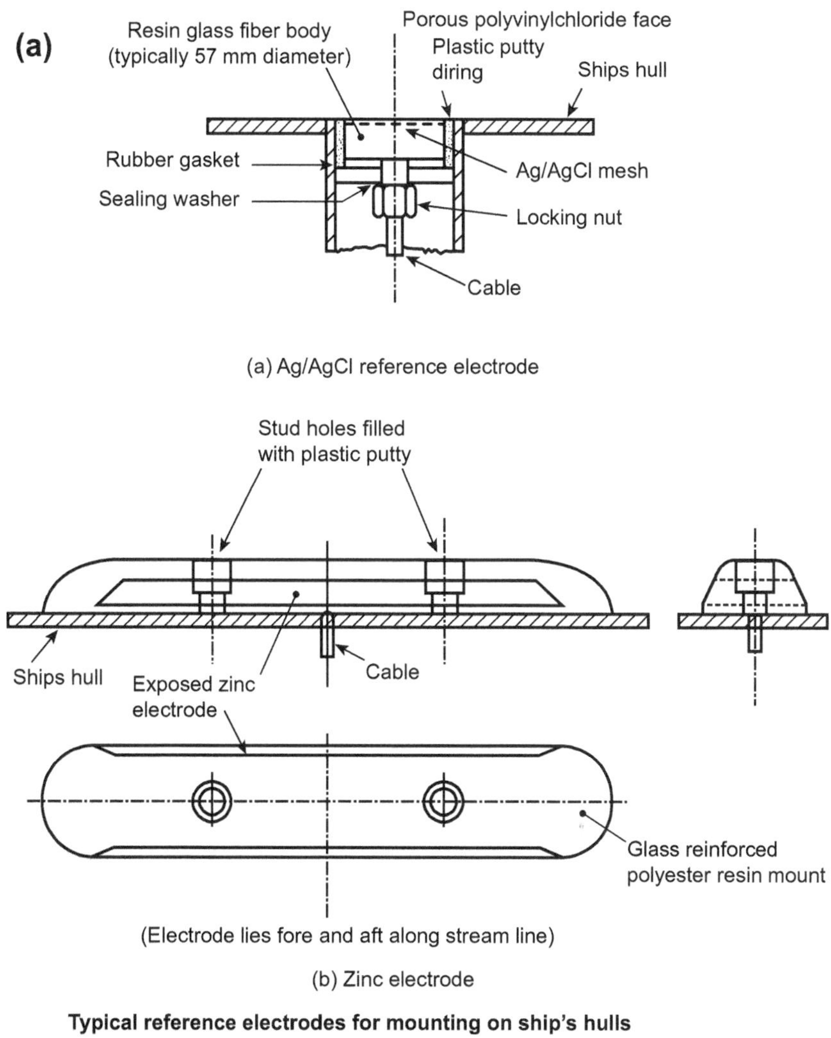

Figure 5.12 Typical reference electrodes for mounting inside plants and ship hulls.

In view of the above, high-resistance or high-sensitivity voltmeter is required for measuring potentials between pipeline and reference electrode. High-sensitivity voltmeter has a resistance of 50,000 ohm/V or higher. DC voltmeters with sensitivity of 200,000 ohm/V with various ranges of full-scale voltage are normally used for pipe to electrode potential measurement. This is a sufficiently sensitive instrument that keeps error in potential reading low enough for practical field conditions and the observer can be satisfied that the potential reading taken is sufficiently accurate.

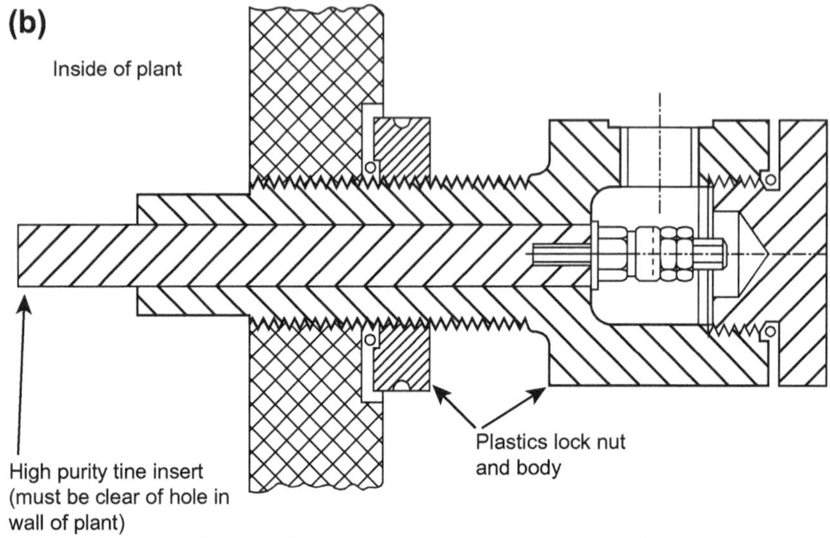

"Screw-in" zinc electrode for measurement of metal/electrolyte potential inside plant

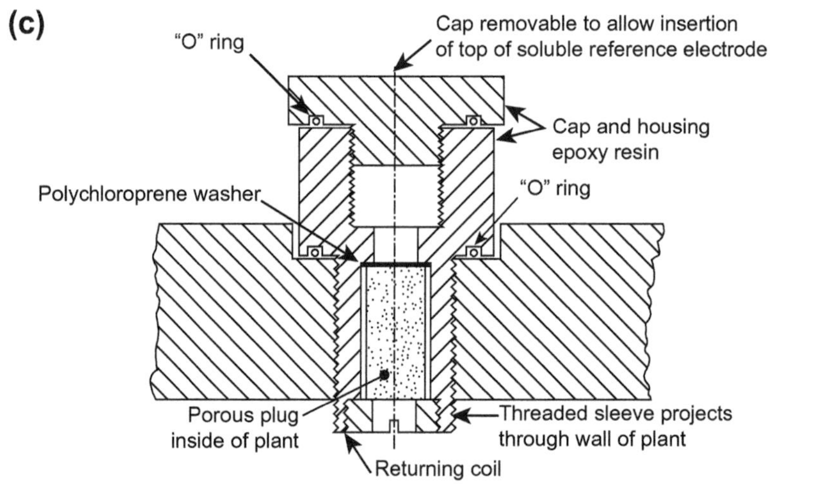

Porous plug reference point for measuring metal/electrolyte potential inside plant

Figure 5.12 *(continued).*

5.15.2 Potentiometer–Voltmeter

This instrument is used for measuring pipe to electrolyte potential in areas having high-resistivity soil. By using this instrument, the effect of high resistance in the measuring circuit will be reduced to a negligible amount. This instrument operates by

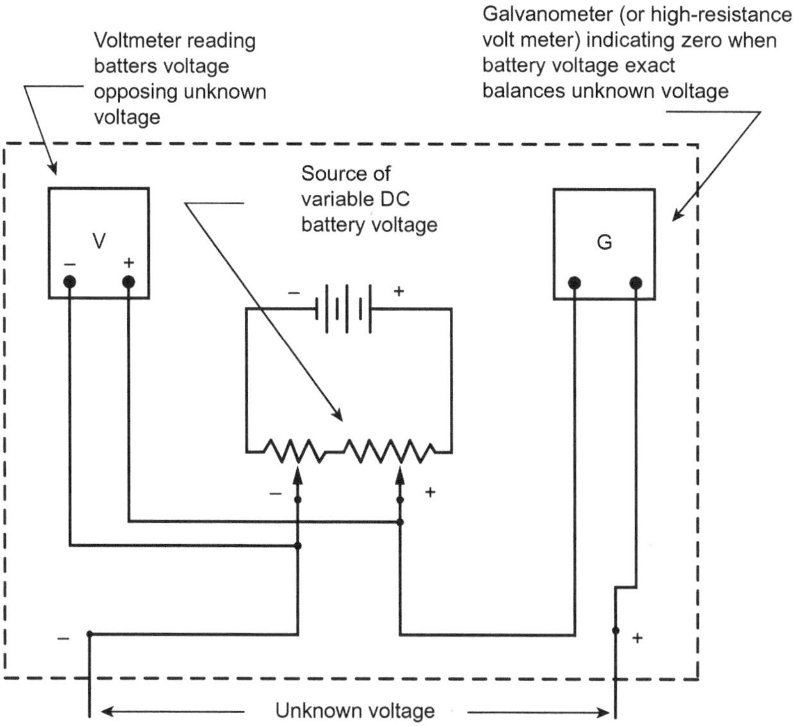

Figure 5.13 Potentiometer–voltmeter circuit diagram.

matching the unknown potential to be measured with a variable-voltage battery supply. A sensitive galvanometer built in the instrument indicates the matching of the two potentials. The indicating voltmeter then measures the matched battery voltage. By this means, the operating current for the voltmeter is taken from the low-resistance battery circuit rather than from the high-resistance measuring circuit. Figure 5.13 illustrates the fundamental circuitry of potentiometer–voltmeter.

5.15.3 Potentiometer

There is a difference between potentiometer and potentiometer–voltmeter. The resistance across the battery terminals of the potentiometer is calibrated so that the voltage taken from it to balance the unknown voltage can be read from indicating dials or solid wires, eliminating the need for an indicating voltmeter. The balance is indicated by a sensitive galvanometer the same way as potentiometer–voltmeter. Circuitry of a potentiometer is illustrated in Figure 5.14.

In some cases, despite the accuracy and advantage, a potentiometer or potentiometer–voltmeter may be impractical to be used. For instance, in stray current areas, potential changes are rapid and it is difficult to keep these instruments on null balance and as such it is impractical to use potentiometer or potentiometer–voltmeter. In this

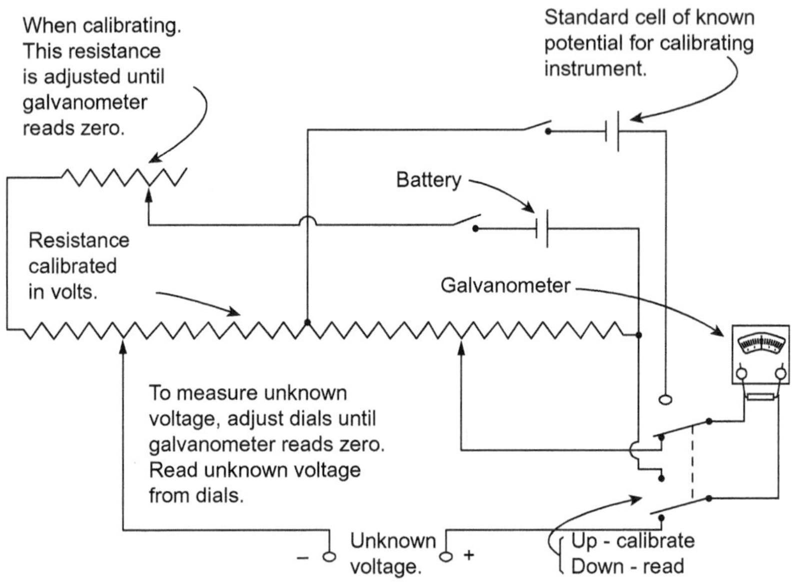

Figure 5.14 Potentiometer schematic diagram.

case, direct-reading voltmeter can be used provided that correction factor of the meter is established for the particular range and multiplied by the reading taken to obtain true potential value.

To obtain external circuit resistance and establish correction factor, potential should be measured on two ranges of specific voltmeter with a known internal resistance. Assume that comparative readings are −0.92 V on 2 V range and −0.85 V on the 1 V range using a voltmeter having a sensitivity of 62,500 ohm/V. Internal voltmeter resistance will then be 125,000 ohm on 2 V range and 62,500 ohm on the 1 V range. The external circuit resistance in ohms (R) and correction factor can then be obtained as follows:

$$0.92 \times \frac{125,000 + R}{125,000} = 0.85 \times \frac{62,500 + R}{62,500}$$

From the above, equation, R (external circuit resistance) = 11,200 ohm m the above and correction factors:
for 2 V range

$$\frac{125,000 + 11,200}{125,000} = 1.085$$

for 1 V range

$$\frac{62,500 + 11,200}{62,500} = 1.179$$

Using the above correction factors, true potential will be
On 2 V range

$$0.92 \times 1.085 = -1.00\ \text{V}$$

On 1 V range

$$0.85 \times 1.179 = -1.00\ \text{V}$$

6 Test Methods

6.1 Test Methods for the Short Circuit Point

Cathodic protection (CP) testing is a very effective method for preventing metal corrosion and deterioration. Some testing needs to be regularly done to avoid such mishaps. Since underground storage tanks, metals, pipes, connectors, or tubes are located in the ground, they frequently come in close contact with soil. This is where the corrosion appears. If not treated at the earliest, it can prove to deliver harmful results. With the appropriate use of direct electricity current in order to curtail any sort of corrosive action, procedures for CP testing take care of metal deterioration in a manner that makes the underground storage tank the cathode of an electrochemical cell.

In order to constantly battle these issues or avoid them, companies must search and locate an efficient and favorable compliance testing provider. Since, compliance testing providers undergo frequent fuel inspections, leak detections, and provide remedies for fuel infiltration, users can be assured that their petroleum systems will operate very smoothly. Corrosion experts make sure that there are no interruptions with cathodes and confirm that the CP procedures followed are foolproof and beneficial.

A regular system check every 60 days is a must. This enforces consistency on all the products and system performance. CP testing enables your company to initiate a technique of upkeeping your equipment.

CP testing involves certain necessary methods of operation such as regular surveys in order to ensure that the system installed is genuine. It also provides the most cost-effective system to save up on maintenance expenditure. Other surveys for inspection of leak detection and corrosion of the underground storage tanks are also undertaken. If petroleum companies can work toward a regular check on all their devices, products, and other equipment, they can assure themselves of having good quality storage tanks. If tackled with urgency and through inspection, these hindrances can be prevented. All one needs to do is to find a suitable and fine compliance testing provider to counter the occurring problems in time.

6.1.1 Locating Casing Short Circuit

Tests can be carried out to locate the point of short circuit. Fig. 6.1 illustrates the test points and tests setup.

As shown in Fig. 6.1, a battery current, which is measured by an ammeter, is passed between the pipe and the casing. This current will flow along the casing to the

Cathodic Corrosion Protection Systems. http://dx.doi.org/10.1016/B978-0-12-800274-2.00006-5

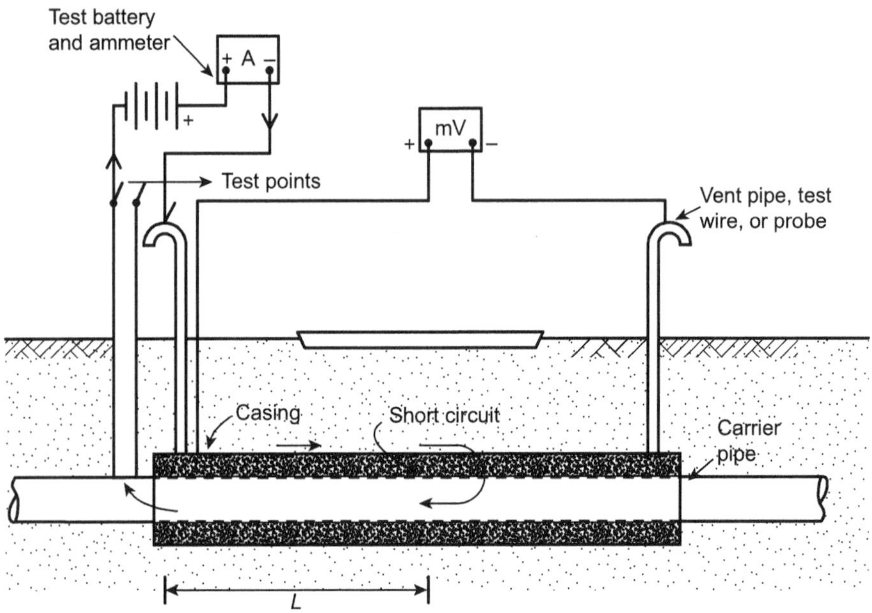

Figure 6.1 Locating casing short circuit.

point of short circuit where it will transfer to the carrier pipe and return to the battery. A millivoltmeter connected between the two ends of the casing will indicate a value dictated by the measured current flowing through the casing resistance, IR drop within the casing length which the current actually flows through. From the voltage and current value, one can obtain the resistance of the casing span, which is subject to current flow. The resistance corresponds to the length of the casing between the point of the short circuit and millivoltmeter connection at the left end of the casing.

From the size and thickness (or weight per unit length) of the casing, one can estimate its resistance per unit length using Table 6.1.

Therefore, the length of the casing span that is subject to current flow can be obtained using the following formula:

$$\text{Length (L in meter)} = \frac{\text{Millivolt observed} \times 10^{-3}}{\text{Current observed in Amps} \times \text{Resistance per meter of casing pipe in ohms}}.$$

The length obtained is the distance between the point of the short circuit and millivoltmeter connection at the left end of the casing.

The short circuit may not be in the casing itself. It may be due to contacts between the test point wires and the casing vent (or the end of the casing itself) or between the test wires and test point conduit mounted on the casing vent. Therefore, inspection should include a thorough examination for these possible contacts.

Table 6.1 Steel Pipe Resistance (Based on a Steel Density of 7.83 gr/cm^3 and Steel Resistivity of 18 micro-ohm-centimeter)

Pipe Size, in	Outside Diameter, in	Wall Thickness, mm	Weight Per Meter, kg	Resistance of 1 m in ohms $\times 10^{-6}$
2	2.375	3.91	5.43	259.84
4	4.5	6.02	16.07	87.78
6	6.625	7.11	28.27	49.89
8	8.625	8.18	42.56	33.15
10	10.75	9.27	60.27	23.4
12	12.75	9.53	73.81	19.1
14	14	9.53	81.25	17.36
16	16	9.53	93.16	15.14
18	18	9.53	105	13.44
20	20	9.53	116.97	12.06
22	22	9.53	128.73	10.95
24	24	9.53	140.87	10.02
26	26	9.53	152.69	9.24
28	28	9.53	164.59	8.57
30	30	9.53	176.65	7.99
32	32	9.53	188.4	7.49
34	34	9.53	200.31	7.04
36	36	9.53	212.21	6.65

Resistance of 1 m in ohms

$$\mathbf{R} = \frac{78.38 \times \text{Resistivity in microohm–centimeter}}{\text{Weight per meter}} \times 10^{-6}$$

6.1.2 Insulation Tests

6.1.2.1 Locating of Insulation Defects

If an insulating flange is found to be defective, a step-by-step check should be made as follows:

An insulated bolt sleeve may be broken down. The shorted bolt may be removed and insulation replaced. To determine which bolt is defective, it is necessary to check each bolt electrically. To check each bolt, an ohmmeter may be used for checking the resistance between the pipeline (flange face) and the bolt. Bolts that are shorted will have zero resistance to the pipe.

If all bolts have a high resistance to the pipe, examine the outer surface of flange properly, once more to ensure that there is nothing that could cause the short. If there is nothing on the outer surface, then check for any possible by passing pipe, since if they are not insulated it would give the indication of a shorted flange.

If the results of the examination in the above step give no indication of a shorted flange, the gasket is shorted, and it will be necessary to take the line out of service to replace the gasket.

In the case of a shorted insulated flange having bolts insulated only on one side, the ohmmeter test would not be applicable because all bolts will have metallic contact

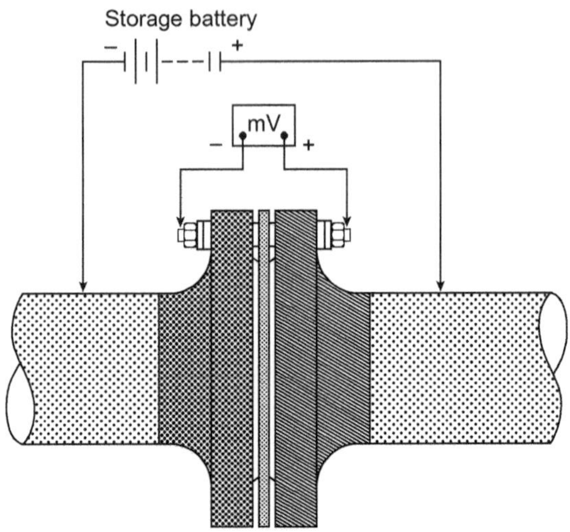

Figure 6.2 Checking the insulated flange for shorted bolts.

and low resistance to the pipe. One method of checking the bolts on such a flange is to remove one bolt at a time and inspect the insulation for damage. This is a laborious and time-consuming job on a large flange. It may be done electrically as illustrated in Fig. 6.2.

With a millivoltmeter connected between the two ends of each bolt in turn, a heavy current (such as from a storage battery as shown in Fig. 6.2) is momentarily passed through the shorted flange. Any shorted bolts will carry this current and show a deflection on the millivoltmeter. Satisfactory insulated bolts will show no deflection.

6.1.2.2 Measuring the Percentage of "Leakage"

Where desired, a test can be conducted to obtain the percentage of "leakage" of an insulating device. This is shown in Fig. 6.3.

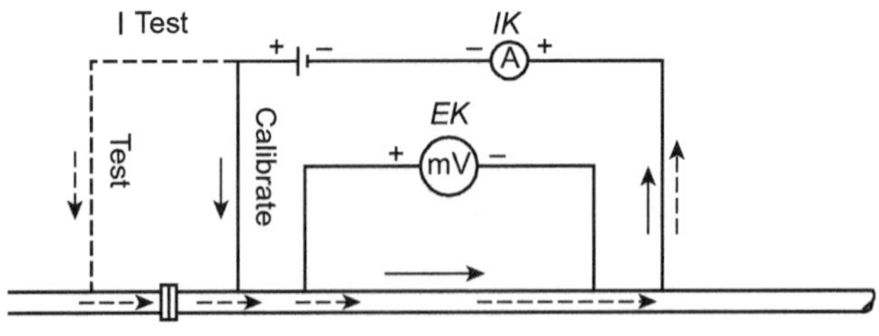

Figure 6.3 Leakage test.

- **Leakage test**

Insulating joints or fittings may become partially or completely shorted due to lightning or other causes. The integrity of insulating fittings must be tested by some reliable method. The performance of these devices is generally critical to the operation of CP systems. One way to measure the percentage of leakage is illustrated below:

- **Example:**

Given the following data,

1. Calculate the calibration factor $\boldsymbol{K} = \frac{IK}{\Delta EK} = \frac{\text{Amps}}{\text{mV}}$
2. **Calculate the percent leakage:**

$$\%\,\text{Leakage} = \frac{K \times E(Test) \times 100}{I(Test)}.$$

Calibration	$IK = \mathbf{+38.0\ A}$
	$EK = \mathbf{+33.5\ mV}$
Test	$I\ Test = \mathbf{+6.0\ A}$
	$E\ Test = \mathbf{+4.40\ mV}$
Calculate	$\boldsymbol{K} = \frac{IK}{\Delta EK} = \frac{38.0\ \text{A}}{33.5\ \text{mV}} = \mathbf{1.13\ A/mV}$
% Leakage	$= \frac{K \times E\ Test}{I\ Test} \times 100 = \frac{1.13\ \text{A/mV} \times 4.40\ \text{mV} \times 100}{6.0} = \mathbf{82.9\%}$

6.2 American Society for Testing and Material Standard Test Method for Half-Cell Potentials of Uncoated Reinforcing Steel Bar in Concrete

This test method is under the jurisdiction of American Society for Testing and Materials (ASTM) Committee C-9 on Concrete and Concrete Aggregates and is the direct responsibility of Subcommittee C 09.03.15 on Methods of Testing the Resistance of Concrete to its Environment. Current edition approved May 29, 1987. Published July 1987. Originally published as C 876-77. Last previous Edition C 876-77.

6.2.1 Scope

This test method covers the estimation of the electrical half-cell potential of uncoated reinforcing steel in field and laboratory concrete, for the purpose of determining the corrosion activity of the reinforcing steel.

This test method is limited by electrical circuitry. A concrete surface that has dried to the extent that it is a dielectric and surfaces that are coated with a dielectric material

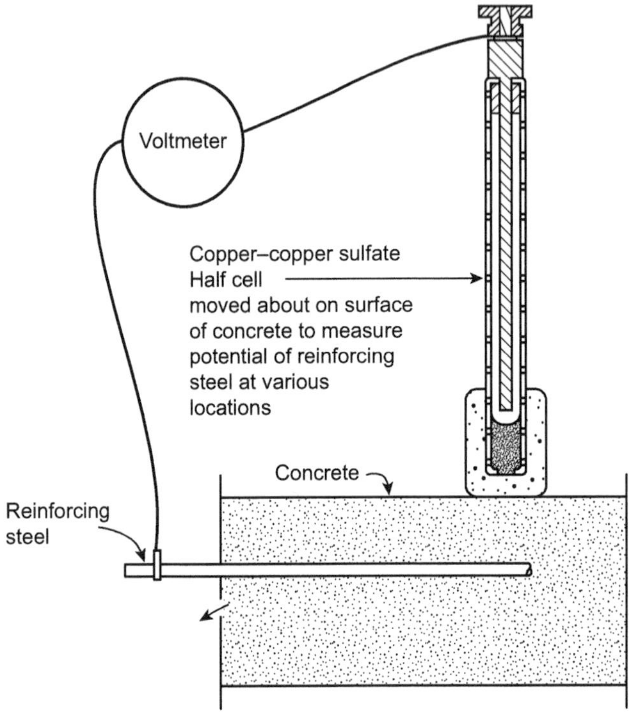

Figure 6.4 Copper/copper sulfate half-cell circuit.

will not provide an acceptable electrical circuit. The basic configuration of the electrical circuit is shown in Fig. 6.4.

The values stated in inch–pound units are to be regarded as the standard. This section may involve hazardous materials, operations, and equipment. This section does not purport to address all the safety problems associated with its use. It is the responsibility of the user to establish appropriate safety and health practices and determine the applicability of regulatory limitations before use.

6.2.2 Significance and Use

This test method is suitable for in-service evaluation and for use in research and development work. This test method is applicable to members regardless of their size or the depth of concrete cover over the reinforcing steel. This test method may be used at any time during the life of a concrete member. The results obtained by the use of this test method should not be considered as a means for estimating the structural properties of the steel or of the reinforced concrete member.

The potential measurements should be interpreted by engineers or technical specialists experienced in the fields of concrete materials and corrosion testing. It is often necessary to use other data such as chloride contents, depth of carbonation, delamination survey findings, rate of corrosion results, and environmental exposure conditions, in addition to half-cell potential measurements, to formulate conclusions concerning corrosion activity of embedded steel and its probable effect on the service life of a structure.

6.2.3 Apparatus

The testing apparatus consists of the following:

6.2.3.1 Half Cell

A copper/copper sulfate half cell (see Note) consists of a rigid tube or container composed of a dielectric material that is nonreactive with copper or copper sulfate, a porous wooden or plastic plug that remains wet by capillary action, and a copper rod that is immersed within the tube in a saturated solution of copper sulfate. The solution should be prepared with reagent grade copper sulfate crystals dissolved in distilled or deionized water. The solution may be considered to be saturated when an excess of crystals (undissolved) lies at the bottom of the solution.

The rigid tube or container should have an inside diameter of not <25 mm (1 in); the diameter of the porous plug should not be <13 mm (½ in); the diameter of the immersed copper rod should not be <6 mm (¼ in), and the length should not be <50 mm (2 in).

Present criteria based upon the half-cell reaction of $Cu \rightarrow Cu^{++} + 2e$ indicate that the potential of the saturated copper/copper sulfate half cell as referenced to the hydrogen electrode is −0.316 V at 22.2 °C (72 °F). The cell has a temperature coefficient of about 0.0005 V more negative per degrees centigrade for the temperature range from 0 to 49 °C (32–120 °F).

Note: While this test method specifies only one type of half cell, that is, the copper/copper sulfate half cell, others having a similar measurement range, accuracy, and precision characteristics may also be used. In addition to copper/copper sulfate cells, calomel cells have been used in laboratory studies. Potentials measured by cells other than copper/copper sulfate half cells should be converted to copper/copper sulfate equivalent potential. The conversion technique can be found in Practice G3, and it is also described in most physical chemistry or half-cell technology textbooks.

- **Electrical junction device**

An electrical junction device should be used to provide a low electrical resistance liquid bridge between the surface of the concrete and the half cell. It should consist of a sponge or several sponges prewetted with a low electrical resistance contact solution. The sponge may be folded around and attached to the tip of the half cell so that it provides electrical continuity between the plug and the concrete member.

- **Electrical contact solution**

In order to standardize the potential drop through the concrete portion of the circuit, an electrical contact solution should be used to wet the electrical junction device. One such solution is composed of a mixture of 95 mL of the wetting agent (commercially available wetting agent) or a liquid household detergent thoroughly mixed with 5 gal (19 L) of potable water. Under working temperatures of less than about 10 °C (50 °F), approximately 15% of either isopropyl or denatured alcohol must be added to prevent clouding of the electrical contact solution, since clouding may inhibit the penetration of water into the surface to be tested.

- **Voltmeter**

The voltmeter should have the capacity of being battery operated and have a $\pm 3\%$ end-of-scale accuracy at the voltage ranges in use. The input impedance should not be <10 mega-ohm when operated at a full scale of 100 mV. The divisions on the scale used should be such that a potential difference of ≤ 0.02 V can be read without interpolation.

- **Electrical lead wires**

The electrical lead wire should be of such a dimension that its electrical resistance for the length used will not disturb the electrical circuit by >0.0001 V. This has been accomplished by using not more than a total linear of 150 m (500 ft) of at least AWG No. 24 wire. The wire should be suitably coated with a direct burial-type solution.

6.2.4 Calibration Standardization

6.2.4.1 Care of the Half Cell

The porous plug should be covered when not in use for long periods to ensure that it does not become dried to the point that it becomes a dielectric (upon drying, pores may become occluded with crystalline copper sulfate). If cells do not produce the reproducibility or agreement between cells, cleaning the copper rod in the half cell may rectify the problem. The rod may be cleaned by wiping it with a dilute solution of hydrochloric acid. The copper sulfate solution should be renewed either monthly or before each use, whichever is the longer period. At no time should steel wool or any other contaminant be used to clean the copper rod or half-cell.

6.2.5 Procedure

6.2.5.1 Spacing Between Measurements

While there is no predefined minimum spacing between measurements on the surface of the concrete member, it is of little value to take two measurements from virtually the same point. Conversely, measurements taken with a very wide spacing may neither detect corrosion activity that is present nor result in the appropriate accumulation of data for evaluation. The spacing should therefore be consistent with the member being investigated and the intended end use of the measurements (see Note).

Note: A spacing of 1.2 m (4 ft) has been found to be satisfactory for the evaluation of bridge decks. Generally, larger spacings increase the probability that localized corrosion areas will not be detected. Measurements may be taken in either a grid or a random pattern. Spacing between measurements should generally be reduced where adjacent readings exhibit algebraic reading differences exceeding 150 mV (areas of high corrosion activity). Minimum spacing generally should provide at least a 100-mV difference between readings.

6.2.5.2 Electrical Connection to the Steel

Make a direct electrical connection to the reinforcing steel by means of a compression-type ground clamp, or by brazing or welding a protruding rod. To ensure a low electrical resistance connection, scrape the bar or brush the wire before connecting to the reinforcing steel. In certain cases, this technique may require removal of some concrete to expose the reinforcing steel. Electrically connect the reinforcing steel. In certain cases, this technique may require removal of some concrete to expose the reinforcing steel. Electrically connect the reinforcing steel to the positive terminal of the voltmeter.

Attachment must be made directly to the reinforcing steel except in case where it can be documented that an exposed steel member is directly attached to the reinforcing steel. Certain members, such as expansion dams, date plates, lift works, and parapet rails may not be attached directly to the reinforcing steel and, therefore, may yield invalid readings. Electrical continuity of steel components with the reinforcing steel can be established by measuring the resistance between widely separated steel components on the deck. Where duplicate test measurements are continued over a long period of time, identical connection points should be used each time for a given measurement.

6.2.5.3 Electrical Connection to the Half Cell

Electrically connect one end of the lead wire to the half cell and the other end of this same lead wire to the negative (ground) terminal of the voltmeter.

6.2.5.4 Prewetting of the Concrete Surface

Under certain conditions, the concrete surface, an overlaying material, or both must be prewetted by either of the two methods with the solution to decrease the electrical resistance of the circuit.

A test to determine the need for prewetting may be conducted as follows:

Place the half cell on the concrete surface and do not move.

Observe the voltmeter for one of the following conditions:

1. The measured value of the half-cell potential does not change or fluctuate with time.
2. The measured value of the half-cell potential changes or fluctuates with time.

If condition (1) is observed, prewetting the concrete surface is not necessary. However, if condition (2) is observed, prewetting is required for an amount of time such that the voltage reading is stable (±0.02 V) when observed for at least 5 min. If prewetting cannot obtain condition (1), either the electrical resistance of the circuit is too great to obtain valid half-cell potential measurements of the steel, or stray current from a nearby direct current (DC) traction system or other fluctuating DC, such as arc welding, is affecting the readings. In either case, the half-cell method should not be used.

6.2.5.5 Method A for Prewetting Concrete Surfaces

Method A is used for those conditions where a minimal amount of prewetting is required to obtain condition (1). Accomplish this by spraying or otherwise wetting either the entire concrete surface or only the points of measurements with the solution described. No free surface water should remain between grid points when potential measurements are initiated.

6.2.5.6 Method B for Prewetting Concrete Surfaces

In this Method, saturate sponges with the solution and place on the concrete surface at locations. Leave the sponges in place for the period of time necessary to obtain condition (1). Do not remove the sponges from the concrete surface until after the half-cell potential measurements; place the electrical junction device firmly on top of the prewetting sponges for the duration of the measurement.

6.2.5.7 Underwater, Horizontal, and Vertical Measurement

Potential measurements detect corrosion activity, but not necessarily the location of corrosion activity. The precise location of corrosion activity requires a knowledge of the electrical resistance of the material between the half cell and the corroding steel. While underwater measurements are possible, results regarding the location of corrosion must be interpreted very carefully.

Often, it is not possible to precisely locate points of underwater corrosion activity in saltwater environments because potential readings along the member appear uniform. However, the magnitude of readings does serve to indicate whether or not active corrosion is occurring. Take care during all underwater measurements so that the half cell does not become contaminated and that no part other than the porous tip of the copper/copper sulfate electrode (CSE) half cell comes into contact with the water.

Perform from horizontally and vertically upward measurements exactly as vertically downward measurements. However, additionally ensure that the copper/copper sulfate solution in the half cell makes simultaneous electrical contact with the porous plug and the copper rod at all times.

6.2.6 Recording Half-Cell Potential Values

Record the electrical half-cell potentials to the nearest 0.01 V. Report all half-cell potential values in volts and correct for temperature if the half-cell temperature is outside the range of 22.2 (72 ± 10 °F).

6.2.7 Data Presentation

Test measurements may be presented by one or both methods. The first, an equipotential control map, provides a graphical delineation of areas in the member where corrosion activity may be occurring. The second method, the cumulative frequency diagram, provides an indication of the magnitude of the affected area of the concrete member.

- **Equipotential contour map**

On a suitably scaled plan view of the concrete member, plot the locations of the half-cell potential values of the steel in concrete and draw contours of equal potential through points of equal or interpolated equal values. The maximum contour interval should be 0.10 V.

- **Cumulative frequency distribution**

To determine the distribution of the measured half-cell potentials for the concrete member, make a plot of the data on normal portability paper in the following manner:

Arrange and consecutively number all half-cell potentials by ranking from the least negative potential to the greatest negative potential.

Determine the plotting position of each numbered half-cell potential in accordance with the following equation:

$$\mathbf{f_x} = \frac{r}{\sum n + 1} \times \mathbf{100}, \tag{6.1}$$

where $\mathbf{f_x}$ is the plotting position of the total observations for the observed value in percentage; r is the rank of the individual half-cell potential; and Σn is the total number of observations.

Label the ordinate of the probability paper "Half-Cell Potential (volts, CSE)." Label the abscissa of the probability paper "Cumulative Frequency (percentage)," Draw two horizontal parallel lines intersecting the −0.20 and −0.35 V values on the ordinate, respectively, across the chart.

After plotting the half-cell potentials, draw a line of best fit through the value.

6.2.8 Interpretation of Results

Laboratory testing of reinforced concrete specimens indicates the following regarding the significance of the numerical value of the potentials measured. Voltages listed are referenced to the copper/copper sulfate (CSE) half cell.

If potentials over an area are more positive than −0.20 V CSE, there is a >90% probability that no reinforcing steel corrosion is occurring in that area at the time of measurement.

If potentials over an area are in the range of −0.20 to −0.35 V CSE, corrosion activity of the reinforcing steel in that area is uncertain.

If potentials over an area are more negative than −0.35 V CSE, there is a >90% probability that reinforcing steel corrosion is occurring in that area at the time of measurement.

In laboratory tests in which potentials were more negative than −0.50 V, approximately half of the specimens cracked due to corrosion activity.

Positive readings, if obtained, generally indicate a poor connection with the steel, insufficient moisture in the concrete, or the presence of stray currents and should not be considered valid.

6.2.9 Report

The report should include the following:

- Type of cell used if other than copper/copper sulfate.
- The estimated average temperature of the half cell during the test.
- The method for prewetting the concrete member and the method of attaching the voltmeter lead to the reinforcing steel.
- An equipotential contour map, showing the location of reinforcing steel contact, or a plot of the cumulative frequency distribution of the half-cell potentials, or both.

The percentage of the total half-cell potentials that are more negative than −0.35 V. The percentage of the total half-cell potentials that are less negative than −0.20 V.

6.2.10 Precision and Bias

The difference between two half-cell readings taken at the same location with the same cell should not be >10 mV when the cell is disconnected and reconnected.

The difference between two half-cell readings taken at the same location with two difference cells should not be >20 mV.

The ASTM takes no position in respecting the validity of any patent rights asserted in connection with any item mentioned in the relevant standard. Users of the relevant standard are expressly advised that the determination of the validity of any such patent rights, and the risk of infringement of such rights, are entirely their own responsibility.

6.3 Subsea Pipeline CP Survey Method

6.3.1 Introduction

This Section provides a narrative description of the CP survey method (CPSS) Subsea Pipeline Cathodic Protection Survey System. It describes the components of the system that will be provided, operations, and results. Fig. 6.5 illustrates the system.

6.3.2 System Components

The system provided by CPSS consists of the following components:

- Probe,
- Remote reference electrode,
- Digitizer,
- Data receiver/monitor,

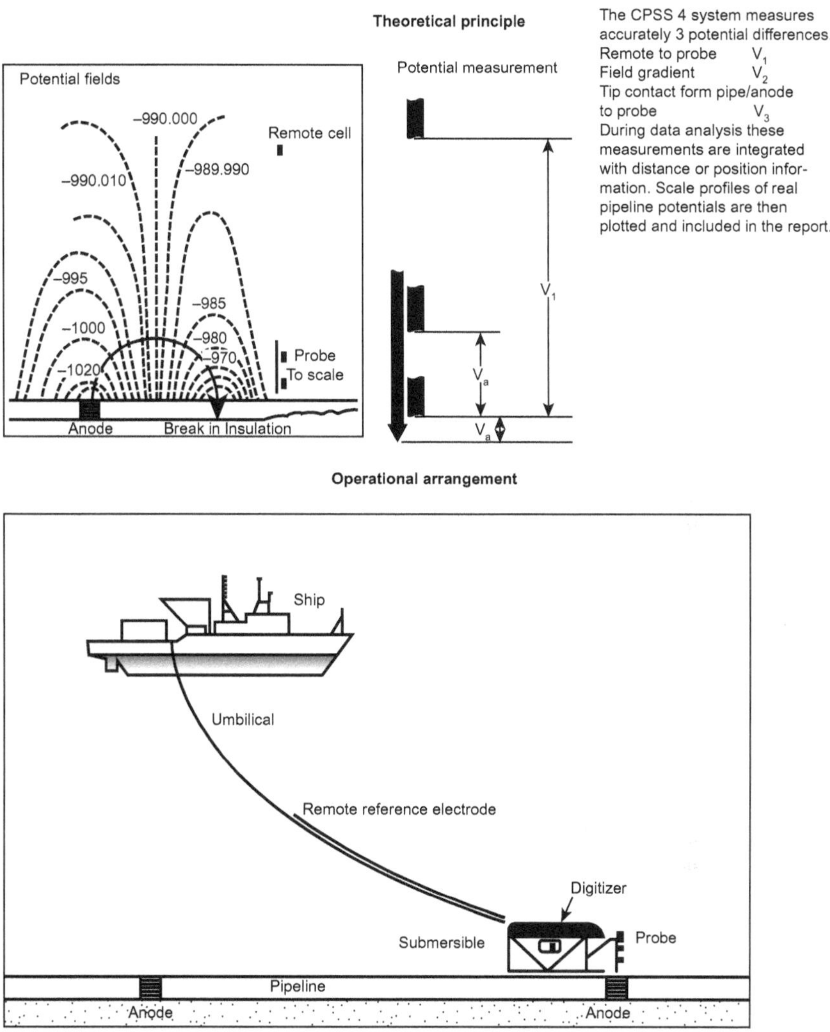

Figure 6.5 Subsea pipeline cathodic protection survey system.

- Compute with software-controlled data recording,
- Survey engineers.

In the following paragraphs, each component is described in terms of what it does and how it links with the rest of the system:

6.3.2.1 Probe

The probe is the primary data-gathering instrument for the survey. Its design, with two precisely located reference electrodes allows the measurement of a radical field gradient.

A spiked tip permits the measurement of pipeline potential and facilitates positive contact with anodes or pipe metal from which calibration measurements may be taken.

A shock-absorbing bracket attaches the probe to the manipulator arm of the submersible. The bracket design ensures stable contact during the measurement of anode and pipe metal potentials and absorbs the stresses placed on the probe during accidental strikes attendant to normal operation.

An armored electrical cable connects the probe to the digitizer. A sleeve on the probe and recesses on the digitizer protect the underwater connectors from accidental damage.

6.3.2.2 Remote Reference Electrode

The remote reference electrode provides a "remote earth" or reference zero against which to measure the data obtained by the probe. During operation, it is positioned 50–100 m from the probe, well away from any electrical fields caused by anodes, the pipe, or the submersible. An electrical cable connects the electrode to the digitizer. Both the electrode and its cable are clipped to the submersible's umbilical, as shown in Fig. 6.5.

6.3.2.3 Digitizer

The digitizer converts the information from the probe and the remote electrode to a digital signal and transmits it through the submersible's umbilical to the data receiver on board the ship. A digital signal is used because it is relatively unaffected by electrical noise interference and a digital signal can be transmitted without any loss of accuracy over much greater distances than an analog signal. The digitizer is powered by a low voltage DC via a cable connected to the umbilical junction box. It is secured firmly to the submersible with clamps to prevent movement.

6.3.2.4 Data Receiver/Monitor

The data receiver/monitor receives the signal from the digitizer via the submersible umbilical. Information arrives at the rate of five times per second, giving a fast response to rapidly changing potentials, and the digital data are transmitted to the computer for high-accuracy recording. The signal channels for remote electrode and field gradient are, in addition, converted to analog form and feed a two-pen chart in the recorder monitor. This chart provides a convenient, real-time, pictorial record of changing pipeline potential and radial field gradient.

The analog signal for the potential of the pipeline versus remote electrode is also routed via an offset correction circuit to a digital voltmeter, and thence, it may be taken to the Customer's video writer if desired. This digital voltmeter gives an approximate indication of pipe potential during a dive.

6.3.2.5 Computer and Software-Controlled Data Recording

The computer is used to process and record all information relevant to each dive. Information arrives along the link from the data received, from the navigation system, from the internal clock, the pipe tracker, if fitted, and from the computer's keyboard. Under software control, this large volume of data is compressed before recording on cassette tapes to maximize the effective capacity of each cassette.

The record of potentials and gradients is of an extremely high accuracy in order that the eventual analysis will be of the highest quality. All records are linked to time so that any additional information, such as corrected navigation fixes recorded during the dive, may be added later.

In addition to its data recording function, the computer provides a software-controlled display of all relevant data channels on its internal digital readout.

To permit a 24 h operation, two Engineers should operate the survey. During operation, some space in a dry environment must be provided for the computer, monitor, and at least one Engineer. The Engineer must be able to communicate easily with the Submersible Operator at all times.

6.3.3 Survey Operation

During the survey, the submersible travels along the pipeline at a speed of 0.5–2 kilometers per hour. The probe is held vertically in the manipulator arm so that its tip is approximately 20 cm above the pipe's center line. For buried pipelines, the submersible must have a pipe-track system to keep the probe above the center line of the pipe. Burial depth information is also required. Potential readings are taken at the rate of five per second.

Changes in the potential trace on the chart usually give advance warning that anodes, damaged areas, or other points of interest are being approached. The Engineer can then warn the Submersible Operator to slow the vehicle down to prepare for more careful examination of those areas. When the point of interest is reached, the manipulator arm is moved so that the probe spike can be stabbed into the relevant area. A good contact is confirmed by a stabilized signal, and the reading and navigation fix number are recorded.

Occasionally, damage to the pipe coating beneath the concrete will show up in the potential readings but will not be readily identifiable visually. In this, and similar instances, the Submersible Operator can be advised to perform additional visual inspection to try to determine the nature and severity of the damage.

Following these general procedures, the survey thus proceeds to completion:

6.3.4 Survey Results

On a real-time basis during the survey, the system provides the following results:

- Pipeline CP potential.
- Anode potentials and approximate output currents.
- Position identification of coating damage.

As mentioned earlier, these results are sufficiently detailed to enable the immediate identification of problem areas.

After returning to their office, Engineers will analyze the recorded data utilizing the comprehensive, detailed, and accurate information to produce a report covering the following:

- General pipeline condition.
- Analysis of pipeline CP levels corrected for ohmic drops in seawater and mud using the combination of high-accuracy field gradient measurement with potential measurement.
- Analysis of bracelet anodes, with a table showing anode locations, corrected potentials, and output currents.
- Analysis of areas of particular importance or unusual interest (coating damage, spool pieces, anode sleds, clamps, risers, insulating flanges, etc.) supported by a table of relevant data.

Pipeline CP levels will be plotted on a chart against distance along the pipe. This will be based on the position fixed and recorded during the dive and analysis of start and stop times. If more accurate, processed navigation fix information is available, this can be used to produce a plot of potential against the corrected position.

6.4 Line Current Survey Test Method

This survey helps to determine the distribution of current along a cathodically protected pipeline. Permanent test leads provided on the pipeline are used to determine the direction and measure the amount of current flowing in a particular length and span of pipe (Fig. 6.6). This survey should be annually conducted on pipelines. Test procedures to carry out this inspection are outlined hereunder:

Test procedure with the test point consisting of two wires:

The two wires bridge a span of a known length. At each point of measurement, while making a line current survey, the potential drop across the span is observed and

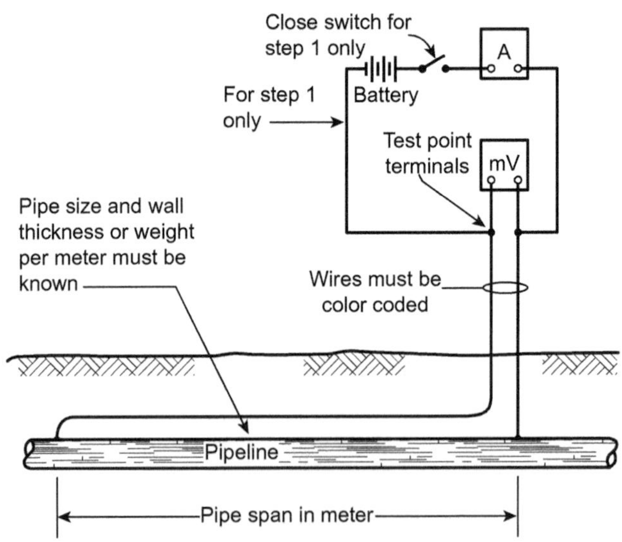

Figure 6.6 Current measurement, 2—wire test point.

recorded together with the polarity of instrument connections to indicate the direction of current flow. Currents may be calculated by Ohm's law after determining the span resistance of the pipe being surveyed (from Table 6.1) and the resistance of the test circuit.

The following is a procedure for the test:

Step 1
Measure the circuit resistance of the test leads and pipe span by passing a known battery current through the circuit and measuring the resulting voltage drop across the test point terminals.
Step 2
Measure the voltage drop across the test point terminals caused by the normal current flowing in the pipeline. Usually, since an instrument with a range suitable for these measurements will have a low internal resistance, the resistance of the lead wires may induce a substantial error in the reading obtained.
This will be in millivolts or in fractions of a millivolt.
Instrument resistance and the value of lead wire resistance must be known and correction made for the resistance of the external circuit (measured in Step 1). Note the polarity of meter connection to test point terminals and indicate the direction of the current flow (+ to −) along the pipeline:

$$\text{R.B Correction factor} = \frac{\text{Resistance of external circuit} + \text{Resistance of voltmeter}}{\text{Resistance of voltmeter}}. \tag{6.2}$$

Step 3
Using pipeline resistance in Table 6.1, determine the resistance of the pipeline span.
Step 4
Calculate the pipeline current flow by Ohm's law:

$$\text{Current in milliamps} = \frac{\text{Corrected millivolt drop (from Step 2)}}{\text{Pipe span resistance in ohms}}. \tag{6.3}$$

Table 6.1 for steel pipe resistance may be used as a general guide to pipeline resistance. This table is based on a Steel resistivity of 18 microhm-centimeter. Steel resistivity varies between 15 and 23 microhm-centimeter, and 18 microhm-centimeter is used as an average.

The following is an example of determining the line current flow on a 75 cm (30 in) pipe, 61-m (200 ft) span with a 9.5-mm (0.375 in) wall thickness (or weighing 177 kg per meter with a pipeline direction of EAST–WEST).

Step I
Battery current = 1.2 A.
The resulting voltage drop across the test point terminals = 0.108 Vikt.
External circuit resistance 0.108/1.2 = 0.09 ohm.
Step II
The potential drop across the test point terminals caused 0.16 mV by a normal current (protective current) flowing in the pipeline (on a 2-mV range of an instrument with a

resistance of 1000 ohm per volt. or 2 ohm per 2 mV).
Correction factor = 20.09 2 + 1.045.
Corrected potential drop across the test point terminals caused by a normal current flowing in the pipeline: 0.16 × 1.045 = 0.17 mV.
Polarity of the meter of the West-end terminal: +ve.
Step III
Resistance of pipeline span (200 ft, 30, 0.375 wt) = 2.44 × 10 − 6 × 200 = 0.00049 ohm
Step IV
Pipeline current (protective current) flow 0.17 mV/0.00049 ohm = 346 mA
Direction of current flow → (west to east).

6.4.1 Test Procedure with the Test Point Consisting of Four Wires

In a four-wire test point for line current measurement, two wires are connected at each end of the test span. The four wires have different colors to be distinguished from one another. The two inner leads serve for measuring current and for measuring external circuit resistance; the two outer leads are used for calibrating the test span. The four-wire calibrated line current test point permits more accurate pipeline current measurements. The reason is that each span can be calibrated accurately. This avoids errors in the length of pipe span and pipe resistance, which is possible when using the two-wire test point.

The general arrangement for a pipeline current measurement is shown in Fig. 6.7.

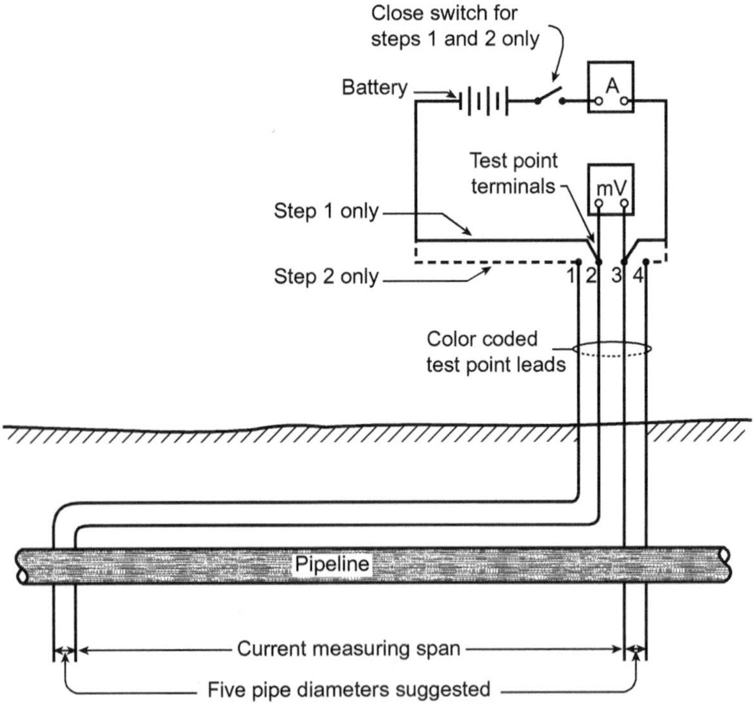

Figure 6.7 Current measurement, 4—wire test point.

The following is the procedure for the test:

Step 1
Measure the circuit resistance of the test leads and pipe span (between terminals 2 and 3) by passing a known battery current through the circuit and measuring the resulting voltage drop across terminals 2 and 3.
Calculate the resistance in ohms by the application of Ohm's law (resistance = AmpsVolt).
Step 2
Calibrate the span by passing a known amount of battery current between the outside leads (terminals 1 and 4) and measure the change in the potential drop (corrected for the effect of circuit resistance) across the current-measuring span (terminals 2 and 3). Divide the current flow in amperes by the change in potential drop in millivolts to express the calibration factor in (amperes per millivolt). Normally, this calibration needs to be done only once for the same location. The calibration factor may be recorded for subsequent tests at the same location. However, on pipelines in which the operating temperature of the pipe changes considerably (with accompanying changes in resistance), more frequent calibration is necessary; pipelines carrying hot fuel or gas trunk line at the outlet of the compressor station are typical examples of lines that need frequent calibration.
Step 3
Measure the potential drop in millivolts across the current-measuring span (terminals 2 and 3) caused by the normal pipeline current. Apply the correction factor for circuit resistance. Calculate current flow by multiplying the corrected potential drop by the calibration factor determined in Step 2.
Also note the direction of the current flow.

The following is a sample determination of the current flow using a four-wire test point. In this sample, the same pipeline section is used as in the example of a two-wire test point. Steps to be taken for determining line current in the four-wire procedure are as follows:

Step I
Circuit resistance between terminals 2 and 3 measured as 0.09 ohms (refer to Step I of the example of determining the line current for a two-wire test point under B3.1.4).
Step II
Battery current passed between terminals 1 and 4 = 10 A.
Corrected potential drop (with current ON) = 5.08 mV.
Corrected potential drop (with current OFF) = 0.17 mV.
Change in potential drop V = 4.91 mV.
Calibration factor 10 A/4.91 mV = 2.04 A/mV.
Step III
The corrected potential drop across current-measuring span (terminals 2 and 3) caused by normal pipeline current (protective current) = 0.17 mV.
Polarity: West-end terminal is +ve.
Pipeline current (protective current) flow = 0.17 mV × 2.04 A per mV = 0.34 A.
Direction of current flow: West to east.

6.4.2 *Test Procedure Using the Null Amp Test Circuit for Line Current Measurement*

This is an effective method of measuring the line current. In this method, a null Ampere test circuit is used. The circuit is illustrated in Fig. 6.8.

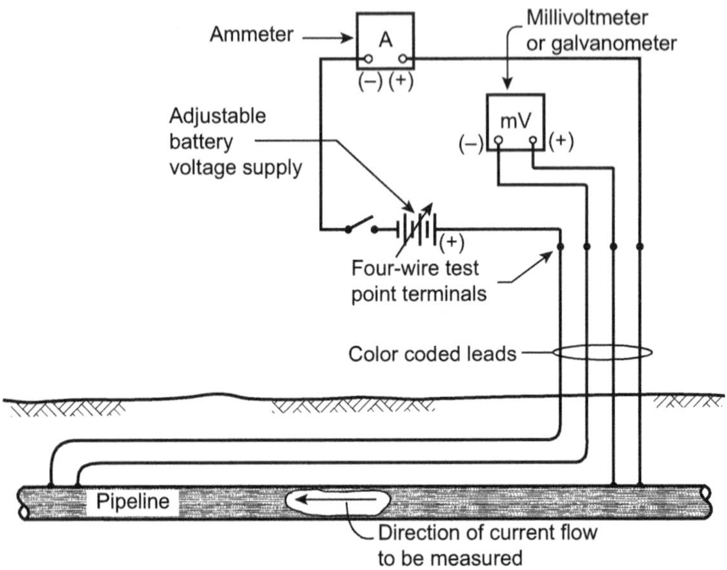

Figure 6.8 Null ammeter circuit.

As shown in Fig. 6.8, a four-wire test point is used in this method, with a sensitive galvanometer connected between the inner pair of wires. Current from the battery is forced to flow between the outer pair of wires in opposition to the protective current flowing in the pipe. With the battery output adjusted to give a zero deflection on the galvanometer, the ammeter in series with the battery will read the protective current originally flowing in the pipe.

This method is most effective with a permanent four-wire test point on buried lines where all pipe connections are at essentially the same temperature to avoid thermal potentials.

The sensitivity of the test will depend on the sensitivity of the galvanometer. The galvanometer need not be calibrated, but it should preferably have a sensitivity of ≤ 1 mV full scale to permit measuring a reasonably small amount of current in large pipelines.

6.5 Computer Modeling of Offshore CP Systems Utilized in CP Monitoring

Offshore structures with large dimensions and structural complexity combined with high-current density requirements for protection represents a combination in which traditional potential measurements have proved to be inadequate.

Many computerized modeling techniques have been developed for the analysis of offshore CP systems. The following are some references on computer modeling that may be used in CP analysis.

6.5.1 Computerized Modeling Techniques

The finite difference and finite element methods are numerical discretization procedures for the approximate analysis of complex boundary value problems. Development of these methods have to a large extent followed the rapid development of electronic computers and have been applied successfully for many years in various fields of engineering such as structural stress analysis and heat conduction. In 2014, there is a growing interest in the computerized modeling of CP systems, for both improving CP designs and analysis of readings and for the "diagnostics" of CP performance as indicated above.

Iterative solutions of the finite difference from of the Laplace's Equation for electrochemical systems were applied in 1964 by R.N. Fleck. The Finite Difference Method (FDM) has been employed in the modeling of offshore CP systems and CP generally in the 1970s. This includes the direct solution of the linear equations using simple Gaussian elimination combined with an iteration procedure for adaptation to nonlinear boundary conditions.

Despite high efficiency, that is, low computer time for solving the linear equations when employing the FEM technique, the finite element method has a become dominant technique in numerical analysis, largely due to the fact that it is more easily adapted to complex geometry. Multipurpose program packages now available are therefore most often based on the FEM technique.

In 2014, there has also been considerable development in the application of integral equations and Boundary Element methods in potential theory and similar problems. Achievements are reduction in the size of the numerical problem. As far as is known, such methods have not yet been used in CP modeling.

6.5.2 Short Theoretical Background

Based on the requirement for continuity of electrical charges in an electrolyte, it can be shown that variations in the electrochemical potential, E obey Laplace's equation:

$$\mathbf{V}^2 E = 0. \tag{6.4}$$

Calculation of the potential variation along cathodically protected structures involves the solution of this equation, which for cylindrical coordinates in three dimensions is written as follows:

$$\frac{\partial^2 E}{\partial r^2} + \frac{1}{r}\frac{\partial E}{\partial r} + \frac{\partial^2 E}{\partial \theta^2} + \frac{\partial^2 E}{\partial Z^2} = 0, \tag{6.5}$$

where r is the radius; θ is the angle; and Z is the direction of the axis.

To solve this equation, appropriate boundary conditions must be specified. These are given by the geometry of the structure, that is, dimensions and position of the anodes, of steel members, or areas of exposed steel in the case of a coated structure with coating defects, position of the structure in water relative to the mud line, etc.

Further, the reaction kinetics for anodes and protected steel must be specified. For a homogeneous electrolyte, a relationship exists between the current density ($\mathbf{i_s}$) at the electrode surfaces (anodes and exposed steel) and the potential gradient:

$$\mathbf{i_s} = -\sigma \frac{\partial E}{\partial n_s}, \tag{6.6}$$

where n_s is the normal to the exposed electrode surface and $\sigma = \boldsymbol{L}/\rho$ is the conductivity of the seawater.

In the case of a coated and insulated surface, the current density is reduced to zero:

$$\frac{\partial E}{\partial n_s} = \mathbf{0}. \tag{6.7}$$

The boundary conditions at the electrode surfaces are given by the reactions taking place. These reactions are irreversible, that is, displaced from equilibrium. One of the half reactions is dominating on the electrode surface, and the external behavior is written in the following form:

$$\mathbf{i_{electrode}} = \mathbf{i_{corr}} \left(10^{\eta/\mathrm{ba}} - 10^{-\eta/\mathrm{bc}} \right), \tag{6.8}$$

where $\mathbf{i_{corr}}$ is the corrosion current density for the electrode; ba and bc are anodic and cathodic Tafel slopes; and the overvoltage, η is defined as follows:

$$\eta = E - E_{\mathbf{corr}}. \tag{6.9}$$

Here, $E_{\mathbf{corr}}$ is the corrosion potential of the electrode.

It should be observed that the two Tafel slopes, ba and be refer to different reactions on the same electrode, and that these further may differ locally on the electrode surfaces, that is, on protected steel and on the sacrificial anodes. Changes with time are also observed in the polarization characteristics. Consideration deviations from the theoretical relationship of Eqn (6.6) are found frequently. In addition, E_{corr} for anodes and cathodic areas may differ as well. It has therefore been found necessary to tabulate data obtained from testing and offshore monitoring:

$$\mathbf{i_{electrode}} = \mathbf{f}(E). \tag{6.10}$$

For cathodically protected steel in seawater, $\mathbf{i_{electrode}}$, in addition to being a function of E depends on environmental conditions in the sea such as O_2 content, temperature, salinity, and water flow rate. As already indicated above, the current density will further change with time reflecting the build-up of scale, calcareous deposits, and marine growth on the steel.

Descaling of the steel structures during winter storms in harsh offshore areas may finally result in much higher-than-average current density requirements locally on the protected steel. The importance of including such local variations in the analysis has been proved.

6.5.3 Numerical Solutions

Numerical solutions to this problem using the FDM have been presented in the literature both for two-dimensional and three-dimensional systems. By using the Taylor expansion, the differential Eqn (6.5) is transformed to a set of linear equations of the following form:

$$[c][E] = [I], \tag{6.11}$$

where :

- I is the current,
- E is the potential, and .
- c is the conductivity matrix.

Using a simple physical analogy, one can show that this corresponds to replacing the continuum, that is, the seawater by a network or a mesh of elements. Neighbor elements are connected by conducting bars, and each bar of electrical resistance is equal to the resistance of the electrolytic element in the same direction. Utilizing ohm's and Kirchhoff's laws, one can obtain a set of equations, in the form of the above equation, one equation per element.

The finite element technique has been described previously. The application of this method involves a numerical procedure in which the differential, and the appropriate boundary conditions are handled simultaneously by a functional. Minimizing this functional is equivalent to solving an equation for the appropriate boundary conditions. The problem is discretized by dividing the electrolyte/sea water into a number of finite elements (in a similar fashion as for FDM), for example, tetrahedral, and approximating the potential in each element by a simple function.

Obviously, the CP modeling of complex offshore structures requires a high flexibility in regard to mesh refinement, for adaptation of local boundary conditions, geometry, etc. Running the programs further requires large computers with a considerable capacity to handle the large numerical problems involved in an offshore CP analysis.

6.5.4 Analysis of Existing CP Systems

Computerized modeling can be used in different ways to take the full advantage of readings obtained with various CP monitoring techniques. Once a problem of unsatisfactory CP performance has been detected, the cause(s) to the problem and the requirements for rectification need to be established. Of primary importance is to define additional current requirements for the satisfactory protection of the structure. In this regard, computer models have been used for the analysis of current density and potential readings to accurately estimate the following:

- Current output from sacrificial anodes,
- Current consumption on exposed steel at different protective levels (different potentials).

Such data provide a sound basis for estimating additional current requirements and for the redesign of the CP system. Other applications have included the following:

- Analysis and definition of typical potential profiles at nodes and other confined and critical areas. Such profiles are subsequently used for predication of the potential in the subject area on the basis of a few potential readings at specified reference positions. The number of readings to be taken, and thereby the diving time, is effectively reduced.
- Analysis of interference effects caused by impressed current anodes of high current output or between cathodically protected structures.

6.5.5 Performance of Sacrificial Anodes

Data on operational status of the sacrificial anodes are very useful for different purposes. Of vital interest is first to check that the anode really is operating, and second to obtain current drain, for example, as a basis for CP redesign, for estimating the remaining life and to obtain data on current output versus potential as a means of checking its capabilities to increase the current output in periods of high loads.

Over the last few years, there has been an increasing application of equipment for electric field strength/current density monitoring in CP surveys of offshore Structures. Probes for Remote Control Valves (RCV) and diver operations are used to measure the electric field strength at typical stand-off anodes, at sacrificial bracelet anodes on pipelines, etc.

As is obvious, there is a strong reduction in the field strength and in the local current density with increasing distance from the anode surface. Accordingly, there is a need to relate the reading, obtained at a specified distance from the anode, back to the anode surface. This is achieved by comparison of measured field strength values with figures obtained from computer modeling of the same anode geometry. By modeling of the anodes for different conditions, that is, for different current output levels, the output from an anode is found simply by comparison of the reading with tabulated figures for a known current output.

In monitoring of platform anodes, readings are obtained at well-defined positions, that is, at a specified distance from the anode surface. For this purpose, the probe is provided with a support or spacing piece. Preferably, this should be manufactured to keep the probe at a minimum distance of 10–15 cm from the anode surface, to avoid too strong effects of local variations over the anode surface in the anodic current density. When monitoring long, rod-shaped offshore anodes, two to three readings per anode may therefore be required to obtain results with a good accuracy. Still efficiency is maintained with a typical figure of 15–20 anodes monitored per hour.

Surveys of submarine pipelines are normally conducted using RCVs and manned submersibles with the probe carried in front of the vehicle. The electric field strength variations along the pipeline are continuously monitored and logged on magnetic tape. Although the sensor-to-pipe distance can be measured, such data are not always monitored.

Results from the numerical modeling of a bracelet anode are some curves. The various curves represent variations in the radial field strength along the pipe axis and for different radial distances, and for a current output corresponding to 1000 mA/m^2. As observed from these curves, the field strength in the radial direction is strongly reduced with increasing distances from the pipe. However, it is also obvious that the shape of the field strength curves is strongly dependent on the radial distance. By computerized analysis, also making use of a curve-fitting procedure this is utilized:

- To estimate the radial distance between the sensor and the pipe on passing the anode bracelet.
- To estimate the current output from the anode (a technique has been developed to obtain potential profile data for pipelines, anode output, and remaining life).

6.5.6 Analysis of Attenuation Curves at Sacrificial Anodes

By computer modeling a section of a structure, including one or a few sacrificial anodes, the potential profile may be obtained. In this case, the number of anodes, surface area of exposed steel, and boundary conditions generally have been fitted to cause the anodes to supply an output corresponding to 1000 mA/m^2. This typical profile, also named the attenuation curve, mainly reflects the current output from the anodes. (This may not always be true, as, for instance, at nodes and complex areas where, e.g., "shadow" effects may add considerable IR drops in the seawater.)

The accuracy of such estimates is expected to be slightly inferior to that of readings obtained using electric field strength/current density monitor equipment. Errors may be introduced by inaccurate reference electrodes as well as by unsatisfactorily defined polarization characteristics for the exposed steel. An error of $\pm$10 mV in the ΔE figure for the anode would in this case represent an error of 7–9% in the current output estimate. If we compare the readings obtained with the two methods, we see that they seldom differ by >20% for any anode.

6.5.7 Potentials in Nodal Areas—Improved Efficiency in Potential Surveys of Nodes

Potential profiles for nodes and similar shielded areas on a structure have often been found to be critical, that is, such areas exhibit the least satisfactory potential levels. An example of such a nodal area with a total of eight members meeting at the joint has been modeled. Anodes are attached outside the nodal area. As is observed, an IR drop of 50–60 mV is obtained across the nodal area in this particular case. Figures for such IR drops in excess of 60 mV may not be unusual.

However, with the increasing use of RCVs in the inspection of steel structures, it has often been proved inconvenient or even impossible to obtain potential readings in such narrow corners of the steel members, and readings in the potentially most critical areas are lost.

Under such conditions, computer modeling has been used to provide estimates of the (relative) potential profile at the nodes. Potential and field strength readings are used in the mapping of boundary conditions in a few selected areas. These data are subsequently used in the modeling of a large number of nodes of different geometries, to provide similar potential profiles. These profiles are subsequently utilized in future potential surveys:

- It is now sufficient to take potential readings at a few selected reference positions, each located for convenient access 2–3 m outside the node. These reference readings are used for the calibration of the (relative) potential profiles above, to obtain the complete profile for the node area each time a survey is conducted.

In an early test of this technique, potential readings at a node were found to fit the predicted profile with maximum deviations of 10 mV.

6.6 Coating Resistance Measurement Method

The electrical resistance of the coating is expressed as the resistance per average square meter of the coating. This may also be expressed as conductance in microhms per average square meter (which is the reciprocal of the electrical resistance multiplied by one million). The coating resistance of a buried or submerged pipeline may be determined by the following methods:

6.6.1 Current–Voltage Change Method

This is the most practical method used for determining the effective resistance or conductance of the coating in a pipeline section. This method is based on calculating coating resistance directly from current and voltage change measurement obtained from field tests.

Test setup and arrangement are illustrated in Fig. 6.9.

The procedure for determining the coating resistance or conductivity using this method is as follows:

To perform the test, batteries will be sufficient as a source of power supply. A current interrupter automatically switches the circuit on and off (usually 30 s on and 15 s off). The interrupter also assures the inspector when at remote locations that the battery installation is operating properly and draining current from the pipeline as long as his potential and line current measurement continue to change in accordance with the established on–off cycle.

Test data may be taken for a section of 8 km. Testing can be continued section by section in each direction from the power source until changes in the observed currents and potentials (as the current interrupter switches on and off) are insignificant. The length of the section that can be maintained at −0.85 V or better will be established at the same time.

On coated pipeline systems provided with test points for potential and line current measurement, a survey will proceed rapidly. The survey can be carried out by a single inspector. For maximum accuracy, however, two engineers in radio communication can observe data simultaneously at each end of each section tested. This becomes essential if the pipeline under testing is affected by variable stray current.

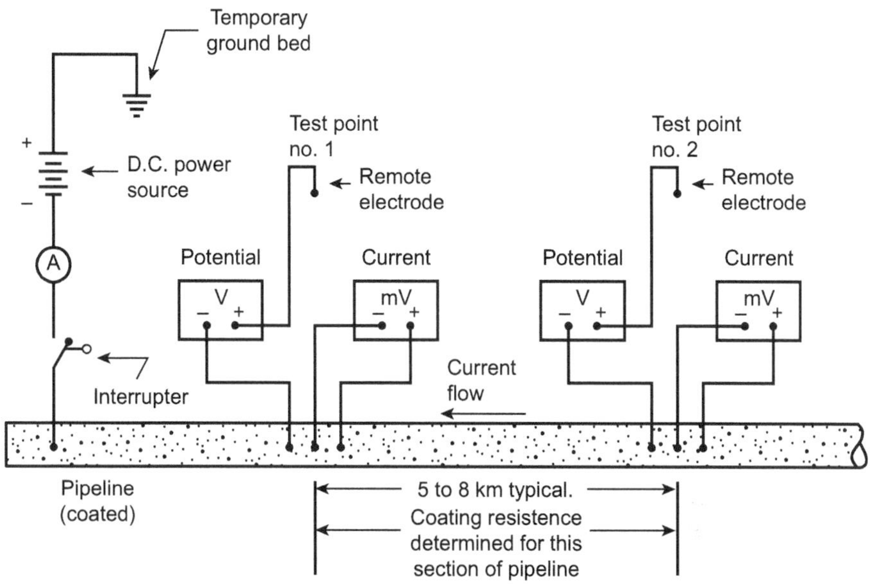

Figure 6.9 Coating resistance and CP current requirement tests.

To obtain data for the calculation of coating resistance, readings are taken for pipeline at each end of each test section as follows:

Potential readings to remote CSE with interrupter ON and OFF.

6.6.2 Pipeline Current with Interrupter On and Off

From these readings, the change in pipe potential (ΔV) with the change in line current (I) at each end of the section can be determined. The difference in the two $\Delta\Delta I$ values will be the test battery current collected by the line section when the current interrupter is switched on. The average of the two V values will be the average change in the pipeline potential within the test section caused by the battery current collected, namely, Δ.

The average V in millivolts divided by the current collected in milliamperes will give the resistance to the earth, in ohms, of the pipeline section tested. If one knows the length and the diameter of a pipe in the section tested, one can calculate its total surface area in square meters to be Δ.

Multiplying the pipe section to the earth resistance by the area in square meters will result in a value of ohms per average square meter. This is the effective coating resistance for the section tested.

Some inspectors express coating conditions in terms of coating conductivity in mhos or microhms. This is simply a matter of conversion. The reciprocal of the resistance per average square meter is the conductivity in mhos. The reciprocal times

106 is the conductivity in micromhos. A sufficient number of readings should be taken to ensure acceptable precision.

The following is an example of how this test method can be implemented:

- **Test section**

The Test Section is a 5000-m-diameter coated line that lies between Test Point 1 and Test Point 2. The wall thickness of the pipe is 9.52 mm (0.375 in).

Test data taken at test Point No. 1 are as follows:

Pipe/soil potentials (Reference CSE) are as follows:

−1.75 V ON,
−0.89 V OFF,
$\Delta V = -0.86$ V.

Potentials drops across test section are as follows:

+0.98 mV ON,
+0.04 mV OFF.

Span calibration is 2.30 Amperes per millivolt.

Pipeline current is as follows:

+2.25 A ON,
+0.09 A OFF,
$\Delta I = +2.16$ A.

Test data taken at test Point No. 2

Pipe/soil potentials (Reference CSE) are as follows:

−1.70 V ON,
−0.88 V OFF,
$\Delta V = -0.82$ V.

Potentials drops across the test section are as follows:

+0.84 mV ON,
+0.02 mV OFF.

Span calibration is 2.41 A per mV.

B.6.3.2.4 Pipeline current

+2.03 A ON,
+0.05 A OFF (Negative current indicates current flow in the opposite direction),
$\Delta I = +2.08$ A.

Calculation of coating resistance

Average $\Delta V = \frac{-0.86+(-0.82)}{2} = -0.84$ V.

Current collected is 2.16 – 2.08 = 0.08 A.

Pipe-to-earth resistance = 0.84V/0.08 A = 10.5 ohms

Surface area of the test section 4785 m^2.

Effective coating resistance 10.5 × 4785 = 50,242 ohms per average square meter.

Coating conductance is 10ˆ6/50,242 = 19.9 microhms per average square meter.

6.7 Test Method and Calculation for "Attenuation Constant"

Coating effectiveness of the pipeline may be evaluated using the attenuation method, which involves a limited number of field measurements.

The measurements lead to a figure that is termed "Attenuation Constant" per kilometer. From this figure, the spread of the protective current can be evaluated, and the approximate distance between drain points can be estimated.

When the original design fails to provide complete CP due to a sharp drop in the pipe/soil potential along the pipeline, the estimated distance would tell the designer where to put the rectifiers for supplementary protection. "Attenuation Constant" is derived from the attenuation equation. The equations describe the interrelation of factors affecting the degree of protection achieved at any specific location on a cathodically protected pipeline. The factors are as follows:

- Total current drain.
- Radial resistance of the path from the pipe surface to the remote earth. This factor includes both the resistance of the path through the coating on the pipe and the resistance through the soil.
- Diameter and the wall thickness of the pipe.
- Distance of the location from the drain point.

The basic formulas for attenuation are as follows:

$$\mathbf{dV_x = dV_0 e^{-ax}}, \tag{6.12}$$

where $\mathbf{dV_x}$ is the change in potential as distance × kilometers from the drain point in millivolts; $\mathbf{dV_o}$ is the change in potential at the drain point with the cell directly over the pipe in millivolts; **e** is a constant 2.72 (base of natural logarithms); **a** is the attenuation constant in per kilometers; and **x** is the distance from the drain point in kilometers.

$$\mathbf{Ln\ dV_x = Ln\ dV_o + Ln\ e^{-ax}}, \tag{6.13}$$

$$\mathbf{Ln\ dV_x = Ln\ dV_o - ax}, \tag{6.14}$$

$$\mathbf{a} = \frac{\mathbf{Ln\ dV_o - Ln\ dV_x}}{X}. \tag{6.15}$$

Equation (6.5) is the basis for constructing the attenuation graph for each value of attenuation constant. In addition to the above, it can also be shown that the following relationships apply:

$$\mathbf{a} = \frac{R_S}{R_L}, \tag{6.16}$$

where R_S is the longitudinal resistance in the pipe wall (ohms per kilometer); R_L is the leakage resistance or radial resistance from the pipe surface to the remote earth (ohms-kilometer).

$$R_S = \frac{\rho L}{A}, \tag{6.17}$$

where ρ is the resistivity of the steel pipe usually estimated at 18 microhm-centimeterm unless actual resistivity is known; L is the unit pipe length, in this case 1 km or 1 × 105 cm; A is the cross-sectional area of the pipe in square centimeters.

$$R_L = \frac{\rho}{2nL}\mathbf{Ln}\,\frac{D}{r}, \tag{6.18}$$

where ρ is the average resistivity of the soil around the pipe to remote earth in ohms centimeters; D is the distance from the pipe surface to effectively "remote" earth; r is the pipe radius in the same units as for D.

The following relationships are also applicable:

$$\mathbf{a} = \frac{R_S}{R_K}, \tag{6.19}$$

where R_S is as shown in Eqn (6.17); and R_K is the characteristic resistance of the entire line looking in one direction only from the drain point.

For a line uniform in both directions from the drain point,

$$\mathbf{P_K = 2R_G}, \tag{6.20}$$

where $\mathbf{R_G}$ is the resistance of the entire line looking in both directions from the drain point.

$$\mathbf{R_G} = \frac{\mathbf{dV'_o}}{dI}, \tag{6.21}$$

Here, $\mathbf{dV'_o}$ is the change in the potential at the drain point related to a remote cell; dI is the current–drain change that causes $\mathbf{dV'_o}$.

From this,

$$R_K = \frac{\mathbf{2dV'_o}}{dI}, \tag{6.22}$$

$$\mathbf{a} = \frac{R_S dI}{\mathbf{2dV'_o}}. \tag{6.23}$$

For a line on which there are rectifiers spaced at a distance 2AL kilometers apart, the relationship is as follows:

$$\mathbf{dV_m = dV_o cosh\ al}, \tag{6.24}$$

where $\mathbf{dV_m}$ is the change in potential at the midpoint between rectifiers.

Practically, the same value for $\mathbf{dV_m}$ is obtained if the following two values are added:

$$\mathbf{dV_{AL} = dV_{Ao} e^{-aAL}}, \tag{6.25}$$

$$\mathbf{dV_{BL} = dV_{Bo} e^{-aBL}}, \tag{6.26}$$

where values ($\mathbf{dV_{AL}}$), ($\mathbf{dV_{Ao}}$), and (**aA**) all relate to rectifier (A).

$$\mathbf{dV_m = dV_{AL} + dV_{BL}}. \tag{6.27}$$

6.7.1 Significance of Attenuation Constant

To put the various levels of attenuation constant in perspective, a graph showing the effective spread of protection for various values of attenuation constants (from 0.05 to 0.5 per kilometer) has been prepared (Fig. 6.10).

6.7.1.1 Method of Calculation

It is assumed that the potential change at the drain point is to be 1000 mV from the static pipe/soil potential.

Thus, if the static pipe/soil potential is about 500 mV, the maximum pipe/soil potential would be 1500 mV (all potential values are with reference to the CSE).

To construct the graph, the value of the potential change was calculated for the 10-km distance for each value of attenuation constant, and the value so calculated was plotted on the 10 km line.

A straight line was then drawn from $\Delta = 1000$ at km zero through the point plotted at km 10 for the particular value of "a." V.

For example,

If $V = 1000\ \Delta$,
$X = 10$ km,
$a = 0.50$ per km.

$$\mathbf{Ln\ \Delta V_x = Ln\ \Delta V_o - a_x = 6.90 - (0.5 \times 10) = 1.90}, \tag{6.28}$$

$$\mathbf{\Delta V_x = 6.7}.$$

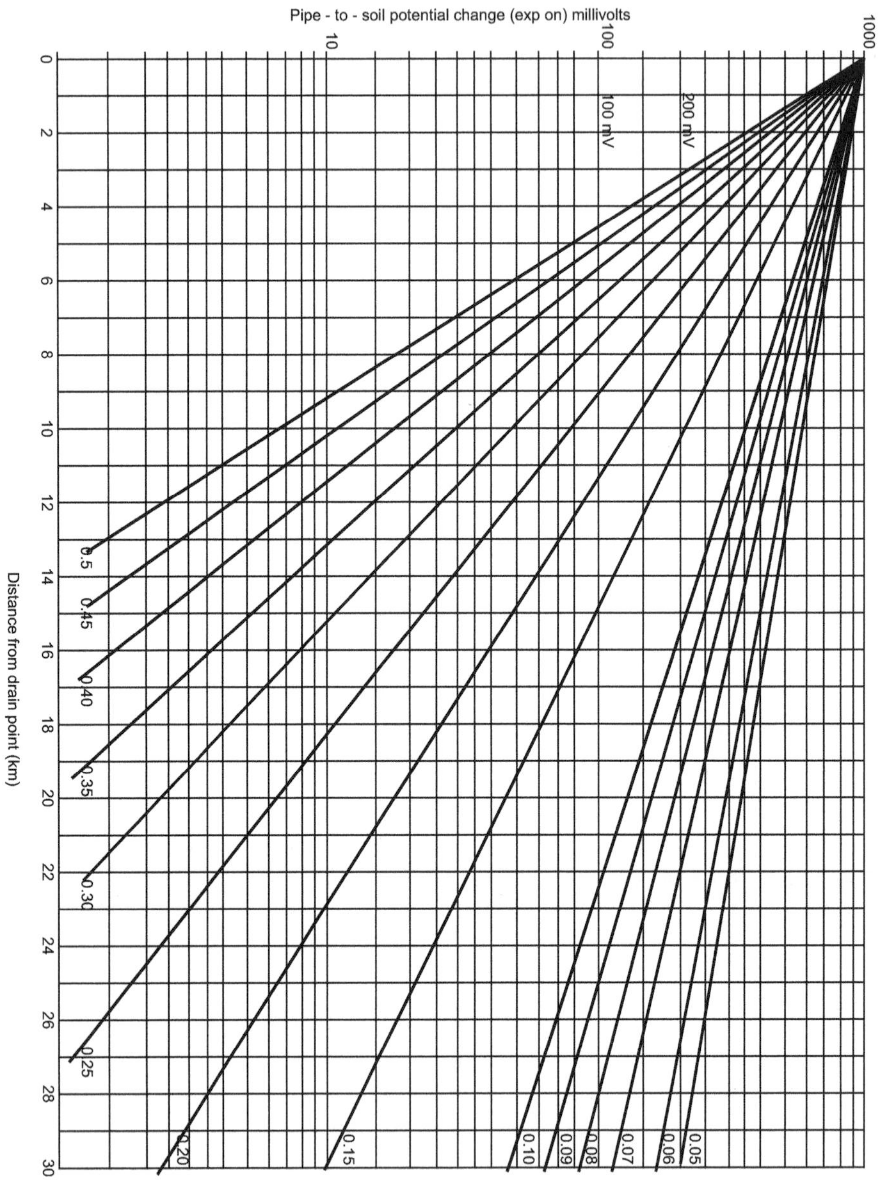

Figure 6.10 Graph spread of protection relating to the attenuation constant.

6.7.2 Example Use of Graph Relating to Attenuation Constant

If the static potential is assumed to be −500 mV to the CSE and the desired minimum potential is −900 mV to copper sulfate (allowing a 50-mV safety factor), then the minimum potential change due to each rectifier at the midpoint (between rectifiers) would have to be as follows:

$$\frac{900 - 500}{2} = \mathbf{200\ mV}. \tag{6.29}$$

From the graph, it will be seen that the distance from the drain point at which this potential change (200 mV) is just achieved is a function of the attenuation value.

For instance, if the attenuation constant is 0.45 per km, the distance at which the potential change attainable is 200 MV is 3.6 km. To achieve protection to the 900-mV minimum level with this value of attenuation constant, it would be necessary to space the rectifiers 2 × 3.6 or 7.2 km apart. To put the various levels of attenuation constant in perspective, a rating table has been prepared. This table can be used to describe the relative effectiveness of any coating, as installed, in terms of attenuation constant per kilometer of the line concerned. Based on the attenuation constant value, the approximate distance between drain points can be estimated, and the location for installation of a new rectifier to supplement ineffective CP system may be decided.

6.8 Coating Inspection by the Pearson Method

The Pearson survey is an above-ground survey technique used to locate coating defects in buried pipelines and is named after J.M. Pearson who developed the technique. The survey compares the potential gradients along the pipeline measured between two movable electrical ground contacts. The potential gradients result from an injected alternating current (AC) signal leaking to the ground at coating defects or metallic objects within the pipeline trench.

6.8.1 Equipment

The equipment required for the survey comprises the following:

- Transmitter to provide an AC signal of approximately 1000 Hz for conventional pipeline coatings, for example, enamel, tape, extruded coating, and a reduced frequency of 175 Hz for thin film coating, for example, fusion-bonded powder epoxy. The transmitter is powered from internal batteries or for long surveys from an external high-capacity battery.
- Receiver, hand-held, self-contained, battery operated with pick-up sensitivity controls, audible warning, earphone output, and, in some cases, recording capability. The receiver is tuned to the transmitter frequency.
- Earth contact set of boot cleats, studded boots, or modified aluminum ski poles.
- Connecting cable harness between earth contacts and receiver.
- Earth spike and connecting cables for the transmitter.

The above equipment is normally all contained in a portable box for easy transportation. The total weight is approximately 25 kg.

Optional equipment available from some suppliers enables the instrument to be used as a general-purpose pipe/cable locator and comprises the following:

- Signal level meter on receiver.
- Pipe location antenna.
- Pipe depth indication.
- Signal recorder and playback chart recorder or interface to a portable computer.

6.8.2 Procedure

The equipment is set up as shown diagrammatically in Fig. 6.11. The transmitter is electrically connected with one lead to the pipeline, usually by connecting to a CP test lead or an accessible part of the pipeline, and the other lead to a good remote earth and then energized.

Using the receiver in the pipe locating mode, or a separate pipe locator, one can locate and identify the section of pipe to be tested so as to enable the survey operators to follow the route of the pipe exactly above the pipe. For record purposes, it may be useful to insert pegs at measured intervals.

The survey may be carried out with an impressed current CP system energized. However, any sacrificial anodes, bonds to other structures, or similar are best disconnected before commencing the survey to ensure that they do not mask defect areas or drastically reduce the length that may be surveyed from one injection point. With the line located, the receiver is then connected via the cable harness to the earth contacts worn or held by the two operators such that at all times, earth contact is made by each operator. The connecting cable provides for a separation of 6–8 m between the operators.

Surveying should commence at a sufficient distance from the transmitter and earth spike to minimize interference from the transmitter and/or return current flow in the earth.

The two operators walk over the top of the pipeline to locate coating defects. When the front contact approaches a defect, an increased signal level is indicated in the earphones by an increase in volume or by a higher reading on the receiver signal level meter. As the front contact passes the defect, the signal fades and then peaks again as the rear contact passes over the defect.

The defect is logged on the record sheet at a measured distance from a reference point (by triangulation if possible) and/or may be indicated with a marker or nontoxic paint. The signal is recorded automatically for later interpretation if the receiver is fitted with recording equipment. Where the signal is not easily interpreted or where there may be more than one defect within the span of the operators, this may be clarified by surveying at right angles to the pipeline, that is, one operator walks over the pipeline and the second walks parallely to the pipeline at 6–8 m from the pipeline. In this mode, each defect is indicated as the operator over the pipeline traverses the fault. This is utilized when using recording equipment.

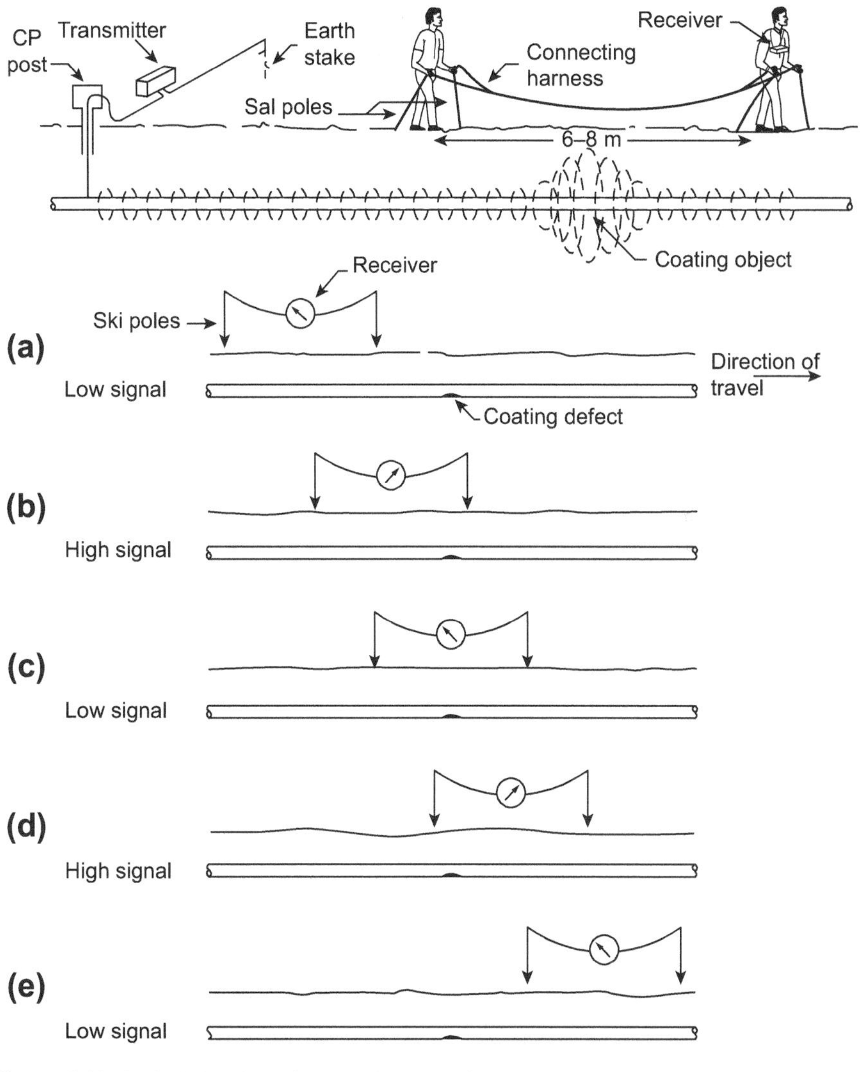

Figure 6.11 Typical signal level on passing a coating. (a) Low signal (b) High signal (c) Low signal (d) High signal (e) Low signal.

The above procedures are general, and in all cases, the equipment manufacturer's instructions are to be followed.

6.8.3 Data Obtained

The information obtained from a Pearson survey is the change of signal intensity at probable defect locations.

For instruments without a signal level meter, only the location and the operator's aural interpretation of the signal strength can be recorded. Where a signal level meter is fitted, further data concerning the rate of increase and decrease and the maximum signal level may be recorded, together with intensity variations around the defect location, which can assist in analyzing the magnitude and disposition of the defect. This is done automatically when the receiver is of the recording type.

6.8.4 Presentation of Data

Defect indications are either by audible tone or by the signal meter level. The accuracy of recording signal level changes either manually or automatically, and the locations where they occur are very important to enable further investigations to be carried out.

Locations of probable defects can be measured to fixed points so that they may be returned to at a later date.

In addition to recording the probable defect locations, valuable information may be gained by recording various observations of the signal levels on the meter, which will assist in the evaluation of the probable defect. For example, if the signal level rises to a peak rapidly and then falls away, or if it rises steadily and remains high for a distance before decaying, these characteristics may be recorded. This is done automatically if the receiver is of the recording type.

The survey record sheet may provide space to adequately note all pipeline features, reference points, other services crossing, signal intensity levels, and characteristics, etc. to enable the results of the survey to be analyzed and determine areas where further investigations or remedial measures are required.

Where repeat surveys are carried out, these records can be compared and may show further deterioration of the condition of the coating.

6.9 Coating Inspection by the C-Scan System

This system introduces the latest technique of above-ground coating inspection. The system locates coating defects on buried pipelines and provides an assessment of external coating quality. In addition to the coating condition, the system can also be used to locate contact or connection points between buried services, which can be of great importance in the proper functioning of pipeline CP. The system is based on current attenuation survey. The current attenuation is the only technique that gives a value (attenuation) that is a direct indication of the pipeline coating.

Excavations confirmed almost all defect locations detected by C-Scan. The C-Scan located the defects found by the Pearson survey and also located defects that the Pearson survey did not find. At the end of a survey (or whenever required), the detector unit may be plugged into any computer or printer, and the survey report can be printed out in full. Alternatively, the data can be displayed on a computer monitor or stored on disk or tape for further analysis.

The coating condition of a pipeline can be monitored over its working life. This is done by repeating the general survey at intervals of three to five years. The results of the repeat surveys are compared with those from the first survey to monitor any deterioration of the coating with age (aging effect on coating condition).

6.9.1 C-Scan System Features

Operation of the C-Scan detector is largely automatic. When required, prompt messages, instruction messages, and warnings are displayed to the operator to guide him through the survey sequence.

The operating frequency of the system has been selected to minimize interference from commonly occurring sources. The filtering system and the directional nature of the detector antenna help to eliminate almost all suspicious signals. The C-Scan system includes the following items:

- Signal generator with built-in charger.
- Generator/pipe connector lead.
- Generator/earth connector lead provided with extension. Set of earth spikes, complete with connector leads.
- Detector unit containing power pack, antenna, and computer.
- Spare power pack for the detector unit.
- Charger for the detector power packs.

6.9.2 Performing Survey by C-Scan

The C-Scan system uses a generator to apply an AC signal to the pipe, and this is done by connecting one of the generator output leads to the pipeline at a CP test point, valve, or access point; the other lead of the generator is attached to a remote earth.

A constant AC signal (set by the operator) is applied to the pipeline. The applied AC signal will flow along the pipeline in both directions, decreasing in magnitude as the signal leaks to the earth through the coating. At coating defects, where the pipeline is in direct contact with the soil, the signal leakage to earth is greater.

A detector is used to measure the strength of the AC signal on the pipe. The operator starts making measurement. The first measurement is made at least 100 m from the point of injection, usually at a fence or other identifiable feature. The detector will automatically take 1000 readings of the magnetic field and will compute and display to the operator the depth to the center line of the pipeline and the signal level (strength of the remaining signal current). If required, the detector unit will also record (store) these data and give them a (marker) number for subsequent reference.

This procedure is then repeated at the next survey point, usually the next road crossing or the next cathodic test point. The attenuation is calculated from these values and gives a direct indication of the coating condition in that section.

An initial survey, called the general survey, may be carried out on a pipeline by taking measurements at road crossings and points of easy access with intervals of approximately 1 km. This will identify the sections of the pipeline between measurement points where coating defects exist. The defect location may be defined by progressively halving the distance in the suspect section until it has been reduced to a practical length where a detailed close interval current loss survey can be carried out.

6.9.3 Advantages of the System

The main advantages of this system are as follows:

- Rapid assessment of the system.
- Accurate defect location and depth determination.
- Repeating the survey at regular intervals to monitor coating condition during the working life of the pipeline.
- Survey can be carried out in plant areas where the pipe is covered by concrete or tarmac.
- No need to walk the entire length of the pipeline and areas of difficult access, such as rice fields, etc.
- Survey includes road, rail, and river crossings.

6.9.4 Theoretical Background

An electrical current applied to a well-wrapped buried metal pipeline will decrease gradually with increasing distance from the current injection point, as the current escapes to the earth through the coating.

If the coating has a uniform thickness and separates the pipe from the surrounding soil at all points, the strength of the signal current on the pipe will decline logarithmically, and the rate of decline will be dependent primarily on the effective resistance of the coating in use (resistance per average square meter of coating), and the area of coating in contact with the soil per unit length of pipe (i.e., for a given coating, the decline is proportional to the circumference of the pipe).

It should be borne in mind that because of the relative magnitude of the resistance involved, local changes in soil resistivity can usually be ignored.

If there is a low resistance electrical path from the pipe direct to the soil at any point, there will be a substantial local increase in the rate of loss of signal current. Such a low resistance path could arise from the following:

- Improper applied coating.
- Mechanical damage to the coating before, during, and after application.
- Decay of the coating due to soil stress or bacterial effects.
- Disbonding of the coating from the pipe (provided that ground water has penetrated into the gap to create an electrical path to the earth).
- A leak in the pipe itself causing the coating to fail at the leak point.

Because the resistance of such an electrical path to the earth is likely to be several orders of magnitude less than the resistance of the undamaged coating, the resultant loss of current, even from a single small fault of a few square millimeters, can usually be detected by a significant increase in the apparent rate of current decline over quite a long length of pipe.

In practice, the existence of one or two small faults on a section of pipeline several hundred meters in length can usually be tolerated because CP system is expected to prevent corrosion resulting from these faults. In this case, it may not be necessary to locate the specific faults immediately, but the rate of the logarithmic decline of the current between two specific points can be logged for future reference so that any deterioration of the condition of the pipeline coating can be monitored.

The logarithmic rate of decline of the current (which is known as attenuation level) is measured in terms of millibels (1/100 of decibel) per meter. This rate of decline of the current is effectively independent of the applied current so that it is virtually an absolute indication of the average condition of the coating between two given points at the date on which the survey is made.

If the attenuation level over a given section of pipeline is particularly high, serious faults (or a large number of small faults) in the protective coating are indicated. Intermediate readings of attenuation levels can be taken to identify the worst sections. The survey can be made extensively over the suspect section at close intervals (3–5 m), and readings of the actual current levels in this section can be taken, recorded, and plotted. By so doing and noting the place where the rate of current decline is the steepest, the precise location of faults (generally to an accuracy of 1–2 m) can be determined. This locates the defect to the extent that excavation for visual check and repair is justified.

6.9.5 Operation Principle

As mentioned earlier, this system of coating evaluation is based on current attenuation level. The main operating components of the system are "The Signal Generator" and "The Detector Unit."

The Signal Generator is attached to the pipeline and also to an appropriate earth point, and produces a constant AC signal that passes along the pipe.

The detector unit measures the electromagnetic field radiating from the pipeline and uses this to:

1. Locate the pipeline.
2. Determine the depth of the pipe.
3. Determine the residual strength of the signal current at the observation point.

The above data can be stored in the memory of the detector unit's computer and are used to compute and store current attenuation rates between any two given observation points.

The complete record of the survey (with time and data) may be printed out at the end of the survey to provide a permanent record.

6.9.5.1 Signal Generator

The C-Scan signal generator contains the following:

Power supply consisting of six 12-V batteries. Capacity of each battery is 5 A/hr.

A built-in charger suitable for a mains input of 110 or 240-V single phase with a frequency of 50 or 60 Hz, respectively.

The inverter–frequency changer that inverts the output of the signal generator to a sine wave signal at a frequency of 937.5 Hz, with a nominal maximum voltage of 60 V.

The maximum current obtainable is approximately 1 A. The Root Mean Square value of the current being produced is displayed on the digital ammeter of the instrument. The value is expressed in milliamperes.

The actual level of current produced is set by the surveyor using the instrument's controls. Once set, this current level will be maintained at a constant level (despite any transient changes in soil resistivity), until the batteries are exhausted. In tests, the unit when fully charged has maintained a constant output of 750 mA for up to 16 h of continuous operation.

6.9.5.2 Detector Unit

The detector unit consists of two vertical tubes joined by two smaller crosstubes, the upper one of which serves as a carrying handle. The taller vertical tube contains parts as given hereunder:

- The antenna system.
- The keyboard and liquid crystal display (LCD) that are mounted on the top of the tube.

The other vertical tube contains the computer, the power supply, and the printer lead. The antenna consists of five large-diameter air cored coils. Two of these are at the bottom of the tube with their axes horizontal and at right angles to each other (c and d). Another identical pair (a and b) is at the top of the tube approximately 80 cm above the first pair, and a fifth coil with its axis vertical.

6.9.5.3 Principle of Operation

In operation, the computer calculates the vector sum of the field strength measured by the two pairs of coils (ab and cd) and indicates to the operator that it has acquired the signal radiating from the buried pipe. The LCD gives an indication of the field strength and guides the operator toward the pipe. A comparison of the field strengths at the top and bottom of the antenna, with the strength of the field measured by the vertical axis (e) coil, is used to inform the operator when he is close to the pipeline.

When the field measured by the (e) coil falls below a threshold value, the instrument informs the operator to stop because is overhead. Once the instrument is stationary in the overhead position for a few seconds, it will automatically collect a sample of 200 readings of the field strength values at the top and bottom of the antenna over a period of approximately 4 s. Provided the standard deviation of the sample is below a threshold value, the computer will calculate the depth of the pipe (below the center of the bottom coil pair) and the strength of the residual signal current on the pipe.

The depth (in meters) and current value (in milliamperes) are shown on the LCD for a few seconds, and if the operator takes no action, the instrument will repeat the sample and calculation routine.

Using the keyboard and, following the prompts appear on the display, the operator can store the display data (which is mathematically given a reference number), and request the computer to calculate the logarithmic attenuation of the signal from any previous location stored in its memory. This information is displayed on the LCD and may also be stored if required. The computer can store up to 100 complete sets of data (location reference number, depth of pipe, strength of signal current, distance from a previous location reference, and logarithmic attenuation of the signal), in millibels per meter, between the two points. At the beginning of each survey, the computer will also automatically note the time and data using its internal clock. This information will also appear on the subsequent printout. Current readings obtained in the course of a close interval survey to locate specific faults are not stored for printout, as they are unlikely to be relevant to subsequent surveys. These readings are generally noted down for plotting purposes by the operator during the course of the operation.

At the end of a survey (or whenever required), the detector unit may be plugged into a standard computer printer, and the survey report may be printed out. Alternatively, the data may be displayed on a computer monitor or stored on disk or tape for further analysis.

The detector unit is powered by a nickel–cadmium rechargeable power pack providing enough power for continuous running for over 20 h.

In addition, there is a separate built-in lithium power source to maintain the data stored in the memory and the operation of the clock, when the power pack has been unplugged. This power source has an estimated life of 10 years.

The C-Scan system is not immune to electromagnetic interference, but it incorporates a number of features that are designed to keep such interference to a minimum. The feature include the following:

1. The generation of a pure sine wave so that no complex harmonics are produced.
2. The use of a carefully selected frequency with virtually no harmonics in common with other frequencies in general use.
3. The collection of 200 signal "samples" over a period of approximately 4 s to eliminate transient interference.

The instruments are not affected by the presence of mains AC current, nor does any residual ripple of CP rectifier affects the operation of the system.

6.10 Coating Evaluation by Electromagnetic Current Attenuation Survey

The electromagnetic current attenuation survey is an above-ground survey technique used to locate coating defects on buried pipelines and to provide a comparative assessment of coating quality.

An AC signal is applied between the coated pipe and remote earth. When the signal current flows along a straight conductor (pipeline), it creates a magnetic field, cylindrical in shape, around the pipeline. The shape of this field is not affected by the presence of pipeline coating, by different types of soil, or nonreinforced concrete,

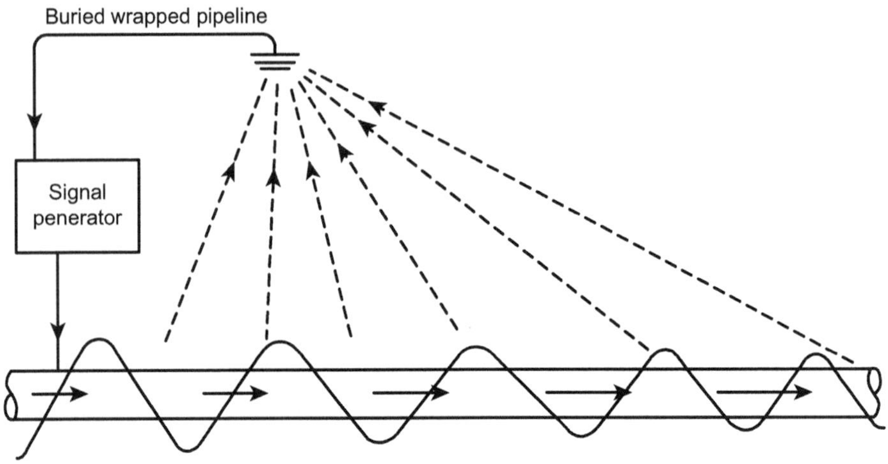

Figure 6.12 Current attenuation.

tarmac, etc. It is possible by electromagnetic induction to detect and measure the intensity of this signal with an antenna.

The applied AC signal will flow along the pipeline in both directions, decreasing in magnitude as the signal leaks to the earth through the capacitive and resistive effects of the coating. At coating defects, where the pipeline is in direct contact with the soil, the signal leakage to the earth will be more significant (Fig. 6.12).

Thus, signal losses over a section of the pipeline are due to the following:

- Dielectric and resistive losses due to the nature and quality of the coating.
- Capacitive losses of the AC signal to earth.
- Discrete points of coating damage resulting in steel/soil contact.
- Metallic contact with other buried conductors.

Where the current is observed to attenuate at a constant rate over a long section of the line, these losses are due to the intrinsic characteristics of the pipeline coating rather than resulting from evenly spaced holidays of a similar area. Where a metal-to-soil contact exists, the signal loss greatly increases at this point, clearly indicating the presence of a coating defect.

6.10.1 Equipment

The equipment required for the survey comprises the following:

- Transmitter to provide an AC signal of approximately 1000 Hz. Thin-film coatings may require a reduced frequency in the order of 200 Hz. The transmitter is powered from internal batteries. Longer surveys may require the use of an external high-capacity battery.
- Receiver, hand held with integral antenna for monitoring and indicating the location of pipeline, depth of pipeline, and signal strength of the electromagnetic field.
- Earth spike and connecting cables for the transmitter.

The above equipment is normally contained in a plastic or aluminum box. Its total weight is between 15 and 30 kg, dependent on the manufacturer's guidelines.

Some equipment is fitted with a microprocessor control that enables current flow and attenuation rates to be computed and data to be stored and printed out on completion of the survey. Optional equipment available may enable the instrument to be used for general-purpose pipe/cable locating, Pearson survey, etc.

6.10.2 Procedure

The equipment is set up as shown in Fig. 6.13. The transmitter is electrically connected with one lead to the pipeline, usually by connecting to a CP test lead or any accessible part of the pipeline, and the other lead is connected to a good remote earth. The transmitter is energized and set to the appropriate output.

The survey may be carried out with an impressed CP system energized. However, any sacrificial anodes, bonds to other structures, or similar are best disconnected before commencing the survey to ensure that they do not mask defect areas or cause excessive signal loss, thereby reducing the length that may be surveyed from one transmitter location.

The first measurement is made at a sufficient distance from the transmitter and earth spike to minimize interference from the transmitter and/or return current flow in the earth. The receiver "location" mode is used to accurately position it directly above the pipeline.

The depth to the center line of the pipeline and the signal level, corrected for depth, is measured and recorded.

This procedure is then repeated at the next measurement point. Subtraction of the two values gives the signal loss between the two measuring points. To calculate the

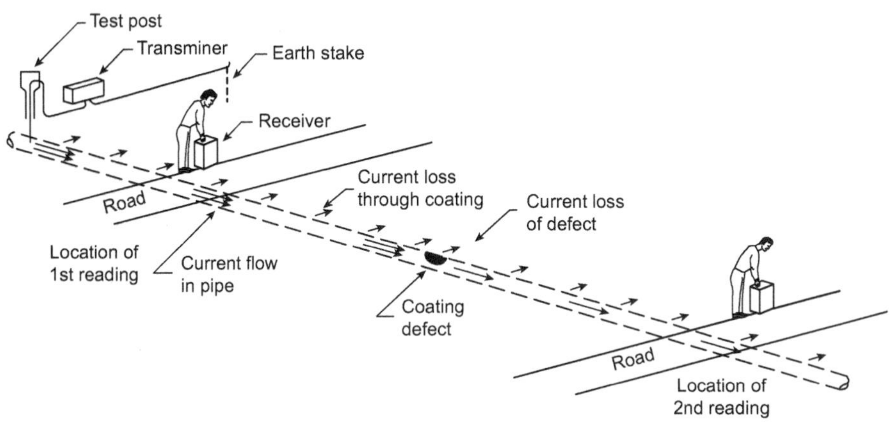

Figure 6.13 Principle of electromagnetic current attenuation survey.

attenuation rate, the signal loss is divided by the distance between the two measuring points. This may be computed manually or by an in-built microprocessor.

As the survey proceeds, measurements are taken at intervals until the intensity of the signal is insufficient to be monitored accurately. At this point, the transmitter is moved, and the above procedures are repeated for the next section of the pipeline.

The length of the pipeline surveyed from any one transmitter location may vary from <200 m to in excess of 20 km depending upon the quality of the pipe coating, the size of the pipeline, the resistivity of the soil, etc.

A survey cannot be conducted over the last 2–3 km of a well-coated pipe or 0.5 km where high losses are present working toward an electrical isolating device. Either the transmitter should be located at the isolating device or the device should be temporarily bonded across.

An initial survey may be carried out on a pipeline by taking measurements at road crossings and points of easy access with intervals of approximately 1 km. This will identify sections of pipeline between measurement points where coatings defects exist. The defect location may be defined by progressively halving the distance in the suspect section until it has been reduced to a practical length where detailed measurements would then be applied.

Measurements are to be avoided at locations where pipe bends or other metallic services are present as there will be a distortion of the magnetic field giving erroneous readings.

The above procedures are general, and in all cases, the equipment manufacturer's instructions are to be followed.

6.10.3 Data Obtained

The data recorded from an electromagnetic current attenuation survey would be as follows:

- Location reference.
- Chainage.
- Distance between measurements.
- Depth.
- From these data, the following may be calculated:
  - Signal level attenuation.
  - Signal level attenuation rate.

6.10.4 Presentation of Data

The results of the survey may be presented in one of the following formats:

1. Tabular.
2. Graphical—plot of signal level against distance (Fig. 6.14).
3. Histogram—plot of section attenuation or attenuation rate against distance (Figs 6.15 and 6.16).

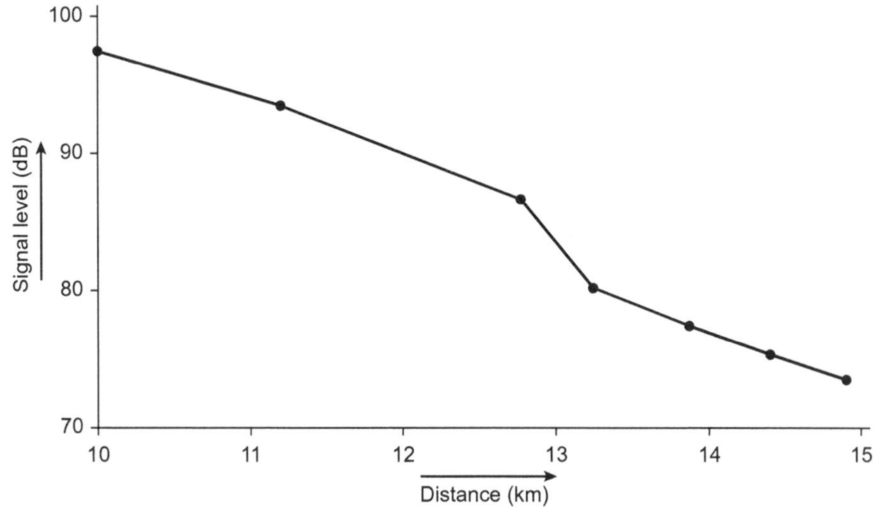

Figure 6.14 Plot of signal level vs distance.

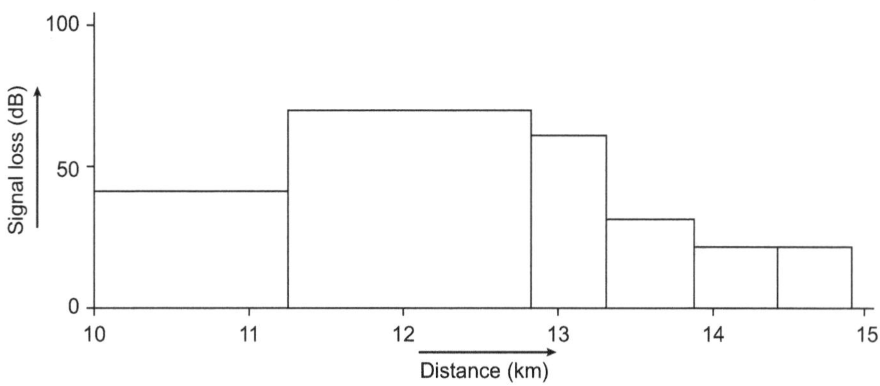

Figure 6.15 Plot of signal loss vs distance.

6.10.5 Criteria and Interpretation

The criteria for determining the coating quality are not precise and differ from one pipeline to another.

The attenuation rate is primarily dependent on the overall quality of the coating and the coating area per unit length of pipeline, that is, it is a function of quality and circumference. "Quality" is related to a combination of the following: type of coating, general age and condition, and the presence or absence of discrete faults due to mechanical damage or other factors. Attenuation rates are not affected by changes in wall thickness and are only marginally affected by changes in soil resistivity except in

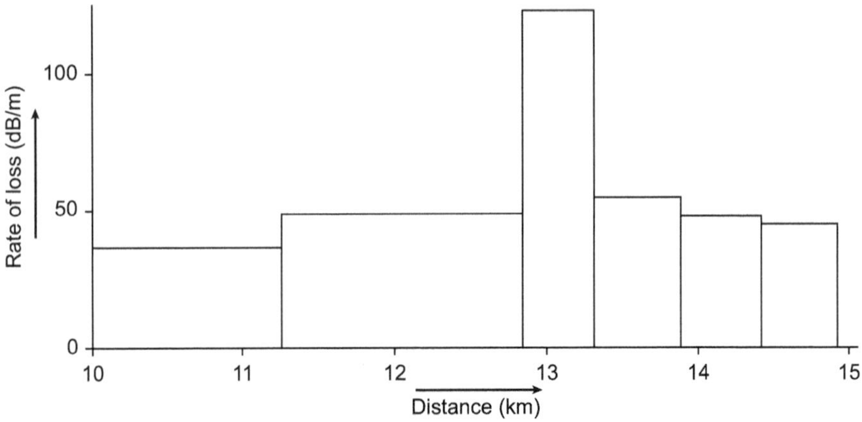

Figure 6.16 Plot of rate of loss vs distance.

extreme situations such as in deserts or where there is very poor or nonexistent coating over long lengths of pipe.

From the data, the average attenuation rate of the pipeline can be determined. Any significant change of gradient on the graph or increase in the section attenuation rate on the histogram may be treated as a suspected defect.

Where intervals of measurement are irregular or are very small, the histogram plot of attenuation rate against distance may be misleading, and a plot of attenuation against distance may be preferred.

A high signal loss may not be present at a substantial coating defect in high-resistivity soil. Because of high soil resistivity, the loss of current will be restricted. However, the corrosion risk at such an undetected defect will be reduced since the corrosion current would also be restricted by the high soil resistivity. Signal level is corrected for depth.

6.11 Close Interval Pipe-to-Soil Potential Survey

The close interval pipe-to-soil potential survey is a survey technique to provide a detailed profile of the potential difference between the steel and the soil. This profile can be used to determine detailed information on the performance of the CP system, the coating system, interaction effects, etc.

The performance of CP systems has normally been monitored on a sample basic, that is, pipe-to-soil potential measurements taken at test stations installed at intervals of approximately 1 km along the route of the pipeline.

Based on this information, a judgment is made of the performance of the CP system and also, by inference, the performance of the pipe coating system.

If a defect of any kind occurs on a pipeline at a location remote from a test station, it is probable that this will remain undetected. It has therefore become apparent that a much greater sample of pipe-to-soil potential measurements, representative of the pipeline as a whole, needs to be collected to ensure that the corrosion prevention systems are performing satisfactorily.

To achieve this, it is necessary to place the reference electrode at a much greater number of locations than the number of test stations. Close interval potentials are generally measured at between 1- and 5-m spacing to provide a "continuous" pipe-to-soil potential profile of the structure.

Equally essential is the need to reference, record, and store the data collected and to present the larger number of measurements in tabular and/or graphical form.

The methods and measurements utilized are based on methods used in everyday routine monitoring procedures. The advent of portable microprocessor-based measuring equipment and computer-assisted data handling, together with accurate, synchronized high-speed current interruption devices, have allowed these surveys to provide practical data collection methods that are of value when assessing the performance of the corrosion-prevention systems applied to buried pipelines.

The potentials may be recorded with the CP system permanently energized as in routine "on" surveys. However, these measurements will include the IR drop error caused by the flow of CP currents in the soil.

To eliminate this error, cyclic timers are fitted to the transformer rectifiers or to other DC power sources to interrupt the CP current. Potential measurements are then recorded with the CP switched on and immediately after switching off (IR drop error-free or instantaneous off measurement) over the entire length of the pipeline. It is of utmost importance to ensure that all power sources that can influence the pipe-to-soil potential at the point being measured are switched in unison by the use of synchronized timers. Surveys measuring the "Instantaneous Off" potentials are commonly referred to as "polarized" potential surveys. This method is not practical for use with distributed sacrificial anode systems.

In areas of stray current interaction where the pipe-to-soil potentials are not stable, stationary data collectors may also be used.

The foregoing covers the general requirements for a close interval potential survey. However, when planning to carry out a survey, the following practical considerations may be kept in mind:

- The need for mobile operation to traverse the pipeline length and obtain data at a reasonable speed.
- The requirements for current interruption to switch the CP current sources at high speeds, at a ratio that will not depolarize the CP levels and with a minimum of synchronization error.

The need for polarized potentials to be collected at such intervals to provide an accurate profile of IR drop error-free measurements.

- The need for switching cycle and distance between readings to be so arranged that the measurement equipment records accurately with respect to instrument stabilization and synchronization errors.
- The method by which equipment operates and data are recorded, that is, manual, semi-automatic and automatic.

6.11.1 Measurement Intervals

The measurement intervals between pipe-to-soil potentials may be determined by the pipe diameter and the depth of cover over the pipeline.

For pipeline diameters of $\leq$600 mm, a minimum spacing of 1 m is appropriate, although specific circumstances may necessitate closer intervals in order to determine particular characteristics.

For pipeline diameters in excess of 600 mm, a 1-m spacing is appropriate, but this may be increased. The measurement interval should not, however, be >1.5 times the normal pipeline depth of cover or 5 m whichever is the smallest.

In practice, polarized measurements are commonly taken at

- Every meter,
- Every 5 m,
- Ratio of 4 "on" readings and 1 "off" reading, each reading being taken at meter intervals.

6.11.2 Switching Frequency

The switching frequency of the CP and associated equipment required to enable pipe-to-soil potential measurements to be made, may be determined by the following factors:

- The degree of synchronization that can be achieved between the timers.
- The measurement response time of the data collector.
- The measurement interval of pipe-to-soil potentials in automatic time-dependent surveys.

An important consideration in the selection of time cycles is the provision of long-term switching (during the survey period) without any degradation of the polarized potential level due to the "off" period. Depolarization will cause inaccurate and unreliable potentials to be measured. For this reason, a minimum on/off ratio of 4:1 is common practice. This assumes that the quality of the measurement circuit will provide for a stabilized accurate measurement during the "off" period (response time) and that the synchronization error between switching devices does not conflict with the measurement period.

Switching of the CP equipment may be by devices interrupting either the AC or DC side of the transformer rectifier or other power sources and may be either built in or fitted for the duration of the survey. The current interrupts must have the facility to switch at high speeds, cause minimum transient switching surges, have a selection of commonly used time cycles, and have <100-m s synchronization error between devices and maintain this for a minimum of one working day. An image retaining oscilloscope or high-speed recorder may be used to verify that no significant depolarization occurs with the time cycle selected and that timers are fully synchronized.

If switching is carried out on the AC supply, the polarized potential measurements may adversely be affected by the unbroken connection of the pipeline to the ground bed via the DC power source, allowing some currents still to flow.

6.11.3 Distance Measurement

The survey needs to have some means of distance measurement with respect to the mobile potential data collector. Systems are available to accurately record the distance from a fixed point that can be related to pipeline chainage. Basic systems operate with annotating pipeline chainage at the start and finish of each survey section.

For more details, a pedometer or chainer can be used, and the distance is keyed into the data collector. With fully automatic systems, a wire measurer is often used, which can give an accuracy to within a few meters per kilometer.

All systems need to allow for chainage to be entered, together with any physical features found along the route of the pipeline so that defects indicated on the plots may be physically located.

6.11.4 Stationary Measurements

In areas where stray interaction currents or variable potentials may be present during the mobile survey, stationary data collectors may be employed at fixed locations on the pipeline as close as possible to the anticipated point of maximum interaction. Stationary measurements may be obtained at one or more of the following locations:

- Start and finish of surveyed section.
- Anticipated locations of maximum interaction effect.
- Locations close to the mobile unit.

The data collected from the "mobile" unit can then be compared with the data from the "stationary" unit. If fluctuations of potential on the "mobile" plot coincide with similar fluctuations on the "stationary" plot, this is indicative of interaction or stray currents and not necessarily due to inadequacies of the corrosion protection system. The data collected are therefore to be accurately time related in order to make this comparison.

6.11.5 Equipment

The equipment required for the survey varies from very simple manual equipment to complex microprocessor/computer equipment.

In general, the equipment comprises the following:

- Synchronized timers for switching the transformer rectifiers or other DC power supplies (not required for "on" surveys).
- Data collector with input impedance in excess of 10 Megaohms with an adequate level of AC rejection which may be
- Manual—voltmeter/multimeter.
- Semiautomatic—recording voltmeter.
- Microprocessor data collector with manual entry.
- Automatic—microprocessor data collector with fully automatic recording of data.
- Cu/CuSO4 reference electrode(s) fitted to pole(s).
- Insulated fine wire that may be reusable or disposable, wound on reels.

- Distance measurer that may be
- Chainer.
- Pedometer.
- Wire measurer that automatically measures the length pulled off the fine wire reel and records the value in the data collector.
- Pipeline location equipment.
- Computer hardware and software for processing the data.
- Plotter for producing the profile plots.

6.11.6 Procedure

Before commencing the survey, the exact route of the pipeline is located and marked out so that the survey operator may follow the route of the pipeline exactly over the pipe. Alternatively, a pipe location operator may immediately precede the survey operator to minimize the working period over the pipeline route.

If polarized potentials are to be measured, timers are synchronized and installed in all transformer rectifiers or other CP power supplies, bonds, or other connections that may influence the potentials over the section of pipeline to be surveyed.

If the survey is subject to interaction or fluctuating potentials, stationary recording units would be installed.

With the equipment connected and the pipeline located, the survey operator, carrying the mobile data collector and $Cu/CuSO_4$ reference electrode(s), traverses the pipeline paving out the fine wire and placing the reference electrode at the required interval over the pipeline and maintaining it in that position long enough for the data collector to measure and store the pipe-to-soil potentials. Alternatively, two reference electrode(s) may be fitted to poles such that one reference electrode is always in contact with the ground. Distance traversed and physical features, such as roads, hedges, and streams, are recorded to assist with locating specific areas after processing the data.

As each section between pipeline connection points is completed, the fine wire is either re-reeled for further use or recovered and properly disposed of, to prevent a hazard to livestock. Where the fine wire crosses roads, paths, or fields, with livestock present, the wire may be pegged down.

The data are usually transferred from the data recorder to a computer for producing data printouts and plots of the pipe-to-soil potential profiles.

The above procedures are general, and in all cases, the equipment manufacturer's instructions are to be followed.

6.11.7 Data Obtained

The data obtained from a close interval pipe-to-soil potential survey include the following items:

- Time.
- Date.

- Distance (pipeline chainage).
- Locations.
- Energized potential.

Polarized potential.

- Physical route features.
- CP power supply outputs.
- Bond currents.

6.11.8 Presentation of Data

The data collected during the survey are normally processed by a computer and presented in tabular and graphical form.

The raw data are normally processed immediately following the completion of each day's survey to verify the full functioning of the survey equipment and identify if any areas may require resurveying due to interaction, equipment malfunction, etc.

Plotted profiles are produced at a later date when all the data have been processed.

The data presentation is normally computer assisted. This allows for data to be organized in the correct chainage order regardless of survey direction and section order.

6.11.9 Criteria and Interpretation

The criteria used for closed interval pipe-to-soil potential surveys are the same as those used for routine potential surveys.

The close interval survey enables the potential profile for the whole length of the pipeline to be plotted. With the CP system switching "on" and "off," and polarized potentials, free of IR drop error, measured, a profile of the true pipe-to-soil potential can be determined.

It is the measurement of the polarized potential that determines the effectiveness of the CP current in both qualitative and quantitative terms.

Any areas on the profile that fall outside the selected potential criteria are not being provided with adequate CP and could therefore be corroding. Interpretation of the pipe-to-soil potential profile may highlight the following:

- Coating defects.
- Underprotected sections of pipe.
- Interaction by stray currents.
- Overprotected sections of pipe.

6.12 An Example for Cathodic Field Interference Test Method

To illustrate the interference test method, assume that a CP system incorporating a rectifier as the power source has been applied to a section of coated pipeline and that there are several pipeline crossings in the section as shown in Fig. 6.17.

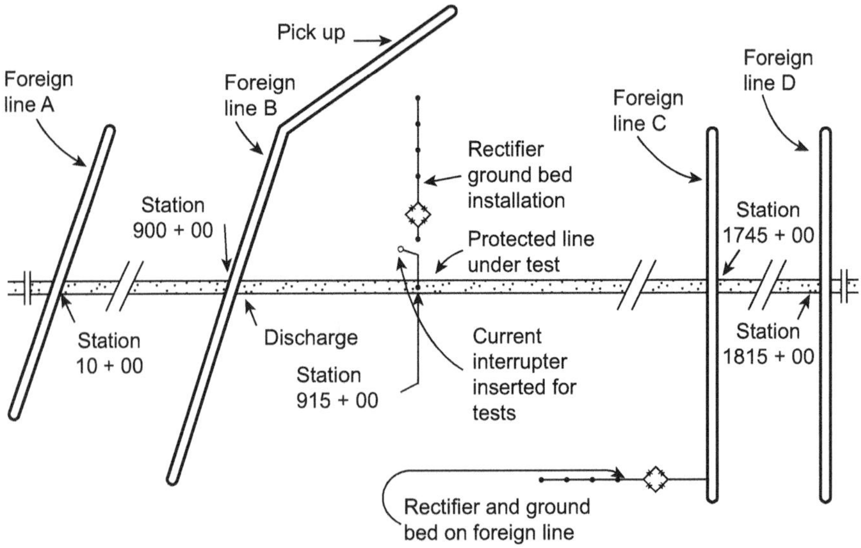

Figure 6.17 An example for cathodic field interferences.

Assume that at each foreign line crossing a test point has been installed, as shown in detail in Fig. 6.18, with two color-coded leads brought to the test point terminal from each point.

With the current interrupter operating, each foreign line crossing is visited and the potential of each line is measured under both current "ON" and current "OFF" conditions. For these tests, the CSE is placed directly over the point of crossing. If there is any question as to the crossing location, use a pipe locator to determine where

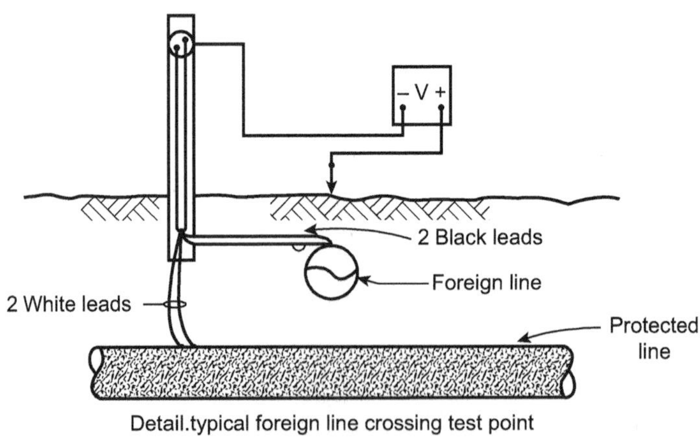

Figure 6.18 Testing foreign pipeline crossings for stray current interference from CP installations.

Table 6.2 Survey Notes on Stray Current Investigation

Foreign Line Designation and Location		Potential vs Close Copper/Copper Sulfate Electrode					
		Own Pipeline, V			Foreign Pipeline, V		
		With Current			With Current		
Name	Pipeline Station	On	Off	ΔV	On	Off	ΔV
A	10 + 00	−1.03	−0.98	−0.05	−1.02	−1.04	+0.02
B	900 + 00	−1.98	−1.02	−0.96	−0.32	−0.68	+0.36
C	1765 + 00	−0.68	−0.64	−0.04	−0.78	−0.78	0
D	1815 + 00	−1.10	−1.06	−0.04	−0.68	−0.68	0

it is. Data taken at the several pipeline crossings of Fig. 6.18 may be recorded as shown in Table 6.2:

Data entered have been selected to illustrate various types that may be encountered. In addition to the data shown for this illustration, field data sheets should include full information on the line protected; this includes data, current output of the interrupted rectifier, and other pertinent facts.

The following are some conclusions that can be reached from the data:

Crossing A

Pipeline under test is fully protected but can be expected to have a substantial coating holiday in the vicinity of the crossing. This is based on the fact that the potential of the foreign line (which is also cathodically protected) decreases when the interrupted rectifier switches from "OFF" to "ON." This indicates that there is appreciable current flowing to the line under test, creating more negative soil locally around the foreign line.

The foreign line in this instance is cathodically protected as indicated by the fact that although its potential at the crossing is reduced by the operation of the rectifier on the line under test, the reduction is not sufficient to indicate any loss of protection. On the basis of the data shown, no corrective measures would be required at Crossing A.

Crossing B

Pipeline under test is fully protected. The foreign pipeline is not cathodically protected because its potential is well below 1.0 V (with the rectifier "OFF" on the line under test). With the rectifier ON, the foreign line potential is shifted severely in a less negative direction, indicating the probability of severe corrosion damage to the foreign pipeline. Corrective measures will be required.

Crossing C

The line under test is receiving inadequate CP, apparently because of interference from the cathodically protected foreign line. If the potential of the line under test to a

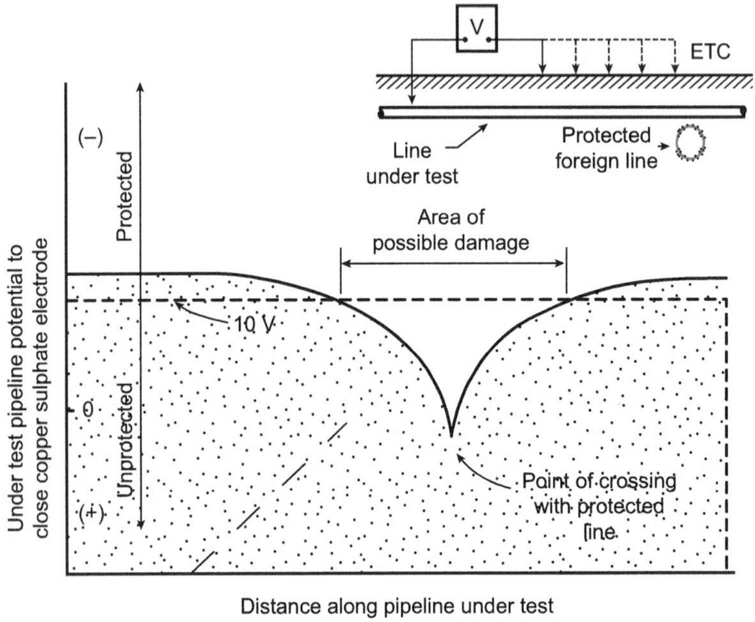

Figure 6.19 Pipe-to-earth potentials on under test pipeline passing through the area of influence around a cathodically protected foreign line.

remote electrode with respect to both lines and this potential is found to be representative of normal protective potentials (above −1.0 V), the poor potential at the crossing is shown to be a local condition probably caused by the foreign line crossing. Corrective measures will be required. The length of the line under test, which is below −1.0 V, can be determined by taking readings to close electrode, directly above the line under test, in each direction from the point of crossing. If the data when plotted, give a curve similar to that shown in Fig. 6.19, interference from the protection system on the foreign line is confirmed. In this case, the foreign pipeline is not affected adversely; it is the line under test that is adversely affected by the foreign line.

Crossing D

The line under test is protected adequately. The foreign line does not have full CP but is not affected by the CP system on the line under test. No corrective action is required.

6.13 Tests for Electrical Continuity

Any disconnections or high-resistance joints will become apparent when structure/electrolyte potential measurements are made at various points along the length of the structure. Where this method is being used to locate high-resistance joints, sufficient current should be used to give a large structure/electrolyte potential depression at the point of negative cable connection, for example, −1.0 to −1.5 V. Where there is an abrupt decrease in the structure/electrolyte potential depression, there is either a resistive joint or a connection to a massive buried structure between the point of test and the negative cable connection point.

In the case of structures such as buried pipelines, when such discontinuities are suspected, a constant voltage (up to 12 V) can be applied between two points of pipeline some distance apart. Discontinuities will be indicated by a change in the potential gradient, that is, a sudden voltage drop. It is possible to locate these points

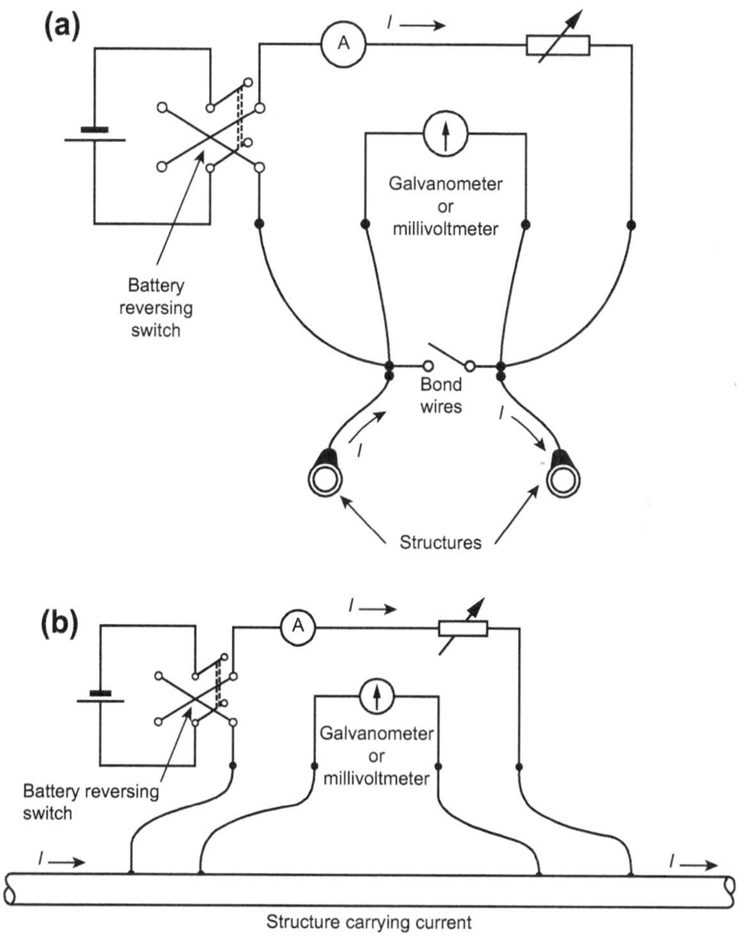

Figure 6.20 Tests for electrical continuity. (a) Zero resistance ammeter circuit used to measure bond current (bond disconnected). (b) Zero resistance ammeter circuit used to measure current flowing in bond or structure (circuit not disconnected).

accurately by using two wander probes, each making contact with the pipe, when a sensitive voltmeter is connected between them.

The comparative longitudinal resistance of a pipe joint caused by the passage of a known impressed current of the order of 10–20 A, the galvanometer being calibrated as a millivoltmeter. The four connections to the pipe can be made using probes with hardened points; the probes connected to the millivoltmeter should be insulated. The impressed current should be reversed, and the test should be repeated to confirm that the potential drop measured is due solely to the applied test current (Fig. 6.20).

The following are to be noted:

Probing should not be carried out without prior consultation with other authorities concerned, to ensure that there is no possibility of causing damage to their buried services.

Another method of locating discontinuities, which can be applied to buried cables and pipelines, uses an audio frequency current impressed on the structure from which a signal is pitched up by means of a search coil and head phones. Points of electrical discontinuity are indicated by a sudden decrease in the signal strength.

This method has proved satisfactory in open country but needs care in urban areas where the presence of other services in the ground will often cause misleading results. If may also be necessary to determine whether the armor of a buried cable is satisfactorily in contact with the sheath. Evidence of discontinuity is afforded if there is an appreciable potential difference between the sheath and armor, particularly when CP is applied by means of connection to the sheath.

7 Installation of Cathodic Protection Systems

This chapter covers the minimum requirements for installation, testing, and commissioning of cathodic protection (CP) systems (impressed current and galvanic) for buried and immersed steel structures such as buried pipelines, distribution networks, in-plant facilities, and marine structures, which includes installation, startup, measurements, testing, commissioning, and inspection procedures.

This chapter is generally applicable to buried structures. For specific structural systems (Installations that require special attention, techniques, and materials are not covered. Each such installation requires special considerations based on many influencing factors and cannot be covered adequately in a single standard), this chapter could also be used in conjunction with the project specification and drawings for those structures.

Because of the inaccessible nature of much of the CP equipment in service, it is necessary to confirm, prior to shipment to site and prior to installation, that materials and equipment comply with the appropriate standard specification to avoid unnecessary and protracted delays while replacements are sought or repairs are undertaken.

7.1 Galvanic Anodes

Anodes should be inspected for the following:

1. Freedom from electrical damage.
2. Electrical security and continuity of connections.
3. Anode-to-core continuity.
4. Correct metal mass.
5. Correct profile.
6. Compliance of anodes (including anode backfill) with "standards."

Insulation of all cable tails should be inspected for the presence of nicks, cuts, or other forms of damage.

Packaged anodes should be inspected and steps taken to ensure that backfill material completely surrounds the anode. The individual container for the backfill material and anode should be intact. If individually packaged anodes are supplied in waterproof containers, that container must be removed before installation. Packaged anodes should be kept dry during storage. Electrical continuity between anode and lead wire should be tested without compromising the integrity of the package.

Cathodic Corrosion Protection Systems. http://dx.doi.org/10.1016/B978-0-12-800274-2.00007-7

Other galvanic anodes, such as unpackaged "Bracelet" type or ribbon, should be inspected to ensure that dimensions conform to design specifications and that any damage during handling does not affect application. If a coating is used on bands and the inner side of "Bracelet" anode segments, it should be inspected and, if damaged, repaired before the anodes are installed.

When a separate suspension such as rope is used to support the weight of an anode, the suspension system should be inspected for damage, and all defects should be repaired.

In the case of weld-on type of galvanic anodes, steel cores should be inspected for conformance to specifications. If anode cores have welded joints or connections, these should be inspected to ensure compliance with structure-welding specifications.

When galvanic anode suspension cables are used for the lead wire, the cables should be inspected for strength and good electrical contact with the anode. Where separate suspension cables are used, care should be taken to ensure that anode lead wires are not in such tension so as to damage the lead wires or connections.

If coatings are specified for galvanic anode supports or suspension cables, they should be visually inspected and the coatings repaired if damaged.

7.2 Impressed Current Anodes

Impressed current anodes should be inspected for conformance to standard specifications concerning correct anode material and size, length of lead wire, and secure cap, if used. Care should be taken to avoid cracking or damaging anodes during handling and installation.

Lead wire should be carefully inspected to detect defects in insulation. Care should be taken to avoid damage to insulation on wires. Defects in the lead wire must be repaired, or the anode must be rejected. Anode backfill material should conform to standard specifications. Cables should be inspected to ensure that cable runs can be achieved, preferably in one take-off from a reel or drum and that the cable is of correct construction for the intended application. Insulation of all cables should be inspected for the presence of nicks, cuts, cracks, abrasions, and excessive thinning below a specified thickness or other forms of damage.

7.3 Transformer/Rectifier Equipment

Testing should be carried out before acceptance of a transformer/rectifier unit, to confirm compliance with the standard specifications and to ensure that the equipment is suitable for the intended purpose.

The following tests should be carried out on transformer/rectifier equipment:

1. Visual inspection to ensure that all rectifier and surge protection equipment and all specified current outputs have been provided.
2. Polarity check to ensure that output terminals are correctly identified.

3. A step-by-step check of the unit output against the calculated load, to ensure that a uniform control pattern is available.
4. Insulation resistance tests conducted and recorded on all transformer/rectifier(s) in accordance with company-approved testing method to ensure that the equipment has neither deteriorated nor has been damaged during shipment.
5. The contractor to ensure that Oil-cooled transformer/rectifier(s) be filled to the normal liquid level before being placed in operation. Five samples of insulation oil should be tested for dielectric strength and have the results and the average recorded if required by the Company.
6. Functional tests of time switches to be installed.
7. Functional tests of other special equipment to be fitted.

Where appropriate, each insulating joint should be electrically tested, pressure tested, and finally electrically retested. Where supplied for welding into position, the associated pipe pieces should be of a sufficient length to prevent damage to the joint insulation by heat transfer during the welding process. During welding, the manufacturer's recommendations on cooling rate should be followed.

7.4 Excavation and Backfilling

All necessary excavating, shoring, sheathing, bracing, pumping, and backfilling required to install ground beds, cables and connections as specified need to be provided. When excavation is carried below grade, the fill-to-grade material should be well tamped. In no case should any frozen earth be used for backfilling, nor should any backfilling be placed on or against frozen earth. Trenches under roads and paved areas should be backfilled with coarse sand to meet the approval of the Company. Excavated material should not be used.

Any earth excavating procedure presents safety hazards related to the presence of unstable soils, water, released products, and moving equipment. Personnel involved in excavation, equipment installation, and backfilling should be knowledgeable about and should follow the safety standards.

The excavation should provide adequate space for the installation of anodes, cables, and ancillary equipment. Special attention should be given to sloping or shoring the sides of the excavation to make them stable. Metallic pipelines should be located through the use of a line locator and mechanical probe. Excavations within 600 mm of the pipe should be done by hand. Nonmetallic lines in the immediate proximity of excavations should be exposed by hand. Damage to pipelines, coatings, conduit, cable, or other buried equipment as a result of excavation should be repaired in accordance with standards at a cost to the contractor before backfilling.

All electrical and control equipment marred by shipment or erection should be touched up, using the same color and type of finish as the original, according to the standard for painting. The transformer/rectifier cabinet must not be coated with mastics, tars, or any other similar materials.

7.5 Installation of CP Systems for Buried Pipelines

This section specifies the minimum requirements for the installation of CP systems that will control corrosion of the buried pipelines.

7.5.1 Installation of Impressed Current Systems

- **Ground beds**

Because anodes are often brittle, care should be taken to ensure that they are not damaged by handling. Unless specially designed, they should not be suspended or lowered by their cable tails because connections are essentially electrical and not mechanical. Proper implements, tools, and facilities should be provided and used for the safe and convenient performance of the work.

All materials should be examined carefully for damage and other defects immediately before installation. Defective materials should be marked and held for inspection by the Company, so that they may prescribe corrective repairs or reject the materials.

Anodes should be installed in the center of any backfill, and the backfill should be gently tamped into place around the anode. Care should be taken to prevent anode breakage.

On completion of the installation of a ground bed, the resistance of the ground bed to the remote earth should preferably be measured by using an alternating current (AC) earth tester. Measured resistance should be compared with the design resistance.

Resistance in a ground bed may be lowered by permanently adding water to each anode by using plastic water piping and drip-irrigation fittings. However, where ground bed resistance is still too high, the ground bed will need to be extended.

The ground beds should be of the following forms as will be specified by the design documents.

- **Horizontal ground bed**
  Horizontal ground beds should be constructed in locations as determined in the design drawings and with the following considerations:
  - Anodes should be installed horizontally in a group and connected in parallel in the trench at a minimum depth of 2000- and at 4500-mm centers, unless otherwise specified by the design documents.
  - A bedding of the trench should be made 600 mm wide and with a depth and length as specified on the design drawings. The trench walls should be vertical throughout, and the bedding should be tamped to provide a uniform surface.
  - Anodes should be installed with a minimum of 150-mm compacted metallurgical grade coke breeze encapsulating the circumference and a minimum of 2250-mm coke breeze extending beyond each end. The anode lead wires should then be brought out of the coke breeze and spliced, taped, and coated to the positive header cable.
  - The coke breeze should be thoroughly and properly tamped; for maximum coupling between the anode and the earth. Care should be taken during backfilling to avoid

damage to the anode. Loose backfill can give disappointingly high resistances and shorten the anode life.

- The process of tamping down should be achieved in stages after every 10-cm layer of coke breeze has been poured into the trench. The tamping down process while having to be very thorough should in no way damage the anodes.

  The ground bed excavation should then be backfilled with fine soil by hand until a minimum cover of 200 mm over coke breeze is achieved. Power equipment should then be used to restore the excavation to the original ground level.

  Before backfilling the trench, vent pipes should be placed at their predetermined locations on each anode and filled with gravel. If the backfilling operation does not produce sufficient compaction to eliminate the possibility of future settling, a berm should be installed over the backfill such that original elevations will be met.
- The header cable should then be laid on a 100-mm layer of fine sand covered with a further 100 mm of sand. The remaining space of the trench should be backfilled with earth to the ground level. For the protection of the cable, protective tiles or bricks should then be installed on top of the sand as shown in the standard drawing. The remaining space of the cable trench should be filled with backfill.

The following are to be noted:

1. In horizontal installations, ditch width at anode depth should be that of the design width of the carbonaceous backfill layer. Where this is not possible because of trenching conditions, form boards may be used to restrict the backfill. After the carbonaceous material and anodes have been placed inside the form boards and tamped earth outside, the form boards must be withdrawn. The coke breeze should be retamped to fill the space occupied by the form boards.
2. Top soil should be stripped and stockpiled at the commencement of excavation and redistributed over the excavated area upon completion.
3. Maximum anode loading should be determined by employing good engineering practices.
4. Due consideration to the use of anode irrigation equipment should be given by the contractor. Where applicable, complete details of the proposed equipment and installation methods should be provided for approval of the relevant experts.

- **Vertical ground bed**

  Vertical ground beds should be constructed in locations as determined in design drawings and with the following considerations:
  - Anodes should be installed vertically in a group (at straight line) in separate holes and connected in parallel.
  - The anode hole should be made so that it is at least 1300 mm (4 ft) deeper than the length of the anode rod and 200 mm (8 in) larger in diameter than the diameter of the anode.
  - The bottom of the hole should be filled to a depth of 300 mm (1 ft) of metallurgical grade coke breeze, and tamped until well packed. Tamping will reduce the anode-to-soil resistance, and will thereby increase the efficiency of the installation. The anode must be centered carefully in hole, and the backfill material should be poured into the hole to cover the anode. The backfill should be gently tamped into place around the anode. When tamping with power tampers (preferred) or by hand, particular care must be taken to prevent damage to the anode or to the anode lead wire.
  - This procedure should be repeated until the anode is covered by at least 300 mm (1 ft) of backfill. After making the electrical connection of the anode lead wire-to-header cable the vent pipe should be placed in its predetermined location and filled with gravel.

The following should be noted:

The purpose of the gravel is to provide a gas vent for the oxygen, chlorine and, in some special cases, hydrogen, which may evolve under various conditions from the anode area.

- The hole should then be backfilled with the excavated earth to the ground level. The header cable should be laid on a 100-mm layer of fine sand covered with a further 100 mm of sand. For the protection of the cable, protective tiles or bricks should be installed on top of the sand as shown in the standard drawing. The remaining space of the cable trench should be filled with backfill.

- **Deep-well ground bed**

  Deep-well ground beds should be constructed in locations as determined in the design drawings, and with the following considerations:
  - The anode bed for a deep-well ground bed should be drilled with a rotary rig (using mud or air) or by cable tools where applicable.
  - Depending on the type of the drilling rig used, the following possibility may be considered: the use of the rig itself for placing the anode system in the hole where a well is being drilled in soft formations that cannot be depended on to hold an open hole without collapsing; a rotary rig can continue circulating drilling mud in the hole, after reaching the design depth, until just prior to placing the anode system.
  - All drilling procedure and installation of casing and well head completions should be in accordance with public laws.
  - Casing should be set prior to installation of anodes to prevent damage to the lead wires.
  - Downhole components should not be bundled or fastened with materials that will cause gas entrapment or backfill bridging.
  - The lead wire-to-anode connection resistance should be checked before installation.
  - The lead wire insulation must be protected from abrasion and sharp objects. Prior to installation, lead wire insulation should be visually inspected for flaws or damage.
  - Further assurance of lead wire insulation integrity may be achieved by conducting suitable wet tests using proper safety precautions.
  - When installing a suspended anode, where separate suspension is required, care should be taken to ensure that the lead wire is not in sufficient tension to damage the anode lead wire or connections.
  - The deep well should be either of the dry (closed hole) or the wet (open hole) type as specified by design documents.

The following are to be noted:

1. When possible, ground beds should be located at a minimum distance of 30 m from any buried metal structures.
2. Salt should be added to deep-well ground beds for the purpose of lowering the resistance of the ground bed is absolutely forbidden.

- **Closed well (closed hole)**
  - Individual anodes should be centered in the well with a suitable device that will allow passage of backfill material, will not entrap gases, and will not damage lead wire insulation or preclude proper placement of anodes.
  - Before pumping backfill material, all anodes should be placed at the predetermined depth and the vent pipe should be set from the bottom anode to the top of the well.
  - The following should be considered: Before pouring or shoveling backfill material from the top of the well, displace the drilling mud with clear water, and place the vent pipe and

two deepest anodes at their predetermined depth. Pour backfill material into the well to cover the first anode, place the third anode, and repeat the procedure for each following anode.

- The backfill material should be wetted, as may be required to prevent bridging the well.
- The following should be considered: If strata resistivities permit moderate vertical shifting of anode position, the release of anode lead wire tension to provide slack may prevent excessive loading of the lead wire or the lead wire-to-anode connection in the event of caving or settling of the backfill material.
- Type 3 coke breeze should be used as a backfill unless specified otherwise by experts.
- All deep ground bed installations (rectifier, well, and venting location) should be marked with adequate signs so as to advise all personnel to vent the installation properly before commencing work and to keep fire away.
- One plastic vent pipe should be used to aid in dissipating gases to the atmosphere.
- The following criterion should be kept in mind: The plastic vent pipe that extends below the anodes normally has a series of small holes on 15- to 30-cm centers drilled in the immediate vicinity of the anodes. These holes should be of such a small diameter as to prevent the entry of the backfill material into the vent pipe.
- The vent pipe should be capped at both ends during the backfilling operation to minimize filling with backfill material or mud.
- The following should be considered: A threaded fitting installed at the surface end of the vent pipe will facilitate water or air injections that may be required to eliminate gas blockage. The use of a screened bushing on the threaded fitting will prevent the entry of insects and foreign objects. The connection of a hose to the vent pipe with the end inserted in an open water container will provide a visible test of gas venting.
- Vent pipes should be located so as to preclude the entry of corrosive gases into the test box and rectifier. All lead wire conduits should be sealed.
- A uniformly low resistivity backfill should be installed in the well until the top anode is adequately covered. Suitable backfill will decrease the anode resistance to the electrolyte, increase anode life, prevent caving, and facilitate gas venting. The remainder of the well should be filled with a nonconductive, nonabrasive permeable backfill material (sand or pea gravel).
- Backfilling may be accomplished by pumping, shoveling, or pouring. Backfilling method usually is determined by the characteristics of the strata and the backfill material used.
- The backfill material should be presoaked with water as is recommended to minimize the possibility of bridging. A wetting agent may be used.
- Backfilling of wells containing drilling mud and/or water may be accomplished by pumping the backfill material (in slurry form) to the bottom of the well and allowing the well to fill from the bottom up to displace the drilling mud and/or water.
- Observations of the change in anode resistance to the earth should be used to determine if the backfill material has been placed around the anode.

- **Open well (open hole)**
  - Each anode should be suspended, placed in position, raised, lowered, or removed for inspection by individual polypropylene ropes.
  - Anodes should be centered in the casing, considering the required spacing between them, within the aqueous electrolyte.
  - The final depth of the well will depend on the subsurface strata and the number and length of the anodes. The approximate depth should be defined by the designer.

Necessary precautions should be taken to prevent deleterious modification of ground water quality.

- Each anode should be provided with an individual insulated lead wire, or a cable sufficiently long should be connected to the anode lead wire with a cable connector (line tap) and in-line (two-way) splicing kit.
- Each cable group of each deep well should be brought inside the positive test box (type 2) located at the deep-well head.
- The test box should be installed as per the standard requirement for individual termination of anodes and rectifier positive lead wires.
- A shunt should be installed in each anode circuit to monitor the current output.
- Resistors should be installed in individual anode circuits to balance anode outputs.
- Anode wires should be sealed to prevent capillary action between insulation layers, as may be necessary to prevent corrosive elements from entering the test box.
- Lead wire entry should be sealed as may be necessary to prevent the entry of gases.

7.5.2 Installation of Transformer/Rectifier Equipment

It is essential that transformer/rectifier units be installed by suitably qualified personnel and be installed in accordance with the requirements of the National Electrical Code NFPA-70, Latest Edition and this Standard.

The manufacturer's installation and operating manuals should be available at the site before installation of the transformer/rectifier. The instructions contained therein should be adhered to.

Air-cooled transformer/rectifier(s) should be installed pole mounted by means of using four roll and plug-type connectors and in a free place for cooling purposes.

Oil-cooled transformer/rectifier(s) should be installed in nonhazardous areas and away from any equipment that creates heat. Oil-cooled transformer/rectifier(s) should be installed on a concrete plinth in accordance with the details specified by standards as required by the job.

Transformer/rectifiers should not be installed in series or in parallel in the same CP circuit. Transformer/rectifiers should be installed in nonhazardous area. If this is not possible, the construction of the rectifier units should fulfill the requirements of the hazardous area classification applicable for the site.

The following is to be noted:

When electrical work is carried out in hazardous areas, requirement of IEC 79.14 should be adhered to in conjunction with the area classification drawings and the standards.

If installed outdoors, the enclosure should have a minimum degree of protection IP 54 in accordance with IEC 529.

Transformer/rectifier foundations should contact tank support beams only.

Transformer/rectifier foundations should allow space below the tank bottoms to permit painting.

If the proposed rectifier site is in an area where flooding may be a problem, the maximum high water level should be ascertained and the transformer/rectifier should be mounted so that it will be above this level.

The transformer/rectifier should be firmly secured to the plinth with holding down bolts to be supplied by the Contractor to the approval of the Engineer.

The transformer/rectifier manufacturer's instructions should be followed completely.

The AC and direct current (DC) cabling should be installed through steel conduits to connect the transformer/rectifier.

The AC current cables and DC current cables should be placed in separate conduits.

After the installation of cables the ends of the steel conduit should be fitted with a suitable blanking disc and coated with waterproof sealing compound (plastic inserts should be used in conduit ends to protect cables).

The electricity supply should be taken from the nearest existing electricity pole or a new one to be installed and brought to a pole mounted electricity meter by underground cables. The T/R unit should then be supplied from this meter.

Before connecting the supply to the unit, it should be checked whether it is the correct voltage as stated on the rating plate of the transformer/rectifier.

The connections of DC cables to the transformer/rectifier must be mechanically secure and electrically conductive. Before the transformer/rectifier is energized, it must be verified whether the negative conductor is connected to the structure to be protected and the positive conductor is connected to the anodes at the power source output terminals.

Caution: The negative lead of the rectifier must be attached to the structure to be protected. If the structure is mistakenly attached to the positive lead, it will serve as an anode and rapid corrosion failure can result.

The transformer/rectifier should be connected into either the existing earthing circuit or should be separately earthed to a new earthing system according to design specifications.

When the metal work of the transformer/rectifier unit is bonded to the earthing terminal, precautions should be taken to ensure that there is no possibility of a metallic connection, even for a short period, between the earthing system and the ground bed of the CP installation.

After erection of a unit, it is important that the following be checked:

1. Oil level is correct, and if the unit is oil cooled.
2. Fuse ratings are correct.
3. Input and output cables are properly identified prior to connection to the electricity supply.

Transformer/rectifier(s) should not be energized until all check-out and commissioning tests have been completed.

The following is to be noted:

When electricity is connected, correct polarity and ground bed resistance should be verified by energizing the unit.

7.5.3 Cabling

All cabling should be routed and installed in accordance with the design drawings and to meet the approval of the engineer.

Sufficient information should be given in the design drawings to indicate the general routes of cables. Final routes are to be determined on site and changes made only where absolutely necessary and with the approval of the Engineer.

Cables for connection between the transformer/rectifier and pipe and ground bed should conform to the dimensions and characteristics indicated in the drawings and/or materials specifications.

Cables run between the ground bed and transformer/rectifier and between the transformer/rectifier and structure(s) should be continuous and free of splices.

To avoid kinks and knots, all cables should be carefully unreeled and laid directly into the prepared trench. Where cables are reeled on drums, the drums should be mounted on jacks.

Trenches should be kept away from buried pipes containing hot fluids and from pipes liable to temperature rise owing to steaming out.

The bottom of the trench receiving direct buried cables should contain relatively smooth, undisturbed, and well-tamped earth. Care should be taken to ensure that there are no sharp rocks or other objects in the cable trench bottom that could damage cable insulation.

Cables should be laid with sufficient "Slack" to avoid breaking during or after backfilling and to allow for shifting and settling. When connections are made to a pipe, the cable should be wrapped around the pipe twice and taped down. Each wire terminated in the test box should have at least 15 cm of slack coiled.

Cable runs under roads and areas subject to vehicular traffic should be installed in a steel or aluminum conduit of a minimum size of 20 mm.

The conduit used should be reamed carefully after cutting to length to remove all sharp edges. Bushings should be installed on both ends of the conduit.

The positive cable anode lead is especially critical to the operation of the system. It is imperative that insulation remain intact. Extreme care should be taken to ensure that the entire cable and all connections are waterproof. Care should be taken to ensure that there are no short circuits between the positive cable and the structure or conduit.

Cables should enter the rectifier, ground bed test box, and where applicable, other enclosures, in properly sized rigid conduit extending 450 mm below the ground surface. Plastic inserts should be used in conduit ends to protect cables.

Cables should be installed as follows:

- Cables should be laid in prepared trenches.
- Before cables are placed, the trench bottom should be leveled and backfilled with a layer of soft sand of a 10-cm thickness.
- This soft sand should be leveled and the cable placed thereon. The laying of cables should be carefully done to avoid any damage to insulation. After laying and before covering, all cables should be examined for cuts, nicks, and any other damage. All damaged cables must be repaired before burying.
- The cable should then be covered with a layer of fine sand that is 20 cm deep. The sand should be lightly tamped. Machine compaction should not be used.

- A protective covering warning device (bricks, tiles, or red concrete slab) should then be applied. The protective covering should be placed without disturbing the sand fill while pouring.
- The remainder of the trench should then be backfilled and compacted with soil such that existing elevations are met.

All cable main runs should be tagged at each end with waterproof identification tags as per cable schedules. Tagging method should be approved by an engineer prior to being carried out.

Color code of the cables should be as follows:

Between T/R and positive test box: Red
Between T/R and negative test box: Black
Between positive test box and ground bed: Red
Between negative test box and structure: Black
Between test point and pipeline: Black
Between test point and casing: Red
Between test point and foreign line: Red
Between test point and insulating joint/flange: Red

All cable runs should be identified with cable markers of the type shown in Standard drawings, installed at 50-m intervals and turning points. Markers should be installed at one edge of the trench.

The following information should be marked on each marker plate, with a steel die stamp:

- Direction of cable runs.
- Location of trench with respect to the marker.

Plates should have a blank space of approximately 15 × 50 mm for the company's use.

The following are to be noted:

1. The distance between the top of a cable and the surface under which it is installed (depth of burial) should be sufficient to protect the cable from damage imposed by expected surface usage.
2. The top surface of the cable in the trench should be a minimum of 70 cm below the finished grade.
3. Burial depth may be increased where necessary to meet underground conditions.
4. In areas where frost conditions could damage cables, greater burial depths than indicated above may be desirable.
5. Lesser depths than indicated above may be used in rocky terrains. Supplemental protection should be provided. Supplemental protection should be sufficient to protect the cable from damage imposed by expected surface usage.
6. Where the surface is not to final grade, under which a cable is to be installed, the cable should be placed so as to meet or exceed the requirements indicated above, both at the time of installation and subsequently thereto.
7. The horizontal separation between direct buried cables and other underground structures should not be <300 mm to permit access to and maintenance of either facility without damage to the other.

8. Where a cable crosses under another underground structure, the structure should be suitably supported to prevent transfer of a harmful load onto the cable system.
9. Where a cable crosses over another underground structure, the cable should be suitably supported to prevent transfer of a harmful load onto the structure.
10. Adequate support may be provided by installing the facilities with sufficient vertical separation.
11. Adequate vertical separation should be maintained to permit access to and maintenance of either facility without damage to the other. A vertical separation of 300 mm is, in general, considered adequate, but the parties involved may agree to a lesser separation in special cases.
12. Plowing in of cable in soil containing rock or other solid material should be done in such a manner that the solid material will not damage the cable, either during the plowing operation or afterward.
13. The design of cable plowing equipment and the plowing-in operation should be such that the cable will not be damaged by bending, side-wall pressure, or excessive cable tension.
14. At low temperatures, some plastics are so brittle that they may crack when bending the cable, and therefore, no cables should be installed during freezing weather.

7.5.4 Electrical Connections

- **Attaching cables**

The thermit-welding process (cad welding) should be used for attaching test leads, and bonding lead wires to structures.

The thermit-welding process should be such that copper penetration into the pipeline material should not be >1 mm and that the hardness should remain within the original pipeline requirements.

Thermit welding should not be used for austenitic stainless steel and duplex steel pipelines.

Thermit welding should not be used for structures that contain or have contained flammable or combustible liquid.

The following are to be noted:

1. Connections of test lead wires to the structure must be installed so that they remain mechanically secure and electrically conductive. Care should be taken to ensure that cables and connections are not damaged during backfilling. Sufficient cable slack should be provided to avoid strain.
2. All cable attachments to structures should be coated with an electrically insulating material provided or approved by the Company. This coating should be compatible with the structure coating and cable insulation, and have good adhesion to both.
3. The following welding process, as an alternative to thermit welding, for the cable connections may be required:

- **Welding**

A metal plate, of a minimum size of 50×50 mm, provided with a welded M 10 threaded stud bolt, should be welded to the pipeline by two continuous welds in the circumferential direction of the pipe only. The plate should be made of the same material as that of the pipeline.

The cables should be connected to the threaded stud bolt using crimped or brazed cable lugs, nuts, and serrated washers.

- **Stud welding**

Stud welding may be done using an electrical (resistance welding) or mechanical (friction welding) process that should be approved by the Company.

The stud material and consumables should be compatible with the pipeline material. The process should not influence the pipeline material properties to fall outside the original specifications.

The size of threaded studs should be $\geq$8 mm to suit the cable size. The cables should be connected to the stud using crimped or brazed cable lugs, nuts, and serrated washers.

- **Pinbrazing**

The pinbrazing process should use specially designed cable lugs and brazing pins to braze the cables to the pipeline and should be approved by the Company.

The brazing materials should be compatible with the pipeline material. Penetration of copper and/or other brazing metals into the pipeline should not be $>$1 mm, and the hardness should remain inside the original pipeline requirements.

Pinbrazing should not be used on austenitic stainless steel and duplex stainless steel pipelines.

- **Glued connections**

Where welding, brazing, or thermit welding is not possible, for example, for safety reasons, the contractor may design glued electrical connections using metal plates bonded with electrically conductive epoxy resin. This method should not be used for current carrying cables (drain cables, bond cables). The materials to be used and the installation procedure should be approved by the Company.

- **Splicing cables**

Cable splicing plays a very important role in a good CP system. Cable splices should be properly insulated to preclude current leak.

Anode lead wire-to-header cable connections and header cable splices should be made by using a split bolt connector (line tap). An epoxy resin-splicing kit should then be applied over the tightened zero resistance connection in accordance with the manufacturer's recommendations.

The following are to be noted:

1. For good insulation results, the manufacturer's instruction for epoxy resin-splicing kit installation should be rigidly followed.
2. The resin should not have exceeded its specified shelf life.
3. Buried connections must be protected with extreme precautions against the entrance of any moisture, because any discharge of current to the earth from the cable will destroy it in a matter of days or hours.

4. Proper cleaning (degreasing and abrading) of the insulation is necessary to ensure that a watertight bond is achieved between the insulation and the cable-jointing compound. Where repairs are carried out, a minimum of 50 mm of cable insulation, on each side of the repair, should be contained within the repair.
5. Random checks should be made during installation of joints in accordance with the manufacturer's instructions. Where applicable, these checks should ensure that
   a. the joint area is dry;
   b. the resin compound has not overrun its expiry date;
   c. sheath abrasion, if specified, is properly carried out;
   d. the connector stagger and other dimensions are observed;
   e. the preparation, installation, and tightening of conductor connectors are correct;
   f. the appropriate tools, particularly for compression connectors, are used by the installer;
   g. the cable is laid straight and the box and cable are well supported so that movement when pouring the encapsulant is not likely;
   h. the cold pour encapsulant is thoroughly mixed;
   i. the encapsulant fills the mold and does not distort its shape significantly;

The electrical testing of the installation is deemed to be completed by final system installation/commissioning tests.

- **Cable-jointing procedure**
  - The instructions and procedures given in this Clause should be observed at all times during the preparation and installation of a cable joint.
  - Joint kits should be inspected before use and any defect should be corrected.
  - When using resin compounds, good housekeeping practices should at all times be observed in accordance with the manufacturer's instructions. The following precautions should be taken when handling jointing materials used in the preparation of cable joints.
    - Avoid using in a confined unventilated area.
    - Avoid breathing the vapors.
    - Wear protective clothing at all times when handling cold pour resins.
    - Avoid contact with the skin and eyes.
      The following are to be noted:
      In the case of accidental contact with the skin, treat the affected area with copious quantities of water (or with the reaction agent recommended by the resin manufacturer). For the eyes, follow the same treatment and immediately obtain medical aid.
    - Containers of resin compounds should be kept closed at all times except when actually in use.
    - Smoking should be prohibited.
    - Accidental spillage should be cleared immediately.
    - After use, all containers should be disposed of strictly in accordance with the manufacturer's instructions.

  d. The following equipment may be required for use:
    - a basic jointer's tool kit including consumable materials;
    - a tent or some effective means of protecting the jointing operations from moisture, rain, or excessive cold or heat;
    - those tools supplied or recommended by the splicing kit manufacturer, for example, compression tools;
    - special equipment, for example, fire extinguishers and pumps.

e. The equipment used in making cable joints should be regularly maintained in accordance with the manufacturer's instructions.

f. At all times, every effort should be made to ensure dirt does not become entrapped in a joint. Tools should be laid out in an orderly manner and when not in use replaced in their chosen place. Waste products, for example, trimmings from sheaths or insulation, should be placed in a receptacle provided for the purpose.

- **Jointing application**
  - Before starting the joint, the jointer should ensure that all the correct materials are available.
  - The cables to be joined should be lined up approximately in the position required for the joint.
  - The outer covering(s) of the cable(s) should be removed to the dimensions given in the jointing instructions, followed by the removal of other cable materials to expose the cores. The core insulation(s) should then be removed or partly removed over sufficient lengths to take the connectors.
  - Polyvinyl chloride (PVC) sheaths should be removed with a sharp knife or special tool designed for the purpose, by making a circumferential cut for cable ends or tow cuts at joint positions, plus one longitudinal cut. The circumferential cut(s) is made first and the PVC is cut through about two-thirds of its thickness to avoid damaging the cable component below the sheath. The longitudinal cut is then made with the knife blade almost tangential to the cable. The PVC sheath is completely penetrated when making this cut. The sheath can then be removed by tearing it away at the circumferential cut(s).
  - Polymeric insulation should be removed with a knife, and care should be taken not to damage the conductor.

    The following are to be noted:

    Polymeric material is easily cut when slightly warm, but care should be taken not to overheat it.
  - The conductors and connectors should be cleaned before the connections are made. It is important that the cleaning of conductors be strictly in accordance with the manufacturer's instructions.
  - Mechanical connectors should be tightened in accordance with the manufacturer's instructions.
  - The joint mold should be presented to the joint to ensure adequate clearances. When satisfactory, secure the mold and fill with encapsulating compound where appropriate.

The following is to be noted:

In low-temperature conditions, cold pour compounds can be harder to mix and will have longer curing times. Every effort should be made to store the compound at an ambient temperature >5 °C, and at all times, the manufacturer's storage instructions should be observed.

7.5.5 Installation of Test Stations (Test Points)

The contractor should install the CP test points that should be as indicated in the Standard drawings. The contractor should supply the necessary materials for installation of test point when required by the Company.

The contractor should install CP test points at locations specified in the design drawings. Precise location of test point connections to the structure should be subject to the engineer's approval prior to their attachment.

Unless specified otherwise, CP test facilities should be installed at distances of a maximum of 1000 m along the pipeline and at 250–300 m in urban or industrial areas and, in addition, at all foreign pipeline crossings, insulating flanges/joints, cased crossings, on both sides of river crossings, and at any location where interference with other buried installations is found at the time of starting up of the CP system in accordance with the design drawings.

Care should be taken to avoid damage to structure coating during excavation and backfill.

If pipelines are running in parallel, but are not in the same trench, each pipeline should be provided with separate potential-monitoring facilities. Test points should be installed not more than 2.5 m away from the pipeline.

Cables necessary for the connections between structure and test point should be as specified in the design drawings. Cables should be laid on a padding of soft earth at least 10 cm thick in trench at least 0.80 m deep and should be covered with at least 15 (15) centimeters of soft earth. Cables should be so placed that they will not be subject to excessive strain and damage during backfill operation. All test point cables should be installed with a sufficient slack.

The structure and test lead wires should be clean, dry, and free of foreign materials at points of connection when the connections are made. Connections of test lead wires to the structure must be installed so that they will remain mechanically secure and electrically conductive.

The test lead connections should be properly bonded to the structure by the thermit-welding process.

The thermit weld on the structure should be made after installation of the structure. In any case, the contractor should ensure that the cables are maintained intact. Splicing of the cable should not be permitted.

All test lead wire attachments and all bared test lead wires should be coated with an electrically insulating material. If the structure is coated, the insulating material should be compatible with the structures coating and wire insulation.

Conductor connections at bonds to other structures or across insulating joints should be mechanically secure, electrically conductive, and suitably coated. Bond connections should be accessible for testing.

CP test points attached to the structure should be tested for electrical continuity between structure and test connection, before commissioning of the CP system. Any cable not passing the final tests should be replaced.

All test point cable leads should be color coded or otherwise fitted with identification tags adjacent to the cable lug. Damage to wire insulation should be avoided. Test leads should not be exposed to excessive heat and sunlight.

Each test point should be clearly labeled and/or marked with a specific number as follows:

- For above-ground test points, this should be achieved by stamping a plate attached to the test point.
- For grade level test points, marker plates should be installed on the nearest adjacent building or the wall in-built up areas.

- **Types of test points**

  Type A—Single test points

  - Above-ground test points for desert or rural area installation: this type should be installed along the main branch and/or crosscountry lines in accordance with standard drawings. It utilizes a combined line marker and a terminal box with a screw-on cover with test wire coiled and left in the box through a conduit with its ends taped to avoid contact.
  - Grade level test points for urban area installation: this type should be installed in sight holes (embed into the ground) in the pipeline axis, off limits of road crossings in accordance with standard drawings. It utilizes a street sight hole with a cover with test wires coiled and left in the terminal board. Ample wire slack should be left in the housing below the terminal panel to allow for backfill settlement and for withdrawing the terminal panel.

  Type B—Crossings and parallelism with existing pipelines

  This type will consist of two separate cables attached to each individual pipeline, terminating in a test box (type 1) with suitable facilities to install direct or resistive bonds. The cables to each pipeline should be identified by color coding or tags.

  Type C—Casing test point

  This type should be installed in accordance with standard drawings. It utilizes a combined line marker and a terminal box with a screw-on cover with test wires coiled and left in the box through a conduit with their ends taped to avoid contact and clearly labeled.

  If the casing is >30 m, the test point should be installed at both ends of the casing. Shorter casings should be provided with a test point at one end only.

  In each test point, one test cable should be connected to the pipeline, and one test cable should be connected to the casing. Both cables should be terminated in the test point.

  Type D—Insulating joint/flange test point

  This type should be installed across each insulating joint/flange in easily accessible locations. Two cables should be connected to each side of the joint or flange. All cables should be separately terminated in a common test box (type 1) with suitable facilities to install direct or resistive bonds. The cables to each side of the insulating joint/flange should be identified by color coding or tags.

  Type E—Line current measurement test point

  This type should consist of two pairs of cables, each pair connected to the pipeline 30–60 m apart. All cables should be separately terminated in a common test box (type 1) with suitable facilities. The cables should be identified by color coding or tags.

7.5.6 Installation of Test Box(es)

The test box(es) internally equipped with a copper bus bar, copper links, copper terminals, and a proper rotary resistor should be installed for the following purposes:

1. Connection of anodic cables (header cable and positive cable) between ground bed and positive pole of transformer/rectifier, and control of the ground bed current through the rotary resistor circuit (as a positive test box).
2. Connection of cathodic cables (negative cables) between the structure and negative pole of transformer/rectifier, and control of the CP system (as a negative test box).
3. Bonding between different cathodic circuits.

The box(es) should be installed in accordance with standard drawings.

7.5.7 Earthing of CP Equipment

The object of electrical earthing is to ensure effective operation of the transformer/rectifier in the event of earth fault current, which might otherwise cause damage to property, and protect against danger to life through shock due to the installation metal work being maintained at a dangerous potential relative to earth.

Local earthing circuit should be installed at the CP station(s), in accordance with 6.2.8.3 and as detailed on the standard drawings.

Each earthing system will be composed of the following:

- Earthing pits for connection and inspection of the copper rods.
- Copper rods inserted in the earth.
- Bonding header cables between the pits.
- Earthing cables from header cable to CP equipment and fence.

The requirements for the connection of metal works of CP station(s) are specified in the following:

BS 7430 (1991) "Code of Practice for Earthing"
Formerly C.P. 1013 (1956)
BS 6651 (1985) "Code of Practice for Protection of Structures Against Lightning"

The following are to be noted:

1. Earthing should be installed fully underground.
2. Earthing should be carried out at locations where the soil resistance is the lowest. Sandy soil should be avoided.
3. The grounding resistance should be kept as low as possible by adding salt, coke, or any other kind of backfilling.

If CP station(s) to be installed inside the area with individual earthing system, such as compressor station, valve station, and city gate station. The CP equipment should be adequately bonded together and connected to the existing earthing system.

In the absence of earthing drawings, CP equipment should be adequately bonded together and connected to the earth electrodes.

7.5.8 Fencing

Fencing should be erected by competent laborers, experienced in industrial-type fence erection.

Particular care should be taken during fence erection so that no underground piping, cable, or other appurtenances are touched or damaged.

On completion of the work, all excess and waste materials resulting from fence construction should be removed from the site by the contractor.

7.5.9 Parallel Power Lines

If the pipeline runs in the vicinity of high-voltage power lines, the contractor should investigate whether high AC voltages can be present on the pipeline by induction or otherwise and whether devices have to be installed for protection of the pipeline and personnel.

The contractor should show (by calculation or otherwise) that no harmful voltages will be present or design additional facilities to prevent excessive voltages. Such facilities may consist of dedicated pipeline earthing and/or the installation of polarization cells or surge arrestors across isolating joints/flanges and across the output terminals of DC voltage sources.

7.5.10 Lightning Protection

In areas of lightning activity, the contractor should install suitable lightning protection to protect the pipeline isolation and CP equipment. This should consist of suitably rated surge arrestors. Surge arrestors should be mounted across isolating joints/flanges and across the output terminals of DC voltage sources.

7.5.11 Surge Arrestors

Surge arrestors required to prevent elevated voltages due to faults in adjacent electrical power systems or lightning should be of the spark gap type and should be such that the following are criteria met:

- The impulse breakdown voltage of the electrodes is lower than that of the isolating joint across which they are mounted.
- The spark gap is capable of discharging the expected lightning currents without sustaining damage.
- The spark gaps are fully encapsulated to prevent sparks in open atmosphere and to protect the spark gaps from moisture.

7.6 Installation of Galvanic Anode Systems

Anodes should be installed according to design specifications and drawings. Before the anode is buried, it is important that any waterproof wrapping material be removed. Typical galvanic anode installations should be of the following types:

7.6.1 Single Packaged Anode

Anodes should be installed at a minimum distance of 1.5 m from the pipeline and at least 30 cm (1 ft) deeper than the pipeline.

The native earth should be thoroughly tamped around the anode, watered, and then backfilled to the surface (after making all anode lead connections and insulating them).

Anodes should be placed 2 m away from any secondary buried structure, so that the secondary structure does not lie between the anode and the primary structure.

In distribution systems, where space limitations are extremely critical and where soil resistivities and auguring conditions permit, anodes should be placed in auger holes alongside the pipe with the hole being deep enough that reasonable spacing between pipe and anode is obtained.

7.6.2 Multiple Galvanic Anodes

In multiple galvanic installation, anodes should be placed in a straight-line configuration for the lowest resistance to the earth. The line of the anodes may be either perpendicular to the pipeline, or may be along a line parallel to the pipe.

A Parallel line of magnesium anodes should be about 5 m away from the pipeline, with zinc; this distance should be about 3 m for optimum performance.

Where anodes and backfill are provided separately, anodes should be centered in the backfill, and the backfill should be compacted before any additional backfill soil is added. The backfill should be thoroughly wetted before burial is completed.

The connection to the pipe should be made before more than one anode is installed; it will then be possible to observe the current output of successive anodes as they are connected, and installation should be halted before the average output per anode falls below 150% of the designed value.

One 0.01-ohm measuring shunt should be installed; in each lead wire, current limiting resistors are not permitted.

The anodes thus installed should be permitted to operate unrestricted for a period of three weeks or more. This will permit adequate polarization and stabilization of current output. After this time, a current output and pipe-to-soil potential survey should be made. Resistors should be installed where needed, and the current should be reduced to the designed value. It is particularly important to check the potential at the midpoints between stations (if they are unequal in size, then at the low point). If these potentials should all be found to be >0.85, then the installation is said to be complete.

7.6.3 Extruded Ribbon Anodes

Extruded ribbon anodes (of either magnesium or zinc), should be plowed-in parallel to the pipeline along sections of bare or poorly coated lines where continuous local protection is required.

Connections between the pipeline and anode core wire should be made at intervals to complete the protection circuit. The crossconnections should be made at test points at a convenient location, to measure current flow periodically and estimate the rate of anode material consumption. Intervals between crossconnections should not be >300 m.

Spacing between the ribbon anode and pipeline is not critical. To remain clear of the pipe during plowing-in operations, a spacing of 1.5 m may be used.

The anode strip should be deep enough to be in continuously moist soil (at least 0.6 m).

Extruded ribbon anodes of magnesium (or zinc) are furnished bare. Using anodes in the earth without a special backfill involves the risk of anode passivation and inadequate amounts of current. The anodes should be plowed-in with suitable special backfill according to the design specification. An adequate allowance for satisfactory dispersion around the anode is 32 kg of backfill per 30 m (100 ft) of ribbon anode.

7.6.4 Connection of Galvanic Anodes to the Pipeline

The anodes should be connected to the pipeline using the combined marker, test point, and bond box. This equipment should be made for the following purposes:

- Pipe-to-anode ground bed connection;
- Pipe-to-soil potential measurement;
- Installation of a rotary resistor between anodes and the pipeline to allow the anode current control;
- Marking the location of the anodes.

Anode lead wire should be connected to a loop shaped cable (called header cable), using a suitably sized split bolt (line tap) or compression-type connectors and a proper branch type (3 way) splicing kit. The splicing compound should be applied over the tightened zero resistance connection.

The coated splice should be insulated by taping with at least one half-lapped layer of rubber tape and one half-lapped layer of electrical insulating tape, with the joint insulation overlapping the wire insulation by a minimum of 50 mm.

The current-carrying cable is composed of two sections in black color: One section will connect the header cable to terminal No. 1 of "Combined Marker, Test Point, and Bond box," the other section connects the pipeline to terminal No. 2.

A test wire should be connected between the pipeline and terminal No. 3 at "Combined Marker, Test Point, and Bond box."

Thermit welding (cad welding process) should be used to connect the anode lead wire to the pipeline.

The copper wire connection to the steel main is the most critical insofar as insulation is concerned. At this point, all copper at the connection must be coated completely to avoid the possibility of a shielded copper–steel corrosion cell.

All connections must be permanently of a low resistance. Any gradual development of joint resistance can reduce the anode output.

Insulation of underground connections on galvanic anode installations should be well done to prevent current wastage. The connection should be waterproofed completely to prevent the possible development of resistance within the joint.

Care should be taken so that lead wires and connections are not damaged during backfill operations. Lead wires should have enough slack to prevent strain. Anodes should not be carried or lowered into the excavation area by the lead wire.

The following are to be noted:

1. The chemical backfill in packaged galvanic anodes will take up moisture slowly even if wetted with water after placing in the auger hole and before completing the earth fill. For this reason, the anode will not attain full output immediately. Depending on the amount of moisture in the earth, it may be a matter of days or even weeks before the full output is attained.
2. When bare galvanic anodes are placed in auger holes and backfilled with a separate chemical backfill, it is the usual practice to install the backfill dry. There will be a time lag before the full current output is attained as in the case with packaged anodes. It is possible to mix the chemical backfill with water and pour the slurry into the auger hole to surround the anode. The full output will be attained immediately. There is, however, the danger of shrinkage as the excess water leaves the slurry. This shrinkage may operate to cause the ultimate reduction in current output. Backfill installed dry, on the other hand, tends to swell upon taking up moisture to develop maximum coupling between the anode and the surrounding earth. For this reason, the use of dry backfill is considered the best practice.

7.7 Installation of CP Systems for Compact Buried Structures

This section outlines the procedures for the installation of CP systems for the of external surfaces of compact buried structures, including tank farms, service station tanks, tower, footings, steel pilings (in soil), short well casings, compressor and pump stations, refineries, petrochemical plants, and associated pipework.

The installation of CP systems for compact buried structures is basically similar to the installation of buried pipelines, so many of the requirements outlined in the previous section in respect of buried pipelines are applicable to compact buried structures, with the following exceptions:

1. Before any work is carried out on or near an insulated flange, the area should be checked for hazardous atmospheres.
2. To avoid the risk of electric shock and the possibility of sparking, it is advisable that insulating joints be crossbonded before being disassembled. This precaution is essential for hydrocarbon product lines.

3. Galvanic anodes should preferably be sited on a line normal to the long axis of the tanks at a distance of about 5 m from the outside surface of the tank; if two anodes are used, one should be positioned on each side of the tank. For a well-coated tank, the siting of the anodes is not critical, and they may be sited to suit conditions, at a distance of approximately 3–6 m from the tank.

   The anodes should be buried at a depth that places them in permanently moist soil if possible.

   The lifting lugs situated at either end of the tank provide convenient points of attachment for anode cables. The lugs should be scraped carefully to expose the bare metal, and the cable end should be attached by a bulldog clamp or by thermit welding; the coating should then be made good.

   For tanks that are already buried, the cable can be connected to the vent pipe.

   The cables from the tanks should preferably be connected to the cables from the anodes via a test box, including a measuring wire from the tank to enable periodic checks of the steel-to-soil potential to be made, as well as current measurements of the anodes.
4. Impressed current ground beds should be arranged symmetrically around a tank or group of tanks. Dependent upon the space available, the ground beds should be located not less than one tank diameter from the tank periphery to provide optimum current distribution over the tank bottom. If this is not possible, consideration should be made to distribute a number of anodes or ground beds evenly around the periphery of the tank or to install borehole ground beds. The top anode of a borehole ground bed should be at a minimum depth of 10 m to facilitate current distribution.

   If flammable liquids are being stored in the tanks, the preferred siting of the ground beds is outside of the bund walls. Where this is not possible, the ground beds and all connections should either be totally buried or, if above ground, comply with the requirements of the electrical classification of the hazardous area. This should also apply to any negative drain point connection to the tank. If borehole ground beds are used, any steel casing should be finished below the ground level to ensure that any spark hazard due to inadvertent contact between the casing and protected steelwork cannot occur.

7.7.1 Structure Preparation (to be Considered by the Structural Constructor)

The tank foundation mound should as far as possible be constructed so that it will distribute protection current uniformly to the whole of the underside of the tank. This means that the use of rubble, rock fill, etc., should be avoided and the mound should consist of fine-grained and well-compacted material, to a minimum depth of 150 mm.

Storage tank bottoms are generally constructed by lap welding individual plates and are therefore electrically continuous. Where groups of tanks are to be cathodically protected, provision should be made for bonding between individual tanks.

If it is desired to confine the protection current to the tanks; isolating joints should be installed in all pipelines and fittings connected to the tanks including electrical and instrumentation connections.

If flammable liquids are being stored, such joints should be located outside the tank bund. Earthing electrodes connected to the tank should be of zinc, stainless, or galvanized steel.

7.7.2 Installation of Permanent Reference Electrodes

If the installation of the metallic structure is likely to obstruct correct electrode placement, permanent reference electrodes should be installed immediately prior to construction. For large structures, consideration should be given to installation of reference electrodes and associated cabling prior to the laying of foundations. Cabling should be laid with sufficient free play to allow for foundation movement and structural loading.

Reference electrodes should be installed as close as possible to the buried structure without touching or shielding the surface. The backfill around the electrode should have a resistivity not greater than that of the soil surrounding the buried structure. Allowance should be made for foundation settling when locating reference electrodes.

Where reinforced concrete foundations are to be laid, care should be taken to ensure that all reference and test point cabling and equipment are electrically isolated from metallic reinforcement materials.

Reference electrodes, associated cabling, and connections should all be checked for damage prior to installation. Correct operation and electrical isolation of the system should be confirmed prior to the final reinstatement of backfill material.

The actual location of permanent reference electrodes and cabling should be accurately documented on the as-built drawings.

7.7.3 Installation of Insulating Flanges, Joints, and Couplings

All insulating flanges, joints, and couplings should be installed in accordance with the requirements outlined in the standards.

The assembly of an insulating flange requires particular care, to ensure that insulation is not lost due to mechanical failure of the components.

It is to be noted that the use of resistance methods to determine the integrity of insulating flanges in the field can produce unreliable results.

Completed flanges should be coated in accordance with design specifications.

Insulating joints should be checked for insulation integrity by measurement of structure-to-soil potential on each side of the joint, with the reference electrode in the same location. Different potential readings indicate adequate insulation. If the potential readings are the same, a CP current (or changed CP current) should be applied to one side of the joint, and the potential should be remeasured. If the potentials remain the same on both sides, the joint is not adequately insulating.

7.8 Installation of CP Systems for Internal Surfaces

This Clause outlines procedures for the installation of CP systems for internal surfaces of pipes and structures, including heat exchangers, hot water systems, clarifiers, ballast and water storage tanks, cooling conduits, and reservoirs, that are in contact with natural waters including sea water and waters of near neutral pH.

Full construction details and installation procedures of the CP system for each specific type of structure will be specified in design specifications and drawings.

Many of the requirements outlined in previous sections in respect of buried pipelines are applicable to internal surfaces.

The installer should be thoroughly familiar with the specifications for the work, and should ensure that all work is completed in accordance with good industrial practice and the relevant specifications. Departures from design specifications should be approved by the designer and/or Company and permanently recorded for future reference.

Care should be taken to ensure that cables and other components are protected from damage during installation. All cable connections need to provide reliable long-term low-resistance electrical contact.

7.8.1 Materials and Equipment Acceptance (or Compliance)

Impressed current anodes should be provided with individual lead wires to the rectifier for control and measurement of current output from each individual anode.

Because anode cables may be subject to attack from a high chlorine environment found near some anodes, it is important that the cable insulation and sheathing be resistant to such an environment, or otherwise be suitably oversheathed or protected.

7.8.2 Installation of Impressed Current Systems

Impressed current anodes should be installed in accordance with design specification and drawings.

Impressed current anodes should not be directly attached to the internal part of the structure. They are required to be insulated from the structure and, in all cases, the electrical connection is to the positive terminal of the DC power source.

Because anodes are often brittle or have thin film electrodeposited coatings, care should be exercised to ensure that they are not damaged during handling.

Certain anodes are specifically designed for suspension by their cable tails, and may be lowered into position by the cable. Other anodes generally of the direct immersion type may require to be lowered into position by separate ropes, as their cable tails are designed for electrical purposes only and not for mechanical suspension. The installation drawings should be checked before the commencement of anode installation.

The following are to be noted:

1. Anodes that are in close proximity to a coated steel structure should be provided with an adequate dielectric shield, designed so that the potential at the periphery of the shield does not exceed −1.2 V with reference to a copper/copper sulfate electrode.
2. In the case of cantilever anodes, which are generally rod shaped and project from the structure, obstruction of the active anode surface can be avoided by using an adequate shroud length to prevent the build-up of a calcareous deposit on the structure surface.

3. For safety reasons, suspended anodes, other than light anodes of platinized titanium or mixed metal oxides that are specifically designed to be suspended by their cable tails, should be supported by a suitable rope of polypropylene, to prevent the anode cable from bearing the anode weight.

Cable supports should be corrosion resistant and located so that the cable insulation does not become abraded due to cable movement from wind or electrolyte forces. Cable routes should also avoid areas of likely damage from physical operations on the structure.

Cable joints should be completely waterproofed using an appropriate cable-jointing compound. Waterproofing is particularly important on the positive side of an impressed current system to prevent localized rapid corrosion and subsequent failure of the corrosion protection system.

The following are to be noted:

Proper cleaning (degreasing and abrading) of the insulation is necessary to ensure that a watertight bond is achieved between the insulation and the cable-jointing compound. Where repairs are carried out, a minimum of 50 mm of cable insulation should be applied to each side of the repaired cable joint.

Anode-to-cable tail encapsulation for immersed anodes is generally fitted at the factory. Prior to installation, the encapsulation should be carefully inspected for any faults or handling damage during transit.

Anodes that project from support pipes or require centering through insulating sleeves may require inspection after installation.

The following are to be noted:

Of special importance to be inspected during the installation is to ensure that the anode material and size are in accordance with the relevant standard, where applicable and/or to the approved specifications.

Warning:

Where underwater diving inspection or maintenance is likely, structures should have warning notices displayed advising of the danger of electrical gradients near the anodes and the need to switch off the system prior to diving.

Caution:

Signs should be displayed indicating the presence of any immersed cables or anode support ropes that are not physically protected.

Anodes and their support cables on structures located in flowing fluids should be designed to withstand vibration and impact.

Requirements of this standard and local authorities should be observed during the installation of a transformer/rectifier especially with regard to AC input, cabling, and positioning.

After installation of a unit, it is important that the following be checked:

1. The input and output terminals are correctly identified, and the structure cable is connected to the negative output terminal prior to connection to the electricity supply.
   The following is to be noted:
   When electricity is connected, correct polarity and loop resistance should be verified by energizing the unit, and checking that the structure potential is shifted in the negative direction.

2. The oil level is correct (if the unit is oil cooled).
3. The fuse ratings are correct.

7.8.3 Safety Precautions

Precautions must be taken to avoid the following:

1. The effects of lightning, both on the protected structure and via the electricity distribution system (personnel protection aspects should also be included), should be taken into account.
2. Electrical gradients resulting from impressed current systems occurring in water around fully and partially submerged anodes and in waterways adjacent to anode installations should be considered.
   The following are to be noted:
   Paralysis and respiratory failure may result if a person comes into contact with electric field strengths >3 V/m in water. Should the design result in a possible electric field strength exceeding this value in waters located close to impressed current anodes, warnings should be given and access to such areas prevented by shielding or by other means.
3. Sparks in the presence of flammable substances and explosive gas mixtures that may be present around oil treating vessels should be avoided.
4. The cable-to-anode connections in impressed current systems should never be disconnected, nor should the anode be removed while the rectifier is in operation.
5. Usual precautions to prevent fire or explosion must be taken before a CP system can be installed or repaired in a vessel handling water mixed with oil or gas.
6. The rectifier case, external AC disconnect switch box, and any related metallic equipment must be properly grounded using recognized safe grounding practices.
7. Special gaskets capable of withstanding high temperatures should be used to mount anodes in fired vessels, particularly if the gaskets are located near fire tubes.

7.8.4 Installation of Galvanic Anode Systems

Anodes should be installed according to design specification and drawings.

The common methods of installation of galvanic anodes are as follows:

1. by direct attachment to the internal part of the structure; or
2. by suspension in the electrolyte from the structure using a cable or a rigid metal support; the cable is connected to the structure above electrolyte.

The following are to be noted:

For safety reasons, suspended anodes should be supported by a suitable rope of polypropylene to prevent the anode cable bearing the anode weight.

Anodes that are to be installed flush with the structure may be attached to the structure by either of the following methods:

1. Welding of the anode core to the structure.
2. The use of structure studs nuts to attach the anode core.

In all cases, the anode should be in reliable long-term low-resistance metallic contact with the structure. This may be achieved by the use of fusion joints, bolted connections, or by direct screwing into the structure surface. Ensure that corrosion resistant materials are used and the joints are effectively insulated (wrapped).

Before immersion of the anodes, it is necessary to remove any material wrapped around them. The anodes should not be painted and, where necessary, should be protected from accidental paint application.

Adequate support of anodes is necessary to avoid possible cable failure.

7.8.5 Permanently Installed Reference Electrodes

If permanent reference electrodes are installed for the measurement of the structure-to-electrolyte potential, it is important that they be continually immersed when in use.

Reference electrodes should be located in accordance with the design requirements. Each reference cell should be wired to a termination position by a separate and isolated conductor, insulated from the structure and the electrolyte, and protected by continuous conduit. Separate conductors can be installed together using a multicore cable.

It is essential that reference cell wiring be electrically shielded between the structure exit point and the termination position.

The most convenient method of mounting reference electrodes inside a plant is by means of a "screw-in" assembly such that the electrode can easily be withdrawn for inspection and replacement of either the entire unit or the electrode material. The electrodes can be wired to central monitoring and control equipment. A disadvantage lies in the difficulty of checking the accuracy of the electrodes, once installed.

If it is impossible to use "screw-in" mountings, reference electrodes can be attached by suitable nonmetallic fixings to the protected surface and the insulated connecting leads brought out through the plant wall through a suitable gland.

At least one reference electrode should be installed for each cathodically protected compartment. The reference electrode should be installed at the position where corrosion is most likely, for example at junctions of ferrous and nonferrous materials and/or remote from anodes.

The following are to be noted:

1. Care must be taken in placing the reference electrode in the treating vessel. For potential measurements the electrode must be as far from the anodes as possible. In pressure vessels, the electrode is "Lubricated" (introduced into the vessel against existing vessel pressure) through a gate valve installed in the vessel for that purpose.
2. Contamination of the reference electrode with oil or sediments such as iron sulfide must be avoided. A salt bridge may be used to prevent contamination of the reference electrode.
3. Location of the reference cell near an anode may indicate a higher potential than elsewhere in the vessel.

7.9 Installation of CP Systems for Marine Structures

This section specifies general construction requirements for the installation of CP systems that will control corrosion of the submerged zones of marine structures and the buried parts of integral offshore/onshore structures.

Full construction details and installation procedures of the CP system for each specific type of marine structure will be specified in design specifications and drawings.

Many of the requirements outlined in Clause 6 in respect of buried pipelines are applicable to submarine pipelines.

CP systems installed onshore to protect submarine pipelines should comply with Clause 6 of this Standard.

The contractor should be thoroughly familiar with the specifications for the work, and should ensure that all work is completed in accordance with good industrial practice and the relevant specifications. Departures from design specifications should be approved by the design engineer and/or company and permanently recorded for future reference.

It is necessary that precautions be taken in combustible atmospheres to prevent sparking due to potential differences between protected and unprotected structures. Any insulated devices should be crossbonded before being separated, and the CP system switched off.

Care should be taken to ensure that cable and other components are protected from damage during installation. All cable connections need to provide reliable long-term low-resistance electrical contact.

7.9.1 Immersed Structures

7.9.1.1 Installation of Impressed Current Systems

The installation should be done under the supervision of a corrosion specialist to verify that the installation is made in accordance with design specification and drawings.

Impressed current anodes should be installed in accordance with design specifications and drawings. Special care should be taken to avoid damage to anodes and their lead wires during installation. Careful supervision of this phase is most essential to proper long-term performance of the CP system.

Impressed current anodes may be installed by one or more of the following methods:

1. Anodes may be lowered in a casing and are allowed to extend below a termination fitting at the bottom. This method provides a means of anode retrieval or replacement without diver assistance.
2. Anodes may be installed on platform members using offset steel structural supports attached to the platform members. Diver assistance is required for anode replacement.
3. Anodes may be installed on the sea bottom floor, remote from the structure. The anodes may be supported by concrete foundations and buoyancy tanks to minimize the possibility of the anodes becoming covered with mud.

Because anodes are often brittle or have thin-film electrodeposited coatings, care should be taken to ensure that they are not damaged during handling. Certain anodes are specifically designed for suspension by their cable tails and may be lowered into position by the cable. Other anodes, generally of the direct immersion type, may need to be lowered into position by separate polypropylene ropes, as their cable tails are designed for electrical purposes only and not for mechanical suspension. The installation drawings and the recommendations of manufacturer should be checked before commencement of anode installation.

Cable supports should be corrosion resistant and located so that the cable insulation does not become abraded due to cable movement from wind or water forces. Cable routes should also avoid areas of likely damage from physical operations on the structure.

Cable joints should be completely waterproofed using an appropriate cable-jointing compound. Waterproofing is particularly important on the positive side of an impressed current system to prevent localized rapid corrosion and subsequent failure of the CP system.

The following are to be noted:

Proper cleaning (degreasing and abrading) of the insulation is necessary to ensure that a watertight bond is achieved between the insulation and the cable-jointing compound. Where repairs are carried out, the encapsulation should include a minimum of 50 mm of the cable insulation on each side of the repaired cable joint.

Anode-to-cable tail encapsulation for immersed anodes is generally fitted at the factory. Prior to installation, the encapsulation should be carefully inspected for any handling damage during transit. Anodes that project from support pipes or require centering through insulating sleeves may require diver inspection after installation.

Where underwater diving inspection or maintenance is likely, structures should have warning notices displayed advising of the danger of electrical gradients near the anodes and the need to switch off the system prior to diving.

Signs should be displayed indicating the presence of any immersed cables or anode support ropes that are not physically protected.

Of special importance to be inspected during the installation is to ensure that the anode material and size are in accordance with relevant parts of available standard where applicable and/or to the approved specifications.

Conductor cable connections to the rectifier, from the anode(s) and the structure, must be mechanically secure and electrically conductive. Before energizing the power source, verify that the negative (−) conductor is connected to the structure to be protected, that the positive (+) conductor is connected to the anode(s), and that the system is free of short circuits. After the DC power source has been energized by authorization of the supervising corrosion specialist, suitable measurements should be made to verify that these connections are correct in polarity.

Connections between the positive header cable and lead wire(s) from the anode(s) should be mechanically secure and electrically conductive. The connections must be sealed to prevent moisture penetration and ensure electrical isolation from the environment. Submerged connections require seals suitable for the water pressure and environment to which they may be subjected.

When installing a suspended anode, where separate suspension is required, care should be taken so that the lead wire is not in such a tension as to damage the anode lead wire or connections.

Requirements of this Standard and local authorities should be observed during the installation of a transformer/rectifier especially with regard to an AC input, cabling, and positioning. Rectifier or other power source should be installed out of the way of operational traffic and remote from areas of extreme heat or likely contamination by mud, dust, water spray, etc. Where two or more rectifiers are installed, they should be spaced for proper flow of cooling air.

Wiring to rectifiers should comply with any applicable regulatory codes and with the operator's specifications. An external disconnect switch in the AC wiring to the rectifier should be provided.

Testing of the power source should be carried out to ensure adequate electrical connection and that no damage has occurred during installation.

The cables and connections should be carefully inspected to detect insulation defects. Defects should be properly repaired.

7.9.2 Installation of Galvanic Anode Systems

Anodes should be installed according to design specification and drawings.

Various methods for fixation of anodes to the object to be protected may be employed. The method employed should be based on an evaluation of the design requirements to electrical connection, loading, and stresses in the parts to which the anodes are attached.

The common methods of installation of galvanic anodes are as follows:

1. By direct attachment to the structure before structure immersion.
2. By direct attachment to the structure after structure immersion.
3. By placing the anode on the sea bed and connection to the structure by cable, either above or under water level.
4. By suspension in the water from the structure via a cable or a rigid metal support, and connection of the cable to the structure above water.

In all cases, the anode should be in reliable long-term low-resistance metallic contact with the structure. This may be achieved by the use of fusion joints or bolted connections using corrosion-resistant materials followed by effective insulation (encapsulation) of the joints.

Before anode immersion, it is necessary to remove any wrapping material. The anodes should not be painted and, where necessary, should be protected from accidental paint application.

It should be aimed at minimizing the drag forces caused by the sacrificial anode system.

Provisions for in-service installation of future additional current capacity should be made. Such provisions may include spare j-tubes for additional impressed current cables. Other provisions may be "pig tails" on pipelines and various sorts of brackets, guides, etc.

The anodes should be attached to the structure in such a manner that they remain secure throughout the service life.

The anode core should be welded to the structure either directly (e.g., on offshore structures) or by a cad-welded cable between the core and the structure is used (e.g., for bracelets around pipelines).

The distance between the anode and the structure depends on the condition of the structure. For coated steel, the minimum distance is zero; for a bare structure, the minimum is 25 cm. The maximum distance is not critical provided the ohmic resistance of the interconnection is small compared with the anode resistance in the medium.

Underwater installation of anodes may be performed with mechanical fixing devices or by welding. Where the latter is done, welding should be performed in a dry environment provided by a hyperbaric chamber. Wet welding should only be allowed on members where cracks and defects will be harmless. Mechanical fixing devices may not give reliable electrical connections for >5 years.

Where separate suspension is required, care should be taken when installing a suspended anode to ensure that the lead wire is not in sufficient tension to damage the anode lead wire, its insulation, or connections.

All galvanic anode installations should be tested to ensure that electrical continuity exists between the anode and the structure.

7.9.3 Electrical Connections

Electrical connections between anodes and steel structures should be made by manual welding or by thermit welding.

For pipelines and risers, attachment welding should be placed at least 150 mm off other welds.

Doubler plates should be used for attachment of anode supports to pressurized parts and highly stressed structural members. Anodes should not be located in areas with high stress concentrations, for example, anode joints.

Doubler and/or gusset plates should be installed on anode supports at the time of anode installation. If installed as part of the anode fabrication, these plates are subject to serious damage during anode hauling and handling.

Suspended galvanic anodes should be installed after the platform is set on location offshore, and the anodes should be tested for good electrical contact to the structure after installation.

Welding of doubler plates and anode supports directly on to load carrying members and pressurized parts should be performed with a qualified welding procedure by qualified welders. These welds should be nondestructively examined as required for the welding of these components.

Attachments of electrical connections by thermit welding should be made using a qualified procedure proved to give sufficient bonding and negligible Cu-penetration along grain boundaries.

The size and the shape of the mold should suit the diameter of the pipe and the anode cable size.

Qualification of the thermit-welding procedure should be based on the visual examination and mechanical testing of three test welds.

The test welds should be sectioned and examined for bonding and possible excessive Cu-penetration using a microscope with a magnification of at least 100×. The Cu-penetration should normally be <0.3 mm for procedures to be used on risers, while it will be a maximum of 0.8 mm for procedures to be used on pipelines.

The hardness in the heat-affected zone should be determined on the macrosections and should be within the normal limit specified for the pipeline system.

Welds made between anode cores and structural members for offshore facilities should have the approval of the welding engineer. Procedure testing will often be required. Wet welding is not permitted.

Other methods used to connect anodes to structures are often used during retrofit exercises when welding is impossible. These are clamping, clamping plus hard-tipped bolting, flash stud welding in mini habitats, stud shooting, etc.

7.9.4 Corrosion Control Test Stations, Connection, and Bonds

Test leads to pipelines associated with offshore structures must be mechanically secure, electrically conductive, and should be readily accessible.

Both the pipe and the test lead wires should be clean, dry, and free of foreign material at the points of connection when the connections are made. The completed connection should be coated to prevent atmospheric corrosion.

Conductive connections to other pipelines or across insulating joints should be installed. All bond connections should be readily accessible for testing.

Steel piles should be electrically connected by means of a continuous copper cable embedded in the concrete deck, and connected to each pile by a welding process equivalent to cadweld or thermoweld. Fender piles should be electrically connected to the main pier structure by a flexible insulated cable.

Current continuity between sections of sheet piling should be provided by the joining of adjacent sections by welding a 25-mm diameter reinforcing bar across the joints at the time of installation.

Bollards should be installed in such a manner so as to prevent any electrical contact between them and the steel pier piling through the reinforcing bars in the concrete deck. This will minimize the possibility of temporarily depleting the CP of the piling when a ship is moored with steel cables.

7.9.5 Installation of Insulating Joints/Flanges and Devices

The assembly of an insulating flange requires particular care to ensure that insulation is not lost or damaged due to the mechanical failure of the components.

The following are to be noted:

The use of resistance methods to determine the integrity of insulating flanges in the field can produce unreliable results.

Completed flanges should be coated in accordance with design specifications.

Insulating joints should be checked for insulation integrity, for example, by the measurement of structure-to-electrolyte potential across the joint, with the reference electrode in the same location.

Different potential readings usually indicate adequate insulation. If the potential readings are the same, the CP current (or changed CP current) should be applied to one side of the joint, and the potential should be remeasured. If the potentials remain the same on both sides, the joint is not adequately insulated.

7.10 Submarine Pipelines

7.10.1 Installation of Impressed Current Systems

Impressed current anodes should be installed in accordance with design specifications and drawings.

The installation should be done under the supervision of a corrosion specialist to verify that the installation is made in accordance with design specifications and drawings.

Impressed current anodes submerged in sea water may be installed by one or more of the following methods:

1. Anodes may be lowered in a casing and are allowed to extend below a termination fitting at the bottom. This method provides a mean of anode retrieval or replacement without diver assistance.
2. Anodes may be installed on the sea bottom floor remote from the structure. The anodes may be supported by concrete foundations and buoyancy tanks to minimize the possibility of the anodes becoming covered with mud.

The anodes should not be mounted on sand and mud unless special precautions are taken to prevent them from being submerged as a result of tidal action.

When installing a suspended anode where separate suspension is required, care should be taken to ensure that the lead wire is not in sufficient tension to damage the anode lead wire, its insulation, or connections.

Rectifiers or other power sources should be installed so as to minimize the possibility of damage, vandalism, or unauthorized entry.

Wiring to rectifiers should comply with local and national electrical codes or requirements of utility supplying power. An external disconnect switch on AC wiring should be provided.

The conductor (negative lead wire) should be connected to the pipeline. Conductor connections to the rectifier must be mechanically secure and electrically conductive. Before the power source is energized, it must be verified that the negative conductor is connected to the pipeline to be protected and the positive conductor is connected to the anodes and that the system is free of shorts. After the DC power source has been energized by authorization of the qualified personnel responsible for corrosion control, suitable measurements should be made to verify that the connections and polarity are correct.

Connections between header cable and conductors from anodes should be mechanically secure and electrically conductive. All connections between anode lead wires and header cable should be insulated and sealed to prevent moisture penetration and to ensure electrical isolation from the environment.

Where the cables cross a beach, they should be buried in suitable backfilled trenches, with concrete slabs positioned over the cables to prevent movement or damage, or positioned and fixed in such a manner that cannot be moved or damaged by sea action.

7.10.2 Installation of Galvanic Anode Systems

Galvanic anode systems should be installed in accordance with design specifications and drawings.

It is important that the anodes be mounted in a manner so as to avoid mechanical damage during handling and installation of pipes. Anode bracelets should be fastened securely on the pipe. The two segments may be welded together with steel strips to ensure satisfactory mechanical connection and proper positioning. Each anode should be electrically connected to the pipe by at least two attachments, preferably one from each half bracelet. The reinforcement of concrete weight coating should not be allowed to be in electrical contact with pipe or anode.

Care should be exercised so as not to reduce the design surface area in contact with the electrolyte. This requirement is especially applicable to bracelet anodes where there may be a possibility of anodes being covered by insulating material or anti-buoyancy material.

The contractor should acknowledge safe receipt of anodes in writing and should maintain records that should correlate anode identification with relevant pipe numbers. A copy of these records should be supplied to the Company.

The contractor should ensure that anodes are kept undamaged during all operations. Any damaged anodes should be segregated and reported to the Company.

Before an anode is immersed, it is important that any waterproof wrapping material be removed.

Anodes must not be painted and should be suitably protected during any painting operations.

Electrical connections for anodes are usually incorporated within the mounting arrangement. For bracelet anodes for pipelines, cable connections to the mounting steel framework are provided, and these must terminate on the pipe.

All galvanic anode installations should be tested to ensure that electrical continuity exists between the anode and the pipeline.

Anode bracelets should be installed as follows:

1. Exposed steel portions of the anode should be coated. The primer and dry film thickness of coating should be the same as that used in pipe coating.
2. The anodes should be placed centrally over the pipes and clamped tightly in place. The segments should then be welded or bolted together as indicated on the design drawings. Bolting material should not be high tensile and should be limited to a hardness of 300 vickers.

3. The coating should be removed from the areas where the bonding leads are to be welded to the pipe. The area must be cleaned to bright metal to ensure proper bonding of the weldment.
4. The bonding leads should be welded to the pipe by the thermit welding or an equivalent process. Attachment welds should be made using consumable and procedures qualified under fully representative conditions. The qualification should consist of one trial weld that should be sectioned and subject to a macroexamination and hardness survey.

The sections should show no cracking or copper penetration, and the hardness should not be >260 vickers.

5. Visual damage and holidays should be repaired in the primer coat on the bracelet. The coating over the weld and surrounding area of the pipe should be repaired.
6. The bracelet should be shielded with a light gage sheet metal or by other methods approved by the Company representative while installing the concrete coating. The concrete coating mesh should be cut back so that it will not be within 50 mm of the anode. An ohm meter should be used to demonstrate to the Company representative that the reinforcement steel is not in contact with the anode.
7. Gaps between the anode bracelets and between the anodes and the concrete coating should be filled with concrete or mastic infill to produce a smooth surface across the bracelet with only the exterior curved surface of the anode exposed.

7.10.3 Corrosion Control Test Stations, Connections, and Bonds

Electrical continuity between the test point and the pipeline should be proven by means of a continuity tester, indicating zero resistance.

The test point should, where possible, be constructed prior to the application of the pipeline weight coating system, and care should be taken to ensure that all bare metal is insulated (except for the point of contact used for the test point).

Care should be taken to ensure that the reinforcement in the antibuoyancy weight coating material does not come into contact with the test point and that a minimum of 30-mm clearance between the reinforcement and the test point should be maintained.

Test points should be installed as required, but will be located midway between sacrificial anodes.

Connections of test lead wires to pipelines above the water must be installed so as to be mechanically secure and electrically conductive. Pipe and test lead wires should be clean, dry, free of foreign material, and properly coated.

Conductive connections to other pipelines or across electrical isolating devices should be mechanically secure, electrically conductive, and suitably coated. Bond connections should be accessible for testing.

7.10.4 Reinforcement

Pipeline weight coating reinforcement material should be carefully installed in accordance with standard specifications.

Each pipeline section should be inspected and tested using a 1000-V insulation test set to ensure that the reinforcement is not in contact with the pipe wall. A minimum reading of 1 mega-ohm will be regarded as satisfactory.

After each joint has been completed, the pipeline section reinforcement should be insulation tested from the remote end; this will be applicable for both lay barged and bottom pull pipelines.

7.10.5 Pipeline Crossings

Where two or more pipelines cross, test points should be fitted to the pipeline and positioned to coincide with the crossing.

Where test points for bonding purposes have not been installed, a clamp arrangement with set screw should be utilized. Care should be taken not to damage the coating and the pipeline. The set screw should be tightened only sufficiently to make a good electrical contact but not to damage the pipe wall.

7.11 Electrical Measurements and Tests

Clause 10 indicates the apparatus needed and the techniques for measuring voltage, current, and resistance and for testing for the continuity of structures to ensure the successful commissioning of a CP installation. Some survey techniques are described.

Electrical measurements and inspections are necessary to ensure that the initial protection of the structure has been established in accordance with applicable criteria, and that each part of the CP system is operating satisfactorily.

It is important for subsequent system checks to be carried out to ensure that the structure remains protected and, if changes are noted, that action is taken to return the system to a protected condition.

Whenever be the surface of a structure exposed, the condition of the coating should be noted, and the coating repaired appropriately.

Under no circumstances should any CP system be energized before inspection and testing is completed.

7.11.1 Potential Measurements

- **Instruments**

All instruments used for determining electrical values should be of an appropriate type and of the required accuracy. They should be maintained in good working condition at all times.

Where fluctuations in the electrical measurements are noted, it may be necessary to substitute recording instruments for meters during surveys.

Electrodes other than copper/copper sulfate and silver/silver chloride may be used, provided that their relationship with these electrodes is either known or established prior to each measurement.

7.11.2 Potential Survey of Internal Protection of the Plant

With a fully enclosed plant, it is normally necessary to install permanent measuring points or reference electrodes. Where the positions at which measurement should be made can be predicted, these facilities are preferably installed before commissioning.

Alternatively, potential surveys can be carried out initially with temporary equipment to determine the positions where the potentials are most positive and whether the most negative potentials are acceptable.

- **Permanently installed reference electrodes**

The most convenient method of mounting reference electrodes inside a plant is by means of a "screw-in" assembly such that the electrode can easily be withdrawn for inspection and replacement of either the entire unit or the electrode material.

The electrodes can be wired to central monitoring and control equipment. A disadvantage lies in the difficulty of checking the accuracy of the electrodes, once installed.

For detailed potential surveys, or if it is impossible to use "screw-in" mountings, reference electrodes can be attached by suitable nonmetallic fixings to the protected

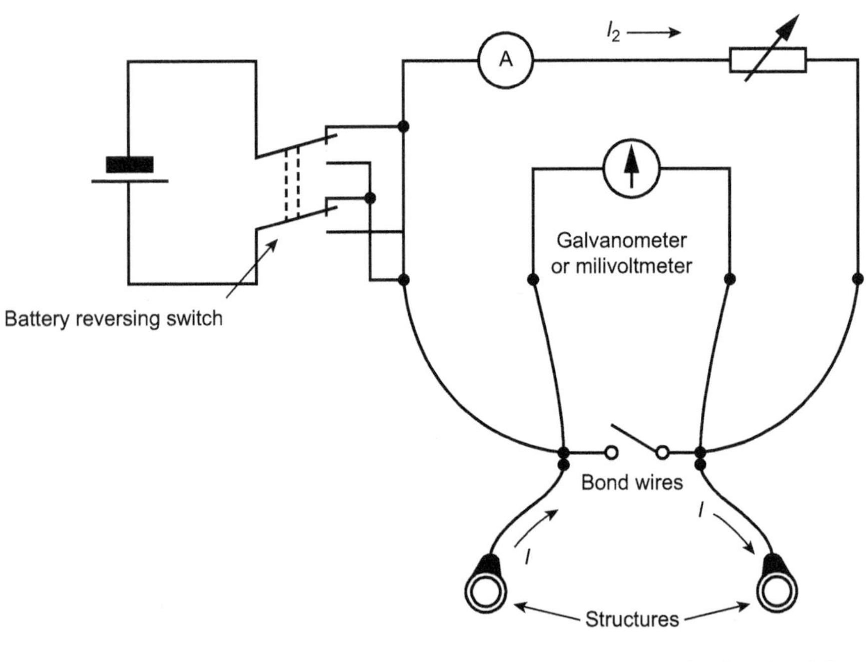

Figure 7.1 Zero resistance ammeter circuit used to measure current flowing in a bond (bond disconnected). With current I_b adjusted to give no deflection on the galvanometer I_b is equal to the bond current.

surface and the insulated connecting leads brought out through the plant wall through a suitable gland.

Generally, it is advisable to install at least one reference electrode for each cathodically protected compartment. The reference electrode should be installed at the position where corrosion is most likely, for example, at junctions of ferrous and nonferrous materials and/or remote from anodes.

7.11.3 Determination of Bond Resistance

It is necessary to determine the value of the resistance that should be connected in series with a bond, to adjust the structure electrolyte potential of a structure to a desired value. This can be done either by insertion of a series of fixed calibrated resistors until a suitable value is found or by adjustment by using a variable resistor, whose resistance is subsequently measured. Alternatively, if the galvanometer shown in Fig. 7.1 is calibrated to indicate voltage, the desired potential conditions on the structure can be obtained by adjustment of the resistor. The necessary resistance value is determined as the ratio of the voltage to the current. This arrangement has the advantage of obviating the need for low-resistance leads. Special milliohm meters are also available for measuring the very low resistance of bonds.

7.12 Tests Prior to Installation of CP on Buried or Immersed Structures

7.12.1 Soil/Water Evaluation

7.12.1.1 Soil/Water Sampling

- Soil samples may be obtained from along the pipeline route with a minimum of one sample from each type of soil noted to exist. Samples are to be ideally between 250 and 2000 g and are placed in sealed, sterile, air-tight containers and should fill the containers completely.
- Where bacteriological analysis is to be undertaken, the soil sample is to be as little disturbed as possible and should completely fill the containers.
- Soil samples are obtained from depths either by excavation or by auguring techniques.
- Water samples may be obtained from rivers, estuaries, and water-logged locations. Samples need to be ideally between 1 and 2 L; placed in sealed, sterile, air-tight containers; and need to fill the containers completely.
- In the case of immersed structures, any analysis of water samples should include measurement of the oxygen content and conductivity. It should be noted that, particularly in the case of estuarine waters, considerable variation can occur depending on the state of the tide and on the season. Moreover, stratification is often present, and the use of a suitable sampling technique is recommended.
- The analysis is to be completed with a minimum delay, from the time of sampling.

7.12.1.2 pH Measurements

- After resistivity measurements, pH measurement is perhaps the most widely used test for corrosivity. Where corrosion could be caused or enhanced by chemical attack, pH measurement may be used to assess this risk.

- The methods available for pH measurement include the following:
  - glass electrode and millivoltmeter;
  - colorimetric;
  - indicator papers.
- Glass electrodes may be used with either potentiometric or high-impedance millivoltmeters. Both types are available as portable, battery-powered units for field use. Apparatus, reagents, and procedures are listed in both ASTM G51 and BS 1377.
- Colorimetric techniques are also described in BS 1377 and may be used as rapid field techniques. However, results can be erratic, and excessive turbidity in the soil may mask end point.
- Indicator papers are a practical site method and are sufficiently accurate for most survey purposes. Dry soils may be wetted with deionized water for this technique.

7.12.1.3 Soluble Salts

Chemical analysis for salts is usually restricted to chlorides, sulfates, carbonates, and sulfides. The latter two are analyzed qualitatively, and chlorides and sulfates are analyzed quantitatively. Quantitative analysis of chlorides and sulfates is undertaken by gravimetric, volumetric, or colorimetric (semiquantitative) analysis. The gravimetric and volumetric analysis of sulfates is detailed in BS 1881. For corrosion purposes, only the water-soluble sulfates are of concern, rather than total sulfates.

- A quick assessment of the resistivity of water may be made from the value of total dissolved solids by the following formula:

$$\frac{6250}{\text{Total Dissolved Solids}} = \text{Resistivity in ohm} - \text{m}.$$

7.12.1.4 Bacterial Analysis

There are a number of microorganisms that thrive in or create conditions conductive to corrosion. These principle organisms are sulfobacteria, ferribacteria and sulfate-reducing bacteria. In soils, the most common form of bacterial corrosion is caused by sulfate-reducing bacteria. These bacteria are most active in anaerobic soils in which the hydrogen ion concentration is near neutral, that is pH 7.0, but are known to grow in the range pH 5.5–8.5. They act by converting sulfates in the soil to sulfides.

Various approaches exist to detect soils in which sulfate-reducing bacteria are likely to thrive. They are as follows:

- Redox potential

Redox potentials are measured in the field by measuring the potential of a platinum electrode using a reference calomel electrode. The reading is pH corrected.

The general accepted criteria for microbial corrosiveness as quoted in BS 7361 are as follows:

Redox potential 100 mV: severe
Corrected to pH7 100–200 mV: moderate

(mV Standard 200–400 mV: slight
Hydrogen Electrode) 400 mV: noncorrosive

This technique is probably the most widely used for assessing microbial activity. The reproducibility of results is however poor, and the equipment can only be used with confidence in relatively soft soils.

- Detection and enumeration

Detection of sulfate-reducing bacteria is undertaken by using one of a number of culture media that include Bars, Postgate, and American Petroleum Institute (API) media. Generally, the culture consists of a nutrient, an indicator, and a redox-poising agent with the pH adjusted to near neutral.

Enumeration is carried out by using a series dilution. After solidification of the culture medium and 2-day incubation, the colonies of bacteria can be counted.

- Chemical tests

These basically cover sulfate content, organic materials content, soluble iron, and hydrogen uptake of the soil.

7.12.1.5 Moisture Content

Moisture content of soils may be determined by one of the methods described in BS 1377.

7.12.2 Structure/Electrolyte "Natural" Potential Survey

A structure/electrolyte potential survey should be carried out to determine the structure/electrolyte potential variation along or over the surface of the structure.

Such a set of potential measurements may indicate those points on the structure where the worst corrosion is likely to be taking place.

With no applied CP, and in the absence of stray currents, the most negative structure electrolyte potentials indicate the corroding areas. On the other hand, if corrosion is due predominantly to stray current in the soil, the more intense corrosion will be associated with the more positive structure/electrolyte potentials.

7.12.3 Stray Electric Currents

Where the presence of stray electric currents is suspected, for example, in proximity to DC electric traction systems or where varying structure/electrolyte potentials indicate the possibility of such currents, it is necessary to determine more accurately the extent of stray current effect on the structure. This can be done by plotting the potential field in the area, using a stationary reference electrode, or a structure, as a reference point.

7.12.4 Tests for Electrical Continuity

Tests should be carried out whenever the continuity of the structure is in doubt, to locate any discontinuities.

7.13 Tests during the Commissioning Period

The structure/electrolyte potentials at various points on a structure will continue to change for some time after protection has been applied. Tests should, therefore, be made at intervals and currents should be adjusted as necessary until conditions become stable with potentials at all points not less negative than the values given in protection criteria.

A comprehensive survey should then be made, and the results should be analyzed to provide a list of conveniently carried out tests by which the continued satisfactory operation of the protection system can be confirmed.

Immediate action should be taken if abnormally positive changes in potential occur, particularly at the point(s) of application of current indicating that one or more transformer rectifiers have been reversed.

More frequent inspections (e.g., at monthly intervals) are recommended in the following situations:

1. the nonoperation of one transformer rectifier would result in a total or partial loss of protection;
2. the nonoperation of the transformer rectifiers is likely, due to factors outside the operators control, for example, known unreliable power supplies, joint operation with a third party, susceptibility to electrical storms; or
3. protection is provided by a single bond from another protected structure.

It is important that commissioning and routine test readings should be permanently recorded. In many instances, comparison with these provides the only information that is available as to the condition and performance of the system. To this end, a routine should be established for the periodic review of the measurements to ensure that the conditions are satisfactory. Consideration should be given to the computerization and graphical presentation of records, with the inclusion of exception reporting for test measurements that fall outside set limits.

7.13.1 Buried Structures

Structure/electrolyte potentials should be measured at a series of points including, particularly, points remote from the ground bed or anode positions.

Outputs should be adjusted to the minimum that gives the desired level of protection.

The period required for the potentials to become stable may vary from a few days for a well-coated structure to a few months for a bare or poorly coated structure.

Where possible, currents from individual sacrificial anodes or transformer-rectifier units should be measured. In the case of a complicated pipe system, it is also useful to measure the current flowing from individual branches or sections.

Once the operating conditions have been established, organizations that might be installing underground equipment in the area in the future should be given sufficient information for them to be aware of possible interaction problems. This will include, for example, ground bed positions and expected currents and, if not already provided, an indication of the routes of the protected structure and of any structures that have been bonded to it to reduce interaction.

7.13.2 Fixed Immersed Structures

Structure/electrolyte potentials should be measured, at the points provided, soon after the protection is switched on. Individual (or group) anode currents should be measured and adjusted to the minimum that gives protection.

The structure potential should be measured by connecting a high input-impedance voltmeter (at least 1 M) to the structure, usually at a test point, and placing a reference electrode, connected to the positive terminal of the voltmeter, as near as practicable to the immersed surface of the structure (Figs 7.2 and 7.3).

Because accurate measurement of the structure potential requires the reference to be located at the surface of the structure, the reference electrode may be located by a diver, a remotely operated vehicle, or be permanently installed at various areas of the structure (e.g., areas of complex geometry or where shielding can occur). Such readings can then be related to readings taken with a reference electrode placed adjacent to the side of the structure.

Care should be taken to ensure that the structure component to which the measuring voltmeter is connected is not carrying a substantial CP current. With impressed current systems, in particular, parts of the structure may be carrying a large current and hence may cause a significant voltage drop error in the measurement.

7.13.3 Internal Protection of Plant

Structure/electrolyte potentials should be measured at the test points before and soon after the installation is switched on (Fig. 7.4). The currents at individual anodes

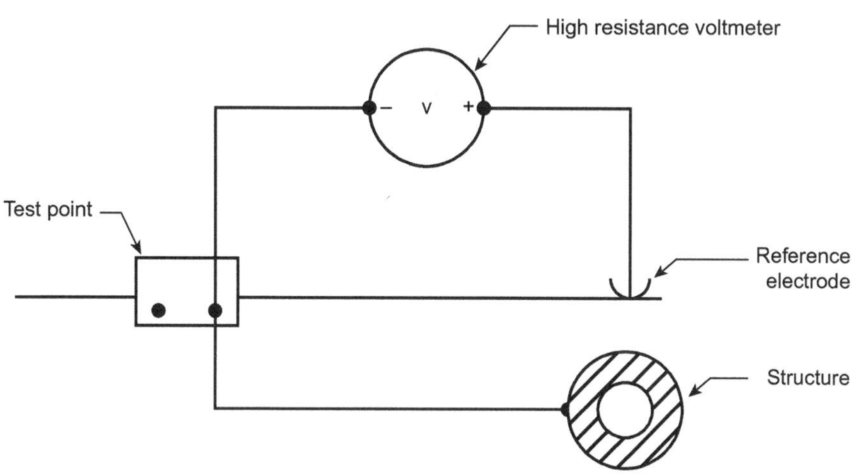

Figure 7.2 Structure potential measurement.

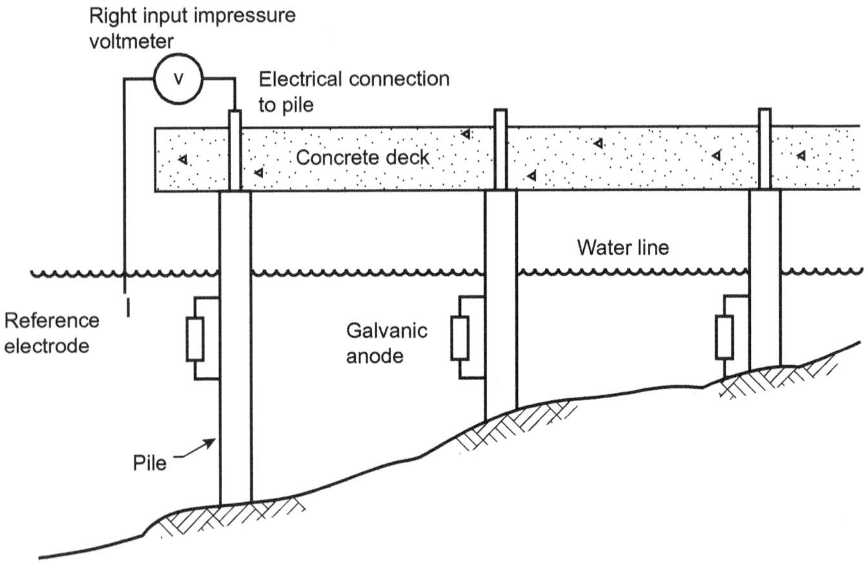

Figure 7.3 Measurement of structure potential on fixed immersed structures.

(or groups of anodes) should be monitored and adjusted as necessary after a further period, for example, one week, then, if no serious departure is observed, again after one month. At each adjustment, the individual and total anode currents should be noted for reference.

Each sensing electrode used for automatic control should be checked against a suitable reference electrode installed close to it. Unless there is experience with a similar plant, reference electrodes should also be installed at a sufficient number of positions in the protected equipment to enable a representative potential distribution curve to be plotted. This will show whether the position of the sensing electrode was chosen judiciously and whether the correct control setting has been selected. If more than one sensing electrode provides the feedback signal to the controller, the readings on each should be compared for incompatibilities before and after switching on the protection. Readings may show differences due to the presence of electropositive materials, and the gradients around anodes. Ideally, all the sensing signals should be within 50 mV when the protection is switched on. Slightly wider tolerances (e.g., 100 mV) may still form an acceptable basis for control.

7.13.4 Internal Surfaces

The structure potential should be measured by connecting the positive terminal of a high-impedance voltmeter (at least 1 mega-ohm) to the structure, usually at a test point. The negative terminal should be connected to a reference cell that is positioned as near as practicable to the immersed surface of the structure.

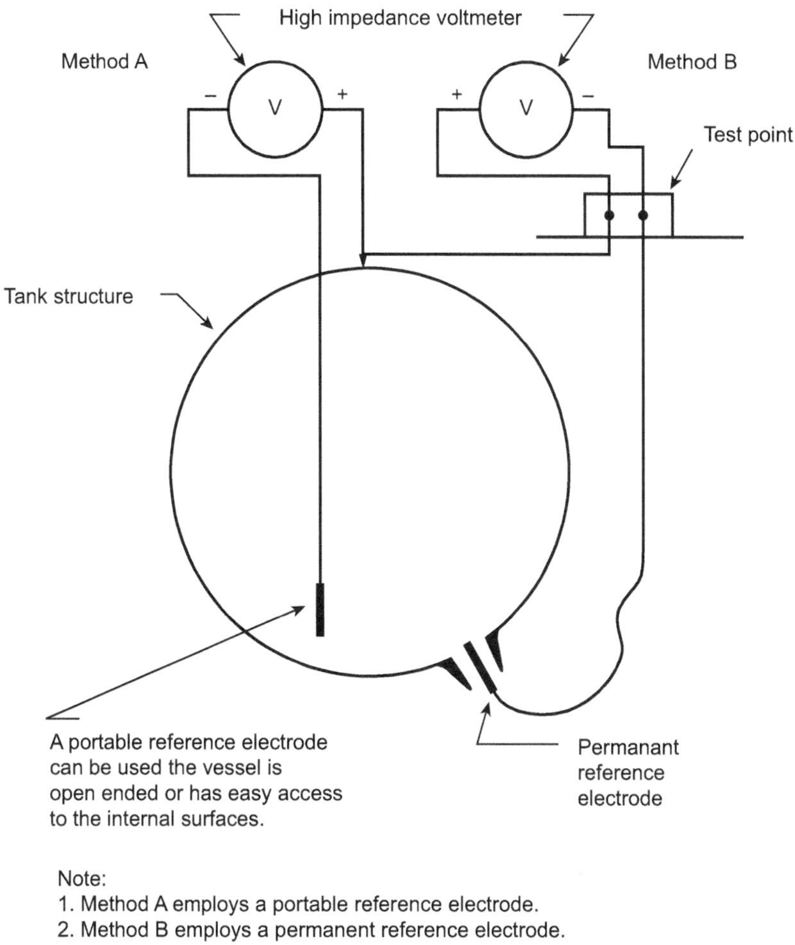

Figure 7.4 Alternative methods for measuring structure potentials.

Because accurate measurement of the structure potential requires the reference to be located at the surface of the structure, the reference electrode may be carried by a remotely operated vehicle or be permanently installed at various areas of the structure (e.g., areas of complex geometry or where shielding can occur). Accurate readings of the structure potential can then be related to readings taken with a reference electrode placed adjacent to the side of the structure.

Care should be taken to ensure that the structure component to which the measuring voltmeter is connected is not carrying a substantial CP current. With impressed current systems, in particular, parts of the structure may be carrying a large current and hence may cause a significant voltage drop error in the measurement.

7.14 Specialized Surveys

There are a number of specialized survey techniques being utilized to provide additional detailed data concerning corrosion prevention systems.

These techniques would normally be carried out by specially trained personnel using purpose-built equipment and instrumentation, often only available from specialist contractors. These surveys are generally time consuming, but the information gained may not be available from other methods.

The surveys covered by this Standard are as follows:

1. Surveys for detecting external pipeline coating defects are
   a. Pearson survey,
   b. Electromagnetic current attenuation survey,
   c. Close interval pipe-to-soil potential survey.
2. Surveys to determine the effectiveness of CP systems are
   a. Close interval pipe-to-soil potential survey,
   b. Current drainage survey.

8 Commissioning of Cathodic Protection Systems

This section indicates the stages at which tests should be made for commissioning of cathodic protection (CP) systems for buried and immersed structures, tanks, and internally protected plant.

The effectiveness of a CP installation depends on applying and maintaining the correct potential difference between the structure and the adjacent environment at all parts of the structure. This is the objective of the procedures described in this section.

The engineer should ensure that the checkout and commissioning procedures are sufficient to demonstrate that the CP system installation satisfies the criteria established by the project tender documents and the associated company specifications and drawings.

8.1 Precommissioning Inspection and Check

Every CP installation should be inspected and tested before commissioning test, this is to ensure as far as practicable that all the requirements of the contract have been carried out and installation is ready for precommissioning. It requires that the test carried out should not in any way be a danger to persons, property, or equipment.

The engineer is entitled to inspect, examine, and test the workmanship during the course of installation; any such inspection should not release from obligations. Any work, in the opinion of the engineer, that is not up to standard should be rectified.

A program should be provided for precommissioning test with the approval of engineer.

All provisions such as testing equipment, special tools, and coordination for availability of power and all pertinent work permit should be envisaged.

The written programs for checkout and commissioning, including detailed check sheets, should be submitted prior to the scheduled date for beginning checkout and commissioning activities.

The approved design drawings and specifications should form the basis for the construction of the CP system. In addition to these drawings, the Company will furnish the drawings for all company-supplied equipment. They should furnish the drawings for all equipment.

As-built corrections should be made to both the supplied drawings, including all changes required during checkout and commissioning.

Detailed check sheets should be developed as part of the checkout and commissioning procedures. The check sheets should have provisions for certifying that the various individual equipment items are properly installed in compliance with the specified performance and safety requirements and are suitable for operation.

Cathodic Corrosion Protection Systems. http://dx.doi.org/10.1016/B978-0-12-800274-2.00008-9

The detailed check sheets should be standardized for each type of CP system and should contain as a minimum

- Checks and tests conducted.
- Record of test results.
- Acceptance signatures.
- Equipment tag number.

The engineer should carry out all precommissioning checks of the completed system. Duties in this connection consist of, but are not limited to, the following:

- **Transformer/rectifiers**

The engineer should check the transformer/rectifier units for correct polarity, supply voltage, fuse rating, and full load test. The engineer should endeavor to keep the time for full load test as minimum as possible. In addition, the testing of the transformer/rectifiers should be carried out to ensure adequate electrical connection and that no damage has occurred during installation.

- **Groundbed**

Groundbed resistance test should be made as soon as the groundbed is installed and backfilled, under the supervision of the engineer.

- **Cable test**

It is necessary to ensure adequate care is taken during installation and backfill to prevent damage to the wiring insulation and test the wiring, after backfilling, for continuity, with a low voltage test. The method and results should be approved by the engineer.

- **Insulating joints**

After the hydrostatic test and before gas commissioning, the engineer should make an electrical test to verify the insulation of the joint. The minimum value should not be less than 1000 ohms.

After commissioning, the effectiveness of the insulation should be tested by CP impressed current. In this test, the current will be drained from one side of the insulating joint to the extent that a negative swing of potential of 400 mV is maintained. In this case, insulating joint is accepted if the potential on the unprotected side of the joint remains natural or becomes more positive. Insulating joints found defective after installation should be replaced by the contractor at his or her expense.

- **Cased crossing**

Test on cased crossings should be made in accordance with standards.

During commissioning of the CP system, tests on the casing insulation will be conducted again by draining the CP current in such a way that a minimum negative swing of 400 mV is achieved on the carrier pipe. In this case, the insulation should be effective if the potential of the casing shifts toward positive.

Results of the above tests should be approved by the engineer.

- **Test stations (test points)**

All installations should be subject to test during construction and on completion of the work. Should any test point prove faulty or damaged in construction and/or installation, the contractor should rectify the defect(s) as soon as discovered and/or instructed by the engineer.

- **Earthing system**

Check completed earthing system for continuity of main earthing system loop and continuity of all taps to equipment.

Check earthing system using a Megger earthing resistance tester or the type that balances out the reference grounds.

Earthing resistance values in all of the above tests should be 5 ohms or less. If greater than 5 ohms, supplemental ground rods should be installed until an acceptable value is obtained.

8.2 Hookup and Commissioning

All work should be executed in strict accordance with the approved drawings and specifications. The written approval is required for any deviations from these drawings and specifications.

It is necessary to bring to attention any areas of the drawings and/or specifications that conflict or do not meet safe and acceptable practices.

The engineer should be responsible for the protection and security of all materials and equipment during all stages of commissioning.

The responsible engineer should not perform any tests that would either void the vendor warranty or damage the materials and equipment.

The engineer should tie-in the alternating current (AC) power source to the rectifier(s) with all work carried out by qualified electricians.

An "adjustive survey" should be provided within few days of system start-up. with the "adjustive survey," the following additional documentation on CP system(s) should be submitted:

1. Copies of accurately dimensioned "as-built" drawings of all installations.
2. Copies of operating and maintenance manuals for all equipment provided and installed by the contractor.

The above information, when available, should be incorporated in the plant data books.

Written programs for commissioning of CP system should be developed. These programs should include detailed check sheets and should be submitted for acceptance prior to the scheduled commencement of the checkout and commissioning activities.

The commissioning program should detail the final tests to be conducted on all equipment after the checkout program is complete. The commissioning program should demonstrate the proper operation of the complete CP system.

Checkout tests may be conducted concurrently with commissioning tests.

8.2.1 Impressed Current Systems

After completion of the installation, checkout, inspection, and testing of the CP equipment, it is necessary to carry out the following procedures.

Prior to energization of the CP systems, the engineer should carry out a natural potential survey all along the length of the structure under protection at all test points and current drainage points.

The transformer/rectifiers then should be energized for a minimum period of 72 h before any measurements are taken.

The engineer should adjust the current by steps to limit the drain point potential according to the criteria for the steel under protection.

During the polarization period with impressed current systems, regular checks of structure potential and transformer–rectifier output should be made in order to avoid gross overprotection.

The frequency of subsequent T/R readings depends on the reliability of the power supply. Readings taken once per fortnight are often chosen to start with.

During the polarization period, the current output of the transformer–rectifiers should be progressively reduced to maintain the steel-to-soil potential at its desired level.

Once all the transformer/rectifiers have been energized, applied potentials should be taken at the same location as for the natural potentials.

Should the interphase potential between two groundbed installations be higher than the minimum criteria laid down, this usually being 0.95 V, the drain point potential should be adjusted to a lower potential to prevent wasteful losses of power at the installations.

Particularly in high-resistivity soils, voltage drop through the soil as a result of applied current for CP can be considerable. As a result, the potentials measured on the surface over the pipe while the current is "on" will not reflect the potential at the pipe surface.

In order to measure a more true potential, the current should be switched off momentarily, the potential (the "off" potential) should be measured immediately after interruption (within seconds). This potential is not affected by any voltage drop in the soil. Transformer–rectifiers should be equipped with automatic and synchronizable interrupters for this purpose.

It is necessary to repeat this test after polarization has taken place, approximately 14 days. This may require further adjustments of the current output of the transformer/rectifier to keep the applied potentials between the minimums and maximums laid down.

Notes:

1. Besides the detailed information related to the design, construction, and commissioning of the installation, records should be kept of current outputs and protective potentials. The data should be analyzed after each survey and corrective action taken soonest but within one month. Whenever a CP system is installed, the necessary instruments should be purchased to enable the engineer-in-charge to make the required measurements.
2. After a protective current has been applied by an impressed current CP system, polarization of the protected structure occurs, causing a gradual fall in current requirements. The rate at which polarization occurs depends on the nature of the medium surrounding the structure

and on the current density; the time elapsing before the current requirement falls to a steady value while maintaining the structure at the desired potential may vary from a few days to many months.

3. The reference electrode should be located in contact with the surface of the earth directly over or not more than five or six pipe diameters away from the structure along the surface of the earth at selected locations as indicated on the drawings. When measuring the potential of a tank, the reference should be placed a half tank diameter away from the structure.
4. Variable resistors should be installed in the negative drain circuit, if required, to balance the current to each of the adjacent pipelines. Each negative circuit should be provided with a suitably sized shunt and diode at the direct current (DC) source. The installation of the diode is required to prevent mutual influence of pipelines during "on–off" surveys.

8.2.2 Sacrificial Anodes

During the polarization period, regular checking of potentials should be carried out to obtain early warning in case the system is inadequate.

The potential of the protected structure can never become more negative than the anode potential itself. The latter is usually well within the range of acceptability.

Current output from the anodes is self-regulating since with further depressed potentials of the structure, the driving force between anode and cathode becomes smaller. This automatically results in a lower current drain. Calculation will show that zinc and aluminum anodes can provide approximately twice the design current at the very start of the polarization period; with the advancement of the polarization the current gradually reaches the equilibrium state (which could be lower than the design current drain).

The sacrificial anode system usually comprises two cables connected to the structure, one for measuring the applied potential and the other to be connected through a terminal panel and a removable link to the sacrificial anode. The removable link is maintained for the measurement of current between the anode and the structure. A sensitive milliammeter with a very low internal resistance is connected between the terminals after the link has been removed and a measurement of the current output is taken. This measurement should be taken as quickly as possible to prevent depolarization between the anode and the structure since increased current will be required should the structure depolarize.

With anodes used suspended under water and accessible, an indication of the probable remaining life can be obtained by periodic lifting and weighing. A comparison of the rate of wastage will also show the extent to which individual anodes are contributing to the protective system.

When sacrificial anodes are used to protect condenser end plates, box coolers, and similar equipment, visual inspection together with potential measurements provide a method of checking.

8.2.3 Interference

The application of CP with impressed current to a buried or immersed structure (referred to as the primary structure) causes DC to flow in the earth or water in its vicinity. Part of the protection current traverses nearby buried or immersed pipes, cables, jetties, or similar structures alongside (termed secondary structures), which may be unprotected and the corrosion rate on these structures may, therefore, increase at points where the current leaves them to return to the primary structure.

Interference with other buried or immersed installations should be measured by the contractor in the presence of the engineer and the owners of the foreign structures, after switching on the impressed current CP system.

The contractor should conduct the tests with the understanding that the criterion is the swing of the potential of the foreign structure from the natural value. The test procedure is detailed hereunder:

1. The foreign structure-to-electrolyte potential should be taken while the transformer–rectifier is turned off.
2. Structure-to-electrolyte potential should be taken while the transformer–rectifier is "on" position.
3. Several readings should be made by the contractor for comparison.

The maximum positive swing of potential at any part of foreign structure resulting from interference should not exceed 20 mV and, in case of interference exceeding a positive change of 20 mV, the contractor should carry out the remedial works.

8.3 Commissioning Survey

The commissioning survey should include the following tests and measurements, where applicable, to ensure that the structure is protected in accordance with the design criteria, and that all equipment is correctly installed and functioning correctly:

1. Measurement of structure potential at all test points, both before and after energization of the CP system.
   Notes:
   a. Following the application of CP, the potential level of the structure will change with time owing to polarization and, to ensure the structure potential is in the desired range, it may be necessary to take several potential measurements over a given time.
   b. Systems requiring large currents may need to be briefly de-energized to permit potential measurement free of voltage drop error.
   c. In tanks containing fluids of relatively high resistivity, it is recommended that "off" potentials are measured to minimize voltage gradient errors for impressed current systems.
2. A check for correctness of polarity of electrical circuits, i.e., positive to anode and negative to structure.
3. A functional test of all test points, to ensure correct installation/operation.
4. Assessment of the effectiveness of the following items:
   a. All insulating devices.
   b. The continuity of bonds.
   c. The isolation of the structure from electrical earths and secondary structures in accordance with the design.
   d. Casing insulation.
   e. Stray current control equipment.
   f. Structure earthing:
      - fortuitous;
      - deliberate; and
      - AC.

5. Measurement of the coating resistance.
6. A check that the anode current distribution is as desired.
7. For impressed current systems, the measurement of back emf and loop resistance.
8. The current required to provide protection.
9. The voltage output of the impressed current system.
10. A test for interference current flow in bonds.

The survey should also identify the following:

a. The variation in output current and structure potential with time. The rectifier output voltage may also be recorded.
b. Locations where future measurements (current, source voltage, structure/electrolyte potential) can be taken to provide a representative view of the operation of the system.
c. Need for any additional test points or CP facilities.

Data from the survey should be recorded and retained for future reference.

Details of tests and testing methods to be carried out during the commissioning periods. The sequence of tests should be commenced as soon as possible after commissioning.

8.4 Commissioning Report

The contractor should submit a commissioning report to the company representative at the conclusion of checkout and commissioning.

Records associated with CP systems should be kept as historical data for future consideration and action. Records are used to demonstrate that a CP system is working.

The commissioning report should include, but does not need to be limited to, the following:

1. Design documentation.
2. Results of periodic survey checks.
3. Results of equipment checks.
4. Agreements made with owners of foreign structures.
5. Location of any test points added to the system.
6. Coating materials and application procedures.
7. Correspondence with regulatory authorities.
8. Information on and location of stray current facilities connected.
9. Commissioning objectives.
10. Description of CP system operation.
11. Original of all check sheets and discrepancy lists.
12. Conclusions and recommendations.

8.5 Installation of Electrical Isolation Equipment

This section deals with the procedures to be used when installing the equipment used for electrically isolating pipelines. The procedures are designed to ensure the following.

A satisfactory degree of electrical isolation is achieved at the time of installation and that the joint is not damaged so as to cause an accelerated degradation rate with time.

The installed equipment is adequately protected against the effects of stray DC or induced AC voltages.

There is adequate provision for test leads to allow for field testing and maintenance. Typical arrangements are shown in Fig. 8.1.

Isolation joints installed at coating/electrolyte changes should always be bonded in test points, to keep the pipeline electrically continuous when the CP is operational.

If CP is to be applied on nonwelded pipelines, the continuity of the pipeline should be ensured. This should be done by the installation of permanent bonds over high-resistance flanges/couplings, using cables and approved attachment methods.

The continuity of nonwelded pipelines should be checked, e.g., by carrying out overlapping potential measurements.

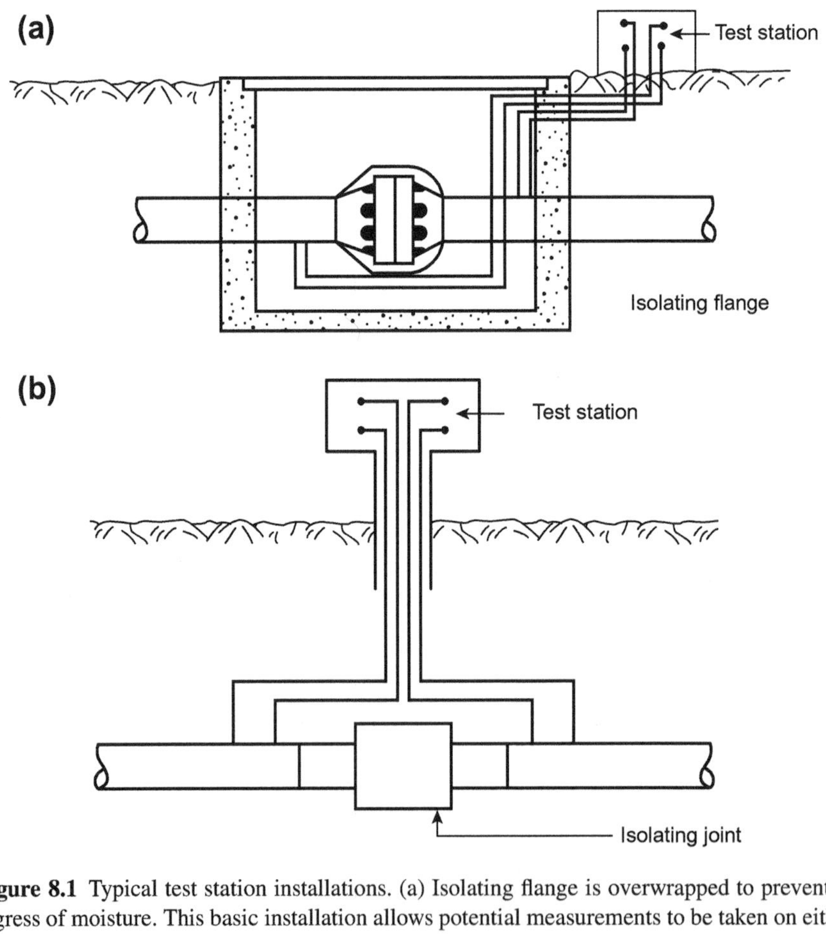

Figure 8.1 Typical test station installations. (a) Isolating flange is overwrapped to prevent ingress of moisture. This basic installation allows potential measurements to be taken on either side of device. (b) Inner cables are of large cross-section and allow for resistance to be bonded across joint as required. Outer cables are for potential monitoring. to allow easy identification, cables should be permanently identified.

Notes:

1. **The buried insulating joints installed on pipelines are occasionally used to open the electric continuity of the pipelines and thus create sections isolated from the rest of the network, when it is desired to**
   a. **measure the electric resistance of pipeline coating;**
   b. **locate pipeline coating defects;**
   c. **limit the zone of influence of the impressed current systems.**
2. **The normal state of buried insulating joints installed on pipelines is to be electrically shunted.**

The aboveground insulating joints installed on district regulators and service lines must never be shunted. Such aboveground insulating joints are not part of the CP installations; they are rather part of the passive protection.

8.5.1 Installation

When installed, all equipment items should be properly supported and aligned so that any forces transferred from adjoining pipe are minimized. This should be considered when equipment locations are selected. The type of equipment chosen should be suitable for the mechanical forces to be encountered at the chosen site. Isolating devices should not be installed in gas systems at locations where there is likelihood of internal moisture accumulation. When possible, the electrical resistance should be checked immediately before and after installation.

Before any work is carried out on or near an isolated flange, the area should be checked for hazardous atmospheres.

To avoid risk of electric shock and the possibility of sparking, it is advisable that insulating joints be cross-bonded before being assembled. This precaution is essential for hydrocarbon product lines.

8.5.2 Insulating Joints

Insulating joints should be ordered with the weld end preparation conforming to the main pipe-laying specifications. The manufacturer's special instructions for installation should always be followed, particularly when welding in joints with short overall lengths, to ensure that the heat generated does not damage the insulating materials used in the joint construction.

The contractor should align, install, and test insulating joints shown in the drawings. The testing should include

1. electrical test to verify the insulating of the joint (the resistance across isolating joints should be measured immediately before welding into the pipeline; the minimum resistance should be 1 mega ohm (106 ohm));
2. hydrostatic test in situ, with the pipeline.

The contractor should perform this work with proper caution so as not to damage the insulating material during the handling, welding, and pipe laying.

In no case should the contractor carry out welding in the vicinity of an insulating joint without shunting the insulation joint. The shunt should be removed by the contractor upon completion of the welding.

When welding the joint into the pipeline, the contractor should take care to ensure that heat should not be conducted along the pipe and cause damage to the internal lining or insulation.

The buried insulating joint should be manually coated as soon as possible after installation, as specified in design specification. The aboveground insulating joints should be painted as specified in design specification.

8.5.3 Isolated Flange Joints

- **Factory preassembled types**

These are supplied with weld ends and may be installed.

- **Flange insulation kits**

The assembly of an insulating flange requires particular care to ensure that insulation is not lost due to mechanical failure of the components. Flanges are welded to the pipeline, and the flange insulating materials are supplied as a kit, designed to provide insulation for one flange of a given size and type, for site installation.

The installation should take place in clean and dry conditions.

Flanges on which insulating gaskets are to be installed should be supplied as matched pairs or reamed on site to ensure correct alignment of bolt holes. The flange faces should be clean and correctly aligned. Misaligned flanges will result in insulating sleeve damage during assembly or subsequent springing of the pipe. Flange faces should be square and free of burrs to allow for correct sealing of nuts, bolts, and washers.

The flange should be assembled using a very high-impedance voltmeter attached across the "open" joint. During assembly, potential difference across the joint must not be lost. It may be necessary to create a potential difference across the "open" joint before commencing assembly.

Note: Using resistance methods to determine integrity of insulating flanges in the field can produce irrelevant results.

Insulated flanges should be assembled and tested before being welded into the pipeline. A voltage of 1500 V DC should be applied across the flange assembly for 1 min without causing breakdown of the insulation or flash over. Subsequently, the resistance across the assembled flange should be measured and should be more than 1 mega ohm (106 ohm).

On existing flanges, the bolt-to-pipe resistance should be measured and the overall effectiveness of the isolated flange determined after CP has been applied. This can effectively be done by measuring the current through the attached pipe using a "Swain-type" current clamp/meter.

Insulated flanges should be protected against ingress of dirt and moisture by the application of flange protectors or protective tape, except when used in sour service conditions.

The isolating gasket should be carefully aligned between the flange faces and the bolt holes. It may be easier to use one size smaller diameter, high-tensile steel bolts, and/or special thin-walled sleeving to assist alignment.

Alignment pins should be inserted to ensure that flange alignment is maintained during the installation of the isolating sleeves.

The isolating sleeves are then positioned in the correctly aligned holes. Isolating sleeves must be of the correct length. If they are too long, they may be damaged when the bolt nuts are finally tightened. If they are too short, they may fail to isolate properly. The length of the isolating sleeve should normally include the two isolating washers, except where alignment allows only one flange to be isolated.

The bolts, complete with isolating washers and steel washers under the bolt and nut heads, are threaded through the sleeves and hand tightened.

Final tightening to the tension recommended for the diameter and pressure rating of the flange should be done in a sequence that provides for equal tensioning without distortion.

The original alignment pins may then the removed and bolts installed, complete with sleeves and washers, as described above.

Complete flanges should be coated in accordance with design specification.

To prevent the failure of the insulating materials and to obtain a satisfactory insulating job with a minimum of effort, the following precautions should be observed in installing insulating flanges:

- Micarta washers should be placed next to the flange and topped by steel washers that rest beneath the nut.
- All micarta washers should be placed on the same half of the flange.
- A backup wrench should be placed on the nut nearest the micarta washer and all tightening of the flange bolt accomplished by turning the nut on the opposite half of the flange.
- The micarta sleeve is designed to extend through both washers. Precautions should be taken to see that it remains in this position during installation work.
- Micarta sleeves should never be hammered or forced into place.

8.5.4 Protection Against External Moisture Ingress

The materials used for the isolating sleeves and washers may absorb water, and the construction of the joint may allow for moisture ingress, both of which will result in a reduction of the electrical resistance of the assembly. It is therefore essential to provide a protective coating. A suitable material may be applied to fill in the crevices and gaps between flange faces, and to mold around the flange faces so that a smooth profile is achieved which, together with the adjacent pipework, may be coated or wrapped to the same standard as the pipeline. Other methods and materials may be used to protect against moisture ingress.

8.5.5 Pipeline Casing Insulators

Pipeline casing insulators should be installed in accordance with the manufacturer's instructions. Special care should be taken to ensure that all subcomponents are correctly assembled and tightened and that no damage occurs during carrier pipe insertion and tightening.

The annulus between the carrier pipe and the casing should be sealed at each end of the casing by means of casing end seals to prevent water from entering the casing.

There must be no inadvertent metallic contact between the casing and the carrier pipe. The spacing of insulators should ensure that the carrier pipe is adequately supported throughout its length, particularly at the ends, to prevent settling and possible shorting.

8.5.6 High Voltage Protection

Isolating devices and supports should be protected against damage from high voltage surges. These surges may be caused by lightning, induced AC from adjacent or overhead high tension cables, or fault conditions.

High voltage surges may permanently damage the isolating materials used in the joint construction.

Isolating devices and supports may be protected with lightning arrestors, electrolytic grounding cells, polarization cells, or combinations of these.

The manufacturer's instructions should be followed strictly when installing protective devices. In particular, they should be physically secured and the connection cables properly sized.

The threshold rating of the protective device should be such that, even allowing for tolerances, the potential applied across the isolating device is below its minimum dielectric strength.

Lightning arrestors and other protective devices may need to be carefully located or housed to prevent dirt and moisture from collecting, which could lead to an external flashover at a relatively low surge voltage. Applicable electrical codes should be consulted.

8.6 Thermit Welding of CP Leads

This Clause 13 covers the connection of CP wire leads to new or in-service carbon steel pipelines under pressure by thermit welding. Connections to pipe less than 3 mm thick should be made using approved clamps or silver soldering.

For the purposes of this standard, thermit welds, thermowelds, and cadwelds are synonymous.

Thermit weld process should be applied only by skilled experienced field personnel.

Thermit welds should not be made on internally plastic-coated pipelines. Internal coatings affected include epoxies, phenolics, nylon, polyethylene liners, etc.

The thermit weld charge should be limited to Thermoweld Cartridge No. 15 F33 (15 g), or equivalent. The maximum size of the electrical conductor should be 25 mm^2 (No. 6 AWG). Where the attachment of a larger conductor is required, a multistrand wire should be used and the strands should be arranged into groups no larger than 25 mm^2 (No. 6 AWG) and each group attached to the pipe separately.

The minimum distance of a thermit weld from a circumferential weld should be 200 mm.

The minimum distance of a thermit weld from a longitudinal weld should be 40 mm.

In attaching one wire to a pipeline, only one charge should be used. If the first thermit weld does not take, a second thermit weld should not be attempted on the same spot.

If a thermit weld is disapproved on the first charge, it should either be removed, the surface cleaned to bright metal, and the process is repeated or a new location on the pipe is selected.

The permissible operating pressure in the pipeline when thermit welding is shown in Table 8.1. If the pipe on which the thermit weld will be attached is not in the table, the permissible operating pressure can be calculated using the following formula:

$$P_p = \frac{2S\,(t-1.59)\times 0.72\times 10^3}{D} \tag{8.1}$$

P_p = Permissible "hot work" pressure (kPa).
S = Specified minimum yield strength of the pipe (MPa).
D = Nominal outside diameter of the pipe (mm).
t = Nominal wall thickness of the pipe (mm).

Note: S, D, and t values are to be obtained from the operating permits for the pipeline.

Thermit welds should not be attached to pipe wall thicknesses less than 3.18 mm while the pipeline is pressurized.

The connection of leads to high-pressure gas lines with wall thickness less than 4.78 mm, but more than 3.18 mm, by thermit welding should be done with the gas flowing.

The use of thermit welds should be avoided in high-stress areas such as elbows, tees, etc.

If more than one weld is required such as two adjacent wires or large conductor split in two to get the required size for a 15 g charge, the spacing between point of connection should not be less than 100 mm (4 in).

A suitable copper sleeve for smaller size wire should be used for a 15 g one-shot mold and the wire bended around the end of the sleeve.

8.7 Pipe Preparation

The pipeline coating should be completely removed including primer and the pipe cleaned with a file to white metal. The pipe should be completely dry.

The wall thickness of the pipe should be checked using a portable ultrasonic wall thickness tester and the permissible pressure of the line checked.

Where possible, thermit welds should be applied to horizontal pipes.

8.8 Thermit Weld Preparation and Procedure

Care should be taken to ensure that the graphite thermit weld mold is completely dry and free from slag or other impurities before proceeding with thermit welding on pressurized pipelines.

Table 8.1 Permissible Hot Work Pressure While Thermit Welding

	Grade		Specified Minimum Yield		OD		WT		Permissible Hot Work Pressure	
NPA	CSA	API	MPa	PSI	mm	in	mm	in	kPa	Psig
2	207	A	207	30,000	60.3	2.375	3.91	0.154	11,529	1664
	241	B	241	35,000	60.3	2.375	3.18	0.125	9170	1330
3	241	A	241	35,000	88.9	3.5	3.18	0.125	6205	900
	290	X42	290	42,000	88.9	3.5	4.78	0.188	15,059	2169
4	241	B	241	35,000	114.3	4.5	3.18	0.125	4826	700
	290	X42	290	42,000	114.3	4.5	4.78	0.188	11,713	1687
6	207	A	207	30,000	168.3	6.625	4.78	0.188	5678	818
	241	B	241	35,000	168.3	6.625	4.78	0.188	6610	955
	290	X42	290	42,000	168.3	6.625	3.18	0.125	3930	570
8	207	A	207	30,000	219.1	8.625	5.56	0.219	5427	783
	241	B	241	35,000	219.1	8.625	8.18	0.322	10,486	1516
	290	X42	290	42,000	219.1	8.625	4.17	0.164	4977	722
10	241	B	241	35,000	273.1	10.750	6.35	0.25	6077	879
	317	X46	317	46,000	273.1	10.750	5.56	0.219	6668	964
	317	X46	317	46,000	273.1	10.750	4.78	0.188	5308	770
12	290	X42	290	42,000	323.9	12.750	6.35	0.25	6166	889
	317	X46	317	46,000	323.9	12.750	6.35	0.25	6740	974
	359	X52	359	52,000	323.9	12.750	4.78	0.188	5060	734
14	290	X42	290	42,000	355.6	14.00	6.35	0.25	5616	810
	317	X46	317	46,000	355.6	14.00	5.59	0.219	5122	743
16	317	X46	317	46,000	406.4	16.00	6.35	0.25	5372	776
	359	X52	359	52,000	406.4	16.00	5.59	0.219	5067	735

CSA-Canadian Standards Association; OD-Outside diameter; WT- Wall Thickness; PSI-Pound per square inch; Psig-Pound per square inch gauge

The copper conductor should be clean and dry and the insulation cut back sufficiently for insertion into the mold.

The copper conductor should be wrapped around the pipe at least once and enough slack provided to allow for pipe in soil movement.

After the completion of the thermit weld on buried pipelines, the bright metal surfaces should be protected and covered by the application of a tape and primer and a "Royston Handy Cap" or equal. The "Handy Cap" should be taped in place using primer and cold applied self adhesive tape to provide a watertight seal on all exposed steel and copper surfaces.

The tape should overlap existing pipe coating to about 25 mm (1 in) minimum.

Manufacturer's instruction for application of "Handy Cap" should be rigidly followed.

Note: The wall thickness of the pipe to be worked on is to be checked with an ultrasonic wall thickness tester after the pipe has been cleaned for welding. The thermit weld is not to be made on a pitted or laminated pipe or where the wall thickness is less than 90% of nominal pipe wall thickness.

8.9 Control of Interference Currents on Foreign Structures

Interference from CP systems arises where a foreign structure intersects the DC path between the anode and cathode. Where the current enters the structure the effect is cathodic. Where it leaves the structure the effect is anodic, and the rate of corrosion at that position may be increased.

Interference may be detected by a change in the potential of the foreign structure when the system current is interrupted. The result of this test indicates whether the foreign structure is being subjected to an increased or a decreased corrosion hazard.

Where a foreign structure is sited adjacent to a protected immersed structure, but not electrically bonded to it, interference can occur. Two common foreign structure types are as follows:

1. A discrete movable structure, such as a moored ship.
2. A buried or immersed pipeline or metal-sheathed cable adjacent to the protected structure or its anode system.

Galvanic anodes used in immersed systems do not usually cause interference to other structures.

Interference problems are more probable with impressed current systems because of the electrolyte voltage gradients usually associated with the anodes.

When marine structures are cathodically protected, adequate precautions should be taken to avoid interference effects when using impressed current and also to ensure that danger does not arise through the production of sparks when ships, barges, etc. make or break electrical contact with the protected structure.

Marine conductors (protective pipes through which wells are drilled) are often closely packed in the conductor bay area. Care should be taken that adequate current densities are available for the protection of the marine conductors. Full electrical continuity may not always be provided and special measures may be required to ensure this.

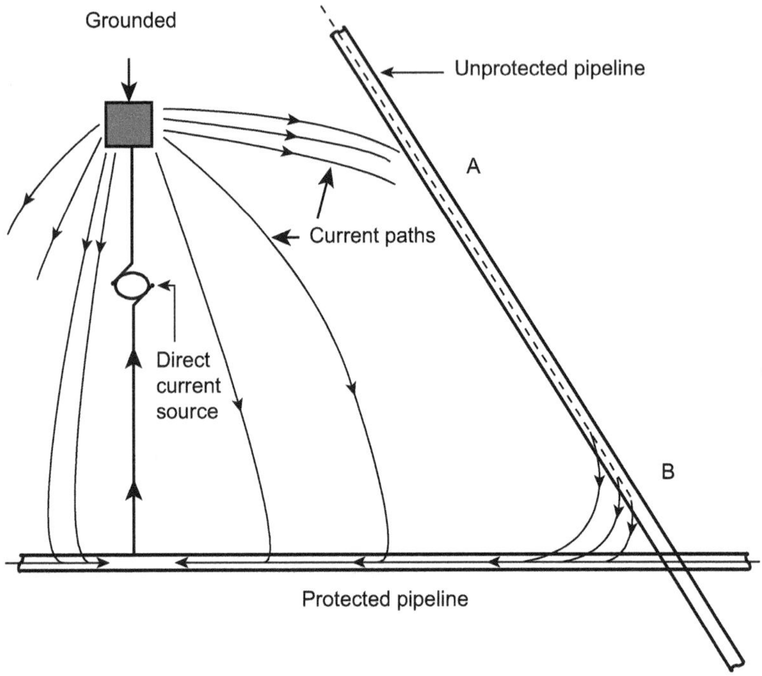

Figure 8.2 Example of current pickup by neighboring structure from CP system, the unprotected pipeline becomes cathodic at the point where the current is picked up (A), and anodic at the point where the current eventually leaves the line (B), resulting in protection at "A" and corrosion at "B" since point "B" becomes anodic, the pipe-to-soil potential measured with a copper/copper sulfate electrode becomes more positive when the rectifier is switched on. A change to positive in potential of more than 50 mV when the rectifier is switched from "off" to "on" indicates an objectionable degree of stray-current corrosion.

Electrical interference effects are negligible with sacrificial anode systems, as these anodes are placed much nearer to the protected structure than to any unprotected steelwork and because their driving voltage is usually much lower than that of impressed current system groundbeds.

Telluric current and induced ACs can be of particular importance on transmission pipelines (see Fig. 8.2).

8.10 Notifying Owners of Other Structures for Interference Testing

It is essential, throughout the installation, testing, commissioning, and operation of a CP system, that notice of actions proposed be given to all organizations and owners having buried metallic pipes, cables, or other structures in the near vicinity of the installation.

These notices are intended to ensure that information becomes available to enable a system to be installed in such a manner that interference is kept to a minimum and that enough information is given to other organizations to enable them to determine whether corrosion interference is likely.

After commissioning tests of the system have been completed, notification should be sent to all organizations who have indicated that they have structures likely to be affected by the operation of the system. The following information should be supplied at least 1 month before the date proposed for interference tests:

1. the anticipated current at which each rectifier or sacrificial anode will be operated during interference tests;
2. an indication of structure/soil potentials along the primary structure before and after the application of protection;
3. dates for the tests.

At the time of the tests, there should be available to all participants, suitably scaled plans showing the layout of the primary and secondary structures at the test locations together with the locations of the CP installations, drawn up from information supplied by both the operator and the other interested parties.

If, at any time after the CP system has been brought into regular service, it is found necessary to alter the system substantially, details of the proposed amendments should be sent to all organizations having buried metallic structures near the revised CP system.

8.11 Interference Testing

8.11.1 Stage at Which Interference Tests Should Be Made

At least 1 month should be allowed, if required, for the owners of nearby structures to examine details of the proposed system and to respond so that the operator may arrange for interference tests. Tests should be made within 3 months of switching on the CP.

8.11.2 Tests to Assess Interference

The changes in structure/electrolyte potential due to interference will vary along the length of the secondary structure and a negative potential change at any point will often indicate the presence of positive changes at other parts of the structure. For most metals, only positive potential changes are liable to accelerate corrosion. The usual object of interference testing is, therefore, to find the areas where the potential change is positive, to locate, by testing a number of positions, points at which the potential change locally reaches a maximum, and to assess each maximum value with sufficient accuracy. In the case of discontinuous structures (such as mechanically jointed pipelines), it is essential that each discontinuous section should be treated as a separate structure for testing.

Quantitative assessment of probable damage is difficult because any current discharge from a foreign structure is difficult to measure, and the surface area from which it discharges is difficult to estimate, particularly in built-up metropolitan areas where there is a multitude of underground services. The extent of foreign structure testing therefore depends on a number of factors, including the following:

1. The relative positioning of the protected structure and foreign structures.
2. Soil resistivity variations.
3. Electrical conductivity per unit length of all structures concerned.
4. Anode current.
5. Condition of coatings on all structures concerned.

Field experience and application of the above factors enables estimation of the likely degree of interference, and the extent of foreign structure testing required. The interference caused by the electrical gradient around the protected structure usually only extends for a radial distance of a few meters from the structure. However, the extent of interference caused by the electrical gradient field around the anode may extend for some hundreds of meters, with the result that foreign structures up to a kilometer or more away may be affected.

In certain cases, negative changes of potential in excess of the level that would be needed for CP may adversely affect the structure or its coating. The current used for the test should be the maximum required during normal operating conditions to give the level of protection required on the protected structure. A test current below the anticipated current required during normal operation may not bring about the maximum changes in potential on secondary structures. The criterion is the magnitude of the change of potential of the secondary structure (see note) with respect to its electrolytic environment that occurs when the CP is switched on or when the sacrificial anodes are connected. This change is usually equal in magnitude to, but has the opposite sign from, the change occurring when the protection is switched off.

Note: When, as is normally the case, the positive terminal of the meter is connected to the reference electrode, the potentials measured are usually negative and a change in the positive direction will be indicated by a reduction in the meter reading.

The change recorded should be that change clearly seen to be due to the switching on of the CP unit, not more than 15 s being allowed for the instrument to indicate the resulting change of structure electrolyte potential before the reading is taken. If there are also fluctuations of potential due to the effects of stray currents from other sources, only those changes caused by the switching on of the CP unit should be recorded. Several observations should be made at each point and compared. In marginal cases, the number of observations should be increased and examined for consistency.

This is of particular importance at positions where tests indicate that the changes in the positive direction on the secondary structure are locally at a maximum. The position of the reference electrode is often important. It is important to synchronize the measurement of structure/electrolyte potential with the switching. This can be done by providing radio, or other, communication between the individual who is making the actual measurement and the one who is controlling the CP unit, the CP being switched on and off alternately by hand. Alternatively, the CP unit can be switched on and off at agreed regular intervals by means of a suitable time switch.

The change in structure/electrolyte potential resulting from the CP should be measured at a sufficient number of points, generally working outward from the anode of the protection system, and with the spacing being sufficiently close, to give an overall picture of the distribution of structure/electrolyte potential change. Detailed attention should be given to crossing points or points of close proximity between the primary and secondary structures and to regions where the change produced has been found to be in the positive direction. Where more than one CP unit is installed on a particular structure, the combined effect should be ascertained. Arrangements should be made for all units that cause an appreciable effect at the position of tests, to be switched on and off or connected and disconnected simultaneously.

The protection current measured at each rectifier during interference tests, and the finally agreed currents to be employed as a result of any remedial measures, should be notified to all organizations attending the tests and all authorities who have indicated that they have structures likely to be affected by the operation of the system.

8.11.3 Tests After Remedial Measures Have Been Applied

Further testing may be required after agreed remedial measures have been applied. If, after providing bonds between two structures or fitting sacrificial anodes in order to reduce interference, the structure/electrolyte potential of the secondary structure is found to be appreciably more negative than that measured with the CP switched off during the initial interference testing, this will normally be sufficient indication that the mitigation procedure is achieving its purpose.

The criterion should be the change of the structure/electrolyte potential between the original condition with the CP switched off and the final remedied condition with the CP operating, switching, and bonding or anode connection being carried out quickly to minimize any effects of variations from other sources.

It may happen that the initial structure/electrolyte potential of the secondary structure is more negative than the potential of the primary structure. For example, a galvanized steel structure without applied CP, even when its structure/electrolyte potential is changed in the positive direction due to the effect of a nearby CP system, may be more negative than a cathodically protected lead or ungalvanized steel structure. Under these circumstances, an adverse effect cannot easily be offset by bonding since, owing to galvanic action between the primary and secondary structures, the structure/electrolyte potential of the latter would be made more positive, the effect being larger than any beneficial effect due to the CP.

Bonding could be made effective only by making the primary structure more negative, for example, by increasing the total protection current or moving one of the groundbeds closer to the point where it is proposed to bond. Alternatively, the secondary structure could be protected by a separate CP system, possibly by installing sacrificial anodes connected to the secondary structure. In exceptional cases it may be found possible, by special agreement between the parties, to accept structure/electrolyte potential changes on the secondary structure greater than the accepted normal limit and thereby avoid the need for remedial action.

8.12 Criteria for Limiting Corrosion Interaction

Any current flow that makes the potential of a metal surface more positive with respect to its surroundings is liable to accelerate corrosion. The structure/electrolyte potential is therefore used as the basis of assessment. Positive structure/electrolyte potential changes are more important.

8.12.1 Limit of Positive Structure/Electrolyte Potential Changes for All Structures

The maximum positive potential change at any part of a secondary structure, resulting from interference, should not exceed 20 mV. Subsequent experience has provided no indication that corrosion damage occurs when this limit is respected. However, in many circumstances, particularly if the existing conditions provide a measure of CP at the relevant part of the structure, higher potential changes could be tolerated.

Some structures are inherently more resistant to stray currents. For example, for cables having a good extruded-plastics coating, the danger of damage to the coating in making test connections for interference testing is probably a greater risk than that due to interference. Any interference associated with cable systems is likely to be manifested at the nearest earthing facility. Should the secondary structure be provided with independent CP, then the owners of the secondary structure may agree to accept a greater positive potential change on this structure, provided that its potential remains more negative than the value given in protection criteria.

8.12.2 Negative Changes of Structure/Electrolyte Potential

Large negative changes may, however, occur if the groundbed of an impressed current CP system is unduly close to a secondary structure. Structure/electrolyte potentials more negative than −2.5 V should be avoided on buried structures.

8.12.3 Control of Interference

Where testing of a CP installation indicates that there is interference at a level that may result in corrosion of the foreign structure, control of the interference may be achieved by taking the following actions:

1. Installing galvanic anodes or an impressed current system on the foreign structure.
2. Bonding the foreign structure to the primary structure through a current controlling resistance, if appropriate.
3. Insulating the foreign structure.
4. Using distributed cathode points to reduce the average potential shift on a poorly coated protected structure.
5. Relocating foreign structures away from the interfering field.
6. Reducing the CP system current.

Where the foreign structure is electrically discontinuous, as may occur on a cast iron pipeline with elastomer ring joints, some bonding of the high-resistance joints may be necessary before the above measures can be adopted.

In practice, it is sometimes found that reducing the system current can reduce the interference on foreign structures to a level acceptable to all concerned, while maintaining a satisfactory level of protection on the protected structures.

Mitigation of interference on a fixed immersed foreign structure is generally achieved by either bonding the two structures, or by installing a separate cathodic system on the foreign structure to protect it in its own right.

With a discrete movable foreign structure, such as a ship berthed at a cathodically protected steel-piled wharf, bonding is usually avoided because of the following factors:

1. A movable structure will usually have its own CP system.
2. A movable structure is usually well coated; when bonded to a large fixed immersed structure system, excessively negative potentials may result on the movable structure, causing coating disbondment.
3. A movable structure will normally only remain near the fixed immersed structure for a short period, i.e., during loading and unloading, and as a result the total interference effect is minimal.
4. The making and breaking of temporary bond connections may result in the generation of sparks or arcs, which can be hazardous at installations handling flammable materials.

8.12.4 Control by the Use of Galvanic Anodes

Interference may be controlled by installing galvanic anodes on the foreign structure, to make the potential at least as electronegative as that which existed prior to the interference.

Note: Advantages and disadvantages of this method are as follows:

1. **Advantages:**
   a. **The owner of the foreign structure has control over the anodes, and thus can be assured of their continued operation.**
   b. **Galvanic corrosion or problems arising from complex stray currents should not occur.**
2. **Disadvantages:**
   a. **If interference is great, the foreign structure is bare or the soil is of high resistivity, the limited driving voltage of the galvanic anodes may not provide sufficient protection current.**
   b. **The performance of the anodes requires to be monitored.**
   c. **Galvanic anodes sited in areas where there are steep potential gradients resulting from a foreign CP system, may accentuate the pickup of stray current, which may discharge at the remote side of the structure and cause corrosion.**

8.12.5 Control by the Use of Impressed Current CP

In special circumstances, interference can also be controlled by installing impressed current CP on the foreign structure.

Note: Advantages and disadvantages of this method are as follows:

1. **Advantages:**
   a. **The owner of the foreign structure has control over the anodes, and thus can be assured of their continued operation.**
   b. **Galvanic corrosion or problems arising from complex stray currents should not occur.**
   c. **If interference is significant, the foreign structure bare, or the soil resistivity high, the high driving voltage of impressed current CP may be required to provide adequate protection.**
   d. **In stray current areas, an impressed current CP system is less likely to accentuate stray current pickup than a galvanic system.**
2. **Disadvantages:**
   a. **Interlocks between the interfering CP system and the second CP system may be required to control adverse effects should failure of the first CP system occur.**
   b. **The interference suppression current will require to be monitored for proper operation.**
   c. **Power supplies for the transformer/rectifier may be difficult to arrange at isolated locations.**
   d. **The impressed current interference suppression system may itself cause additional interference, and it may require registration with, or approval by, the relevant authority.**

8.12.6 Control by Bonding

Interference may be controlled by bonding the foreign structure to the primary structure, or by connecting the foreign structure directly into the impressed current CP system. In the former case, an appropriate resistor is inserted in series with the bond to control the current flow to the level required to just offset interference. In the latter case, a diode and a resistor are inserted into the impressed current circuit.

Diodes are used to prevent the following problems from occurring:

1. Where the foreign structure and the primary structure are of dissimilar metals, interruption to the CP current can lead to galvanic corrosion.
2. Where structures are located in a stray current area, current flow from one structure to the other can affect the overall stray current flow.

Notes:

1. **This is particularly important where stray current drainage from more than one structure is involved.**
2. **Advantages and disadvantages of the bonding method are as follows:**
   a. **Advantages:**
      - **It is an economical solution where the structure access points are close together.**
      - **The existing bond may be capable of automatically coping with an increase in output of the CP unit.**
   b. **Disadvantages:**
      - **Where the structure is remote from the CP installation, bonding may lead to galvanic corrosion and stray current problems should failure of the CP system occur.**

- **The owner of the foreign structure does not have control of both ends of the bond.**
- **The bond requires monitoring to ensure it is not accidentally broken, cut, disconnected, or fused.**

Before bonds are installed, approval must be obtained from the affected parties.

8.12.7 Fault Conditions in Electricity Power Systems in Relation to Remedial and/or Unintentional Bonds

There is a possible risk in bonding a CP system to any metalwork associated with the earthing system of an electricity supply network, whether by intention or not. This is particularly important in the vicinity of high-voltage substations.

Bonds between metalwork associated with an electricity power system (e.g., cable sheaths) and cathodically protected structures can contribute an element of danger when abnormal conditions occur on the power network. The principal danger arises from the possibility of current flow, through the bonds, to the protected structure, due to either earth-fault conditions or out-of-balance load currents from the system neutral.

The current, together with the associated voltage rise, may result in electric shock, explosion, fire, or overheating and also risk of electrical breakdown of coatings on buried structures. Such hazards should be recognized by the parties installing the bond and any necessary precautions taken to minimize the possible consequences. The rise in temperature of conductors is proportional to i 2t, where i is the fault current and t its duration.

Conductors, joints, and terminations should be sufficiently robust, and of such construction, as to withstand, without deterioration, the highest value of i2t expected under fault conditions. For extreme conditions, duplicate bonding is recommended. Precautions should also be taken against danger arising from the high electromechanical forces that may accompany short-circuit currents.

It is difficult to ensure that current-limiting resistances comply with the foregoing requirements; their insertion in bonds through which heavy fault current might flow should therefore be avoided as far as possible. If they are used, it is essential that they be carefully designed for the expected conditions. Bonds and any associated connections should be adequately protected from damage or deterioration.

8.13 Telluric Current

Geomagnetic field variations associated with the ionospheric currents establish large-scale systems of electric currents within the earth by a process of electromagnetic induction. The global pattern of these currents flowing near the surface of the earth is known to be extremely complex due to factors such as the wide range of electrical conductivities of different strata.

The frequency of the fluctuations has been recorded to be a matter of one per several hours.

A pipeline of considerable length being positioned favorably can pick up and discharge telluric currents. If the current picked up is considerable in comparison with the total current applied for corrosion prevention (which can be the case with a very well-coated pipeline in high-resistivity soil), the effect of telluric current on such a system will become noticeable and may have to be corrected.

A number of countermeasures can be taken to combat ill effects from telluric currents. Sectioning the line by insertion of insulation flanges or joints will reduce longline current flow. Installation of discharge points by providing zinc or magnesium anodes at strategic locations will reduce the risks of corrosion at discharge of current at coating imperfections.

It is strongly recommended to obtain expert advice if a case is found suspect.

8.14 AC Effects

ACs induced in pipeline systems running parallel with power lines especially have no influence on the corrosion of the cathodically protected lines but can generate voltages that require mitigation.

It is not uncommon that pipelines and power transmission systems share a right of way. Rules and regulations exist for guidance on earthing of the power transmission system and the distances to be maintained between these and the pipeline(s) in question.

If routing the pipeline close and parallel to an overhead high-voltage system cannot be avoided, a study should be conducted by experts to determine which sections of the pipeline are influenced by a short circuit to earth and to what extent.

Some of the main distances that should be maintained are

- during the construction of a pipeline, it should be separated from the vertical projection of the nearest high-voltage line by at least 10 m for safety reasons;
- valve stations, safety releases, etc. projecting above the ground are not to be installed within 30 m;
- between the pipeline and the earthing pit of the transmission tower the minimum distance should be additional 3 m for a maximum earth-fault current of 5 kA, plus 0.5 m for every additional kA.

Special attention should be paid to the CP of pipelines and the overvoltage protection for rectifiers:

- At valve stations a steel net buried around the valve and electrically bonded to the pipelines may be required for the protection of personnel.
- It is generally advisable to discuss special requirements with the local power company or the labor inspectorate; especially during construction of a pipeline restrictive regulations may be imposed by the local power company.

Figure 8.3 shows some of the current flowing from the anode bed to the protected line collects on the foreign line, flows along it toward the crossing (from both sides), and then discharges through the soil to the protected line. Damage is inflicted on the foreign line in the neighborhood of the crossing.

Figure 8.4 shows the situation when a structure lies in a region of heavy current density, such as the tank shown close to the anode bed, it may pick up current at "A" and discharge it to earth at "B," with resultant damage at the discharge area.

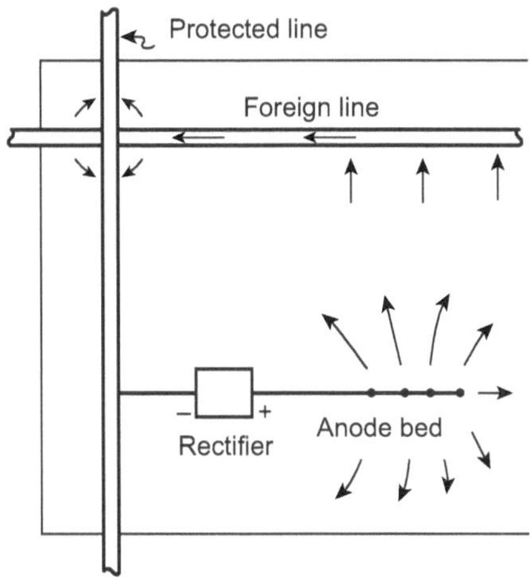

Figure 8.3 Interference.

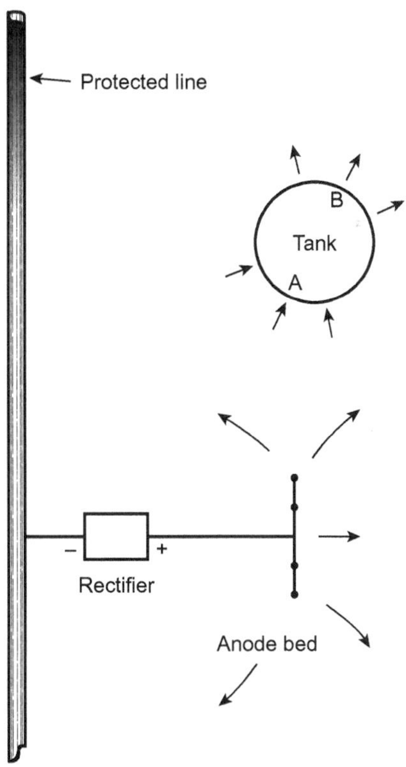

Figure 8.4 Interference (radial current flow).

Sometimes, but not often, an isolated metallic structure lying near a protected line can undergo the same kind of damage.

Figure 8.5 shows the copper sulfate electrode is placed between the two lines at the point of crossing. The resistor is then adjusted by trial and error until there is no change in the potential of the foreign line with respect to the electrode when the rectifier is turned on and off. It is helpful to measure the "short-circuit" current between the two lines first, using the "zero-resistance ammeter" circuit.

Figure 8.6 shows how to determine the current that would flow through a "solid" or zero-resistance bond between two structures, the circuit illustrated is used. The current from the battery is adjusted until the potentiometer (or high-resistance

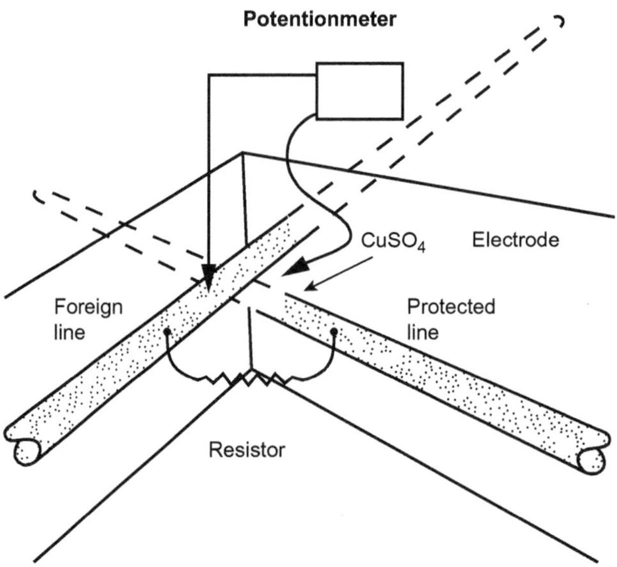

Figure 8.5 Adjustment of crossing bond.

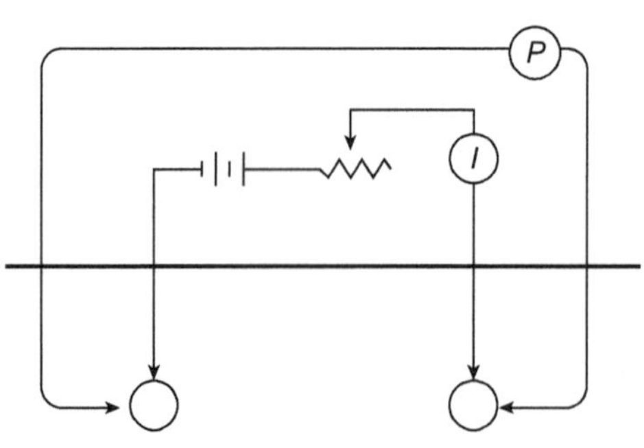

Figure 8.6 Zero-resistance ammeter.

voltmeter) reads zero. Then the current indicated by the ammeter I is the sought for value. There are instruments available that incorporate this complete circuit within themselves.

Figure 8.7 shows much of the current collected on the foreign line flows toward the crossing, where it can be safely handled with a simple bond. Some of it, however, flows in the opposite direction and is discharged over a relatively large and remote area. Such a situation does not arise often and probably does little damage in any case because of the large discharge area. It can be avoided by proper anode bed placement and remedied by the use of auxiliary drainage anodes. The best solution is the installation of one or more magnesium anodes at the point of exposure. The collected current, instead of following the bond back to the protected line, flows to earth by way of the magnesium anodes.

Figure 8.8 shows if the delta (difference between on and off readings) in the position shown in solid lines is appreciably greater than that in the dotted position, then there is current transfer from the foreign line to the protected line. When the point of

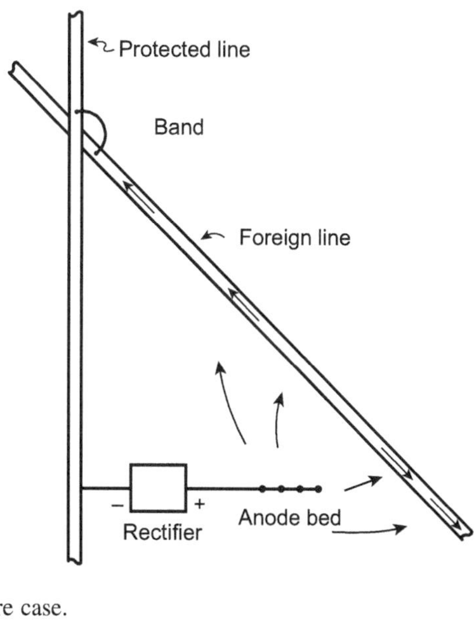

Figure 8.7 An obscure case.

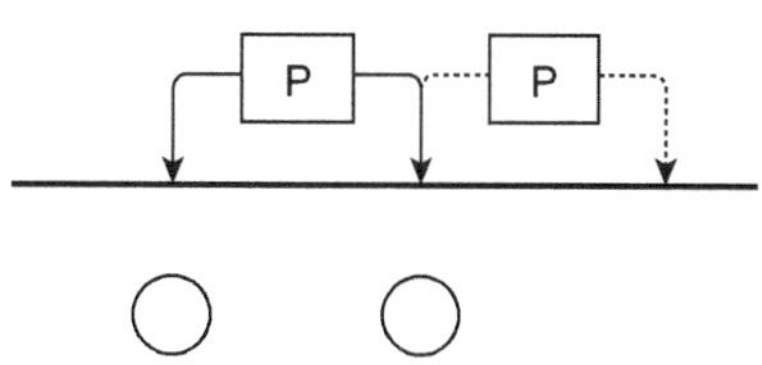

Figure 8.8 Current transfer between parallel lines.

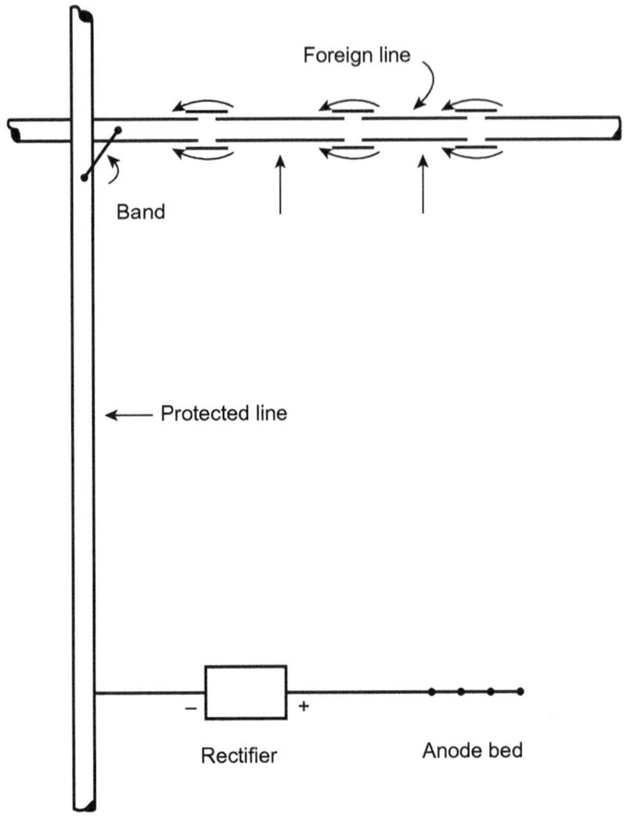

Figure 8.9 Foreign line with dresser couplings.

worst exposure is located, a bond should be installed. A repeat survey must then be made to determine the length of section that the bond will protect, and other bonds installed if required.

Figure 8.9 shows the installation of a bond at the point of crossing will avert the damage there, but there will still be damage done at the mechanical joints, by current bypassing them through the earth. This can be remedied by bonding the joints or by the use of auxiliary magnesium anode drainage.

8.15 Measurement of Soil Resistivity

There are a number of methods for measuring soil resistivity (see Fig. 8.10), the most common is the "Wenner four-pin method."

The equipment required for field resistivity measurements consists of a hand-driven earth tester (Vibroground equipment), four metal electrodes, and the necessary wiring

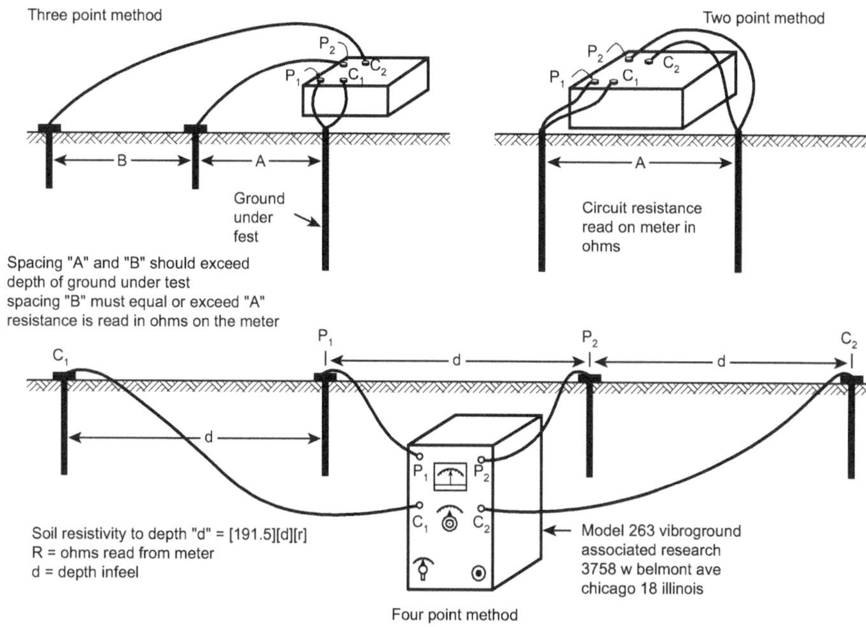

Figure 8.10 Resistivity measurement methods.

to make the connections shown in Fig. 8.10. Terminals should be of good quality to ensure that low-resistance contact is made at the electrodes and the meter.

The Wenner four-electrode method requires that four metal electrodes be placed with equal separation in a straight line in the surface of the soil to a depth not exceeding 5% of the minimum separation of the electrodes. Watering in moderation around the electrodes is permissible to ensure adequate contact with the soil. The electrode separation should be selected with consideration of the soil strata of interest. The resulting resistivity measurement represents the average resistivity of a hemisphere of soil of a radius equal to the electrode separation.

A voltage is impressed between the outer electrodes, causing current to flow, and the voltage drop between the inner electrodes is measured. Alternatively, the resistance can be measured directly. The resistivity is then

$$\rho,\ \textbf{ohm-centimeter} = \mathbf{2\pi aR}(\boldsymbol{a\ in\ cm}) = \mathbf{191.5aR}\ (\boldsymbol{a\ in\ ft}). \quad (8.2)$$

where

$\boldsymbol{a}$ = electrode separation, and
$\boldsymbol{R}$ = resistance, ohm.

Using dimensional analysis, the correct unit for resistivity is ohm-centimeter.

The depth of the electrodes should not exceed the value a/20.

8.16 Field Procedures

Select the alignment of the measurement to include uniform topography over the limits of the electrode span. Do not include large nonconductive bodies such as frozen soil, boulders, concrete foundations, etc. which are not representative of the soil of interest, in the electrode span. Conductive structures such as pipes and cables should not be within "½·a" of the electrode span unless they are at right angles to the span.

Select electrode spacings with regard to the structure of interest. Since most pipelines are installed at depths of 1.5–3 m, electrode spacings of 1.5 and 3 m are commonly used. The "a" spacing should equal the maximum depth of the interest. To facilitate field calculation of resistivities, spacing of 1.58 and 3.16 m, which result in multiplication factors of 1000 and 2000, can be used.

Impress a voltage across the outer electrodes, causing the current to flow. Measure the voltage drop across the inner electrodes and read the resistance directly and record.

Make a record of electrode spacing, resistance, date, time, air temperature, topography, drainage, and indications of contamination to facilitate subsequent interpretation.

It should be recognized that subsurface conditions can vary greatly in a short distance, particularly where other buried structures have been installed. Surface contamination tends to concentrate in existing ditches with surface runoff, appreciably lowering the resistivity below the natural level.

To evaluate contamination effects when a new route is being evaluated, soil samples can be obtained at crossings of existing pipelines, cables, etc. or by intentional sampling using soil augers.

Notes:

1. Other field resistivity measurement techniques and equipment are available. These commonly use two electrodes mounted on a prod that is inserted in the soil-at-grade in an excavation or a driven or bored hole. The two-electrode technique is inherently less accurate than the four-electrode method because of polarization effects, but useful information can be obtained concerning the characteristics of particular strata.

8.17 Frequency of Measurement

The frequency of measurement is dependent on the purpose of the soil survey. Common practice for a general pipeline route survey is to record values at an average of 1 km intervals; however, where the purpose is to locate corrosion "hot spots," measurements every 100–150 m may be necessary.

Where a variation between two successive readings is greater than the ratio 2:1, intermediate readings should be included.

Where pipelines are existing, measurements are taken to one side of the pipeline and at right angles to it at minimum distance equal to the electrode spacing "a" from the pipeline.

For general pipeline route surveys, it is common to undertake a series of readings covering soil above the pipe, soil at pipe depth, soil immediately below the pipe, i.e., 1/2D, D, and 2D where D is depth of the pipe.

For groundbed location surveys, it may be necessary to measure to greater depths and spacings of up to, say, 30 m are recorded depending on the variation of soil resistivity with depth.

In very dry/high-resistance surface soils or where large numbers of measurements or measurements to considerable depths may be required (e.g., for deep well groundbeds), resistivities may be taken using electromagnetic induction techniques.

8.17.1 Presentation of Results

Results are presented in ohm-centimeters and graphically with resistivities (log scale) versus distance (linear) for a pipeline route. Where resistivities have been taken to various depths, the layer resistivity between each depth can be calculated using the "Barnes method."

8.17.2 Criteria and Interpretation

Soil corrosivity assessment may be based on either one or a number of parameters.

BS 7361 quotes a form of assessment based on resistivity as

up to 1000 ohm-centimeter—severely corrosive
1000–10,000 ohm-centimeter—moderately corrosive
10,000 ohm-centimeter and above—slightly corrosive

8.18 Measurement of Electrode Resistance

In some cases, the voltages involved when testing earth electrodes may present a risk of shock and care should be exercised to take the necessary precautions.

Once installed, electrodes are almost certain to be connected, either deliberately or fortuitously, to other items in contact with the general mass of earth. For a new installation, it is generally possible to arrange for a measurement to be made before the electrode is so connected and is still electrically isolated. For existing installations, it is not permissible to disconnect earth electrodes unless the installation is also disconnected from all sources of power. The problem can sometimes be solved by installing multiple electrodes so that, with one disconnected for testing, the remaining electrodes provide an adequately low resistance.

The accuracy of measurement is subject to a number of features, which should be borne in mind when assessing the implications of the value obtained. Apart from seasonal variations and trends in soil resistivity, an electrode is influenced by the presence of other conducting items in the ground, such as cables, pipes, and foundations, as well as other electrodes connected together. A value obtained with existing installations, although not of great accuracy, may nevertheless provide useful information on the stability of the earthing.

For a new installation, a measurement should provide better information than a calculated value based on a measured value for the soil resistivity, because any unknown inhomogeneity in the soil is taken into account.

8.18.1 Measurement of Earth Electrode Resistance

Measurement of the resistance to earth of an earth electrode is not necessarily a simple matter. While certain fairly simple rules can be laid down, circumstances frequently arise that make it necessary to modify them. The resistance of an earth electrode is unique in that only the terminal provided by the electrode itself is definite, the other terminal of the resistance being theoretically at an infinite distance. In practice, a measurement has to be made that includes the greater part, say, 98%, of the total resistance.

There is no point in striving for a high degree of accuracy with such a measurement since, within the volume of such a resistance, there may be considerable nonuniformity in the soil and other disturbing features. An accuracy of 2% is more than adequate, and accuracies of the order of 5% are usually quite acceptable.

The best method of measurement is illustrated in Fig. 8.11. A measured current is passed between electrode X, the one being tested, and an auxiliary current electrode Y. The voltage drop between electrode X and a second auxiliary electrode Z is measured and the resistance of the electrode X is then the voltage between X and Z divided by the current flowing between X and Y. The source of current and the means of metering either the current and voltage or their ratio are often, but not necessarily, combined in one device

The accuracy of the measurement is influenced by the following considerations:

1. Distance between electrodes

The distance between electrodes X and Y has to be such that the resistance area of each, i.e., the area within which roughly 98% of its resistance lies, is independent of the other. If X is a simple rod or plate, Y should be placed 30–50 m from X, with Z

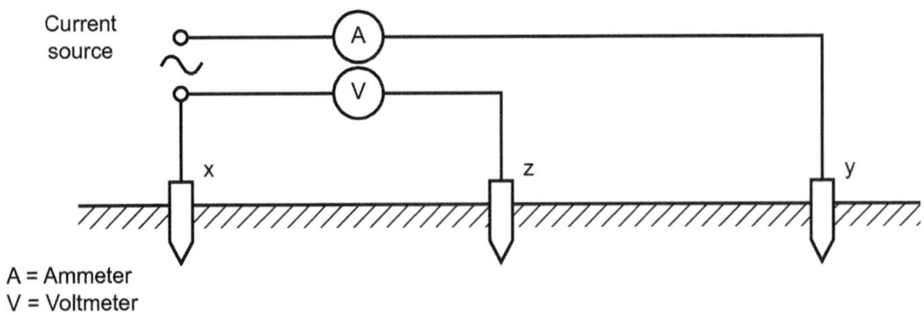

Figure 8.11 Measurement of earth electrode resistance.

about midway between. A reading should be taken, followed by two further readings with Z moved, say, 7 m nearer to X and then 7 m nearer to Y. If the three readings give values for the resistance that agree within the accuracy required, then the mean value can be assumed to be the resistance of X.

If the results do not agree, then Y should be moved further away and the procedure repeated. This whole procedure should be repeated until the three readings do agree.

The above procedure is not satisfactory when X has a low resistance, say, 1 Ω or less. This usually occurs when X is an extended electrode or is composed of a system of electrodes that cannot be measured individually, generally occupying a large area. This problem is usually solved by obtaining earth resistance curves.

To do this, Y should be placed some arbitrary distance, usually some hundreds of meters, away from X and a series of measurements made with Z at various locations along the line X–Y.

If the curve of resistance plotted against the position of Z has a substantially horizontal portion, this will give the required value of resistance. If the curve does not show such a horizontal section, Y has to be moved further away from X and the process repeated until a horizontal portion is obtained (see Fig. 8.12). The horizontal portion does not necessarily occur where Z is midway between X and Y. As an example, in a test on a power station electrode system, ultimately found to have a resistance of 0.05, it was necessary to place Y 700 m away and the horizontal section of the curve was found for potential electrode distances of 70–100 m Ω

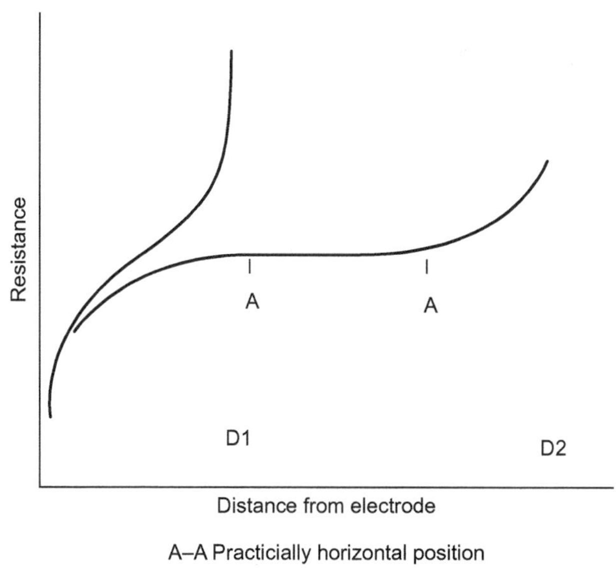

Figure 8.12 Earth resistance curves.

2. Interference by stray earth currents

Soil conduction is an electrolytic phenomenon and hence small DC potentials arise between the electrodes, and stray AC or DC potentials are picked up by the electrodes if there is a traction system in the area. Both of these forms of interference can be eliminated by testing with AC at a frequency different from that of the interfering power currents and their harmonics. This is usually achieved by using a frequency of about 60–90 Hz.

If the source of power for the measurement is a hand-driven generator, the frequency of the measuring current can be varied to obtain the best result. An AC instrument usually incorporates a synchronous rectifier or equivalent in its metering circuits so that it responds only to voltage signals of its own frequency.

3. Resistance of the auxiliary electrodes

The resistance of Y and Z are in series with the measuring and power supply circuits. These electrodes are often, for convenience, single rod electrodes that may have quite a high resistance, depending on the resistivity of the soil in which they are driven. Resistance at Y increases the power supply requirements needed to ensure an accurately measurable voltage between X and Z. Resistance at Z is in series with the voltage measuring circuit and may affect its accuracy. Information on the highest acceptable values of auxiliary electrode resistance are usually provided with instruments designed for earth resistance measurement.

8.18.2 Measurement of Resistance of Earthing Conductor

The following types of instrument may be used to measure the resistance of an earthing conductor:

1. Direct reading DC ohmmeter incorporating a hand-driven generator,
2. Direct reading ohmmeter supplied from a battery, or
3. AC test set generally mains driven and incorporating a suitable transformer providing isolation.

Tests that give no quantitative result (e.g., bell or lamp tests) should never be used to prove the adequacy of earthing conductors.

Of the above instruments, the first two differ only in the magnitude of the test current and they measure only the resistance of the conductor. AC mains testers can provide high test currents, but are usually limited to about 25 A because of the weight of the transformer.

An accuracy of about 5% in the measured value is desirable. To achieve such an accuracy with low-resistance conductors, correct selection of the method of measurement is important and the instrument manufacturer's instructions should be consulted to confirm the conditions under which it can be achieved.

The measured impedance or resistance, except for four-terminal measurements, is usually for a loop consisting of the conductor under test, the return conductor, and probably some test leads. The resistance of the conductor under test is obtained by subtracting a separately measured value for the return conductor and leads from the loop resistance. In order that this process does not introduce too great an error, the resistance of the return conductor and leads should be as low as practicable. For a similar reason, all connections should be made so as to have a low resistance.

The impedance of a ferromagnetic conductor varies with the current; with the sizes of conductor likely to be involved, the highest impedance generally occurs at currents in the region of 25–50 A. Measurement with such a value of current will provide a worst case value, since the magnetic effect decreases as the current increases to fault current values. If measurements are made with DC, and a substantial part of the length of the conductor is of ferrous material, it is recommended that the measured value be doubled to take account of magnetic effects.

8.19 Current Drainage Survey

The current drainage survey is a technique for determining the amount of current required to provide cathodic protection to a buried pipeline and the spread of protection from a single point or the current required to boost CP on a line with deteriorating coating.

To obtain relevant results, pipeline isolation equipment and monitoring facilities (test points) should be installed before current drainage tests are carried out.

If the pipeline has been previously cathodically protected, historical data may be used to determine the current demand.

8.19.1 Method

A temporary CP system should be set up using DC power sources (e.g., batteries, portable rectifiers, etc.) timer units, test points, and one (or more) groundbed(s) in a DC circuit.

After full polarization is achieved, the output should be adjusted to provide the required drain point potential, and then maximizing the output.

Then, the current should be interrupted in a sequence of, e.g., 40 s "on," 20 s "off."

The spread can be measured by taking structure-to-soil potentials away from the drain point until loss of protection occurs.

The location with the least "swing" determines the required minimum output of the final installation in such a way that the minimum acceptable swing is 300 mV negative once the final installation is activated. The ratio swing observed/swing required equals test current applied/final minimum current required.

Notes:

1. The current requirements for coated pipelines is often estimated.
2. Current-drainage tests, although technically reliable, are generally, confined to investigations for complicated structures.
3. This technique may also be used to determine the average coating resistance of a pipeline coating.

An alternative method for determining the current required for CP is the measurement of the structure-to-electrolyte potential with stepped increase of impressed current. A graph is made showing the structure-to-electrolyte potential against the logarithm of the impressed current. The relationship is a straight line with a slight inclination at low currents; after a break point, the curve continues as a straight line with a sharper rise at higher currents.

The break point indicates the current required to provide CP, see Fig. 8.13. This method is often the only feasible one when CP is being considered for structures not fully accessible, e.g., oil well casings. As the measurements are taken in a short period, full polarization does not occur and the structure-to-water or structure-to-soil potential at the break point is not a measure of the potential required to provide protection.

The location of the half-cell in such an experiment is rather critical. A remote location should be found by moving the half-cell further away during which no

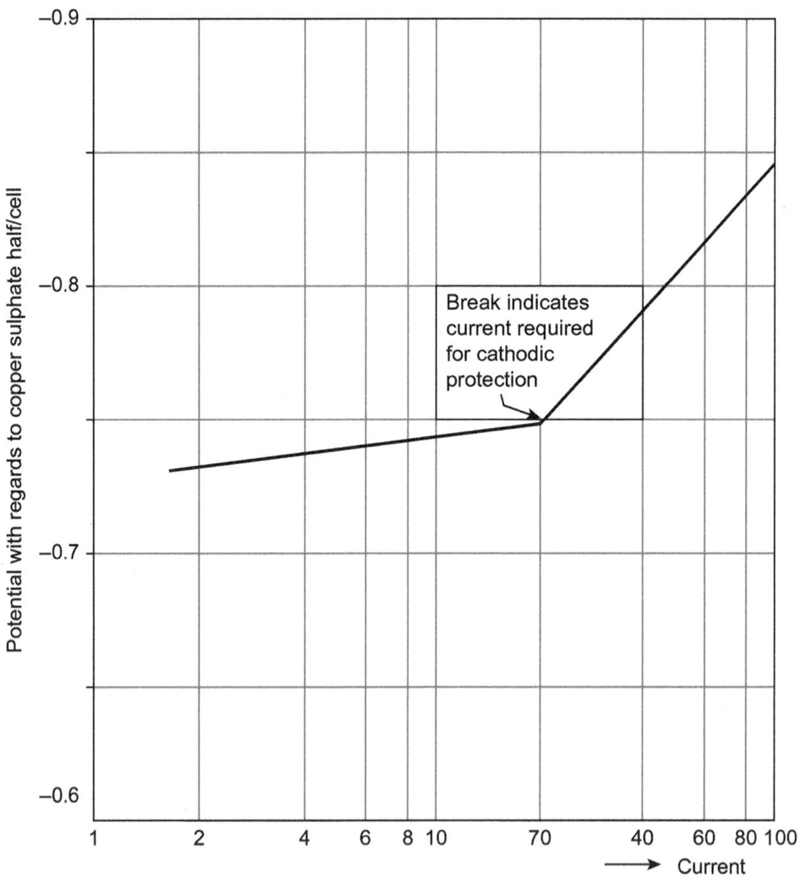

Figure 8.13 Relationship between potential and impressed current.

significant change in potential is observed. This then is the remote potential. The potential after each current step increase should be measured while interrupting the current for approximately 1 s and reading the instant "off" potential. The time over which each current step increase is being applied should be a constant (e.g., 1 or 2 min).

8.20 Determination In situ of the Redox Potential of Soil

This method covers the determination of the redox potential (reduction/oxidation) of soil tested in situ at a selected depth by measuring the electrochemical potential between a platinum electrode and a saturated calomel reference electrode. The test is used to indicate the likelihood of microbial corrosion of metals by sulfate-reducing bacteria, which can proliferate in anaerobic conditions. The redox potential is principally related to the oxygen in the soil, and a high value indicates that a relatively large amount is present. Anaerobic microbial corrosion can occur if a soil has a low oxygen content and hence a low redox potential. This standard requires the use of a calomel reference probe as defined in E.2.2. This is not intended to prohibit the use of the other established portable versions of reference probes, e.g., copper/copper sulfate and silver/silver chloride (see note). In submitting reports, the type of reference probe used and the correction factor applied to convert the measurement to the standard hydrogen electrode should be given. Redox potential may also be measured in the laboratory.

Note: Where the other types of reference probes, e.g., copper/copper sulfate and silver/silver chloride, are used, it is very important to note that their preparation and storage procedures are different from that required for calomel probes and the manufacturer's instructions should be followed. Moreover, copper/copper sulfate probes are not suitable in chloride-contaminated soil or in alkaline environments when silver/silver chloride should be used. Copper/copper sulfate cells are also sensitive to heat, light, and a wide variety of chemicals.

The correction factors for reference probes to convert the measurement to the standard hydrogen electrode are as follows:

- Mercury/mercuric chloride 240 mV at 25 °C
- Copper/copper sulfate 316 mV at 25 °C
- Silver/silver chloride 246 mV at 25 °C

8.20.1 Apparatus

Platinum probe of a design having two separate platinum electrodes embedded in the nose piece. Also a means of protection when not in use. The probe should have a connecting lead permitting the inclusion of each platinum electrode individually in an electrical circuit. Each connection should be separately identified.

Calomel reference probe, having a mercury/mercuric chloride reference electrode, which can be refilled and with a connection to a porous ceramic junction. The calomel reference electrode should be kept clean when not in use by being stored in a sealed container. The precipitation of crystals should be prevented when not in use, particularly at the porous junction, by storing upright and closing the breather hole.

Calibrated millivoltmeter. The total measuring range should be at least 0–2 V DC with a readability at least to 10 mV. The input impedance should be not less than 106 and the polarity (positive or negative) should be marked on the two input terminals.

The instrument should include suitable insulated flexible electric cable and connectors for use with the probes.

The instrument should be recalibrated at intervals not exceeding 2 years.

Installation equipment consists of a soil auger, spade, and trowel to excavate soil to test level, and, where soil is compact, a 1 kg hammer and spike. **pH** measuring apparatus.

Disturbed sample container of glass or dense plastic, that can be hermetically sealed.

Note: When a sample from the test location is required for microbiological examination, a glass container of a size suitable to hold about 500 mL will need to have been cleaned and sterilized by scalding with boiling water beforehand. Alternatively, medically sterilized plastic bags may be used. Fill the container completely and minimize air voids.

8.20.2 Materials

- Saturated solution of potassium chloride in a screw-topped plastic bottle with either pouring lip suitable for filling the reservoir of the calomel reference probe or a separate small dropper or syringe; 500 mL is a suitable quantity.
- Jeweler's rouge.
- Colorless methylated spirits, 70% by volume with 30% by volume distilled water, in a screw-topped wide-mouth bottle; 500 mL is a suitable quantity.
- Distilled water. Two differently marked wash bottles full for cleaning platinum electrodes. About 500 mL is a suitable quantity for each bottle.
- Paper tissues and absorbent-type surgical cotton wool swabs.

8.20.3 Procedure

Assemble from the storage unit according to the manufacturer's operating instructions the calomel reference probe, ensuring that the unit is full of a saturated solution of potassium chloride and that this moistens the porous junction. Remove any air bubbles in the potassium chloride solution by gently tapping the probe and remove excess fluid from the porous junction.

Note: During use on-site, it is important to prevent precipitation of crystals at the porous junction. This may be done by keeping the probe between tests in distilled water in a wide-necked bottle with a rubber bung to ensure that the porous junction is kept moist.

Clean and polish each platinum electrode. Initially smear the surfaces lightly with moist jeweler's rouge and use gentle abrasive action with cotton wool swabs. Follow with a single wash using the methylated spirits. Afterward, wash thoroughly with distilled water. Lastly, dry each electrode with clean paper tissues.

Note: Where the platinum electrode is dipped into distilled water for washing the bottles will need separate identification to select the correct sequence when reused.

Connect the positive terminal of the millivoltmeter with the electric cable to one of the platinum electrodes and the negative terminal to the calomel reference electrode, but leave the circuit open. This circuit should be considered to give positive readings.

Test should always be made below the level of organic growth. A hole not less than 150 mm in diameter is needed to reach the selected level when using separate probes. A combined redox probe may be driven from the surface to the selected level in weak soil, otherwise it may be necessary to auger or dig a hole part way.

If the probes are separate, install them about 100 mm apart in the hole. The platinum probe should penetrate at least 100 mm to ensure full soil contact below any disturbed surface material. A combined redox probe should be pressed into position sufficiently to obtain full soil contact on the electrodes.

Rotate the platinum probe about a quarter turn without letting air reach the probe. Close the electric circuit; then take the reading as soon as the voltage becomes stable. It may be necessary to wait 30 s or more for stable conditions to be reached. Where the probes are separate, turn the platinum probe one revolution under firm hand pressure to ensure good contact. Rotate the combined redox probe a half to one revolution.

Record the reading to the nearest 10 mV when the voltage is steady and record whether it is positive or negative.

Note: Very occasionally the current between the platinum electrode and the reference electrode will be in the reverse direction such as to require the connections to the millivoltmeter to be reversed. In this case, the reading should be considered to be negative.

- Transfer the electric circuit to the other platinum electrode, connecting it again to the positive terminal of the millivoltmeter, and repeat the procedure
- Record the reading to the nearest 10 mV and its polarity.
- If the two readings differ by more than 20 mV, remove the probes, reclean the platinum electrodes and reinstall in a different position at the test site. Do not install the probes in the original position because oxygen will have penetrated and a false reading could result. Remove the probes and clean the electrodes.
- Place a disturbed sample from the position of the test in an hermetically sealed container.
- Determine the **pH** of the sample by the method specified in Clause 9 of BS 1377: 1990 or ASTM G 51.

8.20.4 Calculations and Expression of the Results

The mean of the two acceptable readings and their sign should be recorded as the potential of the platinum probe, **Ep**, to the nearest 10 mV. Calculate the redox potential, **Eh** (in millivolts), to the nearest 10 mV from the equation:

$$\mathbf{Eh = Ep + 250 + 60\ (pH - 7)} \tag{8.3}$$

where

Ep is the potential of the platinum probe (in millivolts) (may be a positive or negative value);
pH is the value of the acidity of an aqueous solution of the soil at the test position;
250 is the correction factor for a calomel reference probe to convert the measurement to the standard hydrogen electrode.

8.20.5 Test Report

The test report should affirm that the test was carried out in accordance with this standard and should contain the following information:

1. The method of test used.
2. The mean value of the potential (in millivolts) of the two platinum probes to the nearest 10 mV.
3. The redox potential (in millivolts) to the nearest 10 mV.
4. The **pH** value.
5. The type of reference probe used in the test.

8.21 Inspection of CP Installations

Adequate inspection during the construction of corrosion control facilities on pipelines can make the difference between first-class performance (to be had if inspection is truly effective) and a system that may perform poorly and that may require relatively high maintenance expenditures if ineffective or no inspection is used.

8.21.1 CP Installations

Inspectors responsible for CP installations must be fully familiar with all details of good CP construction practice as well as with the specific provisions of the installations being made.

In some instances, field modifications may be necessary. This may occur, for example, if vertical anode installations were specified but rock is closer to the surface than expected. A field decision is then necessary to determine whether the type of anode installation may be changed from vertical to horizontal or if best results will be obtained by boring the rock. Occasionally, what appears to be solid rock is actually a relatively thin layer with good soil underneath. The inspector must be qualified to evaluate all such situations when encountered. Where major modifications appear advisable, he or she should check with the designer of the installation to be sure that the system performance will not be affected adversely.

At galvanic anode installations, one of the more particular points to be watched is the anode backfilling operation to be sure that there are no voids in the fill around the anodes. This can be a problem with packaged galvanic anodes placed vertically in augured holes. If the hole is small, it may be difficult to work earth backfill all around the anode package so that no voids backfill material can settle away from the anode, once the container has deteriorated, with probable reduction in anode effectiveness. Where anodes and chemical backfill are installed separately, the inspector must verify that the anode is centered in the hole or trench as specified and that the fill is so placed and compacted that no voids can exist.

All other details of galvanic anode installations must be verified by the inspector as being in accordance with design specification for such construction.

When inspecting impressed current groundbeds, anode placement and backfilling operations must be given careful attention to ensure installation at the design location and to avoid voids in special backfill (carbonaceous), which would tend to increase anode resistance and shorten life. Adequate compacting of carbonaceous backfill materials is important and the inspector must verify that this is done effectively but in a way that will not damage the anode proper or its connecting cable.

The most important single feature of impressed current groundbeds to be verified by the inspector is the insulation on all positive header cable, anode connecting cable, and connections between the two. The inspector must be sure that no damaged cable insulation is backfilled without being repaired and that the insulation of all splices and tap connections is such that they will be permanently watertight and such that no current leakage can occur. Likewise, the inspector must be sure that the cable trench has been so prepared that there are no materials in the trench bottom that can damage the cable insulation and that the backfill in contact with the cable is free of insulation damaging material.

When rectifier installations are involved, the inspector verifies that the rectifier unit has been placed at the specified location and in the specified manner, that all wiring is correct and that requirements of the serving power company and of local jurisdictional codes have been met. Grounding connections should be checked.

Where power supplies other than rectifiers are used, the inspector must be familiar with the details of the specific power source specified in order that he or she may verify that the installation is being made properly.

8.21.2 Test Points, Cased Crossings and Insulating Joints

Construction of test points must be inspected to ensure their installation where called for on construction plans, that connections to the pipeline are sound and well insulated, that color coding is observed, that wires are so placed in the ditch that backfill will not break them, and that the specified terminal panel and housing is installed properly.

Insofar as corrosion control is concerned, cased crossings should be inspected as soon as they are installed. This is necessary to ascertain that the casing is electrically insulated from the carrier pipe and to permit correction of defects before backfilling.

Even though insulating joints may be installed as complete preassembled pretested units, inspection should include verification of installation at the locations indicated on the plans, that the joint is not so placed that it is subject to undue mechanical strains that could cause early failure of insulating material, and that the joint is in fact serving its insulating function once welded into the line.

The inspector charged with inspecting the above features must be thoroughly familiar with the use of test points, cased crossings, and insulating joints. The inspector must be equipped with the necessary instrumentation to verify electrically the satisfactory performance of the various items prior to acceptance. He or she must have the authority to see that corrections are made should defects be found.

8.21.3 Coating Inspection

Pipeline coating inspection, rigorously applied, will result in the best practicable coating performance from the standpoint of maximum effective electrical resistance and maximum stability with time, which is what the money invested in coating is spent for. If pipeline is to be mill coated, inspection starts at the coating plant with close attention given to all phases of pipe cleaning, priming, coating, holiday testing, yard stacking, and loading out for shipment by rail or truck. Additionally, material used are inspected for compliance with the coating specification and the manner in which they are stored and handled is checked for any possible adverse effect on applied coating quality. In the case of hot-applied enamels, the heating kettles are checked for proper temperature range and suitable charging and cleaning cycles. Where other coating systems are used, the application system is checked for continued satisfactory performance under conditions accepted as good practice for that type of coating system. Applied coating thickness is checked for compliance with the specification.

At the job site, inspection practices on mill-coated pipe cover the unloading, hauling, stringing, joint coating, holiday testing, damage repair, lowering in, and backfilling operations.

When coating application is over the ditch, the coating inspector will pay close attention to the pipe cleaning and priming machines to be assured that the prime coat will permit the best practicable bond with hot enamel or other types of field-applied coatings. Where hot-applied enamels are used, inspection of the critical dope kettle cleaning, charging, and heating operations must be thorough. Materials handling techniques are inspected for assurance that materials are kept free of dirt and that wrappers particularly are kept dry. Coating machines are checked for specification coating thickness and smooth application of wrappers under correct tension with correct overlap. As with the mill-coated pipe, holiday testing, damage repair, lowering in, and backfilling operations are checked for adequacy.

With the varied items to be covered and the critical nature of this inspection, it becomes obvious that coating inspection responsibilities should be assigned only to personnel who are fully qualified by experience and training to do the job. Further, their responsibilities should be confined to this one inspection operation and they should have full backing by the owner to obtain immediate and positive correction in the event unsuitable practices are used in any phase of the coating operation. They must have fully adequate specifications covering the entire coating operation.

8.22 Installation in Hazardous Atmospheres

Flammable mixtures of gas, vapor, or dust may develop wherever hydrocarbons or finely divided materials are handled, stored, or processed. Such hazardous atmospheres may be ignited by an electric arc or spark. Unless proper precautions are taken, all electrical installations, including CP systems, will introduce the danger of sparks and ignition.

CP system that are to operate where flammable concentrations of gas or vapor may occur should conform to the statutory and other safety regulations applicable to the particular structure and industry concerned, and approval should be obtained in each individual case as appropriate. Some of the explosion hazards CP may cause and the measures needed to avoid them are described in G.2 to G.10.

8.22.1 Bonds

Intentional or unintentional disconnection of bonds across pipeline joints or any other associated equipment under protection or fortuitously bonded to protected equipment constitutes a hazard.

To avoid the hazard, bonds should be installed outside the hazardous area or in a protected position to avoid an unintentional break. The CP supply should be switched off or disconnected by means of a flameproof switch.

8.22.2 Isolating Joints

Intentional or unintentional short circuit of isolating joints, e.g., by tools, or breakdown due to voltage surges on the protected structure induced by lightning or electrical power faults constitute a hazard.

To avoid the hazard, isolating joints should, if possible, be located outside the hazardous area. Where this is not practicable, measures should be adopted to avoid arcing or sparking. These will include the use of resistance bonds fitted in a flameproof enclosure or located outside the hazardous area, an encapsulated spark gap or surge diverter, zinc earth electrodes connected to each side of the isolating joint, or polarization cell connected across the isolating joint, or to earth. The surfaces of the isolating joint should be insulated to prevent fortuitous short circuiting by tools.

8.22.3 Short Circuits between Points of Different Potentials

Unintentional short circuits by fortuitous bridging of points of different potential, e.g., by metal scraps, odd lengths of wire, mobile plant, constitute a hazard.

This hazard is difficult to foresee but may be limited by bonding all metalwork together to minimize the potential difference between different parts of the structure.

8.22.4 Disconnection, Separation, or Breaking of Protected Pipework

Cathodically protected pipework will have a portion of the protection current flowing through it. Any intentional disconnection, separation, or breaking of the pipework will interrupt the current flow and may produce arcing depending on the magnitude of the current.

During any modification, maintenance, or repair of cathodically protected pipework, transformer–rectifiers that affect that section of pipework should be switched off and a temporary continuity bond attached across any intended break. It is essential that the bond is securely clamped to each side of the intended break and remains connected until the work is completed and electrical continuity restored or until the area is certified as gas-free or nonhazardous.

8.22.5 Electrical Equipment

All electrical equipment installed in a hazardous area should be flameproof and certified for use in that area. To avoid the need for flameproof equipment, the equipment should be located outside the hazardous area. Double pole switches should be provided in each circuit entering a hazardous area to ensure that both poles are isolated during maintenance etc. It is essential that any cables carrying CP current should be installed in such a manner that disconnection cannot take place within the hazardous area without de-energizing the CP system. The cables also should be adequately protected mechanically to prevent accidental breakage.

8.22.6 Test Instruments

The connection and disconnection of instruments used for measuring and testing CP systems may produce arcing or sparking. Where measurements are taken within the hazardous areas, the meter used should be intrinsically safe. The test leads should be connected to the structure before being connected to the meter. Alternatively, the area should be tested and declared gas-free, allowing conventional instruments to be used. Consideration may also be given to the use of permanently installed reference electrodes and test leads with the cables taken outside the hazardous area where conventional instruments can be used.

8.22.7 Internal Anodes

Unintentional short circuiting of impressed current anodes when the liquid level is lowered in plant under internal CP can constitute a hazard.

Arrangements should be made to ensure that the circuit is automatically or manually isolated when the anode is not submerged, i.e., when the anode circuit becomes an open circuit.

8.22.8 Sacrificial Anodes

While the normal operation of sacrificial anode is not considered hazardous, there is a danger of incendive sparking if a suspended or supported sacrificial magnesium or aluminum anode or portion of anode becomes detached and falls onto steel. This risk is not present with zinc anodes.

8.22.9 Instruction of Personnel

In locations where any of the above hazards may occur, it is essential that operating personnel be suitably instructed, and durable warning notices should be authoritatively displayed as appropriate. Suitable written procedures and work authorization permits should be included in the operations manual.

Additional List of Reading on Retrofitting

Al Katheeri, A.B., Kenawy, M.M., Metwali, M.H., Bazuhair, M., 2013. A successful deployment of LWD technique for water shut-off application, an integrated teamwork approach managed to activate horizontal wells economically. In: 18th Middle East Oil and Gas Show and Conference 2013: Transforming the Energy Future, MEOS 2013; Manama; Bahrain; March 10, 2013 through March 13, 2013; Code 98612.

Khanna, A.S., November 28, 2011. Source of the Document Green Corrosion Chemistry and Engineering: Opportunities and Challenges, pp. 339–354.

Lamont, L.A., El Chaar, L., April 2011. Enhancement of a stand-alone photovoltaic system's performance: reduction of soft and hard shading. Renewable Energy 36 (4), 1306–1310.

Peratta, C., Peratta, A., Baynham, J., Adey, R., 2009. Computer modeling of impressed current cathodic protection (ICCP) system anodes. In: ASME 2009 International Design Engineering Technical Conferences and Computers and Information in Engineering Conference, IDETC/CIE2009; San Diego, CA; United States; August 30, 2009 through September 2, 2009; Code 87488.

Tian, Z.-Q., Li, Y.-J., Wang, Z.-F., August 2005. Some problems in catholic protection of underground liquid gas tanks. Corros. Prot. 26 (8), 361–363.

Zhang, F., Shang, M., 2011. The application of SCADA system in an old crude trunk line sustainable solutions for water, sewer, gas, and oil pipelines. In: Proceedings of the International Conference on Pipelines and Trenchless Technology 2011: Sustainable Solutions for Water, Sewer, Gas, and Oil Pipelines, ICPTT 2011; Beijing; China; October 26, 2011 through October 29, 2011; Code 87775.

BSI (British Standard Institute)

BS 4683, 1971. The Construction and Testing of Flameproof Enclosures of, Electrical Apparatus. Part 2.

BS 6651, 1985. The Protection of Structures against Lightning.

BS 7361, 1991. Cathodic Protection.

BS 7430, 1991. Code of Practice for Earthing.

CP 1003. Electrical Apparatus and Associated Equipment for Use in Explosive Atmospheres of Gas or Vapor other than Mining Applications.

DIN (Deutsches Institut Für Normung) DIN 30676 "Design and Application of Cathodic Protection of External Surfaces".

BSI (British Standards Institution)

BS 1377. Methods of Tests for Soil for Civil Engineering Purposes.

BS 6651. The Protection of Structures against Lightning.

BS 7361. Cathodic Protection.

Part 1: Code of Practice for land and Marine Applications BS 7430. Code of Practice for Earthing.

Part 3: Chemical and Electrochemical Tests.

Part 9: In-Situ Tests BS 6651. The Protection of Structures against Lightning.

ASTM (American Society for Testing and Materials)

G 51. Standard Test Method for PH of Soil for Use in Corrosion Testing.

BSI (British Standard Institute)

BS 148. Specification for Unused and Reclaimed Mineral Insulating Oils for Transformers and Switchgear.

BS CP 1003. Electrical Apparatus and Associated Equipment for Use in Explosive Atmospheres of Gas or Vapor other than Mining Application.

BS 7361 Part I. Code of Practice for Land and Marine Applications.

AS (Australian Standard)

AS 2832.1, 1985. Guide to the Cathodic Protection of Metals, Pipes, Cables and Ducts.

AS 2832.2, 1991. Guide to the Cathodic Protection of Compact Buried Structures.

AS 2832.3, 1992. Guide to the Cathodic Protection of Fixed Immersed Structures.

AS 2832.4, 1994. Guide to the Cathodic Protection of Internal Surfaces.

DIN (Deutsches Institut Fur Normung E V)

DIN 30676, 1985. Design and Application of Cathodic Protection of External Surfaces.

DNV Det Norske Veritas-Offshore Standards.

DNV RP B 401, 1976. Rules for the Design, Construction and Inspection of Submarine Pipelines and Pipeline Risers.

DNV RP B 403, March 1987. Monitoring of Cathodic Protection Systems.

DNV RP B 404, 1977. Rules for the Design, Construction and Inspection of Offshore Structures.

DNV Technical Note A5/1Rev.1. Cathodic Protection Evaluation.

NACE (National Association of Corrosion Engineers)

NACE R.P 01–76. Control of Corrosion on Steel Fixed Offshore Platforms Associated with Petroleum Production.

NACE Task Group T.7L.4. Collected Papers on Cathodic Protection Current Distribution.

NACE Task Group T.10.1. Cathodic Protection Criteria – A Literature Survey.

NACE R.P 0169–2002. Control of External Corrosion on Underground or Submerged Metallic Piping Systems.

NACE R.P 0187–96. Recommended Practice-Design Consideration for Corrosion Control of Reinforced Steel in Concrete.

ASTM (American Standards for Testing and Materials)

ASTM C 876–99. Half Cell Potentials of Uncoated Reinforcing Steel in Concrete.

ASTM G 3–89. Standard practice for Conventions Applicable to Electrochemical Measurements in Corrosion Testing. ASTM C-9.

ASTM (American Society for Testing and Materials)

A 256 Standard Method of Compression Testing of Cast Iron.
A 327 Standard Method of Impact Testing of Cast Iron.
A 518M Standard Specification for "Corrosion-Resistant High-Silicon Iron Casting".
B 8 Standard Specification for Concentric-Lay-Stranded Copper Conductors, Hard, Medium-Hard, or Soft.
B 193 Standard Test Method for Resistivity of Electrical Conductor Materials.
B 539 Standard Test Method for Measuring Contact Resistance of Electrical Connections (Static Contact).
D 1047 Standard Specification for Poly (Vinyl Chloride) Jacket for Wire and Cable.
D 1248 Standard Specification for Polyethylene Plastics Molding and Extrusion Materials.
D 2308 Standard Specification for Polyethylene Jacket for Electrical Insulated Wire and Cable.
D 2655 Standard Specification for Crosslinked Polyethylene Insulation for Wire and Cable Rated 0 to 2000 V.
D 3032 Standard Test Methods for Hookup Wire Insulation.
E 10 Standard Test Method for Brinell Hardness of Metallic Materials.
E 94 Standard Guide for Radiographic Testing.
E 802 Standard Reference Radiographs for Gray Iron Casting up to 114 mm ($4^1/_2$ in.) in Thickness.

IEC (International Electrotechnical Commission)

228 Conductors of Insulated Cables.

ISO (International Organization for Standardization)

2859 Sampling Procedures for Inspection by Attributes.

ASTM (American Society for Testing and Materials) ASTM Method "Water Separation Characteristics of Aviation Turbine Fuel. ASTM Standard 2250-66T, Part 17.P. 965" ASTM Standard GS-87; ASTM Standard D-1660.

Aalund, L.R., 1992. Polypropylene system scores high as pipeline anti-corrosion coating. Oil Gas J. 90 (50), 42–45.

Ahmad, Z., 1996. Corrosion phenomena in coastal area of Gulf. Br. Corros. J. 31 (3), 191–197.

Al-Sulaiman, S., Al-Mithin, A.W., Murray, G., Biedermann, A.J., Islam, M., 2008. Advantages and Limitations of Using Field Test Kits for Determining Bacterial Proliferation in Oil Field Waters. NACE – International Corrosion Conference Series, pp. 086551–0865518.

Al-Taq, A.A., Ali, S.A., Nasr-El-Din, H.A., 2009. Inhibition performance of a new series of mono-diamine-based corrosion inhibitors for HCl solutions. SPE J. 14 (4), 627–633.

Altoé, P., Pimenta, G., Moulin, C.F., Díaz, S.L., Mattos, O.R., 1996. Evaluation of oilfield corrosion inhibitors in CO2 containing media: a kinetic study. Electrochim. Acta 41 (7–8), 1165–1172.

Altsybeeva, A.I., Burlov, V.V., Kuzinova, T.M., 2003. Applying corrosion inhibitors to metal surfaces in electrostatic field. Zashch. Met. 39 (2), 217–218.

Amaya, H., Ueda, M., Sagara, M., Noda, K., 2007. Effect of Variation of Local PH on Stress Corrosion Cracking of Stainless Steels in High Concentration Brines. NACE – International Corrosion Conference Series, pp. 070961–0709615.

Andersen, T., Misund, A., April 1983. Pipeline reliability: an investigation of pipeline failure characteristics and analysis of pipeline failure rates for submarine and cross-country pipelines. J. Pet. Technol., 709.

API 941 API (American Petroleum Institute) API 941:2004. Steal for Hydrogen Service at Elevated temperature and Pressure.

API (American Petroleum Institute) API RP 14E. Recommended Practice for Design and Installation of Offshore Production Platform Piping systems–API Publication 941. Recommended Practice for Calculation of Heater Tube Thickness in Petroleum Refinery.

API (American Petroleum Institute) API-RP-38. Recommended Particle for Biological Analysis of Subsurface Injection Waters.

Archer, D., Brovkin, V., 2008. The millennial atmospheric lifetime of anthropogenic CO2. Clim. Change 90, 283–297.

Archer, D., Kheshgi, H., Maier-Reimer, E., 1998. Dynamics of fossil fuel CO2 neutralization by marine CaCO3. Global Biogeochem. Cycles 12, 259–276.

ASME (American Society of Mechanical Engineers) ASME VIII. Boiler and Pressure Vessel Codes.

ASTM (American Society for Testing and Materials) ASTM A353/A353M, 2004. Pressure Vessel Plates, Alloy Steel 9% Nickel, Double Normalized and Tempered. ASTM D664: 2004 "Standard Test Method for Acid No. of Petroleum Products (5.3.4) by Potentiometric Titration".

ASTM (American Society for Testing and Materials) ASTM Corrosion Standard Test Method: G1 to G52 B 117 "Salt Spray (Fog) Testing" B 287 "Acetic Acid-Salt Spray (Fog) Testing".

Atamert, S., Scandella, J.-L., King, J.E., 1995. Corrosion mechanisms of duplex stainless steels in the petrochemical industry. Pipes Pipelines Int. 40 (2), 29–37.

BS 5903. Corrosion Testing of Austenitic Stainless Steel.

Beilin, Yu.A., Nisel'Son, L.A., Begishev, I.R., Filimonov, L.I., Shishkanov, B.A., Ashcheulova, I.I., Podobaev, A.N., Reformatskaya, I.I., 2007. Corrosion pyrophoric deposits as promoters of self-ignition of storage tanks with sour crude oil. Prot. Met. 43 (3), 269–274.

Bhaskaran, R., Palaniswamy, N., Rengaswamy, N.S., Jayachandran, M., Srinivasan, P., 2004. Corrosion auditing in a petrochemical complex. Mater. Perform. 43 (12), 42–44.

Bondos, J.C., Reddy, R.V., Pugh, D.V., Norman, D.A., Pacheco, J.L., Nelson, J.L., 2007. Accurate corrosion prediction through an integrated approach. SPE Prod. Oper. 22 (2), 161–167.

Breton, T., Sanchez-Gheno, J.C., Alamilla, J.L., Alvarez-Ramirez, J., 2010. Identification of failure type in corroded pipelines: a Bayesian probabilistic approach. J. Hazard. Mater. 179 (1–3), 628–634.

Britton, J., April 26–30, 1992. Continuous Surveys of Cathodic Protection System performance on Buried Pipelines in the Gulf of Mexico. Paper no. 422 presented at CORROSION/92, Nashville.

Broecker, W.S., Takahashi, T., 1978. Neutralization of fossil fuel CO2 by marine calcium carbonate. In: Andersen, N.R., Malahoff, A. (Eds.), The Fate of Fossil Fuel CO2 in the Oceans. Plenum, New York, pp. 213–248.

Brondel, D., Edwards, R., Hayman, A., Hill, D., Mehta, S., Semerad, T., 1994. Corrosion in the oil industry. Oilfield Rev. 6 (2), 4–18.

BS (British Standards) BS 5500. Unfired Fusion Welded Pressure Vessels.

Burk, J.D., March 11–14, 1991. Dualnode Field Performance Evaluation – Cathodic Protection for Offshore Structures. Paper no. 309 presented at CORROSION/91. Cincinnati.

DNV Recommended Practice RP 401 Cathodic Protection Design, 1993. Det Norske Veritas Industri Norge AS.

Cavassi, P., Cornago, M., 1999. The cost of corrosion in the oil & gas industry. J. Prot. Coat. Linings 16 (5), 30–40.

Chailes, C.P., 1986. Applied Water Technology. Campbell Petroleum Series USA.
Chandrasekaran, J., Anto, I., Balasubramaniam, K., Venkataraman, K.S., 2009. Higher order modes cluster (HOMC) guided waves for online defect detection in annular plate region of above-ground storage tanks. Insight: Non-Destr. Test. Cond. Monit. 51 (11), 606–611.
Chapra, C.S., Canale, R.P., 1988. Numerical Methods for Engineers, 2nd ed. McGraw-Hill, New York. pp. 734–737.
Chen, Y.Y., Shin, H.C., Liou, Y.M., Wang, L.H., Oung, J.C., 2006. Stress corrosion cracking of type 321 stainless steels under simulated petrochemical conditions containing thiosulfate and chloride. Corrosion 62 (9), 781–794.
Chilingar, G.V., Mourhatch, R., Al-Qahtani, G.D., 2008. The Fundamentals of Corrosion and Scaling for Petroleum and Environmental Engineers. Gulf Publishing Company, TX, USA.
Chitty, G.H., 2002. Corrosion issues with underbalanced drilling in H2S reservoirs. SPE Repr. Ser. 54, 169–176.
Ciaraldi, S., Abdallah, A., Attia, A.M., El-Leil, H.A., Bedair, S., Konecki, M., Lain, A., 1999. Rehabilitation of GUPCO's massive and aged pipeline infrastructure in the Gulf of Suez. Corros. Prev. Control 46 (5), 122–132.
Cochran, J.C., March 22–26, 1982. A Correlation of Anode-to-Electrolyte Resistance Equations Used in Cathodic Protection. Paper no. 169 presented at CORROSION/82. Houston.
Cochran, J.C., April 2-6, 1984. Additional Anode-to-Electrolyte Resistance Equations Useful in Offshore Cathodic Protection. Paper no. 254 presented at CORROSION/84. New Orleans.
NACE Standard RP 0176 Corrosion Control of Steel-Fixed Offshore Platforms Associated with Petroleum Production, 1976. NACE, Houston.
Couper, A.S., Hamel, F.B., June 1970. Process side antifoulants in petroleum refineries, report of work group T-8-2a. Mater. Prot. 9, 29–33.
Craig, B.D., 1998. Predicting critical erosion-corrosion limits of alloys for oil and gas production. Mater. Perform. 37 (9), 59–60.
Cribb, W.R., Grensing, F.C., 2011. Spinodal copper alloy C72900–New high strength antifriction alloy system. Can. Metall. Q. 50 (3), 232–239.
Crook, Paul, 1994. Cobalt-base alloys resist wear, corrosion, and heat. Adv. Mater. Processes 145 (4), 27–30.
Crouch, A.E., Goyen, T., Porter, P., 2004. New method uses conformable array to map external pipeline corrosion. Oil Gas J. 102 (41), 55–59.
Design of Galvanic Anode Cathodic Protection Systems for Offshore Structures, 1998. NACE International Publication 7L198, NACE International, Houston, TX.
Droste, R.L., 1997. Theory and Practice of Water and Wastewater Treatment. John Wiley & Sons, NY, USA.
Duncan, K.E., Gieg, L.M., Parisi, V.A., Tanner, R.S., Tringe, S.G., Bristow, J., Suflita, J.M., 2009. Biocorrosive thermophilic microbial communities in Alaskan North Slope oil facilities. Environ. Sci. Technol. 43 (20), 7977–7984.
Dwight, H.B., 1936. Electr. Eng. 55, 1319.
Emenike, C.O., 1993. Corrosion management during the construction of oil, gas and water lines. Anti-Corros. Methods Mater. 40 (3), 4–6.
Endo, S., Fujita, S., Nagae, M., 1997. New line pipe resists preferential corrosion at welds. Oil Gas J. 95 (11), 92–102.
Eriksrud, E., Sontvedt, T., 1984. Effect of Flow on CO2 Corrosion Rates in Real and Synthetic Formation Waters, Advances in CO2 Corrosion. NACE, Houston, Tx.
Esaklul, K.A., Ahmed, T.M., 2009. Prevention of failures of high strength fasteners in use in offshore and subsea applications. Eng. Failure Anal. 16 (4), 1195–1202.

Evans, N., 1993. Case for corrosion surveillance. Corros. Prev. Control 40 (2), 32–35.

Fatah, M.C., Ismail, M.C., 2011. Empirical equation of CO2 corrosion with presence of low concentrations of acetic acid under turbulent flow conditions. Corros. Eng. Sci. Technol. 46 (1), 49–55.

Fischer, K.P., Finnegan, J.E., April 17–21, 1989. Cathodic Protection Behavior of Steel in Sea Water and the Protective Properties of the Calcareous Deposits. Paper no. 582 presented at CORROSION/89. New Orleans.

Fischer, K.P., Sydberger, T., Lye, R., March 9–13, 1987. Field Testing of Deep Water Cathodic Protection on the Norwegian Continental Shelf. Paper no. 67 presented at CORROSION/87. San Francisco.

Fitzgerald, B.J., Winnik, S., 2004. A corrosion under insulation prevention strategy for petrochemical industry piping. Corros. Manage. 57, 15–19.

Fitzgerald, B.J., Winnik, S., 2005. A strategy for preventing corrosion under insulation on pipeline in the petrochemical industry. J. Prot. Coat. Linings 22 (4), 52–57.

Foss, M., Gulbrandsen, E., Sjöblom, J., 2008. Alteration of wettability of corroding carbon steel surface by carbon dioxide corrosion inhibitors – effect on carbon dioxide corrosion rate and contact angle. Corrosion 64 (12), 905–919.

Foss, M., Gulbrandsen, E., Sjöblom, J., 2009. Effect of corrosion inhibitors and oil on carbon dioxide corrosion and wetting of carbon steel with ferrous carbonate deposits. Corrosion 65 (1), 3–14.

Foster, T., Moores, V.G., March 17-2, 1986. Cathodic Protection Current Demand of Various Alloys in Sea Water. Paper no. 295 presented at CORROSION/86. Houston.

Fulton, W., Harris, J., 1991. Representation Theory. A First Course, Graduate Texts in Mathematics, Readings in Mathematics, vol. 129. Springer-Verlag, USA, New York.

Gaber, M.A.W., 2012. Impact of anti-corrosion liquid epoxy and fusion bond epoxy primer on mechanical testing of three layers polyethylene pipeline coating. J. Appl. Sci. Res. 8 (11), 5349–5359.

Gafarov, N.A., Goncharov, A.A., Kushnarenko, V.M., Shchepinov, D.N., Chirkov, Yu.A., 2003. An analysis of the failures of equipment and pipelines at the Orenburg oil-gas condensate deposit. Prot. Met. 39 (3), 294–296.

Ruiz, Galindez, N.G., Craig, Bruce, D., 1995. Effect of crude oil on corrosion of alloys in severe sour environments. Mater. Perform. 34 (4), 50–53.

Ge, L., Chen, C.-F., Zheng, S.-Q., Cai, X.-W., 2009. Corrosion behavior of nickel based alloy 825 in high temperature and high pressure H2S/CO2 environments. Corros. Protect. 30 (10), 708–710 + 716.

Ghanbarzadeh, A., Neshati, J., Bagherzadeh, M.R., Ghanizadeh, S., 2013. Atmospheric corrosion map of an oil refinery. Anti-Corros. Methods Mater. 60 (2), 106–114.

Gillepsi, B.G., August 1971. A New Process Antifoulant Test Correlates Better with refinery experience. paper 40, NCE 27th Conference. March, 1971, see also Mater. Prot. Perform. 10 (8), 21–25.

Glazov, N.P., 2001. Enhancement of corrosion protection of steel pipelines. Zashch. Met. 37 (5), 464–470.

Grimes, Keith, 1995. Stress corrosion crack in-line pig shows promise in tests. Pipeline Gas Ind. 78 (3), 37–40.

Groysman, A., 2005. Anticorrosion techniques for aboveground storage tanks. Mater. Perform. 44 (11), 40–43.

Hallaji, M., Shekarchi, M., Pargar, F., Taheri, S.R., Ghassemzadeh, F., 2012. Corrosion damage diagnosis of a 44-year-old ammonium nitrate prill tower in a petrochemical complex. J. Perform. Constr. Facil. 26 (4), 532–543.

Han, Y., Zhao, X.H., Bai, Z.Q., Wei, B., 2011. Review of sour gas field on corrosion factors and development progress. Adv. Mater. Res. 201–203, 438–447.

Hanzalek, Liu, S., Olson, D.L., Smith, C., 2000. Am Bureau of Shipping. Houston.

Harris, S., 2007. How to avoid the burning issue of sulphuric acid plant corrosion. PEI Power Eng. Int. 15 (9), 72–74.

Hartt, W.H., Chen, S., Townley, D.W., 1998. Corrosion 54, p 317.

Hendrix, D.E., 1997. Hydrogen embrittlement of high-strength fasteners in atmospheric service. Mater. Perform. 36 (12), 54–56.

Hinds, G., Turnbull, A., 2010. Novel multi-electrode test method for evaluating inhibition of underdeposit corrosion – part 1: sweet conditions. Corrosion 66 (4), 0460011–04600110.

Horn, R.A., Johnson, C.R., 1991. Topics in Matrix Analysis. Cambridge University Press, UK. Section 6.1.

Improving the Safety of Marine Pipelines, 1994. Committee on the Safety of Marine Pipelines, Marine Board. National Research Council, Washington, D.C.

Ichinose, Hiromi, 1987. Cathodic protection laws and regulations. Corros. eng. 36 (7), 417–421.

Ikeda, M.Ueda, Mukai, S., 1985. Influence of Environmental Factors on Corrosion in CO2 Source Well, Advances in CO2 Corrosion. NACE, Houston, TX.

ISO (International Organization for Standardization) ISO 9000. Quality Systems ISO 8993 Corrosion • Anodized aluminum and aluminum alloys - Rating system for the evaluation of pitting #–Chart method.

Ivanov, E.S., Zavadskaya, V.P., Getmanskii, M.D., 1983. Influence of inhibitors on the corrosion fatigue of steel in hydrogen-sulfide-containing media. Sov. Mater. Sci. 18 (5), 443–445.

Jayaperumal, D., Muralidharan, S., Venkatachari, G., Rengaswamy, N.S., 2000. Inhibition effect of ethanolamines on oil well tubular material in hydrochloric acid. Anti-Corros. Methods Mater. 47 (6), 349–353.

Jayaraman, A., Saxena, R.C., 1995. Corrosion and its control in petroleum refineries – a review. Corros. Prev. Control 42 (6), 123–131.

Johnsen, Roy, 1988. A review of materials for PIPELINES transporting unprocessed well fluid. In: Int. Conf. on Pipeline Protection & Maintenance, Stavanger, Norway, 20–21 Sept. 1.

Jong, C.K., Do, J.S., Kang, O.Y., Jae, B.C., Yoon, S.C., Young, J.K., Song, C.C., Woo, S.K., 2005. Development of corroded gas pipeline assessment program based on limit load solution. Key Eng. Mater. 297–300 (I), 47–52.

Joosten, Michael, W., Fischer, Karl, P., Strommen, Roe, Lunden, K.C., 1995. Internal corrosion monitoring of subsea oil and gas production equipment. Mater. Perform. 34 (4), 44–48.

Kanninen, Melvin, F., Pagalthivarthi, Krishnan, V., Popelar, Carl, H., 1992. Theoretical Analysis for the Residual Strength of Corroded Gas and Oil Transmission Pipelines. ASTM Special Technical Publication (1137), pp. 183–198.

Karami, M., 2012. Review of corrosion role in gas pipeline and some methods for preventing it. J. Pressure Vessel Technol. Trans. ASME 134 (5) art. no. 054501.

Kawakami, M., Ishida, M., Suzuki, S., Sasaki, N., Shinoda, Y., Inao, H., Yamada, T., 2003. Corrosion and cathodic protection condition for buried steel pipelines heated by internal fluids. Zairyo to Kankyo/Corros. Eng. 52 (2), 102–106.

Kermani, M.B., Harrop, D., 1996. The impact of corrosion on the oil and gas industry. SPE Prod. Facil. 11 (3), 186–190.

Kermani, M.B., Morshed, A., 2003. Carbon dioxide corrosion in oil and gas production – a compendium. Corrosion 59 (8), 659–683.

Konstantinov, I.O., Burushkin, O.S., Ryabchenko, V.N., Sheshukov, A.I., 1996. Radiometric corrosion monitoring of the industrial oil pipelines. Zashch. Met. (1), 104–110.

Kryzhanivs'kyi, E.I., Nykyforchyn, H.M., 2011. Specific features of hydrogen-induced corrosion degradation of steels of gas and oil pipelines and oil storage reservoirs. Mater. Sci. 47 (2), 127–136.

Lagad, V.V., Srinivasan, S., Kane, R.D., 2008. Facilitating Internal Corrosion Direct Assessment Using Advanced Flow and Corrosion Prediction Models. NACE – International Corrosion Conference Series, pp. 081311–0813110.

Larsen, K.R., 2011. Ohio university's research facility tackles industry's corrosion challenges. Mater. Perform. 50 (1), 24–27.

Lasebikan, B.A., Akisanya, A.R., Deans, W.F., MacPhee, D.E., 2011. The effect of hydrogen sulphide on ammonium bisulphite when used as an oxygen scavenger in aqueous solutions. Corros. Sci. 53, 4014–4025.

Leth-Olsen, H., 2005. Second SPE International Symposium on Oilfield Corrosion 2005: Corrosion Control in Oil and Gas Production – "Fluids, Chemicals, Materials and More", Proceedings, art. no. SPE 95072 CO2 corrosion in bromide and formate well-completion brines, 67–75.

Li, S.-L., Yu, C.-Y., 2005. Corrosion survey and analysis of equipment in an oil refining complex. Corros. Prot. 26 (7), 310–312 + 314.

Li, X., Fu, D., Dong, C., Chen, H., 2002. Exploitation of expert system of fitness – for – service for petrochemical corrosion. J. Chin. Soc. Corros. Prot. 22 (5), 268–272.

Lin, C.-Y., Pan, J.-Y., 2000. Corrosion characteristics of furnaces burning with emulsified diesel oil containing sodium sulphate. Corros. Prev. Control 47 (3), 83–92.

Liu, H., Wang, Y., 2001. A discussion on the method of evaluating residual corrosion strength of in-service pressure pipeline. Tianranqi Gongye/Nat. Gas Ind. 21 (6), 90–92.

Maeng, W.Y., Macdonald, D.D., 2008. The effect of acetic acid on the stress corrosion cracking of 3.5 NiCrMoV turbine steels in high temperature water. Corros. Sci. 50, 2239–2250.

Mahajanam, S.P.V., Joosten, M.W., 2011. Guidelines for filler-material selection to minimize preferential weld corrosion in pipeline steels. SPE Proj. Facil. Constr. 6 (1), 5–12.

Mandke, J.S., October 29, 1990. Corrosion causes most pipeline failures in the Gulf of Mexico. Oil Gas J., 40.

McCoy, J.E., 1970. Corrosion control by cathodic protection – theoretical and design concepts for marine applications. Inst. Mar. Eng. Trans. 82, 210.

McCoy, J.E., 1970. Trans. Inst. Mar. Eng. 82, 210.

McIntyre, P., 2003. Corrosion management for upstream oil and gas. Corros. Eng., Sci. Technol. 38 (2), 89–94.

McKennis, J., Bae, N., Kimura, M., Shimamoto, K., Sato, H., 2008. A New Chemical Mechanistic Postulate Regarding Annular Environmentally Assisted Cracking (AEAC) – Importance of Cations and Key Contaminants in Packer Fluids in Cracking of Martensitic Stainless Steel. NACE – International Corrosion Conference Series, pp. 084831–0848316.

Milliams, D.E., Kroese, C.J., 1979. In: 3rd International Conference on Internal and External Pipe Protection, paper HI.

Minerals Management Service Data Base, Gulf of Mexico Region, New Orleans, LA.

Mollan, R., Anderson, T.R., March 17-2, 1986. Design of Cathodic Protection Systems. Paper no. 286 presented at CORROSION/86. Houston.

Moon, J., Ha, H.-Y., Lee, T.-H., 2013. Corrosion behavior in high heat input welded heat-affected zone of Ni-free high-nitrogen, Fe–18Cr–10Mn–N austenitic stainless steel. Mater. Charact. 82, 113–119.

Morgan, J., 1960. Cathodic Protection. Macmillan, New York, pp. 140–143.

Morshed, A., 2008. Improving asset corrosion management using KPIs. Mater. Perform. 47 (5), 50–54.

NACE (National Association of Corrosion Engineers) Corrosion Test Methods, TM-01–69. Laboratory Corrosion Testing of Metals for the Process Industries.

NACE RP 0472, 2000. Methods and Controls to Prevent in Service Environment (5.5.3) Cracking of C.S. Weldments in Corrosive Petroleum.

NACE Standard MR-01-75 (1991 Rev.) sulfide stress cracking resistance metallic materials for oil field equipment.

NACE Standard MR-01-75 (1991 rev), 1984. Sulphide Stress Cracking Resistant Metallic Materials for Oil Field Equipment. NACE, Houston, Tx.

NACE (National Association of Corrosion Engineers) NACE RP 0170: Protection of Austenitic Stainless Steels and Other 2004 (5.4.5) Austenitic Alloys from Polythionic Acid SCC during Shut-down of Refinery Equipment.

NACE MR 0175. Material for Use in H2S Containing Environments in Oil ISO 15156: 2003 and Gas Production.

Nathan, C.C., Dulaney, C.L., February 1971. Statistical concepts facilitate evaluation of corrosion inhibitors. Mater. Prot. 10, 21–25.

Neville, A., Wang, C., 2009. Erosion-corrosion of engineering steels-can it be managed by use of chemicals? Wear 267 (11), 2018–2026.

Onyekpe, B.O., Dania, L.P., 1999. Flowline corrosion problems: a case study of Shell Petroleum Development Company, Nigeria. Anti-Corros. Methods Mater. 46 (3), 205–211.

Pipeline Cathodic Protection – Part 2: Cathodic Protection of Offshore Pipelines Working Document ISO/TC 67/SC 2 NP 14489, International Standards Organization, May 1, 1999.

Palacios, C.A., Shadley, J.R., 1991. Performance of Oilfield Corrosion Inhibitors under Single and Two-Phase Flow Conditions, Corrosion 91. NACE, Houston, TX, p. 475.

Panossian, Z., Lira de Almeidaa, N., Ferreira de Sousa, R.M., de Souza Pimenta, G., Marques, L.B.,S., 2012. Corrosion of carbon steel pipes and tanks by concentrated sulfuric acid: a review. Corros. Sci. 58, 1–11.

Papavinasam, S., Revie, R.W., Attard, M., Demoz, A., Michaelian, K., 2003. Comparison of techniques for monitoring corrosion inhibitors in oil and gas pipelines. Corrosion 59 (12), 1096–1111.

Papavinasam, S., Revie, R.W., Friesen, W.I., Doiron, A., Panneerselvam, T., 2006. Review of models to predict internal pitting corrosion of oil and gas pipelines. Corros. Rev. 24 (3–4), 173–230.

Part 1: Sampling Plans Indexed by Acceptable Quality Level (AQL) for Lot-by-Lot Inspection.

Pierson, P., Hartt, W.H., 1999. Galvanic anode cathodic polarization of steel in sea water: part IV – conductor arrays on petroleum production platforms. Corrosion 55, 686.

Powell, D.E., Ma'Ruf, D.I., Rahman, I.Y., 2002. Field-testing corrosion inhibitors in oil and gas-gathering systems. Mater. Perform. 41 (8), 42–45.

Rabartdinov, Z.R., Denislamov, I.Z., 2012. Exploitation of oil fields with H2S in the North-Western part of Bashkortostan. Neft. Khoz. – Oil Ind. (8), 96–98.

Ramstad, K., Rohde, H.C., Tydal, T., Christensen, D., 2009. Scale squeeze evaluation through improved sample preservation, inhibitor detection, and minimum inhibitor concentration monitoring. SPE Prod. Oper. 24 (4), 530–542.

Reddy, R.V., Nelson, J.L., Pacheco, J.L., 2006. An integrated approach to accurate corrosion prediction. JPT, J. Pet. Technol. 58 (5), 76–79.

Ridgwell, A., Hargreaves, J.C., 2007. Regulation of atmospheric CO2 by deep-sea sediments in an Earth system model. Global Biogeochem. Cycles 21, GB2008. http://dx.doi.org/10.1029/2006GB002764.

Roffey, R., 1989. Microbial problems during long-term storage of petroleum products underground in rock caverns. Int. Biodeterior. 25 (1–3), 219–236.
Ropital, F., 2009. Current and future corrosion challenges for a reliable and sustainable development of the chemical, refinery, and petrochemical industries. Mater. Corros. 60 (7), 495–500.
Satoh, K., Yamamoto, K., Kagawa, N., 1984. Prevention of CO2 Corrosion in Gas Gathering Systems, Advances in CO2 Corrosion. NACE, Houston, TX.
Schmitt, G., Simon, T., Hausler, R.H., 1990. CO2 Erosion Corrosion and Its Inhibition under Extreme Shear Stress -1. Development of Methodology, Corrosion 90. NACE, Houston, TX, p. 22.
Schrieber, C.F., Reding, J., April 23–27, 1990. Application Methods for Rapid Polarization of Offshore Structures. Paper no. 381 presented at CORROSION/90. Las Vegas.
Serednyts'kyi, Ya.A., 2000. Polyurethane materials as anticorrosive coatings of pipelines. Mater. Sci. 36 (3), 415–421.
Shao, X.-D., 2008. Application of repair technology of internal protection coatings of pipelines for oilfield and relevant economical analysis. Corros. Sci. Prot. Technol. 20 (3), 229–231.
Silva, V.G., Tavares, S.S.M., Baptista, I.P., De Oliveira, R., 2012. Evaluation of the susceptibility of a superduplex stainless steel welded joint to sulfide stress corrosion by slow strain rate tensile tests in sour solutions. Corrosion 68 (1) art. no. 015006.
Silva, C.C., Miranda, H.C., de Sant'Ana, H.B., Farias, J.P., 2013. Austenitic and ferritic stainless steel dissimilar weld metal evaluation for the applications as-coating in the petroleum processing equipment. Mater. Des. 47, 1–8.
Simon Thomas, M.J.J., De Waard, C., Smith, L.M., 1987. Controlling factors in the rate of CO2 corrosion. In: UK corrosion 87, Brighton 26–28 October, p. 147.
Smart III, John, S., Pickthall, Tom, 1995. Online system developed for measuring internal corrosion. Pipeline Gas Ind. 78 (3), 29–34.
Smart III, John S., 1991. The meaning of API RP-14E formula for erosion corrosion. In: Oil and Gas Production, Corrosion. NACE, Houston, TX. Paper 468.
Smith, L., 1999. Control of corrosion in oil and gas production tubing. Br. Corros. J. 34 (4), 247–253.
Smith, S.N., 2003. Corrosion product analysis in oil and gas pipelines. Mater. Perform. 42 (8), 44–47.
Song, F., Bichon, B., Fassett, R., Boss, T., Lu, A., 2009. Cased pipe segments could be less safe than uncased segments. Oil Gas J. 107 (15), 50–56.
Song, F.M., 2010. A comprehensive model for predicting CO2 corrosion rate in oil and gas production and transportation systems. Electrochim. Acta 55 (3), 689–700.
Stalder, F., 1997. Pipeline failures. Mater. Sci. Forum 247, 139–146.
Stanley, L., Jones, B.L., August 2004. Microbe-assisted external corrosion in oil and pipelines. Pipeline World (4), 5–10.
Stiff, H.A., Davis, L.E., 1952. A method for predicting the tendency of oilfield waters to deposit calcium carbonate. Pet. Trans. AIME 195, 213.
Strommen, R., Rodland, A., 1981. Mater. Perform. 20 (10), 7.
Strommen, R., 1985. Mater. Perform. 24 (3), 9.
Sunde, E.D., 1968. Earth Conduction Effects in Transmission Systems. Dover Publications, Inc., New York.
Townley, D.W., March 9–14, 1997. Unified Design Equation for Offshore Cathodic Protection. Paper no. 97473 presented at CORROSION/97. New Orleans.
Tsaprailis, H., Garfias-Mesias, L.F., 2011. Rapid evaluation of metallic coatings on large cylinders exposed to marine environments. Corrosion 67 (12) art. no. 126002.

Turnbull, A., Zhou, S., 2008. Impact of temperature excursion on stress corrosion cracking of stainless steels in chloride solution. Corros. Sci. 50, 913–917.
Tuttle, R.N., 1997. Corrosion in oil and gas production. SPE Repr. Ser. 46, 62–68.
Ueda, M., 2007. Development of corrosion resistant alloys for oil and gas well environments containing CO2 and H2S. Zairyo to Kankyo/Corros. Eng. 56 (2), 43–50.
Uhlig, H.H., Revie, R.W., 1985. Corrosion and Corrosion Control, 3rd ed. John Wiley and Sons, New York. p. 223.
Uthman, H., 2011. Mathematical modeling and simulation of corrosion processes in Nigerian crude oil pipelines. J. Dispersion Sci. Technol. 32 (4), 609–615.
Vagapov, R.K., 2008. Increasing the corrosion resistance of oil production equipment and pipelines. Prot. Met. 44 (7), 667–672.
Vaideeswaran, R., Natarajan, S., Sivan, V., 2003. The corrosion behaviour of incology 825 weldments used in the process and petrochemical industries. Corros. Prev. Control 50 (2), 64–70.
Velazquez, J.C., Valor, A., Caleyo, F., Venegas, V., Espina-Hernandez, J.H., Hallen, J.M., Lopez, M.R., 2009. Pitting corrosion models improve integrity management, reliability. Oil Gas J. 107 (28), 56–62.
Walsh, T., 2006. Ageing assets and the fatigue challenge. J. Offshore Technol. 14 (1), 22–24.
Wan, L., Meng, Y., Wang, C., Yang, L., 2007. Mechanism of corrosion and scale deposit of tubes in western oilfield. J. Chin. Soc. Corros. Prot. 27 (4), 247–251.
Wang, C., Neville, A., 2008. Inhibitor performance on corrosion and erosion/corrosion under turbulent flow with sand and CO2–an AC impedance study. SPE Prod. Oper. 23 (2), 215–220.
Wang, W., Hartt, W.H., Chen, S., 1996. Corrosion 52, 419.
Wang, W., Free, M.L., Horsup, D., 2005. Prediction and measurement of corrosion inhibition of mild steel by imidazolines in brine solutions. Metall. Mater. Trans. B 36 (3), 335–341.
Wang, X.-F., Dong, Z.-J., Liang, Y.-J., Zhang, Z.-H., Chen, C.-F., 2006. Development of economic steels with low Cr content for anti-corrosion oil tube. Corros. Sci. Prot. Technol. 18 (6), 436–439.
Weldon, C., Kroon, D., December 4–6, 1991. Corrosion control survey methods for offshore pipelines. In: Proceedings International Workshop on Offshore Pipeline Safety, New Orleans, p. 196.
Wu, W., Cheng, G., Hu, H., Zhou, Q., 2013. Risk analysis of corrosion failures of equipment in refining and petrochemical plants based on fuzzy set theory. Eng. Failure Anal. 32, 23–34.
Xiaoyan, L., Jungang, L., Qianya, Z., Jinlai, F., Yingli, L., Jingxin, S., 2009. The analysis and prediction of scale accumulation for water-injection pipelines in the Daqing Oilfield. J. Pet. Sci. Eng. 66, 161–164.
Xu, S.-J., Liu, X.-H., 2010. Research & development of anti-corrosion management system in petrochemical industry. Corros. Prot. 31 (3), 188–192.
Yamamoto, T., 2000. Filed test on the corrosion behavior of internally coated oil pipelines located in Daqing and Sichuan oil field in China. Zairyo to Kankyo/Corros. Eng. 49 (4), 209–214.
Yu, C.-Y., 2007. Analyses of stress corrosion cracking environments of petrochemical equipment. Corros. Prot. 28 (1), 23–28.
Yuan, G., Wang, Q., Shen, R., Yuan, J., Lu, L., Wang, C., 2008. String corrosion and string protection during constructing and operating gas storage facility in bedded salt deposit. SPE Prod. Oper. 23 (1), 63–67.
Zav'yalov, V.V., Moiseeva, L.S., 2004. Chemical, hydrodynamic, and metallurgical factors in West Siberian oil pipeline corrosion failure. Chem. Pet. Eng. 40 (1–2), 45–50.

Zettler, H.U., Müller-Steinhagen, H., Hedges, B., 2003. Modification of carbon steel surfaces to reduce corrosion fouling in oil and gas industry. Pet. Sci. Technol. 21 (3–4), 681–698.

Zhan, Z., Cai, Q.-F., Tang, S., Dong, G., Tang, T.-L., 2008. A corrosion survey and relevent measures for corrosion prevention in Tahe Oilfield. Corros. Sci. Prot. Technol. 20 (2), 152–154.

Zhang, X., Li, Z.-M., Zhang, Z.-H., Dong, B., 2006. Experimental study on H2S corrosion on tube in deep well. Corros. Sci. Protect. Technol. 18 (1), 16–19.

Zhang, L., Tian, S.B., Zhao, S.-L., 2007. Preliminary study on corrositivity of fatty acids. Corros. Sci. Prot. Technol. 19 (5), 358–360.

Zhao, J.-M., Zuo, Y., 2000. Pitting corrosion of steel pipeline in oil-water-gas multiphase flow system. Br. Corros. J. 35 (1), 73–74.

Zhao, X.H., Bai, Z.Q., Lin, K., Han, Y., 2011. The corrosion behavior about two Ni-based alloys in CO2/H2S environments. Adv. Mater. Res. 152–153, 1624–1631.

Zhu, X.Y., Modi, H., Ayala, A., Kilbane, J.J., 2006. Technical note: rapid detection and quantification of microbes related to microbiologically influenced corrosion using quantitative polymerase chain reaction. Corrosion 62 (11), 950–955.

Zou, C.J., Tang, Q.W., Zhao, P.W., Guan, E.D., Wu, X., Ye, H., 2013. Further study on the inclusion complex of 2-phosphonobutane-1,2,4-tricarboxylic acid with B-cyclodextrin: a new insight of high inhibition efficiency for protecting steel corrosion. J. Pet. Sci. Eng. 103, 29–35.

Glossary of Terms

Acceleration corrosion test: Method designed to approximate, in a short time, the deteriorating effect under normal long-term service condition.

Acicular ferrite: A highly substructured, nonequiaxed ferrite formed upon continuous cooling by a mixed diffusion and shear mode of transformation that begins at a temperature slightly higher than the transformation temperature range for upper bainite. It is distinguished from bainite in that it has a limited amount of carbon available; thus, there is only a small amount of carbide present.

Acid embrittlement: A form of hydrogen embrittlement that may be induced in some metals by acid.

Active metal: A metal ready to corrode, or being corroded.

Active potential: The potential of a corroding material.

Activity: A measure of the chemical potential of a substance, where chemical potential is not equal to concentration, that allows mathematical relations equivalent to those for ideal systems to be used to correlate changes in an experimentally measured quantity with changes in chemical potential.

Activity (ion): The ion concentration corrected for deviations from ideal behavior. Concentration multiplied by activity coefficient. Activity coefficient, a characteristic of a quantity expressing the deviation of a solution from ideal thermodynamic behavior; often used in connection with electrolytes.

Addition agent: A substance added to a solution for the purpose of altering or controlling a process. Examples include wetting agents in acid pickles, brighteners or antipitting agents in plating solutions, and inhibitors.

Aeration: (1) Exposing to the action of air. (2) Causing air to bubble through. (3) Introducing air into a solution by spraying, stirring, or a similar method. (4) Supplying or infusing with air, as in sand or soil.

Aeration cell: An oxygen concentration cell; an electrolytic cell resulting from differences in dissolved oxygen at two points. Also see differential aeration cell.

Aerobic: Environment containing oxygen, for instance, normal seawater.

Aging: A change in the properties of certain metals and alloys that occurs at ambient or moderately elevated temperatures after hot working or a heat treatment (quench aging in ferrous alloys, natural or artificial aging in ferrous and nonferrous alloys) or after a cold working operation (strain aging). The change in properties is often, but not always, due to a phase change (precipitation), but never involves a change in chemical composition of the metal or alloy.

Alclad: Composite wrought product comprising an aluminum alloy core having on one or both surfaces a metallurgically bonded aluminum or aluminum alloy coating that is anodic to the core and thus electrochemically protects the core against corrosion.

Alkali metal: A metal in group IA of the periodic system, namely, lithium, sodium, potassium, rubidium, cesium, and francium. They form strongly alkaline hydroxides, hence the name.

Alkaline: (1) Having properties of an alkali. (2) Having a pH greater than 7.

Alkyd: Resin used in coatings. Reaction products of polyhydric alcohols and polybasic acids.

Alkylation: (1) A chemical process in which an alkyl radical is introduced into an organic compound by substitution or addition. (2) A refinery process for chemically combining isoparaffin with olefin hydrocarbons.

Alligatoring: (1) Pronounced wide cracking over the entire surface of a coating having the appearance of alligator hide. (2) The longitudinal splitting of flat slabs in a plane parallel to the rolled surface. Also called fish-mouthing.

Alloy plating: The code position of two or more metallic elements.

Alloy steel: Is one that contains either silicon or manganese in amounts in excess of those quoted in plain carbon steel or that contains any other element, or elements, as the result of deliberately made alloying additions.

Alternate-immersion test: A corrosion test in which the specimens are intermittently exposed to a liquid medium at definite time intervals.

Aluminizing: Process for impregnating the surface of a metal with aluminum in order to obtain protection from oxidation and corrosion.

Ammeter: An instrument for measuring the magnitude of electric current flow.

Amorphous solid: A rigid material whose structure lacks crystalline periodicity, that is, the pattern of its constituent atoms or molecules does not repeat periodically in three dimensions. See also metallic glass.

Anaerobic: Lack of oxygen in the electrolyte adjacent to metallic structure.

Anchor pattern/surface profile: Shape and amplitude of profile of blast cleaned or grooved steel, which influences the bond between metallic or paint of films and the substrate.

Anchorite: A zinc–iron phosphate coating for iron and steel.

Anaerobic: In the absence of air or unreacted or free oxygen.

Anion: Negatively charged ion, which migrates to the anode of a galvanic or voltaic cell.

Annealing: A generic term denoting a treatment, consisting of heating to and holding at a suitable temperature, followed by cooling at a suitable rate, used primarily to soften metallic materials, but also to simultaneously produce microstructure. The purpose of such changes may be, but is not confined to, improvement of machinability, facilitation of cold work, improvement of mechanical or electrical properties, and/or increase in stability of dimensions. When the term is used by itself, full annealing is implied. When applied only for the relief of stress, the process is properly called stress relieving or stress-relief annealing.

Anode: An electrode at which oxidation of the surface or some component of the solution is occurring.

Anode corrosion: The dissolution of a metal acting as an anode.

Anode corrosion efficiency: Ratio of actual to theoretical corrosion based on the total current flow calculated by Faraday's law from the quantity of electricity that has passed.

Anode effect: The effect produced by polarization of the anode in electrolysis. It is characterized by a sudden increase in voltage and a corresponding decrease in amperage due to the anode becoming virtually separated from the electrolyte by a gas film.

Anode film: (1) The portion of solution in immediate contact with the anode, especially if the concentration gradient is steep. (2) The outer layer of the anode itself.

Anode polarization: Difference between the potential of an anode passing current and equilibrium potential (or steady state potential) of the electrode having the same electrode reaction.

Anodic cleaning: Electrolytic cleaning in which the work is the anode. Also called reverse-current cleaning.

Anodic coating: A film on a metal surface resulting from an electrolytic treatment at the anode.

Anodic inhibitor: A chemical substance or combination of substances that prevent or reduce, by physical, physiochemical, or chemical action, the rate of the anodic or oxidation reaction.

Anodic metallic coating: A coating, composed wholly or partially of an anodic metal (in sufficient quantity to set off electrochemical reaction), which is electrically positive to the substrate to which it is applied.

Anodic polarization: The change in the initial anode potential resulting from current flow effects at or near the anode surface. Potential becomes mode noble (more positive) because of anodic polarization.

Anodic potential: An appreciable reduction in corrosion by making a metal an anode and maintaining this highly polarized condition with very little current flow.

Anodic protection: A technique to reduce corrosion of a metal surface under some conditions, by passing sufficient anodic current to it to cause its electrode potential to enter and remain in the passive region.

Anodic reaction: Electrode reaction equivalent to a transfer of positive charge from the electronic to the ionic conductor. An anodic reaction is an oxidation process.

Anodizing: Forming a conversion coating on a metal surface by anodic oxidation; most frequently applied to aluminum.

Anti-fouling: Intended to prevent fouling of underwater structures, such as the bottoms of ships; refers to the prevention of marine organism's attachment or growth on a submerged metal surface, generally through chemical toxicity caused by the composition of the metal or coating layer.

Antipitting agent: An addition agent for electroplating solutions to prevent the formation of pits or large pores in the electrodeposit.

Aqueous: Pertaining to water; an aqueous solution is a water solution.

Artificial aging: Aging above room temperature.

Atmospheric corrosion: The gradual degradation or alteration of a material by contact with substances present in the atmosphere, such as oxygen. carbon dioxide, water vapor, and sulfur and chlorine compounds.

Attenuation: The progressive decrease in potential and current density along buried or immersed pipeline in relation to distance from the point of injection.

Attenuation constant "a": The magnitude of the attenuation constant is directly affected by the longitudinal resistance of the pipe and inversely affected by the resistance across the coating. The attenuation constant describes how much the potential change or current flow decreases with increasing distance from the drain point.

Austenitic stainless steel: Steel containing sufficient amount of nickel, nickel and chromium, or manganese to retain austenite at atmospheric temperature.

Austenite: A solid solution of one or more elements in face-centered cubic iron. Unless otherwise designated (such as nickel austenite), the solute is generally assumed to be carbon.

Austenitic: The name given to the face-centered cubic crystal structure (FCC) of ferrous metals. Ordinary iron and steel has this structure at elevated temperatures; also, certain stainless steels (300 series) have this structure at room temperature.

Automatic welding: Welding in which the welding variables and the mean of making the weld are controlled by machine.

Auxiliary anode: In electroplating, a supplementary anode positioned so as to raise the current density on a certain area of the cathode and thus obtain better distribution of plating.

Auxiliary electrode: The electrode of an electrochemical cell that is used to transfer current to or from a test electrode. An electrode commonly used in polarization studies to pass current to or from a test electrode. It is usually made of noncorroding material:

CUI: Corrosion under insulation
DLA: Double layer activation
ER: Electrical resistance
HAZ: Heat affected zone
LPR: Linear polarization resistance

Backfill: Material placed in a drilled hole to fill space around anodes, vent pipe, and buried components of a cathodic protection system.

Bainite: A metastable aggregate of ferrite and cementite resulting from the transformation of austenite at temperatures below the pearlite but above the martensite start temperature.

Banded structure: A segregated structure consisting of alternating nearly parallel bands of different composition, typically aligned in the direction of primary hot working.

Bandwidth: (Ultrasonics) Range of frequencies around a mean value that constitutes the nominal frequency.

Base: A chemical substance that yields hydroxyl ions (OH) when dissolved in water. Compare with acid.

Base metal: (1) The metal present in the largest proportion in an alloy; brass, for example, is a copper-based alloy. (2) An active metal that readily oxidizes, or that dissolves to form ions. (3) The metal to be brazed, cut, soldered, or welded. (4) After welding, that part of the metal which was not melted.

Beach marks: Macroscopic progression marks on a fatigue fracture or stress-corrosion cracking surface that indicate successive positions of the advancing crack front. The classic appearance is of irregular elliptical or semielliptical rings, radiating outward from one or more origins. Beach marks (also known as clamshell marks or arrest marks) are typically found on service fractures where the part is loaded randomly, intermittently, or with periodic variations in mean stress or alternating stress. See also striation.

Bead: A single run of weld metal on a surface.

Bimetallic corrosion: (Galvanic corrosion) Corrosion resulting from dissimilar metal contact.

Biocide: A chemical that kills other forms of life in addition to bacteria.

Biological corrosion: Biological corrosion is not a type of corrosion; it is the deterioration of a metal by corrosion processes that occur directly or indirectly as a result of the activity of living organisms. These organisms include micro forms such as bacteria and macro types such as algae and barnacles. Microscopic and macroscopic organisms have been observed to live and reproduce in mediums with pH values between 0 and 11, temperatures between −1.1 °C (30 °F) and 82.2 °C (180 °F), and under pressures up to 1020 bar (15,000 lb/in^2). Thus, biological activity may influence corrosion in a variety of environments including soil, natural water and seawater, natural petroleum products, and oil emulsion-cutting fluids.

Biostat: A chemical that retards growth of other forms of life in addition to bacteria.

Bipolar electrode: An electrode in an electrolytic cell that is not mechanically connected to the power supply, but is so placed in the electrolyte, between the anode and cathode, that the part nearer the anode becomes cathodic and the part nearer the cathode becomes anodic. Also called intermediate electrode.

Bituminous coating: Coal tar or asphalt-based coating.

Black liquor: The liquid material remaining from pulpwood cooking in the soda or sulfate paper-making process.

Black oxide: A black finish on a metal produced by immersing it in hot oxidizing salts or salt solutions.

Blast cleaning: Cleaning and roughening of a surface (particularly steel) by the use of metallic grit or nonmetallic grit or metal shot, which is projected against a surface by compressed air, centrifugal force, or water.

Blast peening: Treatment for relieving tensile stress by inducing beneficial compressive stress in the surface by kinetic energy of rounded abrasive particles.

Blister: A raised area, often dome shaped, resulting from (1) loss of adhesion between a coating or deposit and the base metal or (2) delamination under the pressure of expanding gas trapped in a metal in a near-subsurface zone. Very small blisters may be called pinhead blisters or pepper blisters.

Blistering: The formation of swellings on the surface of an unbroken paint film by moisture, gases, or the development of corrosion products between the metal and the paint film.

Blow down: (1) Injection of air or water under high pressure through a tube to the anode area for the purpose of purging the annular space and possibly correcting high resistance caused by gas blocking. (2) In connection with boilers or cooling towers, the process of discharging a significant portion of the aqueous solution in order to remove accumulated salts, deposits, and other impurities.

Blue brittleness: Brittleness exhibited by some steels after being heated to a temperature within the range of about 200–370 °C (400–700 °F), particularly if the steel is worked at the elevated temperature.

Blushing: Whitening and loss of gloss of a usually organic coating caused by moisture. Also called blooming.

Bond: A piece of metal conductor, either solid or flexible, usually of copper, connecting two points on the same or on different structures, to prevent any appreciable change in the potential of the one point with respect to the other.

Bond resistance: The ohmic resistance of a bond including the contact resistance at the points of attachment of its extremities.

Bonderizing: A proprietary custom process for phosphatizing.

Braze welding: The joining of metals using a technique similar to fusion welding and a filler metal with a lower melting point than the parent metal but neither using capillary action as in brazing nor intentionally melting the parent metal.

Brazing: A process of joining metals in which, during or after heating, molten filler metal is drawn by capillary action into the space between closed adjacent surfaces of the parts to be joined. In general, the melting point of the filler metal is above 500 °C, but always below the melting temperature of the parent metal.

Brazing alloy: Filler metal used in brazing.

Breakdown potential: The least noble potential where pitting or crevice corrosion, or both, will initiate and propagate.

Breakway corrosion: A sudden increase in corrosion rate, especially in high-temperature "dry" oxidation, etc.

Brightener: An agent or combination of agents added to an electroplating bath to produce a smooth, lustrous deposit.

Brine: Seawater containing a higher concentration of dissolved salt than that of the ordinary ocean.

Brittle fracture: Separation of a solid accompanied by little or no macroscopic plastic deformation. Typically, brittle fracture occurs by rapid crack propagation with less expenditure of energy than for ductile fracture.

Butt joint: A connection between the ends or edges of two parts making an angle to one another of 135–180° inclusive in the region of the joint.

Calcareous coating or deposit: A layer consisting of a mixture of calcium carbonate and magnesium hydroxide deposited on surfaces being cathodically protected because of the increased pH adjacent to the protected surface.

Calomel electrode: An electrode widely used as a reference electrode of known potential in electrometric measurement of acidity and alkalinity, corrosion studies, voltammetry, and measurement of the potentials of other electrodes.

Calorizing: Imparting resistance to oxidation to an iron or steel surface by heating in aluminum powder at 800–1000 °C (1470–1830 °F).

Carbon dioxide welding: Metal-arc welding in which a bare wire electrode is used, the arc and molten pool being shielded with carbon dioxide gas.

Carbonitriding: A case-hardening process in which a suitable ferrous material is heated above the lower transformation temperature in a gaseous atmosphere of such composition as to cause simultaneous absorption of carbon and nitrogen by the surface and, by diffusion, create a concentration gradient. The process is completed by cooling at a rate that produces the desired properties in the workpiece.

Carburization: The absorption of carbon into a metal surface; may or may not be desirable.

Carburizing: The absorption of carbon atoms by a metal at high temperatures; it may remain dissolved, or form metal carbides; Absorption and diffusion of carbon into solid ferrous alloys by heating, to a temperature usually above Ac in contact with a suitable carbonaceous material. A form of case hardening that produces a carbon gradient extending inward from the surface, enabling the surface layer to be hardened either by quenching directly from the carburizing temperature or by cooling to room temperature then reaustenitizing and quenching.

Case hardening: A generic term covering several processes applicable to steel that change the chemical composition of the surface layer by absorption of carbon, nitrogen, or a mixture of the two and, by diffusion, create a concentration gradient. The outer portion, or case, is made substantially harder than the inner portion, or core.

Cathode: The electrode of an electrolytic cell at which reduction occurs. In corrosion processes, usually the area at which metal ions do not enter the solution. Typical cathodic processes are cations taking up electrons and being discharged, oxygen being reduced, and the reduction from a higher to a lower state of valency.

Cathode depolarization: Removal of protective layers and complexing ions, often caused by high acidity, high oxygen content, accidental shorts to other structures, results is increased corrosion current and corrosion rate. Cathode depolarization is indicated by a positive (+) shift in cathode-to-electrolyte potential.

Cathode film: The portion of solution in immediate contact with the cathode during electrolysis.

Cathode polarization: Build up of protective layers and complexing ions on the cathode electrode. Polarization of the cathode results in decreased corrosion and negative (−) shift in cathode-to-electrolyte potential.

Cathodic area: That part of the metal surface which acts as a cathode.

Cathodic corrosion: Corrosion resulting from a cathodic condition of a structure usually caused by the reaction of an amphoteric metal with the alkaline products of electrolysis.

Cathodic disbondment: The destruction of adhesion between a coating and its substrate by products of a cathodic reaction.

Cathodic inhibitor: A chemical substance or combination of substances that prevent or reduce the rate of cathodic reaction by a physical, physiochemical, or chemical action.

Cathodic pickling: Electrolytic pickling in which the work is the cathode.

Cathodic polarization: Polarization of the cathode; a reduction from the initial potential resulting from current flow effects at or near the cathode surface. Potential becomes more active (negative) because of cathodic polarization.

Cathodic protection: The process to reduce or prevent corrosion of metal structures in contact with an electrolyte by the flow of direct current from the electrolyte into the structure surface.

1. Reduction of corrosion rate by shifting the "corrosion potential" of the electrode toward a less oxidizing potential by applying an external "electromotive force."
2. Partial or complete protection of a metal from corrosion by making it a cathode using either a galvanic or an impressed current. Contrast with anodic protection.

Cathodic protection station: A combination of equipment installed to provide cathodic protection to the structure(s).

Cathodic reaction: Electrode reaction equivalent to a transfer of negative charge from the electronic to the ionic conductor. A cathodic reaction is a reduction process. An example common in corrosion is: Ox + ne s Red.

Catholyte: The electrolyte adjacent to the cathode of an electrolytic cell.

Cation: Positively charged ion that migrates to the cathode in a galvanic or voltaic cell.

Caustic: (1) Burning or corrosive. (2) A hydroxide of a light metal, such as sodium hydroxide or potassium hydroxide.

Caustic dip: A strongly alkaline solution into which metal is immersed for etching, for neutralizing acid, or for removing organic materials such as greases or paints.

Caustic embrittlement: An obsolete historical term denoting a form of stress-corrosion cracking most frequently encountered in carbon steels or iron–chromium–nickel alloys that are exposed to concentrated hydroxide solutions at temperatures of 200–250 °C (400–480 °F).

Cavitation erosion: Cavitation is a particular kind of erosion–corrosion caused by the formation and collapse of vapor bubbles in a liquid contacting a metal surface. The resultant shock forces reach high levels in local areas and can tear out jagged chunks of brittle materials or deform soft metals. Where the environment is corrosive, severity of cavitation damage increases.

Cell: A circuit consisting of an anode and a cathode in electrical contact in a solid or liquid electrolyte. Corrosion generally occurs only at anodic areas.

Cementation coating: A coating developed on a metal surface by a high-temperature diffusion process (as carburization, calorizing, or chromizing).

Cementite: A compound of iron and carbon, known chemically as iron carbide and having the approximate chemical formula Fe_3C. It is characterized by an orthorhombic crystal structure. When it occurs as a phase in steel, the chemical composition will be altered by the presence of manganese and other carbide-forming elements.

Characteristic resistance: The characteristic resistance of the pipe is the electrical resistance between the pipe and remote earth in one direction only from the drain point in ohms.

Chelate: (1) A molecular structure in which a heterocyclic ring can be formed by the unshared electrons of neighboring atoms. (2) A coordination compound in which a heterocyclic ring is formed by a metal bound to two atoms of the associated ligand. See also complexation.

Chelating agent: (1) An organic compound in which atoms form more than one coordinate bond with metals in solution. (2) A substance used in metal finishing to control or eliminate certain metallic ions present in undesirable quantities.

Chelation: A chemical process involving formation of a heterocyclic ring compound that contains at least one metal cation or hydrogen ion in the ring.

Chemical cleaning: Method of surface preparation or cleaning involving the use of chemicals, with or without electrical force, for removal of mill scale, rust, sediments, and paint. These chemicals can also be introduced into some systems on-stream while the system is operating.

Chemical conversion coating: A protective or decorative coating that is produced deliberately on a metal surface by reaction of the surface with a chosen chemical environment. The thin layer formed by this reaction may perform several or all of the following functions: protect against corrosion; provide a base for organic coatings; improve retention of lubricants or compounds; improve abrasion resistance; provide an absorbent layer for rust-preventive oils and waxes.

Chemical potential: In a thermodynamic system of several constituents, the rate of change of the Gibbs function of the system with respect to the change in the number of moles of a particular constituent.

Chemical vapor deposition: A coating process, similar to gas carburizing and carbonitriding, whereby a reactant atmosphere gas is fed into a processing chamber where it decomposes at the surface of the workpiece, liberating one material for either absorption by, or accumulation on, the workpiece. A second material is liberated in gas form and is removed from the processing chamber, along with excess atmosphere gas.

Chemisorption: The binding of an adsorbate to the surface of a solid by forces whose energy levels approximate those of a chemical bond. Contrast with physisorption.

Chevron pattern: A fractographic pattern of radial marks (shear ledges) that look like nested letters "V"; sometimes called a herringbone pattern. Chevron patterns are typically found on brittle fracture surfaces in parts whose widths are considerably greater than their thicknesses. The points of the chevrons can be traced back to the fracture origin.

Chromadizing: Improving paint adhesion on aluminum or aluminum alloys, mainly aircraft skins, by treatment with a solution of chromic acid. Also called chromodizing or chromatizing. Not to be confused with chromating or chromizing.

Chromate treatment: A treatment of metal in a solution of a hexavalent chromium compound to produce a conversion coating consisting of trivalent and hexavalent chromium compounds.

Chromating: Performing a chromate treatment.

Chrome pickle: (1) Producing a chromate conversion coating on magnesium for temporary protection or for a paint base. (2) The solution that produces the conversion coating.

Chromizing: A surface treatment at elevated temperature, generally carried out in pack, vapor, or salt bath, in which an alloy is formed by the inward diffusion of chromium into the base metal.

Clad metal: A composite metal containing two or more layers that have been bonded together. The bonding may have been accomplished by corolling, coextrusion, welding, diffusion bonding, casting, heavy chemical deposition, or heavy electroplating.

Clad metals: Two metals rolled together so that the cheaper one forms a backing for the more costly one.

Cleavage: Splitting (fracture) of a crystal on a crystallographic plane of low index.

Cleavage fracture: A fracture, usually of polycrystalline metal, in which most of the grains have failed by cleavage, resulting in bright reflecting facets. It is associated with low-energy brittle fracture.

Cold working: Deforming metal plastically under conditions of temperature and strain rate that induce strain hardening usually, but not necessarily, conducted at room temperature. Contrast with hot working.

Combined carbon: The part of the total carbon in steel or cast iron that is present as other than free carbon.

Compatibility: The ability of a given material to exist unchanged under certain conditions and environments in the presence of some other material.

Compressive strength: The maximum compressive stress a material is capable of developing. With a brittle material that fails in compression by fracturing, the compressive strength has a definite value. In the case of ductile, malleable, or semiviscous materials (which do not fail in compression by a shattering fracture), the value obtained for compressive strength is an arbitrary value dependent on the degree of distortion that is regarded as effective failure of the material.

Concentration cell: A cell involving an electrolyte and two identical electrodes, with potential resulting from differences in the chemistry of the environments adjacent to the two electrodes.

Concentration polarization: That portion of the polarization of a cell produced by concentration changes resulting from passage of current through the electrolyte.

Conductivity: The ratio of the electric current density to the electric field in a material. Also called electrical conductivity or specific conductance.

Contact corrosion: A term primarily used in Europe to describe galvanic corrosion between dissimilar metals.

Contact potential: The potential difference at the junction of two dissimilar substances.

Continuity bond: A metallic connection that provides electrical continuity between metal structures. A bond designed and installed specifically to ensure the electrical continuity of a structure.

Controlled galvanic system: Cathodic protection system using sacrificial anodes controlled by means of resistors, fixed or variable.

Conversion coating: A coating consisting of a compound of the surface metal, produced by chemical or electrochemical treatments of the metal. Examples include chromate coatings on zinc, cadmium, magnesium, and aluminum and oxide and phosphate coatings on steel. See also chromate treatment and phosphating.

Copper-accelerated salt-spray (CASS) test: An accelerated corrosion test for some electrodeposits for anodic coatings on aluminum.

Copper/copper sulfate reference electrode: A reference electrode consisting of copper in a saturated copper sulfate solution.

Copper ferrule: A ring or cap of copper put around a slender shaft (as a cane or tool handle) to strengthen it or to prevent splitting.

Corrosion: The destruction of a substance; usually a metal, or its properties because of a reaction with its surroundings (environment), i.e., physiochemical interaction between a metal and its environment that results in changes in the properties of the metal and that may often lead to impairment of the function of the metal, the environment, or the technical system, of which these form a part.

Corrosion damage: Corrosion effect that is considered detrimental to the function of the metal, the environment, or the technical system, of which these form a part.

Corrosion effect: A change in any part of the corrosion system caused by corrosion.

Corrosion embrittlement: The severe loss of ductility of a metal resulting from corrosive attack, usually intergranular and often not visually apparent.

Corrosion engineer: The person responsible for carrying out the corrosion monitoring and interpretation of data produced.

Corrosion fatigue: The process in which a metal fractures prematurely under conditions of simultaneous corrosion and repeated cyclic loading at lower stress levels or fewer cycles than would be required in the absence of the corrosive environment.

Corrosion fatigue limit: The maximum cyclic stress value that a metal can withstand for a specified number of cycles or length of time in a given corrosive environment. See corrosion fatigue strength.

Corrosion fatigue strength: The maximum repeated stress that can be endured by a metal without failure under definite conditions of corrosion and fatigue and for a specific number of stress cycles and a specified period of time.

Corrosion inhibitor: An inhibitor is a substance that retards or slows down a chemical reaction. Thus, a corrosion inhibitor is a substance that, when added to an environment, decreases the rate of attack by the environment on a metal. Corrosion inhibitors are commonly added in small amounts to acids, cooling water, oil wells and other environments, either continuously or intermittently to prevent serious corrosion.

Corrosion potential: The potential of a corroding surface in an electrolyte, relative to a reference electrode.

Corrosion product: The chemical compound or compounds produced by the reaction of a corroding metal with its environment.

Corrosion protection: Modification of a corrosion system so that corrosion damage is mitigated.

Corrosion rate: The rate at which corrosion proceeds, expressed by inches of penetration per year (ipy); mils penetration per year (mpy); milligrams weight loss per square decimeter per day (mdd); microns per year (μm/year) or millimeters per year (mmpy). One micron is equal to 0.0395 mils.

Corrosion resistance: Ability of a metal to withstand corrosion in a given corrosion system.

Corrosion system: System consisting of one or more metals and all parts of the environment that influence corrosion.

Corrosive agent: Substance that when in contact with a given metal will react with it.

Corrosive environment: Environment that contains one or more corrosive agents.

Corrosivity: Tendency of an environment to cause corrosion in a given corrosion system.

Couple: A cell developed in an electrolyte resulting from electrical contact between two dissimilar metals. See galvanic corrosion.

Coupons: Specimens of materials exposed to tests or real environments to assess the effect of degradation on the material.

Covered filler rod: A filler rod having a covering of flux.

Covering power: The ability of a solution to give satisfactory plating at very low current densities, a condition that exists in recesses and pits. This term suggests an ability to cover, but not necessarily to build up, a uniform coating, whereas throwing power suggests the ability to obtain a coating of uniform thickness of an irregularly shaped object.

Cracking: When a metal part fails by cracking, it is generally obvious that it cracked, but the exact type of cracking and the cause are less obvious. To determine the type of cracking, microscopic examination is necessary. In some instances, the environment plays a minor role, while in others its role is major.

Cracking (of coating): Breaks in a coating that extend through to the underlying surface.

Crazing: A network of checks or cracks appearing on the surface.

Creep: Time-dependent strain occurring under stress. The creep strain occurring at a diminishing rate is called primary creep; that occurring at a minimum and almost constant rate, secondary creep; and that occurring at an accelerating rate, tertiary creep.

Creep-rupture embrittlement: Embrittlement under creep conditions of, for example, aluminum alloys and steels that results in abnormally low rupture ductility. In aluminum alloys, iron in amounts above the solubility limit is known to cause such embrittlement; in steels, the phenomenon is related to the amount of impurities (for example, phosphorus, sulfur, copper, arsenic, antimony, and tin) present. In either case, failure occurs by intergranular cracking of the embrittled material.

Creep-rupture strength: The stress that will cause fracture in a creep test at a given time in a specified constant environment. Also called stress-rupture strength.

Crevice corrosion: Crevice corrosion is a special type of pitting. The anode of a corrosion cell is fixed by the geometry in a crevice or under a deposit. To function as a corrosion site, a crevice must be wide enough to permit entry of the liquid, but narrow enough to maintain a stagnant zone. Metals or alloys that depend on oxide films or passive layers for corrosion resistance are particularly susceptible to crevice corrosion.

Critical anodic current density: The maximum anodic current density observed in the active region for a metal or alloy electrode that exhibits active–passive behavior in an environment.

Critical pitting potential (E_{cp}, E_p, E_{pp}): The lowest value of oxidizing potential at which pits nucleate and grow. It is dependent on the test method used.

Current density: The direct current per unit area, generally expressed as milliamperes per square meter. Current density required to achieve cathodic protection varies depending on environment and metal being protected.

Current efficiency: The ratio of the electrochemical equivalent current density for a specific reaction to the total applied current density.

Deactivation: The process of prior removal of the active corrosive constituents, usually oxygen, from a corrosive liquid by controlled corrosion of expendable metal or by other chemical means, thereby making the liquid less corrosive.

Dealloying: The selective corrosion of one or more components of a solid solution alloy, usually in the form of ions. Also called parting or selective leaching.

Decarburization: Loss of carbon from the surface layer of a carbon-containing alloy due to reaction with one or more chemical substances in a medium that contacts the surface.

Decobaltification: Corrosion in which cobalt is selectively leached from cobalt-based alloys, such as stellite, or from cemented carbides. See also dealloying and selective leaching.

Decomposition potential (or voltage): The potential of a metal surface necessary to decompose the electrolyte of a cell or a component/substance thereof.

Deep anode bed: Type of ground bed using a drilled vertical hole to contain impressed current anodes. Construction techniques vary widely but objectives are low ground bed resistance, uniform current distribution, less interference, and permanent moisture. Presently, current output is up to 150 A per hole. Typical depth range from 30 to 400 m to reach moist low-resistivity strata.

Deep ground bed: One or more anodes installed vertically at a nominal depth of 15 m (50 ft) or more below the earth's surface in a drilled hole for the purpose of supplying *cathodic protection* for an underground or submerged metallic structure. See also ground bed.

Demineralization: Removal of dissolved mineral matter, generally from water.

Dendrite: A crystal that has a treelike branching pattern, being most evident in cast metals, slowly cooled through the solidification range.

Denickelification: Corrosion in which nickel is selectively leached from nickel-containing alloys. Most commonly observed in copper–nickel alloys after extended service in fresh water.

Density (of solids and liquids): The mass of unit volume of a material at a specified temperature.

Deoxidizing: (1) The removal of oxygen from molten metals by use of suitable deoxidizers. (2) Sometimes refers to the removal of undesirable elements other than oxygen by the introduction of elements or compounds that readily react with them. (3) In metal finishing, the removal of oxide films from metal surfaces by chemical or electrochemical reaction.

Depolarization: The elimination or reduction of polarization by physical or chemical means; depolarization results in increased corrosion.

Depolarizer: A substance that produces depolarization.

Deposit: A foreign substance that comes from the environment, adhering to a surface of a material.

Deposit attack: Pitting corrosion resulting from deposits on a metal surface that cause concentration cells.

Deposit corrosion: Corrosion occurring under or around a discontinuous deposit on a metallic surface. Also called poultice corrosion.

Deposited metal: Filler metal after it becomes part of a weld or joint.

Descaling: Removing the thick layer of oxides formed on some metals at elevated temperatures.

Dezincification: Corrosion in which zinc is selectively leached from zinc-containing alloys. Most commonly found in copper–zinc alloys containing less than 83% copper after extended service in water containing dissolved oxygen; the parting of zinc from an alloy (in some brasses, zinc is lost leaving a weak, brittle, porous, copper-rich residue behind). See also dealloying and *selective leaching*.

Dichromate treatment: A chromate conversion coating produced on magnesium alloys in a boiling solution of sodium dichromate.

Dielectric strength: Degree of electrical nonconductance of a material; the maximum electric field a material can withstand without breakdown.

Diffusion coating: Application of metallic coating, the chemical composition of which was modified by diffusing this at melting temperature into the substrate. Any process whereby a base metal or alloy is either (1) coated with another metal or alloy and heated to a sufficient temperature in a suitable environment or (2) exposed to a gaseous or liquid medium containing the other metal or alloy, thus causing diffusion of the coating or of the other metal or alloy into the base metal with resultant changes in the composition and properties of its surface.

Diffusion coefficient: A factor of proportionality representing the amount of substance diffusing across a unit area through a unit concentration gradient in unit time.

Diffusion-limited current density: The current density, often referred to as limiting current density, that corresponds to the maximum transfer rate that a particular species can sustain because of the limitation of diffusion.

Disbandment: The destruction of adhesion between a coating and the surface coated.

Discontinuity: Any interruption in the normal physical structure or configuration of a part, such as cracks, laps, seams, inclusions, or porosity. A discontinuity may or may not affect the usefulness of the part.

Dislocation: A linear imperfection in a crystalline array of atoms. Two basic types are recognized: (1) an edge dislocation corresponds to the row of mismatched atoms along the edge formed by an extra, partial plane of atoms within the body of a crystal; (2) a screw dislocation corresponds to the axis of a spiral structure in a crystal, characterized by a distortion that joins normally parallel planes together to form a continuous helical ramp (with a pitch of one interplanar distance) winding about the dislocation. Most prevalent is the so-called mixed dislocation, which is any combination of an edge dislocation and a screw dislocation.

Double layer: The interface between an electrode or a suspended particle and an electrolyte created by charge–charge interaction (charge separation) leading to an alignment of oppositely charged ions at the surface of the electrode or particle. The simplest model is represented by a parallel plate condenser of 2×10^{-8} cm in thickness. In general, the electrode will be positively charged with respect to the solution.

Drainage: Conduction of electric current from an underground metallic structure by means of a metallic conductor. Forced drainage is that applied to underground metallic structures by means of an applied electromotive force or sacrificial anode. Natural drainage is that from an underground structure to a more negative (more anodic) structure, such as the negative bus of a trolley substation.

Drainage (current requirement) tests: Tests with current applied for a short period, usually with temporary anodes and power sources in order to determine the current needed to achieve cathodic protection.

Driving EMF (galvanic anode system): The difference between the structure/electrolyte potential and the anode/electrolyte potential.

Dry corrosion: Gaseous corrosion.

Ductility: The ability of a material to deform plastically without fracturing, measured by elongation or reduction of area in a tensile test, by height of cupping in an Erichsen test, or by other means.

Dummy cathode: (1) A cathode, usually corrugated to give variable current densities, that is plated at low current densities to preferentially remove impurities from a plating solution. (2) A substitute cathode that is used during adjustment of operating conditions.

Dummying: Plating with dummy cathodes.

Dynamic equilibrium: The condition of an electrode when the rate of anodic dissolution just balances the rate of cathodic plating.

Embrittlement 885F (475C): Embrittlement of stainless steels upon extended exposure to temperatures between 400 and 510 °C (730 and 930 °F). This type of embrittlement is caused by fine, chromium-rich precipitates that segregate at grain boundaries: time at temperature directly influences the amount of segregation. Grain-boundary segregation of the chromium-rich precipitates increases strength and hardness, decreases ductility and toughness, and changes corrosion resistance. This type of embrittlement can be reversed by heating above the precipitation range.

Edge preparation: Squaring, grooving, chamfering, or beveling an edge in preparation for welding.

Elastic deformation: A change in dimensions directly proportional to and in phase with an increase or decrease in applied force.

Elasticity: The property of a material by virtue of which deformation caused by stress disappears upon removal of the stress. A perfectly elastic body completely recovers its original shape and dimensions after release of stress.

Elastic limit: The maximum stress that a material is capable of sustaining without any permanent strain (deformation) remaining upon complete release of the stress.

Elastomer: A natural or synthetic polymer, which at room temperature can be stretched repeatedly to at least twice its original length, and which after removal of the tensile load will immediately and forcibly return to approximately its original length.

Electrical conductivity: See conductivity.

Electrical grounding: Provides a low-resistance path to ground for fault currents in electrical equipment and distribution networks. Since bare copper is commonly used, current requirement calculations must include the copper as a substantial sink for cathodic protection current to adequately size for cathodic protection system.

Electrical isolation: The condition of being electrically separated from other metallic structures or the environment.

Electrical resistivity: The electrical resistance offered by a material to the flow of current, times the cross-sectional area of current flow and per unit length of current path; the reciprocal of the conductivity. Also called resistivity or specific resistance.

Electrochemical admittance: The inverse of electrochemical impedance.

Electrochemical cell: An electrochemical system consisting of an anode and a cathode in metallic contact and immersed in an electrolyte. (The anode and cathode may be different metals or dissimilar areas on the same metal surface.)

Electrochemical corrosion: Corrosion that is accompanied by a flow of electrons between cathodic and anodic areas on metallic surfaces.

Electrochemical equivalent: The weight of an element or group of elements oxidized or reduced at 100%, efficiency by the passage of a unit quantity of electricity. Usually expressed as grams per coulomb (1 amp/s).

Electrochemical impedance: The frequency-dependent complex-valued proportionality factor (SE/6i) between the applied potential or current and the response signal. This factor is the total opposition of an electrochemical system to the passage of charge. The value is related to the corrosion rate under certain circumstances.

Electrochemical impedance spectroscopy: (AC impedance)—A method to study the impedance of a metal/fluid interface by electrochemical properties by applying a sinusoidal polarization potential to the interface through a range of frequencies.

Electrochemical potential: The partial derivative of the total electrochemical free energy at a constituent with respect to the number of moles of this constituent where all factors are kept constant. It is analogous to the chemical potential of a constituent except that it includes the electric as well as chemical contributions to the free energy. The potential of an electrode in an electrolyte relative to a reference electrode measured under open circuit conditions.

Electrochemical series: Same as electromotive force series.

Electrode: (1) An electronic conductor used to establish electrical contact with an electrolytic part of a circuit. (2) An electronic conductor in contact with an ionic conductor.

Electrode polarization: Change of electrode potential with respect to a reference value. Often the free corrosion potential is used as the reference value. The change may be caused, for example, by the application of an external electrical current or by the addition of on oxidant or reductant.

Electrodeposition: The deposition of a substance on an electrode by passing electric current through an electrolyte.

Electrode potential: The potential of an electrode in an electrolyte as measured against a reference electrode. The electrode potential does not include any resistance losses in potential in either the solution or external circuit. It represents the reversible work to move a unit charge from the electrode surface through the solution to the reference electrode. The potential of an electrode as measured against a reference electrode. The electrode potential does not include any resistance loss in potential in solution due to the current passing to or from the electrode.

Electrode reaction: Interfacial reaction equivalent to a transfer of charge between electronic and ionic conductors. See also anodic reaction and cathodic reaction.

Electrogalvanising: Galvanized by electroplating.

Electrokinetic potential: This potential, sometimes called zeta potential, is a potential difference in the solution caused by residual, unbalanced charge distribution in the adjoining solution, producing a double layer. The electrokinetic potential is different from the electrode potential in that it occurs exclusively in the solution phase, that is, it represents the reversible work necessary to bring a unit charge from infinity in the solution up to the interface in question but not through the interface.

Electroless plating: A process in which metal ions in a dilute aqueous solution are plated out on a substrate by means of autocatalytic chemical reduction.

Electrolysis: Production of chemical changes of the electrolyte by the passage of current through an electrochemical cell.

Electrolyte: A chemical substance or mixture, usually liquid, containing ions that migrate in an electric field. An ionic conductor (usually in aqueous solution). (1) A chemical substance or mixture, usually liquid, containing ions that migrate in an electric field. (2) A chemical compound or mixture of compounds that when molten or in solution will conduct an electric current.

Electrolytic cell: An assembly, consisting of a vessel, electrodes, and an electrolyte, in which electrolysis can be carried out.

Electrolytic cleaning: A process of removing soil, scale, or corrosion products from a metal surface by subjecting it as an electrode to an electric current in an electrolytic bath; process of cleaning, degreasing, of a metal by making it an electrode in a suitable bath.

Electrolytic protection: cathodic protection.

Electromotive force: Electrical potential; voltage.

Electromotive force series: (EMF series)—The potential of an electrode in an electrolytic solution when the forward rate of a given reaction is exactly equal to the reverse rate. (The equilibrium potential can only be defined with respect to a specific electrochemical reaction.).

Electromotive force series (emf series): A list of elements arranged according to their standard electrode potentials (hydrogen electrode is a reference point and given the value zero), with "noble" metals such as gold being positive and "active" metals such as zinc being negative.

Electron flow: A movement of electrons in an external circuit connecting an anode and cathode in a corrosion cell; the current flow is arbitrarily considered to be in an opposite direction to the electron flow.

Electronegative: A qualification applied to a metallic electrode to indicate that its potential is negative with respect to another metallic electrode in the system.

Electroplating: Electrodeposition of a thin adherent layer of a metal or alloy of desirable chemical, physical, and mechanical properties on metallic or nonmetallic substrate.

Electropolishing: A technique commonly used to prepare metallographic specimens, in which a high polish is produced by making the specimen the anode in an electrolytic cell, where preferential dissolution at high points smooths the surface.

Electropositive: A qualification applied to a metallic electrode to indicate that its potential is positive with respect to another metallic electrode in the system.

Electroslag welding: A welding process in which consumable electrodes are fed into a joint containing flux; the current melts the flux, and the flux in turn melts the faces of the joint and the electrodes, allowing the weld metal to form a continuous cast ingot between the joint faces.

Electron-beam welding: Fusion welding in which the joint is made by fusing the parent metal by the impact of a focused beam of electrons.

Electrosmosis: The passage of a liquid through a porous medium under the influence of a potential difference.

Embrittlement: Severe loss of ductility of a metal (or alloy). Loss of load carrying capacity of a metal or alloy; the severe loss of ductility or toughness or both, of a material, usually a metal or alloy. Many forms of embrittlement can lead to brittle fracture. Many forms can occur during thermal treatment or elevated-temperature service (thermally induced embrittlement). Some of these forms of embrittlement, which affect steels, include blue brittleness, 885 °F (475 °C) embrittlement, quench-age embrittlement, sigma-phase embrittlement, strain-age embrittlement, temper embrittlement, tempered martensite embrittlement, and thermal embrittlement. In addition, steels and other metals and alloys can be embrittled by environmental conditions (environmentally assisted embrittlement). The forms of environmental embrittlement include acid embrittlement, caustic embrittlement, corrosion embrittlement, creep-rupture embrittlement, hydrogen embrittlement, liquid metal embrittlement, neutron embrittlement, solder embrittlement, solid metal embrittlement, and stress-corrosion cracking.

Encapsulation: To protect the assembly by inhibited organic sealant, plastic caps, or cast potting compound.

Endurance limit: The maximum stress that a material can withstand for an infinitely large number of fatigue cycles; maximum cyclic stress level a metal can withstand without fatigue failure. See also fatigue strength.

Environment: The surroundings or conditions (physical, chemical, mechanical) in which a material exists.

Environmental cracking: Brittle fracture of a normally ductile material in which the corrosive effect of the environment is a causative factor. Environmental cracking is a general term that includes corrosion fatigue, high-temperature hydrogen attack, hydrogen blistering, hydrogen embrittlement, liquid metal embrittlement, solid metal embrittlement, stress-corrosion cracking, and sulfide stress cracking. The following terms have been used in the past in connection with environmental cracking, but are becoming obsolete: caustic embrittlement, delayed fracture, season cracking, static fatigue, stepwise cracking, sulfide corrosion cracking, and sulfide stress-corrosion cracking. See also embrittlement.

Environmentally assisted embrittlement: See embrittlement.

Epoxy: Resin formed by the reaction of bisphenol and epichlorohydrin.

Equilibrium (reversible) potential: The potential of an electrode in an electrolytic solution when the forward rate of a given reaction is exactly equal to the reverse rate. The equilibrium potential can only be defined with respect to a specific electrochemical reaction.

Erosion: Destruction of metals by the abrasive action of moving fluids accelerated by the presence of solid particles in suspension. When corrosion occurs simultaneously, the term "erosion–corrosion" is often used.

Erosion and erosion–corrosion: Erosion is a strictly mechanical phenomenon, while erosion–corrosion is a combination of mechanical action and chemical or electrochemical reaction. Pure erosion seldom occurs in aqueous systems. Erosion–corrosion is characterized by grooves, gullies, waves, rounded holes, and valleys, and usually exhibits a directional pattern. In copper alloy heat exchanger tubes, the attack frequently results in the formation of horseshoe-shaped depressions. Erosion–corrosion is the acceleration of metal loss because of the relative movement between a fluid and a metal surface. Generally, the movement is rapid, and the effects of mechanical wear are involved. Metal is removed as dissolved ions or as solid corrosion products that are swept from the surfaces.

Erosion–corrosion: Corrosion that is increased because of the abrasive action of a moving stream; the presence of suspended particles greatly accelerates abrasive action.

Etch: To corrode the surface of a metal in order to reveal its composition and structure.

Eutectic: (1) An isothermal reversible reaction in which a liquid solution is converted into two or more intimately mixed solids on cooling, the number of solids formed being the same as the number of components in the system. (2) An alloy having the composition indicated by the eutectic point on an equilibrium diagram. (3) An alloy structure of intermixed solid constituents formed by a eutectic reaction.

Eutectoid: (1) An isothermal reversible reaction in which a solid solution is converted into two or more intimately mixed solids on cooling, the number of solids formed being the same as the number of components in the system. (2) An alloy having the composition indicated by the eutectoid point on an equilibrium diagram. (3) An alloy structure of intermixed solid constituents formed by a eutectoid reaction.

Exchange current: When an electrode reaches dynamic equilibrium in a solution, the rate of anodic dissolution balances the rate of cathodic plating. The rate at which either positive or negative charges are entering or leaving the surface at this point is known as the exchange current.

Exchange current density: The rate of charge transfer per unit area when an electrode reaches dynamic equilibrium (at its reversible potential) in a solution; that is, the rate of anodic charge transfer (oxidation) balances the rate of cathodic charge transfer (reduction).

Exfoliation: Exfoliation is a type of subsurface corrosion that occurs and propagates as cracks approximately parallel to the surface. It leaves the metal in a laminated, flaky, or blistered condition, and appears most frequently in aluminum alloys or cupronickels.

Exothermal (cadweld/thermit) weld: Process of insuring electrical connection of cable to piping or steel structures incendiary mixture of fine aluminum powder with a metallic oxide of iron or chromium, which when ignited yields an intense heat.

External circuit: The wires, connectors, measuring devices, current sources, etc. that are used to bring about or measure the desired electrical conditions within the test cell. It is this portion of the cell through which electrons travel.

Failure: A general term used to imply that a part in service (1) has become completely inoperable, (2) is still operable but is incapable of satisfactorily performing its intended function, or (3) has deteriorated seriously, to the point that it has become unreliable or unsafe for continued use.

Faraday's law: (1) The amount of any substance dissolved or deposited in electrolysis is proportional to the total electric charge passed. (2) The amounts of different substances dissolved or deposited by the passage of the same electric charge are proportional to their equivalent weights.

Fatigue: Subjecting a material to repeated stresses ultimately results in cracking. The environment may have an effect on the fatigue limit of a metal, although this is usually a minor factor. Generally, a fatigue failure is a single fracture, which is transgranular in most common metals. There is normally only a single fracture because stresses on other regions are relieved when the fracture occurs. Characteristic chevron patterns or beach marks can appear on the fracture face.

Fatigue crack growth rate: The rate of crack extension caused by constant-amplitude fatigue loading, expressed in terms of crack extension per cycle of load application.

Fatigue life: The number of cycles of stress that can be sustained prior to failure under a stated test condition.

Fatigue limit: The maximum stress that presumably leads to fatigue fracture in a specified number of stress cycles. If the stress is not completely reversed, the value of the mean stress, the minimum stress, or the stress ratio should also be stated. Compare with *endurance limit*.

Fatigue strength: The maximum stress that can be sustained for a specified number of cycles without failure, the stress being completely reversed within each cycle unless otherwise stated.

Ferritic: Pertaining to the body-centered cubic crystal structure (BCC) of many ferrous (iron-based) metals.

Filiform corrosion: Corrosion that occurs under film in the form of randomly distributed hairlines.

Filler rod: Filler metal in the form of a rod. It may also take the form of filler wire.

Filler metal: Metal added during welding, braze welding, brazing, or surfacing.

Film: A thin, not necessarily visible, layer of material.

Finite line: If the line terminates in an insulated flange or dead ends, it is called a finite line.

Flakes: Short, discontinuous internal fissures in wrought metals attributed to stresses produced by localized transformation and decreased solubility of hydrogen during cooling after hot working. In a fracture surface, flakes appear as bright silvery areas; on an etched surface, they appear as short, discontinuous cracks. Also called shatter cracks or snowflakes.

Flame hardening: Hardening of metal surface by heating with oxyacetylene torch, followed by rapid cooling with water or air jet.

Flame spraying: *Thermal spraying* in which coating material is fed into an oxyfuel gas flame, where it is melted. Compressed gas may or may not be used to atomize the coating material and propel it onto the substrate.

Flow line: Pipeline carrying product from wellhead to gas–oil separator plant (GOSP), typically 100–250 DN, coated or uncoated usually above ground on pipe supports with periodic road crossings.

Flux: Material used during welding, brazing, or braze welding to clean the surfaces of joint, prevent atmospheric oxidation, and reduce impurities.

Fogged metal: A metal whose luster has been reduced because of a surface film, usually a corrosion product layer.

Foreign structure: Any metallic structure that is not intended as part of a *cathodic protection* system of interest.

Fouling: An accumulation of deposits. This term includes accumulation and growth of marine organisms on a submerged metal surface and also includes the accumulation of deposits (usually inorganic) on heat exchanger tubing.

Fouling organism: Any aquatic organism with a sessile adult stage that attaches to and fouls underwater structures of ships.

Fractography: Descriptive treatment of fracture, especially in metals, with specific reference to photographs of the fracture surface. Macrofractography involves photographs at low magnification ($<25\times$); microfractography, photographs at high magnification ($>25\times$).

Fracture mechanics: A quantitative analysis for evaluating structural behavior in terms of applied stress, crack length, and specimen or machine component geometry.

Fracture toughness: A generic term for measures of resistance to extension of a crack. The term is sometimes restricted to results of fracture mechanics tests, which are directly applicable in fracture control. However, the term commonly includes results from simple tests of notched or precracked specimens not based on fracture mechanics analysis. Results from test of the latter type are often useful for fracture control, based on either service experience or empirical correlations with fracture mechanics tests.

Free carbon: The part of the *total carbon* in steel or cast iron that is present in elemental form as graphite or temper carbon. Contrast with *combined carbon.*

Free corrosion potential: *Corrosion potential* in the absence of net electrical current flowing to or from the metal surface.

Free ferrite: *Ferrite* that is formed directly from the decomposition of hypoeutectoid austenite during cooling, without the simultaneous formation of cementite. Also called proeutectoid ferrite.

Free machining: Pertains to the machining characteristics of an alloy to which one or more ingredients have been introduced to give small broken chips, lower power consumption, better surface finish, and longer tool life; among such additions are sulfur or lead to steel, lead to brass, lead and bismuth to aluminum, and sulfur or selenium to stainless steel.

Fretting: A type of wear that occurs between tight-fitting surfaces subjected to cyclic relative motion of extremely small amplitude. Usually, fretting is accompanied by corrosion, especially of the very fine wear debris. Fretting refers to metal deterioration caused by repetitive slip at the interface between two surfaces. When metal loss is increased by corrosion, the term "fretting corrosion" should be used.

Fretting corrosion: Another special case of erosion–corrosion, fretting corrosion, occurs when two heavily loaded metals rub rapidly together, causing damage to one or both metals. Vibration is usually responsible for the damage, but corrosion is also a factor because the frictional heat increases oxidation. In addition, mechanical removal of protective corrosion products continually exposes fresh metal. Fretting corrosion occurs more frequently in air than in water.

Furan: Resin formed from reactions involving furfuryl alcohol alone or in combination with other constituents.

Fusion penetration: Depth to which the parent metal has been fused.

Fusion welding: Welding in which the weld is made between metals in a molten state without the application of pressure.

Fusion zone: The part of the parent metal that is melted into the weld metal.

Galvanic: Pertaining to the current resulting from the coupling of dissimilar electrodes in an electrolyte.

Galvanic anode or sacrificial: A metal, which because of its relative position in the electromotive force (emf) series, provides sacrificial protection to metals that are less negative (lower) in the series, when the two are electrically coupled in an electrolyte. The voltage difference between the anode and the structure causes a current flow in the structure that opposes the corrosion current. The common types of galvanic anodes are rod, bracelet, and ribbon.

Galvanic cell: A cell in which chemical change is the source of electrical energy. It usually consists of two dissimilar conductors in contact with each other and with an electrolyte, or of two similar conductors in contact with each other and with dissimilar electrolytes.

Galvanic corrosion: When two dissimilar metals are in contact with each other and exposed to a conductive environment, a potential exists between them, and a current flows. The less resistant metal becomes anodic, and the more resistant, cathodic. Attack on the less resistant metal increases, while on the more resistant one, it decreases.

Galvanic couple: A pair of dissimilar conductors, commonly metals, in electrical contact.

Galvanic couple potential: *Mixed potential.*

Galvanic current: The electric current that flows between metals or conductive nonmetal in a *galvanic couple.*

Galvanic series: A list of metals arranged according to their relative corrosion potentials in some specific environment; seawater is often used.

Galvanize: To coat a metal surface with zinc using any of various processes.

Galvanizing: The accepted term for the coating of iron or steel with zinc by the immersion of the metal in a bath of molten zinc. Galvanizing comes from "GALVANO."

Galvanneal: To produce a zinc–iron alloy coating on iron or steel by keeping the coating molten after hot dip galvanizing until the zinc alloys completely with the base metal.

Galvanometer: An instrument for indicating or measuring a small electric current by means of a mechanical motion derived from electromagnetic or electrodynamic forces produced by the current.

Galvanostatic: An experimental technique where by an *electrode* is maintained at a constant current in an *electrolyte.*

Gaseous corrosion: Corrosion with gas as the only corrosive agent and without any aqueous phase on the surface of the metal. Also called dry corrosion.

Gamma iron: The face-centered cubic form of pure iron, stable from 910 to 1400 °C (1670–2550 °F).

General corrosion: A form of deterioration that is distributed more or less uniformly over a surface; See *uniform corrosion.*

Gibbs free energy: The thermodynamic function $3G = 5H - TSS$, where H is enthalpy, T is absolute temperature, and S is entropy. Also called free energy, free enthalpy, or Gibbs function.

Glass electrode: A glass membrane *electrode* used to measure pH or hydrogen ion activity.

Grain: An individual crystal in a polycrystalline metal or alloy; it may or may not contain twinned regions and subgrains; a portion of a solid metal (usually a fraction of an inch in size), in which the atoms are arranged in an orderly pattern.

Grain boundary: A narrow zone in a metal corresponding to the transition from one crystallographic orientation to another, thus separating one *grain* from another; the atoms in each grain are arranged in an orderly pattern; the irregular junction of two adjacent grains is known as a grain boundary.

Grain-boundary corrosion: Same as intergranular corrosion.

Graphitic corrosion: Deterioration of gray cast iron in which the metallic constituents are selectively leached or converted to corrosion products leaving the graphite intact. The term *graphic quotation* is commonly used to identify this form of corrosion, but is not recommended because of its use in metallurgy for the decomposition of carbide to graphite; deterioration of gray cast iron in which the metallic constituents are selectively leached or converted to corrosion products leaving the graphite intact.

Graphitization: A metallurgical term describing the formation of graphite in iron or steel, usually from decomposition of iron carbide at elevated temperatures. Not recommended as a term to describe *graphitic corrosion.*

Green liquor: The liquor resulting from dissolving molten melt iron from the kraft recovery furnace in water.

Green rot: A form of high-temperature corrosion of chromium-bearing alloys in which green chromium oxide (Cr_2O_3) forms, but certain other alloy constituents remain metallic; some simultaneous carburization is sometimes observed.

Ground (anode) bed: Commonly a group of manufactured electrodes or scrap steel that serve as the anode for the cathodic protection of pipelines, tanks, or other buried metallic structures. Types of ground beds are surface and deep anode configurations.

Half-cell: A pure metal in contact with a solution of known concentration of its own ion, at a specific temperature, develops a potential that is characteristic and reproducible; when coupled with another half-cell, an overall potential develops, which is the sum of both half-cells.

Halogen: Any of the elements of the halogen family, consisting of fluorine, chlorine, bromine, iodine, and astatine.

Hard chromium: Chromium plated for engineering rather than decorative applications.

Hardenability: The relative ability of a ferrous alloy to form martensite when quenched from a temperature above the upper critical temperature. Hardenability is commonly measured as the distance below a quenched surface at which the metal exhibits a specific hardness (50 HRC, for example) or a specific percentage of martensite in the microstructure.

Hardfacing: Depositing filler metal on a surface by welding, spraying, or braze welding to increase resistance to abrasion, erosion, wear, galling, impact, or cavitation damage.

Hard water: Water that contains certain salts, such as those of calcium or magnesium, which form insoluble deposits in boilers and form precipitates with soap.

Heat-affected zone: That portion of the base metal that was not melted during brazing, cutting, or welding, but whose microstructure and mechanical properties were altered by the heat; refers to area adjacent to a weld where the thermal cycle has caused microstructural changes that generally affect corrosion behavior.

Heat check: A pattern of parallel surface cracks that are formed by alternate rapid heating and cooling of the extreme surface metal, sometimes found on forging dies and piercing punches. There may be two sets of parallel cracks, one set perpendicular to the other.

Hematite: (1) An iron mineral crystallizing in the rhombohedral system; the most important are of iron. (2) An iron oxide, FeO, corresponding to an iron content of approximately 70%.

Hermetic seal: An impervious seal made by the fusion of metals or ceramic materials (as by brazing, soldering, welding, and fusing glass or ceramic), which prevents the passage of gas or moisture.

High-temperature hydrogen attack: A loss of strength and ductility of steel by high-temperature reaction of absorbed hydrogen with carbides in the steel resulting in *decarburization* and internal fissuring.

Holidays: Discontinuities in a coating (such as porosity, cracks, etc.) that allow areas of base metal to be exposed to any corrosive environment that contacts the coated surface.

Hot corrosion: An accelerated corrosion of metal surfaces that results from the combined effect of oxidation and reactions with sulfur compounds and other contaminants, such as chlorides, to form a molten salt on a metal surface that fluxes, destroys, or disrupts the normal protective oxide. See also gaseous *erosion*.

Hot cracking: Also called solidification cracking, hot cracking of weldments is caused by the segregation at grain boundaries of low-melting constituents in the weld metal. This can result in grain-boundary tearing under thermal contraction stresses. Hot cracking can be minimized by the use of low-impurity welding materials and proper joint design. See also *cold cracking, lamellar tearing*, and *stress-relief cracking*.

Hot working: Deforming metal plastically at such a temperature and strain rate that recrystallization takes place simultaneously with the deformation, thus avoiding any strain hardening. Contrast with cold working.

Hot dip coating: A metallic coating obtained by dipping the base metal into a molten metal.

Hot shortness: A tendency for some alloys to separate along grain boundaries when stressed or deformed at temperatures near the melting point. Hot shortness is caused by a low-melting constituent, often present only in minute amounts, that is segregated at grain boundaries.

Huey test: Corrosion testing in a boiling solution of nitric acid. This test is mainly used to detect the susceptibility to intergranular corrosion of stainless steel.

Humidity test: A corrosion test involving exposure of specimens at controlled levels of humidity and temperature. Contrast with *salt fog test.*

Hydrogen-assisted cracking (HAC): *Hydrogen embrittlement.*

Hydrogen-assisted stress-corrosion cracking (HSCC): *Hydrogen embrittlement.*

Hydrogen blistering: The formation of blisters on or below a metal surface from excessive internal hydrogen pressure; formation of blisterlike bulges on a ductile metal surface caused by internal hydrogen pressures. Hydrogen may be formed during cleaning, plating, corrosion, and so forth.

Hydrogen controlled electrode: A covered electrode that, when used correctly, produces less than a specified amount of diffusible hydrogen in the weld deposit.

Hydrogen damage: At moderate temperatures, hydrogen damage can occur as a result of a corrosion reaction on a surface or cathodic protection. Atomic hydrogen diffuses into the metal and collects at internal voids or laminations where it combines to form more voluminous molecular hydrogen. In steels, blisters sometimes occur. At higher temperatures and pressures, atomic hydrogen can diffuse into steel and collect at grain boundaries. Either molecular hydrogen is then formed, or the hydrogen reacts with iron carbides to form methane, resulting in cracking and decarburization. Hydrogen cracking is intergranular and highly branched, but not continuous.

Hydrogen disintegration: Deep internal cracks caused by hydrogen.

Hydrogen embrittlement: A process resulting in a decrease in the *toughness* or *ductility* of a metal due to the presence of atomic hydrogen. Hydrogen embrittlement has been recognized classically as being of two types. The first, known as internal hydrogen embrittlement, occurs when the hydrogen enters molten metal, which becomes supersaturated with hydrogen immediately after solidification. The second type, environmental hydrogen embrittlement, results from hydrogen being absorbed by solid metals. This can occur during elevated-temperature thermal treatments and in service during electroplating, contact with maintenance chemicals, corrosion reactions, cathodic protection, and operating in high-pressure hydrogen. In the absence of residual stress or external loading, environmental hydrogen embrittlement is manifested in various forms, such as blistering, internal cracking, hydride formation, and reduced ductility. With a tensile stress or stress-intensity factor exceeding a specific threshold, the atomic hydrogen interacts with the metal to induce subcritical crack growth leading to fracture. In the absence of a corrosion reaction (polarized cathodically), the usual term used is hydrogen-assisted cracking (HAC) or hydrogen stress cracking (HSC). In the presence of active corrosion, usually as pits or crevices (polarized anodically), the cracking is generally called *stress-corrosion cracking* (SCC), but should more properly be called hydrogen-assisted stress-corrosion cracking (HSCC). Thus, HSC and electrochemically anodic SCC can operate separately or in combination (HSCC). In some metals, such as high-strength steels, the mechanism is believed to be all, or nearly all, HSC. The participating mechanism of HSC is not always recognized and may be evaluated under the generic heading of SCC. Embrittlement of a metal caused by hydrogen; sometimes observed in cathodically protected steel, electroplated parts, pickled steel, etc. Hydrogen embrittlement process results decrease of toughness or ductility of a metal due to absorption of hydrogen.

Hydrogen probes: Probes designed to measure the permeation rate of atomic hydrogen H^+ (measured as hydrogen gas H^2) associated with hydrogen-induced cracking.

Hydrogen stress cracking (HSC): *Hydrogen embrittlement.*

Hydrogen sulfide: Without H_2S: environment containing less than 1 ppm H_2S. With H_2S: environment containing more than 1 ppm H_2S. Low H_2S: greater than 3 ppm but less than 6 ppm H_2S. High H_2S: more than 6 ppm.

Hydrolysis: (1) Decomposition or alteration of a chemical substance by water. (2) In aqueous solutions of electrolytes, the reactions of cations with water to produce a weak base or of anions to produce a weak acid.

Hydrophilic: Having an affinity for water. Contrast with hydrophobic.

Hydrophobic: Lacking an affinity for, repelling, or failing to absorb or adsorb water. Contrast with hydrophilic.

Hygroscopic: (1) Possessing a marked ability to accelerate the condensation of water vapor; applied to condensation nuclei composed of salts that yield aqueous solutions of a very low equilibrium vapor pressure compared with that of pure water at the same temperature. (2) Pertaining to a substance whose physical characteristics are appreciably altered by effects of water vapor. (3) Pertaining to water absorbed by dry soil minerals from the atmosphere; the amounts depend on the physiochemical character of the surfaces, and increase with rising relative humidity.

Immersion plating: Depositing a metallic coating on a metal immersed in a liquid solution, without the aid of an external electric current. Also called dip plating.

Immunity: A state of resistance to corrosion or anodic dissolution of a metal caused by thermodynamic stability of the metal.

Impingement corrosion: A form of erosion–corrosion generally associated with the local impingement of a high-velocity flowing fluid against a solid surface.

Impressed current: Direct current supplied by a power source external to the anode system. Typical power sources are rectifiers solar modules and engine generators.

Impressed current protection: Cathodic protection of structures, where the cathodic polarization of metal is secured by electric currents emitted from an independent source.

Inclusions: Particles of foreign material in a metallic matrix. The particles are usually compounds (such as oxides, sulfides, or silicates), but may be of any substance that is foreign to (and essentially insoluble in) the matrix.

Incubation period: A period prior to the detection of corrosion while the metal is in contact with a corrodent.

Industrial atmosphere: An atmosphere in an area of heavy industry with soot, fly ash, and sulfur compounds as the principal constituents.

Inert anode: An *anode* that is insoluble in the *electrolyte* under the conditions prevailing in the *electrolysis.*

Infinite line: It is a relatively long finite line or a line directly connected to a structure of quite different electrical characteristics such as bare line or a tank farm.

Inhibitor: A chemical substance or combination of substances that, when present in the proper concentration and forms in the environment, prevents or reduces corrosion.

Inorganic: Being or composed of matter other than hydrocarbons and their derivatives, or matter that is not of plant or animal origin. Contrast with *organic.*

Inorganic zinc-rich paint: Coating containing a zinc powder pigment in an *inorganic* vehicle.

Insulated flange: A flanged joint in which flange faces and securing bolts are electrically insulated from each other by insulating sleeves, washers and gaskets. Pipe size and pressure rating must be specified. Insulated flanges are used to electrically isolate pipelines and systems.

Insulated joint: Performs similar function to an insulating but does not incorporate flange gaskets or bolts. Provides electrical isolation via epoxy-sealed gap in a section of pipe.

Insulated spool: It is similar to insulated joint but incorporates internal nonconductive lining where electrically conductive pipeline products are involved.

Interdendritic corrosion: Corrosive attack that progresses preferentially along interdendritic paths. This type of attack results from local differences in composition, such as coring commonly encountered in alloy castings.

Interference bond: A welded metallic connection between two sections of a structure or two or more different structures to carry electrical current. Often used to bypass current in known locations to avoid corrosive current discharge (interference) in unknown or undesirable locations. Then resistance bonds are used, the resistance value must be less than the parallel structure-to- structure soil resistance.

Interference test: A structure-to-electrolyte test to determine the existence of corrosion interaction between two buried or immersed structures where one or both are cathodically protected.

Intergranular: Between crystals or grains. Also called intercrystalline. Contrast with transgranular.

Intergranular corrosion: Metals are composed of grains or crystals that form as solidification occurs. A crystal grows until it meets another advancing crystal. The regions of disarray between crystals are called grain boundaries, which differ in composition from the crystal center. Intergranular corrosion is the selective attack of the grain boundary or an adjacent zone. The most common example of intergranular corrosion is that of sensitized austenitic stainless steels in heat-affected zones at welds. Intergranular corrosion usually leaves the surface roughened, but definite diagnosis must be made by microscopic examination. Corrosion that occurs preferentially at grain boundaries.

Intergranular cracking: Cracking or fracturing that occurs between the grains or crystals in a polycrystalline aggregate. Also called intercrystalline cracking. Contrast with *transgranular cracking*.

Intergranular fracture: Brittle fracture of a metal in which the fracture is between the grains, or crystals, that form the metal. Also called intercrystalline fracture. Contrast with *transgranular fracture*.

Intergranular stress-corrosion cracking (IGSCC): *Stress-corrosion cracking* in which the cracking occurs along grain boundaries.

Internal oxidation: The formation of isolated particles of corrosion products beneath the metal surface. This occurs as the result of preferential oxidation of certain alloy constituents by inward diffusion of oxygen, nitrogen, sulfur, and so forth.

Intumescence: The swelling or bubbling of a coating usually because of heating (term currently used in space and fire protection applications).

Ion: An atom, or group of atoms, that has gained or lost one or more outer electrons and thus carries an electric charge. Positive ions, or *cations*, are deficient in outer electrons. Negative ions, or *anions*, have an excess of outer electrons.

Ion erosion: Deterioration of material caused by ion impact.

Ion exchange: The reversible interchange of ions between a liquid and solid, with no substantial structural changes in the solid.

Iron rot: Deterioration of wood in contact with iron-based alloys.

Isocorrosion diagram: A graph or chart that shows constant corrosion behavior with changing solution (environment) composition and temperature.

K_{ISCC}**:** Abbreviation for the critical value of the plane strain *stress-intensity factor* that will produce crack propagation by *stress-corrosion cracking* of a given material in a given environment.

Killed steel: Thoroughly deoxidized steel, for example, by addition of aluminum or silicon, in which the reduction between carbon and oxygen during solidification is suppressed.

Knife-line attack: *Intergranular corrosion* of an alloy, usually stabilized stainless steel, along a line adjoining or in contact with a weld after heating into the sensitization temperature range.

Kraft process: Wood-pulping process in which sodium sulfate is used in the caustic soda pulp-digestion liquor. Also called kraft pulping or sulfate pulping.

Laminar scale: Rust formation in heavy layers.

Lamellar tearing: Occurs in the base metal adjacent to weldments due to high through-thickness strains introduced by weld metal shrinkage in highly restrained joints. Tearing occurs by decohesion and linking along the working direction of the base metal; cracks usually run roughly parallel to the fusion line and are steplike in appearance. Lamellar tearing can be minimized by designing joints to minimize weld shrinkage stresses and joint restraint. See also cold cracking, hot cracking, and stress-relief cracking.

Langelier saturation index: An index calculated from total dissolved solids, calcium concentration, total alkalinity, pH, and solution temperature that shows the tendency of a water solution to precipitate or dissolve calcium carbonate.

Leakage or coating resistance: Leakage resistance of pipe or resistance of pipe radially to remote earth. This resistance includes the resistance of the coating (if any) and is affected by the resistivity of the environment. Unit is ohm-kilometer.

Ledeburite: The eutectic of the iron–carbon system, the constituents of which are austenite and cementite. The austenite decomposes into ferrite and cementite on cooling below the temperature at which transformation of austenite to ferrite or ferrite plus cementite is completed.

Ligand: The molecule, ion, or group bound to the central atom in a chelate or a coordination compound.

Limiting current density: The maximum current density that can be used to obtain a desired electrode reaction without undue interference such as from polarization.

Linear elastic fracture mechanics: A method of fracture analysis that can determine the stress (or load) required to induce fracture instability in a structure containing a cracklike flaw of known size and shape. See also fracture mechanics and stress-intensity factor.

Linear pipe resistance: The longitudinal pipe resistance of the pipe is the pipe resistance in ohms per unit length, which can be calculated from the specific resistivity of steel or iron and by regarding the pipeline as an annular cylinder. The specific resistivity of steel pipe will normally vary from 15 to 23 microhm-cm depending on its chemistry. In the absence of specific test results it is normal to use a value of 18 microhm-cm.

Linear polarization resistance (LPR): (See polarization resistance.) At small applied polarization potentials, the relationship between applied potential approximates the polarization resistance.

Lipophilic: Having an amenity for oil. See also hydrophilic and hydrophobic.

Liquid metal embrittlement: Catastrophic brittle failure of a normally ductile metal when in contact with a liquid metal and subsequently stressed in tension.

Local action: Corrosion due to the action of "local cells," that is, galvanic cells resulting from inhomogeneities between adjacent areas on a metal surface exposed to an electrolyte.

Local cell: A galvanic cell resulting from inhomogeneities between areas on a metal surface in an electrolyte. The inhomogeneities may be of physical or chemical nature in either the metal or its environment.

Localized corrosion: Corrosion at discrete sites, stress-corrosion cracking.

Longline current: Current that flows through the earth from an anodic to a cathodic area of a continuous metallic structure. Usually used only where the areas are separated by considerable distance and where the current results from concentration-cell action.

Luggin probe (luggin haber capillary): A small tube or capillary filled with electrolyte, terminating close to the metal surface under study, and used to provide an ionically conducting path without diffusion between an electrode under study and a reference electrode.

Macroscopic: Visible at magnifications to 25×.

Macrostructure: The structure of metals as revealed by macroscopic examination of the etched surface of a polished specimen.

Magnetite: Naturally occurring magnetic oxide of iron (Fe_3O_4).

Manual welding: Welding in which the means of making the weld are held in the hand.

Martensite: A generic term for microstructures formed by diffusionless phase transformation in which the parent and product phases have a specific crystallographic relationship. Martensite is characterized by an acicular pattern in the microstructure in both ferrous and nonferrous alloys. In alloys where the solute atoms occupy interstitial positions in the martensitic lattice (such as carbon in iron), the structure is hard and highly strained; but where the solute atoms occupy substitutional positions (such as nickel in iron), the martensite is soft and ductile. The amount of high-temperature phase that transforms to martensite on cooling depends to a large extent on the lowest temperature attained, there being a rather distinct beginning temperature (Ms) and a temperature at which the transformation is essentially complete (Mf).

Mechanical plating: Plating wherein fine metal powders are peened onto the work by tumbling or other means.

Metal-arc welding: Arc welding using a consumable electrode.

Metal dusting: Accelerated deterioration of metals in carbonaceous gases at elevated temperatures to form a dustlike corrosion product; a unique form of high-temperature corrosion that forms a dustlike corrosion product and sometimes develops hemispherical pits on a susceptible metal surface; simultaneous carburization is generally observed.

Metal ion concentration cell: A galvanic cell caused by a difference in metal ion concentration at two locations on the same metal surface.

Metallic glass: An alloy having an amorphous or glassy structure. See also amorphous solid.

Metalizing: (1) The application of an electrically conductive metallic layer to the surface of nonconductors. (2) The application of metallic coatings by nonelectrolytic procedures such as spraying of molten metal and deposition from the vapor phase.

Meteor perforation: Perforation of material in outer space resulting from meteor strikes.

Microbial corrosion: biological corrosion.

Microbiologically influenced corrosion (MIC): Corrosion that is substantially increased as the result of the presence of bacteria (such as sulfate-reducing bacteria, SRB; or acid-producing bacteria, APB).

Microscopic: Visible at magnifications above 25×.

Microstructure: The structure of a prepared surface of a metal as revealed by a microscope at a magnification exceeding 25×.

MIG-welding: Metal inert gas arc welding using a consumable electrode.

Mill scale: An oxide layer on metals or alloys produced by metal rolling, hot forming, welding, or heat treatment. Especially applicable to iron and steel.

Mixed potential: The potential of a specimen (or specimens in a galvanic couple) when two or more electrochemical reactions are occurring. Also called galvanic couple potential.

Molal solution: Concentration of a solution expressed in moles of solute divided by 1000 g of solvent.

Molar solution: Aqueous solution that contains 1 mol (gram-molecular weight) of solute in 1 L of the solution.

Mole: One mole is the mass numerically equal (in grams) to the relative molecular mass of a substance. It is the amount of substance of a system that contains as many elementary units (6.023 exp23) as there are atoms of carbon in 0.012 kg of the pure nuclide C12; the elementary unit must be specified and may be an atom, molecule, ion, electron, photon, or even a specified group of such units.

Monomer: A molecule usually an organic compound, having the ability to join with a number of identical molecules to form a polymer.

Noble metal: A metal that is not very reactive (such as silver, gold, or copper) and may be found naturally in metallic form on earth.

Normalizing: Heating a ferrous alloy to a suitable temperature above the transformation range and then cooling in air to a temperature substantially below the transformation range.

Open-circuit potential: The *potential* of an electrode measured with respect to a reference electrode or another electrode when no current flows to or from it.

Organic: Being or composed of hydrocarbons or their derivatives, or matter of plant or animal origin. Contrast with *inorganic*.

Organic acid: A chemical compound with one or more carboxyl radicals (COOH) in its structure; examples are butyric acid, $CH_3(CH_2)_2COOH$; maleic acid, HOOCCH-CHCOOH; and benzoic acid, C_6H_5COOH.

Organic zinc coating: A paint containing zinc powder pigment and an organic (containing carbon) resin.

Organic zinc-rich paint: Coating containing zinc powder pigment and an *organic* resin.

Overaging: *Aging* under conditions of time and temperature greater than those required to obtain maximum change in a certain property, so that the property is altered in the direction of the initial value.

Overheating: Heating a metal or alloy to such a high temperature that its properties are impaired. When the original properties cannot be restored by further heat treating, by mechanical working, or by a combination of working and heat treating, the overheating is known as *burning*.

Overload: When a metal part has been subjected to a single stress beyond its tensile strength, it can fail by overload. The fracture can be either ductile or brittle, depending on factors such as the metal's hardness and operating temperature. In most cases, a single fracture results.

Overvoltage: The difference between the actual electrode potential when appreciable electrolysis begins and the reversible electrode potential.

Oxyacetylene welding: Gas welding in which fuel gas is acetylene and which is burnt in an oxygen atmosphere.

Oxidation: (1) A reaction in which there is an increase in valence resulting from a loss of electrons. Contrast with *reduction*. (2) A corrosion reaction in which the corroded metal forms an oxide; usually applied to reaction with a gas containing elemental oxygen, such as air.

Oxidized surface (on steel): Surface having a thin, tightly adhering, oxidized skin (from straw to blue in color), extending in from the edge of a coil or sheet.

Oxidizing agent: A compound that causes *oxidation*, thereby itself being reduced.

Oxygen concentration cell: A galvanic cell resulting from difference in oxygen concentration between two locations.

Ozone: A powerfully oxidizing allotropic form of the element oxygen. The ozone molecule contains three atoms (O_3). Ozone gas is decidedly blue, and both liquid and solid ozone are an opaque blue-black color, similar to that of ink.

Parent metal: Metal to be joined, base metal.

Parkerizing: Trade name for process for the production of phosphate coating on steel articles by immersion in an aqueous solution of manganese or zinc acid with phosphate.

Partial annealing: An imprecise term used to denote a treatment given cold-worked material to reduce its strength to a controlled level or to effect stress relief. To be meaningful, the type of material, the degree of cold work, and the time–temperature schedule must be stated.

Parting: The selective attack of one or more components of a solid solution alloy; e.g., dezincification, dealumination, etc. See dealloying.

Parts per million: A measure of proportion by weight, equivalent to one unit weight of a material per million (10^6) unit weights of compound. One part per million is equivalent to 1 mg/g.

Passivation: A reduction of the anodic reaction rate of an electrode involved in electrochemical action such as corrosion.

Passivator: A type of *inhibitor* that appreciably changes the potential of a metal to a more noble (positive) value.

Passive: The state of a metal when it is much more resistant to corrosion than its position in the EMF series would predict. Passivation is a surface phenomenon.

Passive–active cell: (1) A cell, the emf of which is due to the potential difference between a metal in an active state and the same metal in a passive state. (2) A corrosion cell in which the *anode* is a metal in the *active* state and the *cathode* is the same metal in the *passive* state.

Passivity: A metal or alloy that is thermodynamically unstable in a given electrolytic solution is said to be passive when it remains visibly unchanged for a prolonged period. The following should be noted:

- During passivation the appearance may change if the passivating film is sufficiently thick (e.g., interference films);
- The electrode potential of a passive metal is always appreciably more noble than its potential in the active state;
- Passiity is an anodic phenomenon and thus control of corrosion by decreasing cathodic reactivity (e.g., amalgamated zinc in sulfuric acid) or by cathodic protection is not passivity.

Patina: The coating, usually green, that forms on the surface of metals such as copper and copper alloys exposed to the atmosphere. Also used to describe the appearance of a weathered surface of any metal.

Pearlite: A metastable lamellar aggregate of ferrite and cementite resulting from the transformation of austenite at temperatures above the bainite range.

Peen plating: Deposition of the coating metal, in powder form, on the substrate by a tumbling action in presence of peening shot.

Phosphating: Forming an adherent phosphate coating on a metal by immersion in a suitable aqueous phosphate solution. Also called phosphatizing.

Phosphatizing: The forming of a thin inert phosphate coating on a surface, usually accomplished by treating with H_3PO_4 (phosphoric acid).

pH: A measure of the acidity or alkalinity of a solution; The negative logarithm of the hydrogen ion activity; it denotes the degree of acidity or basicity of a solution. At 25 °C (77 °F), 7.0 is the neutral value. Decreasing values below 7.0 indicate increasing acidity; increasing values above 7.0, increasing basicity.

Physical vapor deposition: A coating process whereby the cleaned and masked component to be coated is heated and rotated on a spindle above the streaming vapor generated by melting and evaporating a coating material source bar with a focused electron beam in an evacuated chamber.

Physisorption: The binding of an adsorbate to the surface of a solid by forces whose energy levels approximate those of condensation. Contrast with *chemisorption*.

Pickle: A solution, usually acid, used to remove mill scale or other corrosion products from a metal.

Pickle/Pickling: Form of chemical and electrolytic removal or loosening of mill scale and corrosion products from the surface of a metal in a chemical solution (usually acidic). Electrolytic pickling can be anodic or cathodic depending on polarization of metal in the solution.

Pitting: Highly localized corrosion resulting in deep penetration at only a few spots.

Pitting factor: Ratio of the depth of the deepest pit resulting from corrosion divided by the average penetration as calculated from weight loss.

Plane strain: The stress condition in *linear elastic fracture mechanics* in which there is zero strain in a direction normal to both the axis of applied tensile stress and the direction of crack growth (that is, parallel to the crack front); most nearly achieved in loading thick plates along a direction parallel to the plate surface. Under plane-strain conditions, the plane of fracture instability is normal to the axis of the principal tensile stress.

Plane stress: The stress condition in *linear elastic fracture mechanics* in which the stress in the thickness direction is zero; most nearly achieved in loading very thin sheet along a direction parallel to the surface of the sheet. Under plane-stress conditions, the plane of fracture instability is inclined 45° to the axis of the principal tensile stress.

Plasma plating: Deposition on critical areas of metal coatings resistant to wear and abrasion, by means of a high-velocity and high-temperature ionized inert gas jet.

Plasma spraying: A *thermal spraying* process in which the coating material is melted with heat from a plasma torch that generates a nontransferred arc; molten coating material is propelled against the base metal by the hot, ionized gas issuing from the torch.

Plastic deformation: The permanent (inelastic) distortion of metals under applied stresses that strain the material beyond its *elastic limit.*

Plasticity: The property that enables a material to undergo permanent deformation without rupture.

Polarization: The deviation from the open circuit potential of an electrode resulting from the passage of current.

Polarization admittance: The reciprocal of *polarization resistance (di/dE).*

Polarization curve: A plot of *r current density* versus *electrode potential* for a specific electrode–electrolyte combination.

Polarization resistance: The slope (dE/di) at the corrosion potential of a potential (E) versus current density (i) curve. (It is inversely proportional to the corrosion current density when the polarization resistance technique is applicable.).

Polyester: Resin formed by condensation of polybasic and monobasic acids with polyhydric alcohols.

Polymer: A chain of organic molecules produced by the joining of primary units called *monomers.*

Potential: Any of various functions from which intensity or velocity at any point in a field may be calculated. The driving influence of an electrochemical reaction. See also *active potential, chemical potential, corrosion potential, critical pitting potential, decomposition potential, electrochemical potential, electrode potential, electrokinetic potential, equilibrium (reversible) potential, free corrosion potential, noble potential, open-circuit potential, protective potential, redox potential, and standard electrode potential.*

Potential surveys: Measurement of potential of a structure or pipeline relative to a reference electrode potential pitting.

Potentiodynamic (potentiokinetic): The technique for varying the *potential* of an electrode in a continuous manner at a preset rate.

Potentiostat: An electronic device that maintains an electrode at a constant potential; used in anodic protection devices.

Poultice corrosion: A term used in the automotive industry to describe the corrosion of vehicle body parts due to the collection of road salts and debris on ledges and in pockets that are kept moist by weather and washing. Also called deposit corrosion or attack.

Pourbaix (potential–pH) diagram: A plot of the *redox potential* of a corroding system versus the pH of the system, compiled using thermodynamic data and the *Nernst equation.* The diagram shows regions within which the metal itself or some of its compounds are stable.

Powder metallurgy: The art of producing metal powders and utilizing metal powders for production of massive materials and shaped objects.

Precious metal: One of the relatively scarce and valuable metals: gold, silver, and the platinum-group metals. Also called *noble metal(s).*

Precipitation hardening: Hardening caused by the precipitation of a constituent from a supersaturated solution. See also age hardening and aging.

Precipitation hardening: Hardening caused by the precipitation of a constituent from a supersaturated solid solution. See also *age hardening and aging.*

Precipitation heat treatment: *Artificial aging* in which a constituent precipitated from a supersaturated solid solution.

Precracked specimen: A specimen that is notched and subjected to alternating stresses until a crack has developed at the root of the notch.

Pressure welding: A welding process in which a weld is made by a sufficient pressure to cause plastic flow of the surfaces, which may or may not be heated.

Primary current distribution: The current distribution in an *electrolytic cell* that is free of *polarization.*

Primary passive potential (passivation potential): The potential corresponding to the maximum active current density (critical anodic current density) of an electrode that exhibits active–passive corrosion behavior.

Principal stress (normal): The maximum or minimum value at the *normal stress* at a point in a plane considered with respect to all possible orientations of the considered plane. On such principal planes the shear stress is zero. There are three principal stresses on three mutually perpendicular planes. The state of stress at a point may be (1) uniaxial, a state of stress in which two of the three principal stresses are zero, (2) biaxial, a state of stress in which only one of the three principal stresses is zero, and (3) triaxial, a state of stress in which none of the principal stresses is zero. Multiaxial stress refers to either biaxial or triaxial stress.

Profile: Anchor pattern on a surface produced by abrasive blasting or acid treatment.

Protection current: The current made to flow into a metallic structure from its ectrolytic environment in order to effect cathodic protection of the structure.

Protective potential: The threshold value of the corrosion potential that has to be reached to enter a protective potential range. The term used in cathodic protection to refer to the minimum potential required to suppress corrosion.

Protective potential range: A range of corrosion potential values in which unacceptable corrosion resistance is achieved for a particular purpose.

Quench-age embrittlement: Embrittlement of low-carbon steels resulting from precipitation of solute carbon of existing dislocations and from precipitation hardening of the steel caused by differences in the solid solubility of carbon in ferrite at different temperatures. Quench-age embrittlement usually is caused by rapid cooling of the steel from temperature slightly below AC1 (the temperature at which austenite begins to form), and can be minimized by quenching from lower temperature.

Quench aging: Aging induced by rapid cooling after solution heat treatment.

Quench cracking: Fracture of a metal during quenching from elevated temperature. Most frequently observed in hardened carbon steel, alloy steel, or tool steel parts of high hardness and low toughness. Cracks often emanate from filets, holes, corners, or other stress raisers and result from high stresses due to the volume changes accompanying transformation to martensite.

Quench hardening: (1) In ferrous alloys, hardening by austenitizing and then cooling at a rate such that a substantial amount of austenite transforms to martensite. (2) In copper and titanium alloys, hardening by solution treating and quenching to develop a martensite-like structure.

Quenching: Rapid cooling of metals (often steels) from a suitable elevated temperature. This generally is accomplished by immersion in water, oil, polymer solution, or salt, although forced air is sometimes used.

Radiation damage: A general term for the alteration of properties of a material arising from exposure to ionizing radiation (penetrating radiation), such as X-rays, gamma rays. neutrons, heavy-particle radiation, or fission fragments in nuclear fuel material.

Radiography (X-ray): Use of X-rays to measure thickness or imperfection within solid materials.

Rare earth metal: One of the group of 15 chemically similar metals with atomic numbers 57 through 71, commonly referred to as the lanthanides.

Reactive metal: A metal that readily combines with oxygen at elevated temperatures to form very stable oxides, for example, titanium, zirconium, and beryllium. Reactive metals may also become embrittled by the interstitial absorption of oxygen, hydrogen, and nitrogen.

Recrystallization: (1) Formation of a new, strain-free grain structure from that existing in cold worked metal, usually accomplished by heating. (2) The change from one crystal structure to another, as occurs on heating or cooling through a critical temperature.

Redox potential: The potential of a reversible oxidation–reduction electrode measured with respect to a reference electrode, corrected to the hydrogen electrode, in a given electrolyte.

Reducing agent: A compound that causes *reduction,* thereby itself becoming oxidized.

Reduction: A reaction in which there is a decrease in valence resulting from a gain in electrons. Contrast with *oxidation.*

Reference electrode: A nonpolarizable *electrode* with a known and highly reproducible *potential* used for potentiometric and voltammetric analyses.

Refractory metal: A metal having an extremely high melting point, for example, tungsten, molybdenum, tantalum, niobium, chromium, vanadium, and rhenium. In the broad sense, this term refers to metals having melting points above the range for iron, cobalt, and nickel.

Relative humidity: The ratio, expressed as a percentage, of the amount of water vapor present in a given volume of air at a given temperature to the amount required to saturate the air at that temperature.

Remedial bond: A bond installed between a primary and a secondary structure in order to eliminate or reduce corrosion interaction.

Remote earth: The areas in which the structure-to-electrolyte potential change is negligible with change in reference electrode position away from the structure.

Residual stress: Stresses that remain within a body as a result of *plastic deformation.*

Resistance: The opposition that a device or material offers to the flow of direct current, equal to the voltage drop across the element divided by the current through the element. Also called electrical resistance.

Resistance bond: A bond either incorporating resistors or of adequate resistance in itself to limit the flow of current.

Resistance welding: Welding in which force is applied to surfaces in contact and in which the heat for welding is produced by the passage of electric current through the electrical resistance at, and adjacent to, these surfaces.

Resolution: (Ultrasonics) The ability of a system to give simultaneous separate indications from discontinuities that are close together both in depth and lateral position.

Reverse current switch: The reverse current switch is installed in series with interference bonds where stray current is known to reverse direction. Prevents serious corrosion, where reversed current can discharge to the electrolyte, by interrupting the reversed current. A failed switch becomes an open circuit or a solid bond.

Ringworm corrosion: Localized corrosion frequently observed in oil well tubing in which a circumferential attack is observed near a region of metal "upset."

Riser: (1) That section of pipeline extending from the ocean floor up the platform. Also, the vertical tube in a steam generator convection bank that circulates water and steam upward. (2) A reservoir of molten metal connected to a casting to provide additional metal to the casting, required as the result of shrinkage before and during solidification.

Run: The metal method or deposited during one passage of an electrode, torch, or blow-pipe.

Rust: A visible corrosion product consisting of hydrated oxides of iron. Applied only to ferrous alloys. See also *white rust.*

Rusting: Corrosion of iron or an iron-based alloy to form a reddish-brown product, which is primarily hydrated ferric oxide.

Sacrificial anode: The use of an alloy electrode less noble than the structure to which it is connected, to induce galvanic corrosion on the electrode (anode) in preference to that of the structure.

Sacrificial protection/sacrificial anodes/sacrificial pieces: Pieces of metal that, being anodic to the equipment into which they are introduced, will galvanically corrode and so protect the equipment. Cathodic protection, based on wasting of anodic metal to prevent corrosion of cathodic metal—zinc, aluminum, magnesium, carbon steel, etc.—so protecting steel and other more noble metals.

Safety bond: A bond connecting the metallic framework or enclosure of electrical apparatus with earth, in order to limit its rise in potential above earth in the event of a fault, and so reduce the risk of electric shock to anyone touching the framework or enclosure.

Salt fog test: An accelerated corrosion test in which specimens are exposed to a fine mist of a solution usually containing sodium chloride, but sometimes modified with other chemicals.

Salt spray test: See salt fog test.

Saturated calomel electrode: A reference electrode composed of mercury, mercurous chloride (calomel), and a saturated aqueous chloride solution.

Scaling: (1) The formation at high temperatures of thick corrosion product layers on a metal surface. (2) The deposition of water-insoluble constituents on a metal surface.

Season cracking: An obsolete historical term usually applied to stress-corrosion crackling of brass. A term usually applied to stress-corrosion cracking of brass.

Selective corrosion: The selective corrosion of certain alloying constituents from an alloy (as dezincification), or in an alloy (as internal oxidation).

Selective leaching: Selective leaching describes a corrosion process also called "parting" or dealloying. More specifically, it can be called dezincification in the case of brasses, denickelification in cupronickels, etc. Selective leaching may occur in a plug form or in a more evenly distributed layer type. Stagnant conditions and regions under deposits are conducive to selective leaching. In brasses, it can occur at pH extremes in water; high dissolved solids and high temperature also promote selective leaching. The overall dimensions of a part do not change drastically, but appreciable weakening can occur.

Semiautomatic welding: Welding in which some of the variables are automatically controlled, but manual guidance is necessary.

Sensitization heat treatment: A heat treatment, whether accidental, intentional, or incidental (as during welding), that causes precipitation of constituents at grain boundaries, often causing the alloy to become susceptible to intergranullar corrosion, cracking, or SCC (stress-corrosion cracking).

Sensitization: In austenitic stainless steels, the precipitation of chromium carbide usually at grain boundaries, on exposure to temperatures of about 550–850 °C.

Shear: The type of force that causes or tends to cause two contiguous parts of the same body to slide relative to each other in a direction parallel to their plane of contact.

Shear strength: The stress required to produce fracture in the plane of cross-section, the conditions of loading being such that the directions of force and of resistance are parallel and opposite, although their paths are offset a specified minimum amount. The maximum load divided by the original cross-sectional area of a section separated by shear.

Sherodising: The coating of iron or steel with zinc by heating the product to be coated in zinc powder at a temperature below the melting point of zinc.

Sigma phase: A hard, brittle, nonmagnetic intermediate phase with a tetragonal crystal structure, containing 30 atoms per unit cell, space group P42 mnm, occurring in many binary and ternary alloys of the transition elements. The composition of this phase in the various systems is not the same and the phase usually exhibits a wide range in homogeneity. Alloying with a third transition element usually enlarges the field of homogeneity and extends it deep into the ternary section.

Sigma-phase embrittlement: Embrittlement of iron–chromium alloys (most notably austenitic stainless steels) caused by precipitation at grain boundaries of the hard, brittle intermetallic sigma phase during long periods of exposure to temperatures between approximately 560 and 980 °C (1050 and 1800 °F). Sigma-phase embrittlement results in severe loss in toughness and ductility, and can make the embrittled material susceptible to intergranular corrosion. See also sensitization.

Silver/silver chloride electrode (Ag/AgCl): A reference electrode consisting of silver, coated with silver chloride in an electrolyte containing chloride ions.

Slip: Plastic deformation by the irreversible shear displacement (translation) of one part of a crystal relative to another in a definite crystallographic direction and usually on a specific crystallographic plane. Sometimes called glide.

Slow strain rate technique: An experimental technique for evaluating susceptibility to stress-corrosion cracking. It involves pulling the specimen to failure in uniaxial tension at a controlled slow strain rate while the specimen is in the test environment and examining the specimen for evidence of stress-corrosion cracking.

Slushing compound: An obsolete term describing oil or grease coatings used to provide temporary protection against atmospheric corrosion.

Smelt: Molten slag; in the pulp and paper industry, the cooking chemicals tapped from the recovery boiler as molten material and dissolved in the smelt tank as green liquor.

S–N diagram: A plot showing the relationship of stress, S, and the number of cycles, N, before fracture in fatigue testing.

Soft water: Water that is free of magnesium or calcium salts.

Soil resistivity: The electrical resistivity of the soil. This is important in cathodic protection systems as it affects current distribution through the soil and potentials on the protected structure.

Solder embrittlement: Reduction in mechanical properties of a metal as a result of local penetration of solder along grain boundaries.

Solid metal embrittlement: The occurrence of embrittlement in a material below the melting point of the embrittling species. See also liquid metal embrittlement.

Solid solution: A single, solid, homogeneous crystalline phase containing two or more chemical species.

Solute: The component of either a liquid or solid solution that is present to a lesser or minor extent: the component that is dissolved in the solution.

Solution: In chemistry, a homogeneous dispersion of two or more kinds of molecular or ionic species. Solution may be composed of any combination of liquids, solids, or gases, but they always consist of a single phase.

Solution heat treatment: Heating an alloy to a suitable temperature, holding at that temperature long enough to cause one or more constituents to enter into solid solution, and then cooling rapidly enough to hold these constituents in solution.

Solution potential: Electrode potential where half-cell reaction involves only the metal electrode and its ion.

Solvent: The component of either a liquid or solid solution that is present to a greater or major extent; the component that dissolves the solute.

Sour gas: A gaseous environment containing hydrogen sulfide and carbon dioxide in hydrocarbon reservoirs. Prolonged exposure to sour gas can lead to hydrogen damage, sulfide-stress cracking, and/or stress-corrosion cracking in ferrous alloys.

Sour water: Waste waters containing fetid materials, usually sulfur compounds.

Spalling: The spontaneous chipping, fragmentation, or separation of a surface or surface coating.

Spatter: Globules of metal expelled during welding onto the surface of parent metal or of a weld.

Spelter: A brazing alloy consisting nominally of 50% Cu and 50% Zn.

Spheroidite: An aggregate of iron or alloy carbides of essentially spherical shape dispersed throughout a matrix of ferrite.

Splash zone: That part of external structure between a level 2 m above mean high water springs (MHWS) and a level 3 m below the lowest astronomical tide (LAT).

Sputtering: A coating process whereby thermally emitted electrons collide with inert gas atoms, which accelerate toward and impact a negatively charged electrode that is a target of the coating material. The impacting ions dislodge atoms of the target material, which are in turn projected to and deposited on the substrate to form the coating.

Stabilizing treatment: (1) Before finishing to final dimensions, repeatedly heating a ferrous or nonferrous part to or slightly above its normal operating temperature and then cooling to room temperature to ensure dimensional stability in service. (2) Transforming retained austenite in quenched hardenable steels, usually by cold treatment. (3) Heating a solution-treated stabilized grade of austenitic stainless steel to 870–900 °C (1600–1650 °F) to precipitate all carbon, such as TiC, NbC, or TaC, so that sensitization is avoided on subsequent exposure to elevated temperature.

Standard electrode potential: The reversible potential for an electrode process when all products and reactions are at unit activity on a scale in which the potential for the standard hydrogen half-cell is zero.

Steel:

- **Ferrite:** Ferrite is the name given to the body-centered cubic allotropes of iron, á and ä iron, and to body-centered cubic solid solutions.
- **Austenite:** Austenite is the name given to the face-centered cubic, or ã variety of iron, and to the face-centered cubic solid solutions.
- **Cementite:** Cementite is the name given to the carbide of iron, Fe_3C. This is an extremely hard and brittle constituent.
- **Pearlite:** Pearlite is the eutectoid mixture of ferrite and cementite, and is formed when austenite decomposes during cooling. It consists of alternate thin layers, or lamellae, of ferrite and cementite.
- **Martensite:** This is the name given to the very hard and brittle constituent that is formed when a steel is very rapidly cooled from the austenitic state. It is a ferrite, highly supersaturated with dissolved carbon.
- **Sorbite and troostite:** These are names given to the structures produced when martensite or bainite is tempered, that is, heated to same temperature not exceeding 700 °C for the purpose of reducing brittleness and hardness.
- **Bainite:** This is the term given to the decomposition product that is formed when austenite decomposes by either isotherm transformation, or at a cooling rate intermediate between the very rapid cooling necessary for martensite and the slower rate of cooling at which pearlite is formed.
- **Plain carbon steel:** This is a steel containing up to 1.5% of carbon together with not more than 0.5% of silicon and not more than 1.5% of manganese, and only traces of other elements.
- **Alloy steel:** It is one that contains either silicon or manganese in amounts in excess of those quoted above, or that contains any other element, or elements, as the result of deliberately made alloying additions.

Strain: The unit of change in the size or shape of a body due to force. Also known as nominal strain.

Strain-age embrittlement: A loss in ductility accompanied by an increase in hardness and strength that occurs when low-carbon steel (especially rimmed or capped steel) is aged following plastic deformation. The degree of embrittlement is a function of aging time and temperature, occurring in a matter of minutes at about 200 °C (400 °F), but requiring a few hours to a year at room temperature.

Strain aging: Aging induced by cold working.

Strain hardening: An increase in hardness and strength caused by plastic deformation at temperatures below the recrystallization range.

Strain rate: The time rate of straining for the usual tensile test. Strain as measured directly on the specimen gage length is used for determining strain rate. Because strain is dimensionless, the units of strain rate are reciprocal time.

Stray current: Current flowing through paths other than the intended circuit.

Stray-current corrosion: Stray-current corrosion differs from other forms in that the source of the current causing the corrosion is external to the affected equipment. This cause of metal deterioration is frequently misdiagnosed. Stray-current corrosion can cause local metal loss in buried or submerged metal structures, but it occurs much less frequently in underwater transporting equipment than in underground structures. Stray-current corrosion is almost always associated with direct current. At the anodic areas, metal goes into solution and the electrolyte tends to become acidic. It is most commonly encountered in soils containing water.

Stress: The intensity of the internally distributed forces or components of forces that resist a change in the volume or shape of a material that is or has been subjected to external forces. Stress is expressed in force per unit area and is calculated on the basis of the original dimensions of the cross-section of the specimen. Stress can be either direct (tension or compression) or shear. See also residual stress.

Stress concentration factor (K_t): A multiplying factor for applied stress that allows for the presence of a structural discontinuity such as a notch or hole; K_t equals the ratio of the greatest stress in the region of the discontinuity to the nominal stress for the entire section. Also called theoretical stress concentration factor.

Stress corrosion or stress-accelerated corrosion: Corrosion that is accelerated by stress.

Stress-corrosion cracking: Stress-corrosion cracking is the result of the combined action of static stresses and corrosion. The static stresses may be residual or applied service stresses. The environment plays an important role in this type of cracking. The resulting cracks are branched, and can propagate either transgranularly or intergranularly, and sometimes both ways. Caustic cracking of steel is a case of stress-corrosion cracking that is sometimes called caustic embrittlement. Generally speaking, it is not necessarily the concentration of the corrodent in the bulk environment that causes the cracking, but the increased concentration occurring in crevices or in alternately wetted and dried regions.

Stress-intensity factor: A scaling factor, usually denoted by the symbol K, used in linear-elastic fracture mechanics to describe the intensification of applied stress at the tip of a crack of known size and shape. At the onset of rapid crack propagation in any structure containing a crack, the factor is called the critical stress-intensity factor, or the fracture toughness. Various subscripts are used to denote different loading conditions or fracture toughnesses: K_c. Plane-stress fracture toughness. The value of stress intensity at which crack propagation becomes rapid in sections thinner than those in which plane-strain conditions prevail. K_I. Stress-intensity factor for a loading condition that displaced the crack faces in a direction normal to the crack plane (also known as the opening mode of deformation). K_{Ic}. Plane-strain fracture toughness. The minimum value of K_c for any given material and condition, which is attained when rapid crack propagation in the opening mode is governed by plane-strain conditions. K_{Id}. Dynamic fracture toughness. The fracture toughness determined under dynamic loading conditions; it is used as an approximation of K_{Ic} for very tough materials. K_{ISCC}. Threshold stress-intensity factor for stress-corrosion cracking. The critical plane-strain stress intensity at the onset of stress-corrosion cracking under specified conditions. K_Q. Provisional value for plane-strain fracture toughness. K_{th}. Threshold stress intensity for stress-corrosion cracking. The critical stress intensity at the onset of stress-corrosion cracking under specified conditions. DK. The range of the stress-intensity factor during a fatigue cycle.

Stress-intensity factor range, DK.: In fatigue, the variation in the stress-intensity factor in cycle, that is, $K_{max}–K_{min}$.

Stress raisers: Changes in contour or discontinuities in structure that cause local increases in stress.

Stress ratio, A or R: The algebraic ratio of two specified stress values in a stress cycle. Two commonly used stress ratios are (1) the ratio of the alternating stress amplitude to the mean stress, $A = S_a/S_m$, and (2) the ratio of the minimum stress to the maximum stress, $R = S_{min}/S_{max}$.

Stress-relief cracking: Also called postweld heat treatment cracking, stress-relief cracking occurs when susceptible alloys are subjected to thermal stress relief after welding to reduce residual stresses and improve toughness. Stress-relief cracking occurs only in metals that can precipitation-harden during such elevated-temperature exposure; it usually occurs at stress raisers, is intergranular in nature, and is generally observed in the coarse-grained region of the weld heat-affected zone. See also cold cracking, hot cracking, and lamellar tearing.

Striation: A fatigue fracture feature, often observed in electron micrographs, that indicates the position of the crack front after each succeeding cycle of stress. The distance between striations indicates the advance of the crack front across that crystal during one stress cycle, and a line normal to the striation indicates the direction of local crack propagation. See also beach marks.

Structure-to-electrolyte potential: The potential difference between a metallic structure and the electrolyte surrounding it, measured with a reference electrolyte. Typical examples are pipe-to-seawater potentials.

Submerged-arc welding: Metal-arc welding in which a bare wire electrode is used; the arc is enveloped in flux; some of which fuse to form a removable covering of slag on the weld.

Submerged zone: That part of external structure below the splash zone.

Substrate: The basic metal or nonmetal whose surface is being protected.

Subsurface corrosion: Formation of isolated particles of corrosion products beneath a metal surface. This results from the preferential reactions of certain alloy constituents to inward diffusion of oxygen, nitrogen, or sulfur.

Sulfate-reducing bacteria (SRB): A group of bacteria found in most soils and natural waters, but active only in anaerobic conditions of near neutral pH. They reduce sulfates in their environment, with the production of sulfides. SRB increase cathodic protection current requirements as a result of cathodic depolarization by sulfides.

Sulfidation: The reaction of a metal or alloy with a sulfur-containing species to produce a sulfur compound that forms on or beneath the surface on the metal or alloy.

Sulfide stress cracking: Brittle failure by cracking under the combined action of tensile stress and corrosion in the presence of water and hydrogen sulfide. See also environmental cracking.

Surface anode bed: Type of ground bed using impressed current or galvanic anodes. This type of ground bed is used in relatively low-resistivity soils like marl (low-resistivity limestone and clay) with water-retaining capability.

Surface treatment: Any suitable means of cleaning and treating a surface that will result in the desired surface profile and cleanliness and the required coating characteristics.

Surfactant: A surface-active agent; usually an organic compound whose molecules contain a hydrophilic group at one end and a lipophilic group at the other.

Tafel line, tafel slope, tafel diagram: An electrode when polarized frequently yields a current potential relationship over a region, which can be approximated by: $h = \pm B \log (i/i_0)$ where h = change in open circuit potential, i = the current density, B and i_0 = constants. B is known as the Tafel slope. If this behavior is observed, a plot of the semilogarithmic components is known as the Tafel line and the diagram is called the Tafel diagram.

Tafel slope: The slope of the linear portion of a plot of potential versus the log of the current density for a particular electrode in a corroding system.

Tarnish: Surface discoloration of a metal caused by formation of a thin film of corrosion product.

Temper: (1) In heat treatment, to reheat hardened steel or hardened cast iron to some temperature below the eutectoid temperature for the purpose of decreasing hardness and increasing toughness. The process is also sometimes applied to normalized steel. (2) In tool steels, temper is sometimes inadvisably used to denote carbon content. (3) In nonferrous alloys and in some ferrous alloys (steels that cannot be hardened by heat treatment), the hardness and strength produced by mechanical or thermal treatment, or both, and characterized by a certain structure, mechanical properties, or reduction of area during cold working.

Temper color: A thin, tightly adhering oxide skin (only a few molecules thick) that forms when steel is tempered at a low temperature, or, for a short time, in air or a mildly oxidizing atmosphere. The color, which ranges from straw to blue depending on the thickness of the oxide skin, varies with both tempering time and temperature.

Tempered martensite embrittlement: *Embrittlement* of ultra high-strength steels caused by tempering in the temperature range of 205–400 °C (400–750 °F); also called 350 °C or 500 °F embrittlement. Tempered martensite embrittlement is thought to result from the combined effects of cementite precipitation on prior-austenite grain boundaries or interlath boundaries and the segregation of impurities at prior-austenite grain boundaries.

Temper embrittlement: *Embrittlement* of alloy steels caused by holding within or cooling slowly through a temperature range just below the transformation range. Embrittlement is the result of the segregation at grain boundaries of impurities such as arsenic, antimony, phosphorus, and tin; it is usually manifested as an upward shift in ductile-to-brittle transition temperature. Temper embrittlement can be reversed by retempering above the critical temperature range, then cooling rapidly.

Tempering: To reheat hardened steel or hardened cast iron to some temperature below the eutectoid temperature for the purpose of decreasing hardness and increasing toughness.

Tensile strength: In tensile testing, the ratio of maximum load to original cross-sectional area. Also called ultimate tensile strength.

Tensile stress: A stress that causes two parts of an elastic body, on either side of a typical stress plane, to pull apart. Contrast with *compressive stress.*

Tension: The force or load that produces elongation.

Terne: An alloy of lead containing 3–15% Sn, used as a *hot dip coating* for steel sheet or plate. Terne coatings, which are smooth and dull in appearance, give the steel better corrosion resistance and enhance its ability to be formed, soldered, or painted.

Terne plate: Deposition of lead–tin alloy on iron or steel sheets by the hot dip process.

Test access hole: Provides a means of contacting soil through concrete or asphalt for measuring structure-to-soil potentials. Contains no wires and is usually capped but easily accessible.

Test station: Permanent wires attached to the structure and led to a convenient location for electrical measurements. Used at points where the structure or soil is otherwise inaccessible for electrical testing (underground or underwater).

Thermal electromotive force: The *electromotive force* generated in a circuit containing two dissimilar metals when one junction is at a temperature different from that of the other. See also *thermocouple.*

Thermal embrittlement: *Intergranular fracture* of maraging steels with decreased toughness resulting from improper processing after hot working. Thermal embrittlement occurs upon heating above 1095 °C (2000 °F) and then slow cooling through the temperature range of 815–980 °C (1300F–1800 °F), and has been attributed to precipitation of titanium carbides and titanium carbonitrides at austenite grain boundaries during cooling through the critical temperature range.

Thermal cutting: The parting or shaping of materials by the application of heat with or without a stream of cutting oxygen.

Thermal spraying: A group of coating or welding processes in which finely divided metallic or nonmetallic materials are deposited in a molten or semimolten condition to form a coating. The coating material may be in the form of powder, ceramic rod, wire, or molten materials. See also *flame spraying and plasma spraying.*

Thermocouple: A device for measuring temperatures, consisting of lengths of two dissimilar metals or alloys that are electrically joined at one end and connected to a voltage-measuring instrument at the other end. When one junction is hotter than the other, a *thermal electromotive force* is produced that is roughly proportional to the difference in temperature between the hot and cold junctions.

Thermogalvanic corrosion: Corrosion resulting from an *electrochemical cell* caused by a thermal gradient.

Threshold stress: Threshold stress for *stress-corrosion cracking.* The critical gross section stress at the onset of stress-corrosion cracking under specified conditions.

Throwing power: (1) The relationship between the current density at a point on a surface and its distance from the counter electrode. The greater the ratio of the surface resistivity shown by the electrode reaction to the volume resistivity of the electrolyte, the better is the throwing power of the process. (2) The ability of a plating solution to produce a uniform metal distribution on an irregularly shaped cathode.

TIG-welding: Tungsten inert gas arc welding using a nonconsumable electrode of pure or activated tungsten.

Tinning: Coating metal with a very thin layer of molten solder or brazing filler metal.

Torsion: A twisting deformation of a solid body about an axis in which lines that were initially parallel to the axis become helices.

Torsional stress: The shear stress on a transverse cross-section resulting from a twisting action.

Total carbon: The sum of the free *carbon* and *combined carbon* (including carbon in solution) in a ferrous alloy.

Toughness: The ability of a metal to absorb energy and deform plastically before fracturing.

Transcrystalline: See *transgranular.*

Transcrystalline cracking: See *transgranular cracking.*

Transference: The movement of ions through the *electrolyte* associated with the passage of the electric current. Also called transport or migration.

Transgranular: Through or across crystals or grains. Also called intracrystalline or transcrystalline.

Transgranular cracking: Cracking or fracturing that occurs through or across a crystal or grain. Also called transcrystalline cracking. Contrast with *intergranular cracking.*

Transgranular fracture: Fracture through or across the crystals or grains of a metal. Also called transcrystalline fracture or intracrystalline fracture. Contrast with *intergranular fracture.*

Transition metal: A metal in which the available electron energy levels are occupied in such a way that the d-band contains less than its maximum number of 10 electrons per atom, for example, iron, cobalt, nickel, and tungsten. The distinctive properties of the transition metals result from the incompletely filled d-levels.

Transition temperature: (1) An arbitrarily defined temperature that lies within the temperature range in which metal fracture characteristics (as usually determined by tests of notched specimens) change rapidly, such as from primarily fibrous (shear) to primarily crystalline (cleavage) fracture. (2) Sometimes used to denote an arbitrarily defined temperature within a range in which the ductility changes rapidly with temperature.

Transpassive region: The region of an *anodic polarization* curve, noble to and above the passive *potential* range, in which there is a significant increase in current density (increased metal dissolution) as the potential becomes more positive (noble).

Transpassive state: (1) State of anodically passivated metal characterized by a considerable increase in the corrosion current, in the absence of pitting, when the *potential* is increased. (2) The noble region of potential where an electrode exhibits at higher than passive current density.

Triaxial stress: See principal stress (normal).

Tuberculation: The formation of *localized corrosion* products scattered over the surface in the form of knoblike mounds called tubercles.

u-bend specimen: Horseshoe-shaped test piece used to detect the susceptibility of a material to stress-corrosion cracking.

Ultimate strength: The maximum stress (tensile, compressive, or shear) a material can sustain without fracture, determined by dividing maximum load by the original cross-sectional area of the specimen. Also called nominal strength or maximum strength.

Ultrasonic measurement: The timing of the transmission of ultrasonic sound waves through a material to determine the material's thickness.

Underfilm corrosion: Corrosion that occurs under organic films in the form of randomly distributed threadlike filaments or spots. In many cases, this is identical to *filiform corrosion.*

Uniaxial stress: See principal *stress (normal).*

Uniform corrosion: (1) A type of corrosion attack (deterioration) uniformly distributed over metal surface. (2) Corrosion that proceeds at approximately the same rate over a metal surface. Also called general corrosion. General corrosion is characterized by a chemical or electrochemical reaction that occurs uniformly over the exposed surface. Anodic and cathodic sites shift constantly so that corrosion spreads over the entire metal surface. Identifying general corrosion is usually simple, but determining its cause is often difficult. Chemical dissolution by acids, bases, or chelants frequently results in general corrosion.

Vacuum deposition/vapor depositions/gas plating: Deposition of metal coatings by the precipitation, sometimes in vacuum, or metal vapor on the treated surface. Vapor may be produced by thermal decomposition, cathode sputtering, or by evaporation of the molten metal in air or inert gas.

Valence: A positive number that characterizes the combining power of an element for other elements, as measured by the number of bonds to other atoms that one atom of the given element forms upon chemical combination: hydrogen is assigned valence 1, and the valence is the number of hydrogen atoms, or their equivalent, with which an atom of the given element combines.

Vapor deposition: See chemical vapor deposition, physical vapor deposition, and sputtering.

Vapor plating: Deposition of a metal or compound on a heated surface by reduction or decomposition of a volatile compound at a temperature below the melting points of the deposit and the base material. The reduction is usually accomplished by a gaseous reducing agent such as hydrogen. The decomposition process may involve thermal dissociation or reaction with the base material. Occasionally used to designate deposition on cold surfaces by vacuum evaporation. See also *vacuum deposition.*

Voids: A term generally applied to paints to describe *holidays*, holes, and skips in a *film.* Also used to describe shrinkage in castings and weld.

Vulcanized rubber: Treatment of rubber with sulfur or sulfur compound resulting in a change in physical properties of rubber.

Wash primer: A thin, inhibiting paint, usually chromate pigmented with a polyvinyl butyrate binder.

Weather resistance: Ability of a material to resist all ambient weather conditions. These include changes of temperature, precipitation, effect of wind and humidity, sunlight, oxygen and other gases and impurities in the atmosphere, ultraviolet rays, radiation, and ozone.

Weight coat: Steel mesh-reinforced concrete layer applied over a primary coating system. Provides negative buoyancy for submarine pipelines.

Weld: A union between pieces of metal at faces rendered plastic or liquid by heat or by pressure, or by both. A filler metal whose melting temperature is of the same order as that of the parent metal may or may not be used.

Welding: The making of a weld.

Weld cracking: Cracking that occurs in the weld metal. See also *cold cracking, hot cracking, lamellar tearing*, and *stress-relief cracking.*

Weld decay: *Intergranular corrosion,* usually of stainless steels or certain nickel-based alloys, that occurs as the result of *sensitization* in the heat-affected *zone* during the welding operation.

Weld metal: All metal melted during the making of a weld and retained in the weld.

Weld zone: The zone containing the weld metal and the heat-affected zone.

Wetting: A condition in which the interfacial tension between a liquid and a solid is such that the contact angle is 0–90°.

Wetting agent: A substance that reduces the surface tension of a liquid, thereby causing it to spread more readily on a solid surface.

White liquor: Cooking liquor from the kraft pulping process produced by recausticizing *green liquor* with lime.

White rust: Zinc oxide: the powdery product of corrosion of zinc or zinc-coated surfaces.

Work hardening: Same as *strain hardening.*

Working electrode: The test or specimen electrode in an *electrochemical cell.*

Yeld: Evidence of *plastic deformation* in structural materials. Also called plastic flow or creep.

Yield point: The first stress in a material, usually less than the maximum attainable stress, at which an increase in strain occurs without an increase in stress. Only certain metals—those that exhibit a localized, heterogeneous type of transition from *elastic deformation* to *plastic deformation*—produce a yield point. If there is a decrease in stress after yielding, a distinction may be made between upper and lower yield points. The load at which a sudden drop in the flow curve occurs is called the upper yield point. The constant load shown on the flow curve is the lower yield point.

Yield strength: The stress at which a material exhibits a specified deviation from proportionality of stress and strain. An onset of 0.2% is used for many metals.

Yield stress: The stress level in a material at or above the *yield strength* but below the *ultimate strength*, i.e., a stress in the plastic range.

Zinc reference electrode: A 99.9% metallic zinc rod, with an iron content not exceeding 0.0014%, which contacts directly with the electrolyte around the structure for potential measurement purposes. Some of the zinc alloys used for galvanic anodes are also suitable. The metallic electrode can be made in any convenient form.

Index

Note: Page numbers followed by f indicate figures; t, tables.

A

Absorption theory, 86
Aluminum anode alloys, in seawater
  closed-circuit potential, 168
    apparatus, 168
    circuit, 168, 169f
    procedure, 169
    test anode preparation, 168
  consumption rate determination, 166
    apparatus, 166
    circuit, 166, 167f
    procedure, 167–168
    test anode preparation, 166–167
Anode casting
  graphite anode, 135–136
  High-Silicon–Chromium–Iron anode, 135, 135t
  magnetite anode, 136
Anode core steel works
  aluminum anodes, in seawater, 162–163, 163t
  anode casting, surface irregularities, 163
  anode connections, 163–164
  cast anodes material, cracks, 163
  electrical resistance, 162
  insert material, 160–161
  insert surface preparation, 161
  mechanical resistance, 162
  requirements
    dimensions, 162
    shape, 161
    weight, 162
  welding, 161
Anode field interferences, 242
  measurements of, 242
Anode lead wire
  chlorine-resistant type insulation, 137
  conductor, 137
  standard type insulation, 137
Anode location
  fitting out protection, 128
  full hull protection, 128
  impressed current, 130
  individual anode output determination and calculation, 129–130
  stern only protection, 128
  tank descaling, 129
Anode manufacturing, 134
Anode resistance
  current and potential distribution, 101, 101f
  electrical resistances, 100
  interaction, 101
  Ohm's Law, 100
Anodic index, 4
Anodic protection, 131
Attenuation constant, 281–285
  examples, graph relating to, 285
  significance of, 283
Attenuation formulas, 102–103

B

Bare lines, 109
Bracelet-type galvanic anodes
  alloy composition
    aluminum anodes, 157, 158t
    zinc anodes, 158, 158t
  materials and manufacture, 157
  production testing
    aluminum anode, short-term capacity testing, 159–160
    chemical analysis, 159
    destructive and nondestructive testing, 160
    short-term voltage test, 159
Buried steel pipes, 206
  application, 48–50
  bonding stations, 55
  cased crossing test station, 55f, 56

Buried steel pipes (*Continued*)
casings, 57–58
concrete encased pipe, 58
galvanic anode systems, 51–52
impressed current system
anode materials, 51
cables, 51
connections, 51
deep anode beds, 51
surface anode beds, 50
insulated fitting test station, 56, 57f
isolation
corrosion acceleration, 52
electrical ground isolation, 54
electrolytes, 52–53
insulating devices, 53–54
pipe supports, 54
line crossing test station, 56, 56f
line current measurement, 55
potential measurement, 54–55
protecting existing pipelines, 58
protection criteria, 37
scale detachment, 58
system design, 56

C
Calcareous scales, 99
Carbonaceous backfill
bulk density, 142, 142t
particle size, 142, 143t
requirements, 141–142
resistivity, 142
types, 140
Casings, 57–58
Cast galvanic anodes
anode insert
fabrication, 172–173
material, 172
position, 173
surface preparation, 172
anode types, 170
chemical analysis, 171
chemical composition, 170t, 171
destructive and nondestructive testing, 171–172
electrochemical testing, 171
materials and manufacture, 169–170
Cathode field interferences, 241
measurements of, 241–242
Cathodic protection (CP) system
AC effects, 378–382, 379f
anode identification, 175
anodic protection, 175
chemical environment, 89
with contrast, 86–87
different metals, 87t, 88–89
principles of, 83f–84f, 85–86
process plants, 88f, 89–90
solution concentration, 87
surface area, 87
temperature effect, 89
applications, 1, 10–11
electromotive force series, 36, 38t
offshore structures, 40
water box interiors, 40
well casings, 35, 40–41
buried steel pipes, potential survey of
cased crossings, potential survey at, 217–218
instruments, 213–214, 214f
insulating devices, potential tests at, 218–219
IR drop component, compensation for, 216–217
overprotection, 217
potential measurements, 215–216, 215f
cables
bond conductor, 178
chlorine resistant type cable, 177
conductor, 175, 176t
cross-linked polyethylene, 177
high molecular weight polyethylene, 176
insulation, 176, 178
negative conductor, 178
polyvinyl chloride, 176
PVDF, 177
standard type cable, 177
test conductor, 178
casting quality, 173–174
cathode and anodes, 4, 5t, 10f
cathodically protected structures, 29
cathodic area reaction, 9
cell formation, 7–8, 7f–8f
coatings, 99
commissioning report, 361
commissioning survey, 360–361
corrosion interaction, 19
definition, 21

galvanic anodes, 26
ground beds/anodes, 24
impressed current installations, 26
primary structure, coating quality, 24
reduce measures, 26–28
sea and river terminals, 29
secondary structure, coating quality, 24
soil/water resistivity, 26
structure/electrolyte potential, 21, 24
criteria for, 205–210
aluminum, 207
buried pipes, 206
lead, 207
offshore structures and ship hulls, 207
potential limits, 207
stainless steels, 210
steel in concrete, 210
submerged pipelines, 206
tank/pipe and water box interiors, 207
tanks exteriors, 206
well casings, 207
current drainage survey, 389–391
method, 389–391
current survey
offshore structures, 233
pipelines, 233
data analysis, 239–241
impressed current system, 239–240
rectifiers data analysis, 240–241
sacrificial system, 239
data recording, 238–239
design/modification structures
electrical continuity, 13
factors affect, 19
isolation, 14–15
nonuniformity, 11–12
protective coating, 13–14
structural/electrolyte potential, 11–12, 11f
dimensions and weights, 139–140, 139t–140t
electrical isolation equipment installation, 361–366, 362f
installation, 363
isolated flange joints, 364–365
electric drainage system, 19–21, 20f
electrode resistance, measurement of, 385–389
electromotive force, 10
epoxy internal coatings, 183
factory acceptance tests, 178–179
field procedures, 384
foreign structure, interference currents control on, 369–370
frequency of measurement, 384–385
galvanic anode, 2
galvanic anode installation, inspection and survey for, 235–238
attenuation test method, 236
coating resistance measurement, 235–236
C-scan, coating inspection by, 236
over-the-line potential survey, 238
pearson method, 236
visual inspection, 236–238
ground bed inspection, 231–232
on-land ground bed, 231–232
submerged ground beds, 232
hazardous atmospheres, installation in, 397–399
hookup and commissioning, 357–360
impressed current systems, 358–359
interference, 359–360
sacrificial anodes, 359
ICCP system, 2, 3f
alternating current system, 2–3
vs. galvanic anode, 15, 15t
principle, 4f
rectifier units, 3
impressed DC current, 3, 4f
in-plant facilities
anode beds, 60–61
anode materials, 61
application, 59
bonding stations, 62
cables, 61
connections, 61
current monitoring stations, 62
galvanic anode system, 60–61
hazardous locations, 62
integral impressed current systems, 59–60
potential test stations, 61
reference electrodes, 61
inspection following failure report
circuit resistance, increase in, 234
electrical continuity test, 234
stray electric currents, 234

Cathodic protection (CP) system (*Continued*)
installations, inspection of, 394–395
coating inspection, 396
test points/cased crossings and insulating joints, 395–396
interferences, 241–242
interference testing, 371–373
interference testing structures, notifying owners of, 370–371
jetties and ships, 29
lightning protection system
discharge gaps, 18
isolating joints, 18
vicinity, 18–19
limiting corrosion interaction, criteria for, 374–377
marine structures
atmospheric zone, 74
boosted polarization, 82
calcareous coating, 75
cathodic protection design, 82–83
current density, 71t, 72–73
fixed potential monitoring systems, 84
galvanic anode systems, 81
impressed current systems, 78–80
marine sheet piling and wharves, 75–76
offshore GOSPS and sea islands, 76–77
piers, 76
potential survey, 84
retrofits, 84–85
seawater, 71
splash zone, 74
structural zones, 73, 73f
submarine pipelines, 77–78, 77f
submerged zone, 75
vessels, 78
wellhead and tie-in platforms, 76
metallic corrosion, 5–6
meters and equipment, 242–246
calomel electrode, 246
copper/copper sulfate electrode, 243–244, 244f–245f
pure zinc electrode, 245, 247f
reference electrode, 242–243, 243f
silver/silver chloride electrode, 244–245, 246f
offshore sacrificial anodes inspection, 232
offshore structures
diver-operated unit, potential measurements with, 222–225
extent of, 229–230
methods of measurement, 221–222
reference electrodes, 221
subsea pipeline potential survey, 229
surface voltmeter/cable and measuring electrode, potential measurements with, 225–229
time intervals, 229
oil storage tank bottoms, 62–63
overvoltages, 17–18
passivity, 9
performance requirements
closed circuit potentialk, in seawater, 175
consumption rate, in seawater, 175
electrical resistance, 174
mechanical resistance, 174
periodic inspection
potential survey, 211
procedure, 213
test equipment, 211–213
pipe preparation, 367
polarization, 6–7, 7f
positive structure/electrolyte potential changes, 21–22
potentiometer and voltmeter, 246–251
conventional voltmeter, 246–247
potentiometer, 249–251, 250f
potentiometer–voltmeter, 248–249
precommissioning inspection and check, 355–357
principal methods, 1
properties
alkaline conditions, 94
anodic and cathodic reactions, 91, 92f
diffusion control of oxygen reduction, 92–93, 93f
electrical resistance test, 138
fluoropolymer-insulated anode lead wire test, 138–139
mechanical resistance test, 138
radiographic test, 139, 139t
temperature dependence, 138
protection installations, 17
quality assurance provisions, 178, 184
railway signals and protection circuits, 28–29

rectifiers inspection, 230–231
redox potential of soil, determination in situ of, 391–394
apparatus, 391–392
calculations and expression, 393–394
materials, 392
procedure, 392–393
test report, 394
reinforcement (rebar) in concrete structures
electrodes and probes for, 220–221
potential measurement, 219–220
sacrificial anode, 4, 99
sacrificial metal, 1
safety aspects
buried structure installations, 30
chlorine evolution, 34
fault conditions, in electricity power systems, 31
hazard explosion, 32–34
hydrogen evolution, 31–32
immersed structure installations, 30
secondary effects, 16–17
soil resistivity, measurement of, 382–383
structure/electrolyte potential, 22–24
telecommunication services, 28
telluric current, 377–378
thermit welding of, 366–367
thermit weld preparation and procedure, 367–369, 368t
types, 1, 2f
weight and dimensional tolerances, 174
welding processes, 183
Chlorine resistant type cable, 177
Close interval pipe-to-soil potential survey, 298–303
criteria and interpretation, 303
data obtained, 302–303
distance measurement, 301
equipment, 301–302
measurement intervals, 300
presentation of data, 303
procedure, 302
stationary measurements, 301
switching frequency, 300
Coating deterioration, 108
Coating resistance measurement method
current – voltage change method, 278–279, 279f
interrupter on and off, pipeline current with, 279–280
Conventional voltmeter, 246–247
Corrosion control system, 205
Corrosion interaction, 19
definition, 21
galvanic anodes, 26
ground beds/anodes, 24
impressed current installations, 26
primary structure, coating quality, 24
reduce measures, 26–28
sea and river terminals, 29
secondary structure, coating quality, 24
soil/water resistivity, 26
structure/electrolyte potential, 21, 24
Crossing bond, 380, 380f
C-Scan system, coating inspection by, 288–293
advantages of, 290
features, 289
operation principle, 291–293
performing survey by, 289–290
theoretical background, 290–291
Current density, 93f, 97t–98t, 98–99

D
Data recording, 238–239
Depolarization effect, 109
Dielectric test, 185
Dimensional check, 186
Diver-operated unit, potential measurements with, 222–225
Dulling, 161
Dye penetrant test, 186

E
Electrical continuity, 306–308, 307f
Electric drainage system, 19–21, 20f
Electromagnetic current attenuation survey, coating evaluation by, 293–298
criteria and interpretation, 297–298
data obtained, 296
equipment, 294–295
presentation of data, 296, 297f–298f
procedure, 295–296, 295f
Electromotive force (EMF), 10
External moisture ingress, 365

F

Flange insulation kit
- components, 186
- gaskets shall, 186
- insulating gaskets and sleeves, 188t–189t, 190
- materials, 187–188
- packaging, 190
- quality assurance provisions, 190–191
- requirements
  - dimensions, 188
  - surface defects, 189
  - uniformity, 189
  - workmanship, 189
- sampling, 191

G

Galvanic anode systems
- anode core, 145
- anode dimensions and weights, 147, 147t
- anode identification, 150
- anode-to-core resistance method
  - apparatus required, 151–152, 155–156, 164
  - circuit, 152, 152f, 156, 156f, 164, 165f
  - principle, 155
  - procedure, 153, 156–157, 165–166
- buried steel pipes, 51–52
- cast anodes
  - buried anodes, cable attachment, 148, 149f
  - buried magnesium anodes, cable attachment, 148, 149f
  - lead wire insulation, 148–150
- chemical backfill for
  - anode packaging, 150–151
  - composition and analysis, 150, 151t
  - particle size, 150
- classification, 143
- composition
  - magnesium anodes, 144, 144t
  - zinc anodes, 144–145, 145t
- current output, 117–118
- electrical resistance, 146
- finishing and appearance, 146
- *vs.* impressed current system, 42
- magnesium anodes, 42
- marine structures, 120–121, 121f
  - aluminum alloy anodes, 81
  - magnesium anode, 81
  - zinc anode, 81
- materials and manufacture, 143–144
- mechanical resistance, 146
- production testing, 144t–145t, 145
- requirements, 146
- submerged applications
  - anode identification, 155
  - electrical resistance, 155
  - magnesium anodes, 153–154
  - materials and manufacture, 153–154
  - production testing, 154
  - requirements, 154
  - steel insert quality, 155
  - zinc anodes, 153–154
- symbols, 116, 117t
- workmanship, 146
- zinc anodes, 147–148, 147t–148t

H

Heat shrink anode cap, 136

High voltage protection, 366

Hydrostatic test, 185

I

Impressed current anodes, 133–134, 133f

Impressed current CP (ICCP) system
- advantages, 100
- alternating current system, 2–3
- alternative power sources, 43
- anode beds, 45–47
- auxiliary electrode, 94
- bare and poorly coated lines, 108
- cables, 45
- calculations
  - cable resistance, 111, 111t–113t
  - circuit resistance, 111
  - single cylindrical anode, 112
  - single row vertical anode group, 114, 115t
  - transformer rectifier choice, 116
  - two/more rows of vertical anodes, 114
  - vertical and horizontal anode groups, 115, 116t
- cathodic protection, 94, 95f
- cathodic protection transformer rectifiers, 42
- ceramic anodes, 43
- connections, 47

current requirements, 109, 109t
current–voltage relationships
attenuation and formulas, 102–103
calculations, 103–105, 103t
effect of coating, 107
electrical bonding station, 47
electrical continuity, 47
ferro silicon anode, 43
vs. galvanic anode, 15, 15t
insulating devices, 45
marine structures
graphite, 80
high silicon iron, 78–79
lead silver alloy, 80
platinized anodes, 79–80
measurements, 110
minimum protection point, 107–108
multiple drain points, 106–107, 106f–107f
polymeric anodes, 43
principal anode materials, 45, 46t
principle, 4f
protective coatings and insulation, 47
provision for testing, 48
rectifier sizing, 110
rectifier units, 3
safety, 48
single drain point, 105–106, 106f
site survey, 48
test stations, 47
voltage limitations
bare lines, 109
coating deterioration, 108
depolarization effect, 109
maximum rectifier spacing, 108–109
Installation
buried/immersed structures
electrical continuity, 349
soil/water evaluation, 346–348
stray electric currents, 348–349
structure/electrolyte natural potential survey, 348
for buried pipelines, 312–327
cabling, 317–320
earthing of equipment, 325–326
electrical connections, 320–323
fencing, 326
impressed current systems, 312–316
lightning protection, 326–327
parallel power lines, 326
surge arrestors, 327
test box, installation of, 325
test stations, 323–325
transformer/rectifier equipment, 316–317
commissioning period, 349–353
buried structures, 349–350
fixed immersed structures, 350–351
internal protection of plant, 351
internal surfaces, 352–353
for compact buried structures, 329–332
insulating flanges/joints and couplings, 331–332
permanent reference electrodes, 331
structure preparation, 330–331
electrical measurements and tests, 344–345
bond resistance, determination of, 345
internal protection, potential survey of, 345
potential measurements, 344–345
excavation and backfilling, 311
galvanic anodes, 309–310
of galvanic anode systems
connection of, 328–329
extruded ribbon anodes, 328
multiple galvanic anodes, 327–328
single packaged anode, 327
impressed current anodes, 310
internal surfaces, 332–336
galvanic anode systems, 334–335
impressed current systems, 332–334
materials and equipment acceptance, 332
permanently installed reference electrodes, 335–336
safety precautions, 334
marine structures
corrosion control test stations/ connection and bonds, 340
electrical connections, 339–340
galvanic anode systems, 338–339
immersed structures, 336–338
insulating joints/flanges and devices, 341
specialized surveys, 353–354
submarine pipelines, 341–344
corrosion control test stations/ connections and bonds, 343
galvanic anode systems, 342–343

Installation (*Continued*)
impressed current systems, 341–342
pipeline crossings, 344
reinforcement, 344
transformer/rectifier equipment, 310–311
Internal cathodic protection, 124–126, 125t–126t

L
Lead, 207
Lead alloys, 44
Lightning protection system
discharge gaps, 18
isolating joints, 18
vicinity, 18–19
Line current survey test method, 268–272, 268f
null amp test circuit for line current measurement, test procedure using, 271–272, 272f
test point consisting of four wires, test procedure with, 270–271, 270f
Low pressure leakage test, 185

M
Magnesium anodes, 65
Magnetic particle/ultrasonic test, 186
Magnetite (Fe_3O_4), 44
Marine structures
anode location, 127
anode output formula
brackish water, 123
open-circuit potential, 122, 122t
symbols, 121
design calculations, 118–119, 119f
galvanic anode systems, 120–121, 121f
pipelines and attenuation potential, 124
sacrificial anode
external cathodic protection, 127
internal cathodic protection, 124–126, 125t–126t
materials, 124
single cylindrical vertical anode, 119–120
type and number of anodes, 126–128, 128t
vertical anode spacing, straight line, 120
Maximum rectifier spacing, 108–109
Megger test, 185
Monolithic insulating joints
design, 179–181, 181t
manufacture, 182–183
materials, 181–182, 181t

N
Nonadherent coating, 14

O
Offshore sacrificial anodes inspection, 232
Oxide film theory, 86

P
Passivity, 9
Pearson method, coating inspection by, 285–288
data obtained, 287–288
equipment, 285–286
presentation of data, 288
procedure, 286–287
Pipeline casing insulators, 365–366
Platinum alloys, 44
Polarized electric drainage, 20
Polypropylene rope
cast mold, 194
catalyzed resins properties, 195, 195t
dimensions, 196
odor, 195
shelf life, 196
toxicity, 196
working properties, 195
characteristics, 193
description, 194
packaging, 196–197
quality assurance provisions, 193–194
requirements, 191t, 192–193, 192f
sampling, 197
splicing kit, 194
Potential limits, 207
Potentiometer–voltmeter, 248–249
Protection potentials, 98, 98t
Prototype tests, 186
Pure zinc electrode, 245, 247f

R
Radiographic test, 186
Rectifiers inspection, 230–231
Reference electrodes, 221
Resistivity measurement methods, 382, 383f
Rubber splicing tape
color, 201

materials and manufacture, 201
packaging, 201
requirements, 200t, 201
roll size, 201
sampling, 202

S
Sacrificial anode systems, 96–97, 96t, 97f, 97t, 133f
Sequence model theory, 86
Ship hulls, 40
Short circuit point, 253–257
insulation tests, 255–257, 256f
locating casing, 253–254, 254f
Single cylindrical vertical anode, 119–120
Single drain point, 105–106, 106f
Split bolt connectors, line taps
dimensions, 198
electrical insulating plastic tape
color, 200
conductor corrosion, 200
deformation test, 199
materials and manufacturer, 198, 199t
roll size, 200
storage test, 200
finishing and appearance, 198
materials and manufacture, 197–198, 197t
quality assurance provisions, 198
splice coating materials, 198
workmanship, 198
Stainless steels, 210
Standard type cable, 177
Submarine pipelines, bracelet-type galvanic anodes
alloy composition
aluminum anodes, 157, 158t
zinc anodes, 158, 158t
materials and manufacture, 157
production testing
aluminum anode, short-term capacity testing, 159–160
chemical analysis, 159
destructive and nondestructive testing, 160
short-term voltage test, 159
Submerged pipelines, 40, 206
Subsea pipeline CP survey method, 229, 265f
survey operation, 267
survey results, 267–268
system components, 264–267
Swain-type current clamp/meter, 364

T
Tanks exteriors, 40, 206
Tanks/vessels internal surfaces, 35
Test methods
ASTM, half-cell potentials of uncoated reinforcing steel bar, 257–264
apparatus, 259–260
calibration standardization, 260
data presentation, 263
interpretation of results, 263–264
precision and bias, 264
procedure, 260–262
recording half-cell potential values, 262
report, 264
scope, 257–258
significance and use, 258–259
attenuation constant, 281–285
examples, graph relating to, 285
significance of, 283
cathodic field interference test method, example for, 303–306, 304f
close interval pipe-to-soil potential survey, 298–303
criteria and interpretation, 303
data obtained, 302–303
distance measurement, 301
equipment, 301–302
measurement intervals, 300
presentation of data, 303
procedure, 302
stationary measurements, 301
switching frequency, 300
coating resistance measurement method
current – voltage change method, 278–279, 279f
interrupter on and off, pipeline current with, 279–280
C-Scan system, coating inspection by, 288–293
advantages of, 290
features, 289
operation principle, 291–293
performing survey by, 289–290
theoretical background, 290–291
electrical continuity, 306–308, 307f

Test methods (*Continued*)
electromagnetic current attenuation survey, coating evaluation by, 293–298
criteria and interpretation, 297–298
data obtained, 296
equipment, 294–295
presentation of data, 296, 297f–298f
procedure, 295–296, 295f
line current survey test method, 268–272, 268f
null amp test circuit for line current measurement, test procedure using, 271–272, 272f
test point consisting of four wires, test procedure with, 270–271, 270f
offshore structures, computer modeling of, 272–278
attenuation curves analysis, 277
computerized modeling techniques, 273
existing CP systems, analysis of, 275–276
numerical solutions, 275
potentials in nodal areas, 277–278
sacrificial anodes, performance of, 276–277
short theoretical background, 273–274
Pearson method, coating inspection by, 285–288
data obtained, 287–288
equipment, 285–286
presentation of data, 288
procedure, 286–287
for short circuit point, 253–257
insulation tests, 255–257, 256f
locating casing, 253–254, 254f
subsea pipeline CP survey method, 265f
survey operation, 267
survey results, 267–268
system components, 264–267
Thermit/cad weld powder, 202–203
Time intervals, 229
Transformer rectifiers, 42

U
Underground pipelines, 35
Underground storage tanks, 35

V
Vessel/tank internals
anode distribution, 66
corrosive waters, 64
current density, 66
galvanic anode systems, 65–66
impressed current systems, 64
reference electrode position, 66
storage tanks, 66–67
water circulating systems
box coolers, 70
corrosion rate, 67
galvanic anode systems, 69
heat exchangers, 68
impressed current, 70
water box coatings, 69
Visual inspection, 186

W
Water circulating systems
vessel/tank internals
box coolers, 70
corrosion rate, 67
galvanic anode systems, 69
heat exchangers, 68
impressed current, 70
water box coatings, 69
Well casings, 35, 40–41, 207
Wenner four-electrode method, 383

Z
Zero-resistance ammeter, 380, 380f
z resistant type cable, 177

FSC
www.fsc.org
MIX
Papier | Fördert
gute Waldnutzung
FSC® C083411

Kolibri 360 GmbH
Ferdinand-Jühlke-Straße 7
99095 Erfurt, Deutschland
produktsicherheit@kolibri360.de